STUDENT'S SOLUTIONS MANUAL

MATH MADE VISIBLE

COLLEGE ALGEBRA WITH MODELING AND VISUALIZATION

SIXTH EDITION

Gary Rockswold

Minnesota State University, Mankato

with

Terry Krieger

Rochester Community and Technical College

and

Jessica Rockswold

Copyright © 2018 Pearson Education, Inc.
Publishing as Pearson, 330 Hudson Street, NY, NY 10013

1 16

ISBN-13: 978-0-13-441816-2
ISBN-10: 0-13-441816-6

Table of Contents

Chapter 1: Introduction to Functions and Graphs

1.1: Numbers, Data, and Problem Solving

1. $\frac{21}{24}$ is a real and rational number.

3. 7.5 is a real and rational number.

5. $90\sqrt{2}$ is a real number.

7. Natural number: $\sqrt{9} = 3$; whole number: $\sqrt{9}$; integers: -3, $\sqrt{9}$; rational numbers: -3, $\frac{2}{9}$, $\sqrt{9}$, $1.\overline{3}$; irrational numbers: π, $-\sqrt{2}$

9. Natural number: None; whole number:0; integer: 0, $-\sqrt{4} = -2$; rational numbers: 0, $\frac{1}{3}$, 5.1×10^{-6}, -2.33, $0.\overline{7}$, $-\sqrt{4}$; irrational number: $\sqrt{13}$

11. Shoe sizes are normally measured to within half sizes. Rational numbers are most appropriate.

13. Speed limit is measured using natural numbers.

15. Temperature is typically measured to the nearest degree in a weather forecast. Since temperature can include negative numbers, the integers would be most appropriate.

17. $|5 - 8 \cdot 7| = |5 - 56| = |-51| = 51$

19. $-6^2 - 3(2-4)^4 = -6^2 - 3(-2)^4 = -36 - 3(16) = -36 - 48 = -84$

21. $\sqrt{9-5} - \frac{8-4}{4-2} = \sqrt{4} - \frac{4}{2} = 2 - 2 = 0$

23. $\sqrt{13^2 - 12^2} = \sqrt{169 - 144} = \sqrt{25} = 5$

25. $\frac{4+9}{2+3} - \frac{-3^2 \cdot 3}{5} = \frac{13}{5} - \frac{-27}{5} = \frac{40}{5} = 8$

27. $-5^2 - 20 \div 4 - 2 = -25 - 5 - 2 = -32$

29. 8.2×10^1

31. 3.65×10^{-3}

33. $2450 = 2.45 \times 10^3$

35. $0.56 = 5.6 \times 10^{-1}$

37. $-0.0087 = -8.7 \times 10^{-3}$

39. $206.8 = 2.068 \times 10^2$

41. $10^{-6} = 0.000001$

43. $2 \times 10^8 = 200,000,000$

45. $1.567 \times 10^2 = 156.7$

47. $5 \times 10^5 = 500,000$

49. $0.045 \times 10^5 = 4500$

51. $67 \times 10^3 = 67,000$

53. $(4 \times 10^3)(2 \times 10^5) = 4 \cdot 2 \times 10^{3+5} = 8 \times 10^8; 800,000,000$

55. $(5 \times 10^2)(7 \times 10^{-4}) = 5 \cdot 7 \times 10^{2-4} = 35 \times 10^{-2} = 3.5 \times 10^{-1}; 0.35$

57. $\dfrac{6.3 \times 10^{-2}}{3 \times 10^1} = \dfrac{6.3}{3} \times 10^{-2-1} = 2.1 \times 10^{-3}; 0.0021$

59. $\dfrac{4 \times 10^{-3}}{8 \times 10^{-1}} = \dfrac{4}{8} \times 10^{-3-(-1)} = 0.5 \times 10^{-2} = 5 \times 10^{-3}; 0.005$

61. $\dfrac{8.947 \times 10^7}{0.00095}(4.5 \times 10^8) \approx 42381 \times 10^{15} = 4.2381 \times 10^{19} \approx 4.24 \times 10^{19}$

63. $\left(\dfrac{101+23}{0.42}\right)^2 + \sqrt{3.4 \times 10^{-2}} \approx 87166 + 0.2 \approx 87166.2 \approx 8.72 \times 10^4$

65. $(8.5 \times 10^{-5})(-9.5 \times 10^7)^2 = (8.5 \times 10^{-5})(9.025 \times 10^{15}) \approx 76.7 \times 10^{10} = 7.67 \times 10^{11}$

67. $\sqrt[\eta]{192} \approx 5.769$

69. $|\pi - 3.2| \approx 0.058$

71. $\dfrac{0.3+1.5}{5.5-1.2} \approx 0.419$

73. $\dfrac{1.5^3}{\sqrt{2+\pi}-5} \approx \dfrac{3.375}{-2.732} \approx -1.235$

75. $15 + \dfrac{4+\sqrt{3}}{7} \approx 15.819$

77. From 1985 to 2005: $\dfrac{195-108}{108} \times 100 \approx 80.6\%$

79. From 1995 to 2015: $\dfrac{237-152}{152} \times 100 \approx 55.9\%$

81. (a) $\dfrac{B-A}{A} \times 100 = \dfrac{\$1000-\$500}{\$500} \times 100 = 1 \times 100 = 100\%$

     (b) $\dfrac{A-B}{B} \times 100 = \dfrac{\$500-\$1000}{\$1000} \times 100 = -0.50 \times 100 = -50\%$

83. (a) $\dfrac{B-A}{A} \times 100 = \dfrac{\$1.30-\$1.27}{\$1.27} \times 100 = 0.0236 \times 100 = 2.36\%$

     (b) $\dfrac{A-B}{B} \times 100 = \dfrac{\$1.27-\$1.30}{\$1.30} \times 100 = -0.0231 \times 100 = -2.31\%$

85. (a) $\dfrac{B-A}{A} \times 100 = \dfrac{65-45}{45} \times 100 = 0.4444 \times 100 = 44.44\%$

     (b) $\dfrac{A-B}{B} \times 100 = \dfrac{45-65}{65} \times 100 = -0.3077 \times 100 = -30.77\%$

87. $\$35,000(1+200\%) = \$35,000 \times 3 = \$105,000$

89. $0.14 = 1.4 \times 10^{-1}$ watt

91. The distance Mars travels around the sun is $2\pi r = 2\pi(141,000,000) \approx 885,929,128$ miles . The number of hours in 1.88 years is $365 \times 1.88 \times 24 \approx 16,469$ hours . So Mars' speed is $\dfrac{885,929,128}{16,469} \approx 53,794$ miles per hour .

93. (a) First we will write both numbers in scientific notation. 208 million = 2.08×10^8 , $3,000,000 = 3 \times 10^6$. Then, divide the numbers to find the percentage. $\dfrac{3 \times 10^6}{2.08 \times 10^8} \approx 0.144 = 1.4\%$

     (b) First we will write both numbers in scientific notation. 324 million = 3.24×10^8 , $14,500,000 = 1.45 \times 10^7$. Then, divide the numbers to find the percentage. $\dfrac{1.45 \times 10^7}{3.24 \times 10^8} = 0.04475 = 4.48\%$

95. (a) It would take $\dfrac{1.95 \times 10^{13}}{100} \approx 1.95 \times 10^{11}$ or 195 billion \$100-dollar bills to equal the federal debt. The height of the stacked bills would be $\dfrac{1.95 \times 10^{11}}{250} = 7.8 \times 10^8$ inches or $\dfrac{7.8 \times 10^8}{12} \approx 65,000,000$ feet .

     (b) There are 5280 feet in one mile, so the stacked bills would span $\dfrac{65000000}{5280} \approx 12311$ miles . It would reach farther than the 2500-mile distance between Los Angles and New York.

97. (a) $V = \pi r^2 h \Rightarrow V = \pi(1.3)^2(4.4) \Rightarrow V = 7.436\pi \approx 23.4 \text{ in}^3$

     (b) $1 \text{ in}^3 = 0.55$ fluid ounce $\Rightarrow 23.4 \cdot 0.55 = 12.87$ fluid ounces ; Yes, it can hold 12 fluid ounces.

Extended and Discovery Exercises for Lesson 1.1

1. 2.9×10^{-4} centimeters

3. 0.25 feet, or 3 inches

1.2: Visualizing and Graphing Data

1. (a)

    (b) Maximum: 6; minimum: –2

    (c) $\dfrac{3 + (-2) + 5 + 0 + 6 + (-1)}{6} = \dfrac{11}{6} = 1.8\overline{3}$

3. (a)

    (b) Maximum: 30; minimum: –20

    (c) $\dfrac{-10 + 20 + 30 + (-20) + 0 + 10}{6} = 5$

5.

−30	−30	−10	5	15	25	45	55	61

(a)   The maximum is 61 and the minimum is -30.

(b)   The mean is $\dfrac{-30-30-10+5+15+25+45+55+61}{9} \approx 15.11$ and the median is 15.

7. $\sqrt{15} \approx 3.87$, $2^{2.3} \approx 4.92$, $\sqrt[m]{69} \approx 4.102$, $\pi^2 \approx 9.87$, $2^\pi \approx 8.82, 4.1$

$\sqrt{15}$	4.1	$\sqrt[3]{69}$	$2^{2.3}$	2^π	π^2

(a)   The maximum is π^2 and the minimum is $\sqrt{15}$.

(b)   The mean is $\dfrac{\sqrt{15}+4.1+\sqrt[m]{69}+2^{2.3}+2^\pi+\pi^2}{6} \approx 5.95$ and the median is $\dfrac{\sqrt[m]{69}+2^{2.3}}{2} \approx 4.51$.

9.   (a)

(b)   Mean $= \dfrac{31.7+22.3+12.3+26.8+24.9+23.0}{6} = 23.5$; Median $= \dfrac{23.0+24.9}{6} = 23.95$. The average area of the six largest freshwater lakes is 23,500 square miles. Half of the lakes have areas larger than 23,950 square miles and half have less. The largest difference in area between any two lakes is 19,400 square miles.

(c)   The freshwater lake with the largest area is Lake Superior.

11.   *Answers may vary.* 16, 18, 26; No

13.   (a) $S = \{(-1,5),(2,2),(3,-1),(5,-4),(9,-5)\}$

(b) $D = \{-1,2,3,5,9\}$
$R = \{-5,-4,-1,2,5\}$

15.   (a) $S = \{(1,5),(4,5),(5,6),(4,6),(1,5)\}$

(b) $D = \{1,4,5\}$
$R = \{5,6\}$

17.   (a)   The domain is $D = \{-3,-2,0,7\}$ and the range is $R = \{-5,-3,0,4,5\}$

(b)   The minimum x-value is -3, and the maximum x-value is 7. The minimum y-value is -5, and the maximum y-value is 5.

(c)   The axes must include at least $-3 \le x \le 7$ and $-5 \le y \le 5$.

(d)   See Figure 17.

19.   (a)   The domain is $D = \{-4,-3,-1,0,2\}$ and the range is $R = \{-2,-1,1,2,3\}$.

(b)   The minimum x-value is -4, and the maximum x-value is 2. The minimum y-value is -2, and the maximum y-value is 3.

(c)   The axes must include at least $-4 \le x \le 2$ and $-2 \le y \le 3$.

(d)   See Figure 19.

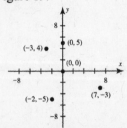

Figure 17

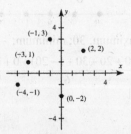

Figure 19

21. (a) The domain is $D = \{-35, -25, 0, 10, 75\}$ and the range is $R = \{-55, -25, 25, 45, 50\}$

    (b) The minimum x-value is -35, and the maximum x-value is 75. The minimum y-value is -55, and the maximum y-value is 50.

    (c) The axes must include at least $-35 \le x \le 75$ and $-55 \le y \le 50$.

    (d) See Figure 21.

23. See Figures 23a and 23b

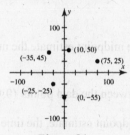

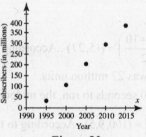

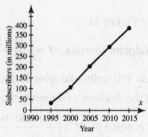

Figure 21   Figure 23a   Figure 23b

25. $a^2 + b^2 = 8^2 + 15^2 = 64 + 225 = 289 = 17^2 = c^2$ Yes

27. $a^2 + b^2 = 7^2 + 22^2 = 49 + 484 = 533 \ne 25^2 = c^2$ No

29. $d = \sqrt{(5-2)^2 + (2-(-2))^2} = \sqrt{3^2 + 4^2} = \sqrt{25} = 5$

31. $d = \sqrt{(9-7)^2 + (1-(-4))^2} = \sqrt{2^2 + 5^2} = \sqrt{29} \approx 5.39$

33. $d = \sqrt{(-9-(-1))^2 + (-20-(-5))^2} = \sqrt{(-8)^2 + (-15)^2} = \sqrt{289} = 17$

35. $d = \sqrt{(-24-12)^2 + (-7-70)^2} = \sqrt{(-36)^2 + (-77)^2} = \sqrt{7225} = 85$

37. $d = \sqrt{(-2.1-3.6)^2 + (8.7-5.7)^2} = \sqrt{(-5.7)^2 + 3^2} = \sqrt{41.49} \approx 6.44$

39. $d = \sqrt{(-3-(-3))^2 + (10-2)^2} = \sqrt{0^2 + 8^2} = \sqrt{64} = 8$

41. $d = \sqrt{\left(\dfrac{3}{4} - \dfrac{1}{2}\right)^2 + \left(\dfrac{1}{2} - \left(-\dfrac{1}{2}\right)\right)^2} = \sqrt{\left(\dfrac{1}{4}\right)^2 + 1^2} = \sqrt{\dfrac{1}{16} + 1} = \sqrt{\dfrac{17}{16}} = \dfrac{\sqrt{17}}{4} \approx 1.03$

43. $d = \sqrt{\left(-\dfrac{1}{10} - \dfrac{2}{5}\right)^2 + \left(\dfrac{4}{5} - \dfrac{3}{10}\right)^2} = \sqrt{\left(-\dfrac{1}{2}\right)^2 + \left(\dfrac{1}{2}\right)^2} = \sqrt{\dfrac{1}{4} + \dfrac{1}{4}} = \sqrt{\dfrac{1}{2}} = \dfrac{\sqrt{2}}{2} = 0.71$

45. $d = \sqrt{(-30-20)^2 + (-90-30)^2} = \sqrt{(-50)^2 + (-120)^2} = \sqrt{2500 + 14,400} = \sqrt{16,900} = 130$

47. $d = \sqrt{(0-a)^2 + (-b-0)^2} = \sqrt{(-a)^2 + (-b)^2} = \sqrt{a^2 + b^2}$

49. $d = \sqrt{(3-0)^2 + (4-0)^2} = \sqrt{3^2 + 4^2} = \sqrt{9 + 16} = \sqrt{25} = 5$

    $d = \sqrt{(7-3)^2 + (1-4)^2} = \sqrt{4^2 + (-3)^2} = \sqrt{16 + 9} = \sqrt{25} = 5$

The side between $(0,0)$ and $(3,4)$ and the side between $(3,4)$ and $(7,1)$ have equal length, so the triangle is isosceles.

51. (a) See Figure 51.

    (b) $d = \sqrt{(0-(-40))^2 + (50-0)^2} = \sqrt{40^2 + 50^2} = \sqrt{1600 + 2500} = \sqrt{4100} \approx 64.0$ miles.

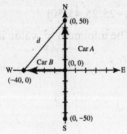

Figure 51

53. Use the midpoint formula $M = \left(\dfrac{24+6}{2}, \dfrac{44+10}{2}\right) = (15, 27)$. According to the midpoint estimate the number of Nintendo Wii units sold after 15 months was 27 million units.

55. Assuming the distance of 0 meters requires 0 seconds to run, the midpoint between the data points $(0, 0)$ and $(200, 19)$ is equal to $M = \left(\dfrac{0+200}{2}, \dfrac{0+19}{2}\right) = (100, 9.5)$. According to the midpoint estimate, the time required to run 100 meters is 9.5 seconds or half the time required to run 200 meters.

57. $M = \left(\dfrac{2016+2020}{2}, \dfrac{7.1+7.8}{2}\right) = (2018, 7.45)$; $M = \left(\dfrac{2020+2024}{2}, \dfrac{7.8+8.0}{2}\right) = (2022, 7.9)$ According to the midpoint estimate, enrollment in two-year colleges will be 7.45 million in 2018 and 7.9 million in 2022.

59. $M = \left(\dfrac{1+5}{2}, \dfrac{2+(-3)}{2}\right) = (3, -0.5)$

61. $M = \left(\dfrac{-30+50}{2}, \dfrac{50+(-30)}{2}\right) = (10, 10)$

63. $M = \left(\dfrac{1.5+(-5.7)}{2}, \dfrac{2.9+(-3.6)}{2}\right) = (-2.1, -0.35)$

65. $M = \left(\dfrac{\sqrt{2}+\sqrt{2}}{2}, \dfrac{\sqrt{5}+(-\sqrt{5})}{2}\right) = (\sqrt{2}, 0)$

67. $M = \left(\dfrac{a+(-a)}{2}, \dfrac{b+3b}{2}\right) = (0, 2b)$

69. $d = \sqrt{(2-5)^2 + (11-7)^2} = \sqrt{(-3)^2 + 4^2} = \sqrt{9+16} = \sqrt{25} = 5$; $M = \left(\dfrac{5+2}{2}, \dfrac{7+11}{2}\right) = \left(\dfrac{7}{2}, 9\right)$

71. $d = \sqrt{(-3-(-8))^2 + (-5-(-2))^2} = \sqrt{5^2 + (-3)^2} = \sqrt{25+9} = \sqrt{34}$; $M = \left(\dfrac{-8+(-3)}{2}, \dfrac{-2+(-5)}{2}\right) = \left(-\dfrac{11}{2}, -\dfrac{7}{2}\right)$

73. $M = \left(\dfrac{7+x}{2}, \dfrac{-4+y}{2}\right) = (8, 5)$; $x = 16-7 = 9$, $y = 10-(-4) = 14$

    $x, y = (9, 14)$

75. $x^2 + y^2 = 25 \Rightarrow (x-0)^2 + (y-0)^2 = 5^2 \Rightarrow$ Center: $(0, 0)$; Radius: 5

77. $x^2 + y^2 = 7 \Rightarrow (x-0)^2 + (y-0)^2 = (\sqrt{7})^2 \Rightarrow$ Center: $(0, 0)$; Radius: $\sqrt{7}$

79. $x^2 + (y+3)^2 = 5 \Rightarrow (x-0)^2 + (y-(-3))^2 = (\sqrt{5})^2 \Rightarrow$ Center: $(0, -3)$; Radius: $\sqrt{5}$

81. $(x-2)^2 + (y+3)^2 = 9 \Rightarrow (x-2)^2 + (y-(-3))^2 = 3^2 \Rightarrow$ Center: $(2, -3)$; Radius: 3

83. $x^2 + (y+1)^2 = 100 \Rightarrow (x-0)^2 + (y-(-1))^2 = 10^2 \Rightarrow$ Center: $(0, -1)$; Radius: 10

85. Since the center is $(1, -2)$ and the radius is 1, the equation is $(x-1)^2 + (y+2)^2 = 1$.

87.　Since the center is $(-2,1)$ and the radius is 2, the equation is $(x+2)^2+(y-1)^2=4$.

89.　$(x-3)^2+(y-(-5))^2=8^2 \Rightarrow (x-3)^2+(y+5)^2=64$

91.　$(x-3)^2+(y-0)^2=7^2 \Rightarrow (x-3)^2+y^2=49$

93.　First find the radius using the distance formula: $r=\sqrt{(4-3)^2+(2-(-5))^2}=\sqrt{50}$.
$(x-3)^2+(y-(-5))^2=(\sqrt{50})^2 \Rightarrow (x-3)^2+(y+5)^2=50$

95.　First find the center using the midpoint formula: $C=\left(\dfrac{-5+1}{2},\dfrac{-7+1}{2}\right)=(-2,-3)$. Then find the radius using

　　the distance formula: $r=\sqrt{(-2-1)^2+(-3-1)^2}=\sqrt{25}=5$.
$(x-(-2))^2+(y-(-3))^2=5^2 \Rightarrow (x+2)^2+(y+3)^2=25$

97.　Find the center using the midpoint formula: $C=\left(\dfrac{5+2}{2},\dfrac{5+1}{2}\right)=\left(\dfrac{7}{2},3\right)$. Then find the radius using the

　　distance formula: $r=\sqrt{\left(\dfrac{7}{2}-2\right)^2+(3-1)^2}=\sqrt{\dfrac{25}{4}}=\dfrac{5}{2}$.
$\left(x-\dfrac{7}{2}\right)^2+(y-3)^2=\left(\dfrac{5}{2}\right)^2 \Rightarrow \left(x-\dfrac{7}{2}\right)^2+(y-3)^2=\dfrac{25}{4}$

99.　See Figure 99.

101.　See Figure 101.

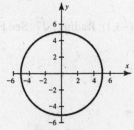

Figure 99

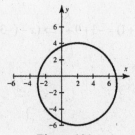

Figure 101

103.　See Figure 103.

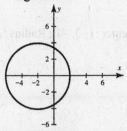

Figure 103

105.　See Figure 105.

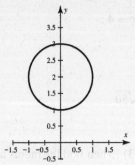

Figure 105

107. See Figure 107.

109. See Figure 109.

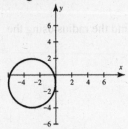

Figure 107

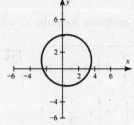

Figure 109

111. $(x^2 + 6x + 9) + (y^2 - 2y + 1) = -1 + 9 + 1 \Rightarrow (x - (-3))^2 + (y - 1)^2 = 9 \Rightarrow$ Center: $(-3, 1)$; Radius: 3 See Figure 111.

113. $(x^2 + 6x + 9) + (y^2 - 2y + 1) = -3 + 9 + 1 \Rightarrow (x - (-3))^2 + (y - 1)^2 = 7 \Rightarrow$ Center: $(-3, 1)$; Radius: $\sqrt{7}$ See Figure 113.

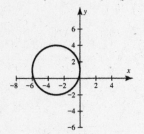

Figure 111

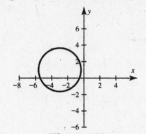

Figure 113

115. $(x^2 + 6x + 9) + (y^2 + 8y + 16) = -9 + 9 + 16 \Rightarrow (x - (-3))^2 + (y - (-4))^2 = 16 \Rightarrow$ Center: $(-3, -4)$; Radius: 4
See Figure 115.

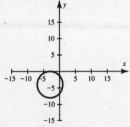

Figure 115

117. $(x^2 - 4x + 4) + (y^2 + 12y + 36) = -4 + 4 + 36 \Rightarrow (x - 2)^2 + (y - (-6))^2 = 36 \Rightarrow$ Center: $(2, -6)$; Radius: 6
See Figure 117.

119. $(4x^2 + 4x + 1) + (4y^2 - 16y + 16) = 19 + 1 + 16 \Rightarrow (2x - (-1))^2 + (2y - 4)^2 = 36 \Rightarrow \left(x - \left(-\frac{1}{2}\right)\right)^2 + (y - 2)^2 = 9$

$\Rightarrow$ Center : $\left(-\frac{1}{2}, 2;\right)$ Radius : 3   See Figure 119.

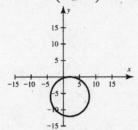
Figure 117

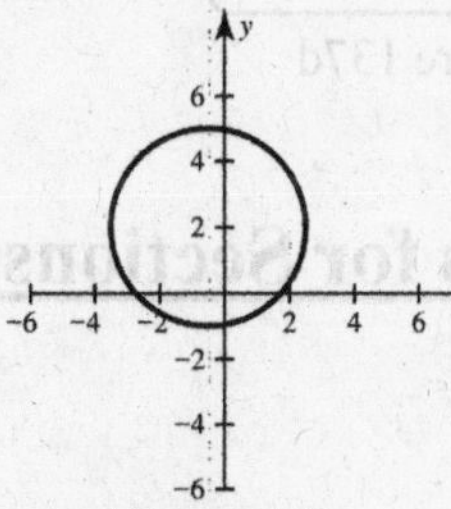

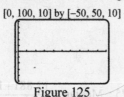
Figure 119

121. $(x^2 + 2x + 1) + (y^2 - 6y + 9) = -14 + 1 + 9 \Rightarrow (x - (-1))^2 + (y - 3)^2 = -4 \Rightarrow$ Not possible

123. x-axis: 10 tick marks; y-axis: 10 tick marks. See Figure 123.

125. x-axis: 10 tick marks; y-axis: 5 tick marks. See Figure 125.

[−10, 10, 1] by [−10, 10, 1]

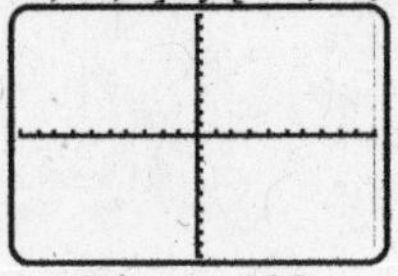

[0, 100, 10] by [−50, 50, 10]

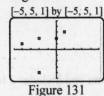
Figure 123

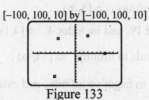
Figure 125

127. Graph b

129. Graph a

131. Plot the points $(1, 3)$, $(-2, 2)$, $(-4, 1)$, $(-2, -4)$ and $(0, 2)$ in $[-5, 5, 1]$ by $[-5, 5, 1]$. See Figure 131.

133. Plot the points $(10, -20)$, $(-40, 50)$, $(30, 60)$, $(-50, -80)$, and $(70, 0)$ in $[-100, 100, 10]$ by $[-100, 100, 10]$. See Figure 133.

[−5, 5, 1] by [−5, 5, 1]

[−100, 100, 10] by [−100, 100, 10]

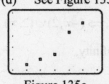
Figure 131

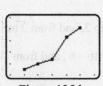
Figure 133

135. (a) The maximum number of Netflix subscriptions is 61.4 million, and the minimum number is 25.1 million. The maximum year is 2015 and the minimum year is 2011 for the given table.

(b) x-min: 2011; x-max: 2015; y-min: 25; y-max: 62. $[2010, 2005, 1]$ by $[5, 20, 5]$. Answers may vary.

(c) See Figure 135c

(d) See Figure 135d

Figure 135c

Figure 135d

137. (a) The maximum U.S. college enrollments who study Chinese is 200 thousand, and the minimum number is 26 thousand. The maximum year is 2015 and the minimum year is 1995.

(b) x-min: 1995; x-max: 2015; y-min: 26; y-max: 200. $[1990, 2020, 5]$ by $[20, 250, 10]$. Answers may vary.

(c) See Figure 137c

(d)   See Figure 137d

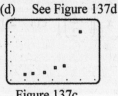
Figure 137c

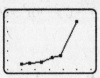
Figure 137d

Checking Basic Concepts for Sections 1.1 and 1.2

1.   (a) $\sqrt{4.2(23.1+0.5^3)} \approx 9.88$

      (b) $\dfrac{23+44}{85.1-32.9} \approx 1.28$

3.   (a) $348,500,000 = 3.485 \times 10^8$

      (b) $-1237.4 = -1.2374 \times 10^8$

      (c) $0.00198 = 1.98 \times 10^{-3}$

5. $M = \left(\dfrac{-2+4}{2}, \dfrac{3+2}{2} \right) = \left(1, \dfrac{5}{2} \right)$

7.   Mean $= \dfrac{13,215+12,881+13,002+3953}{4} = 10,762.75$; Median $= \dfrac{12,881+13,002}{2} = 12,941.5$.

1.3: Functions and Their Representations

1.   If $x \geq 5$, the interval includes 5 and extends to infinity, $\Rightarrow [5,\infty)$.

3.   If $4 \leq x < 19$, the interval extends between 4 and 19, and includes 4, $\Rightarrow [4,19)$.

5.   If $\{x|-1 \leq x\}$, the interval includes -1 and extends to infinity, $\Rightarrow [-1,\infty)$.

7.   If $\{x|x < 1 \text{ or } x \geq 3\}$, the interval extends from 1 to negative infinity, and extends from and includes 3 to infinity, $\Rightarrow (-\infty,1) \cup [3,\infty)$.

9.   If the depicted interval extends from -3 to 5, and includes 5, $\Rightarrow (-3,5]$.

11.   If the depicted interval extends from -2 to negative infinity, $\Rightarrow (-\infty,-2)$.

13.   If the depicted interval extends from -2 to negative infinity, and extends from and includes 1 to infinity, $\Rightarrow (-\infty,-2) \cup [1,\infty)$.

15.   If $\{x|x \neq 2\}$, the interval extends from negative infinity to 2, and from 2 to infinity, $\Rightarrow (-\infty,2) \cup (2,\infty)$.

17.   If $\{x|x \neq -6\}$, the interval extends from negative infinity to -6, and from -6 to infinity, $\Rightarrow (-\infty,-6) \cup (-6,\infty)$.

19.   If $\{x|x \neq -3, x \neq 3\}$, the interval extends from negative infinity to -3, from -3 to 3, and from 3 to infinity, $\Rightarrow (-\infty,-3) \cup (-3,3) \cup (3,\infty)$.

21.   If $\{x|x \neq 1, x \neq 6\}$, the interval extends from negative infinity to 1, from 1 to 6, and from 6 to infinity, $\Rightarrow (-\infty,1) \cup (1,6) \cup (6,\infty)$.

23.   If $f(-2) = 3$, then the point $(-2,3)$ is on the graph of f.

25.   If $(7,8)$ is on the graph of f, then $f(7) = 8$.

27.   See Figure 27.

29.   See Figure 29.

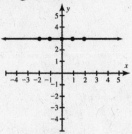

Figure 27

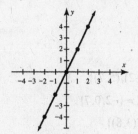

Figure 29

31.   See Figure 31.

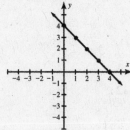

Figure 31

33.   See Figure 33.

35.   See Figure 35.

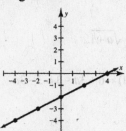

Figure 33

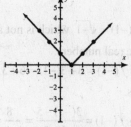

Figure 35

37.   See Figure 37.

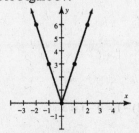

Figure 37

39.   See Figure 39.

41.   See Figure 41.

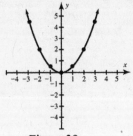

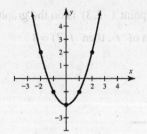

Figure 39   Figure 41

43. (a) $g = \{(-1,0),(2,-2),(5,7)\}$

     (b) $D = \{-1,2,5\}, R = \{-2,0,7\}$

45. (a) $g = \{(1,8),(2,8),(3,8)\}$

     (b) $D = \{1,2,3\}, R = \{8\}$

47. (a) $g = \{(-1,2),(0,4),(1,-3),(2,2)\}$

     (b) $D = \{-1,0,1,2\}; R = \{-3,2,4\}$

49. (a) $f(x) = x^3 \Rightarrow f(-2) = (-2)^3 = -8$ and $f(5) = 5^3 = 125$.

     (b)   All real numbers.

51. (a) $f(x) = x^2 - 3x \Rightarrow f(-3) = (-3)^2 - 3(-3) = 18$ and $f(2) = (2)^2 - 3(2) = -2$.

     (b)   All real numbers.

53. (a) $f(x) = |2-3x| \Rightarrow f(-1) = |2-3(-1)| = 5$ and $f(4) = |2-3(4)| = 10$.

     (b)   All real numbers.

55. (a) $f(x) = \sqrt{-2x} \Rightarrow f(-3) = \sqrt{-2(-3)} = \sqrt{6}$ and $f(5) = \sqrt{-2(5)} = \sqrt{-10}$ which is not a real number.

     (b) $x \le 0$

57. (a) $f(x) = \sqrt{x} \Rightarrow f(-1) = \sqrt{-1}$ which is not a real number, and $f(a+1) = \sqrt{a+1}$.

     (b)   All non-negative real numbers.

59. (a) $f(x) = 6 - 3x \Rightarrow f(-1) = 6 - 3(-1) = 6 + 3 = 9$ and $f(a+1) = 6 - 3(a+1) = 6 - 3a - 3 = 3 - 3a$

     (b)   All real numbers.

61. (a) $f(x) = \dfrac{3x-5}{x+5} \Rightarrow f(-1) = \dfrac{3(-1)-5}{-1+5} = -\dfrac{8}{4} = -2$ and $f(a) = \dfrac{3a-5}{a+5}$.

     (b)   All real numbers not equal to $-5, (x \ne -5)$.

63. (a) $f(x) = \dfrac{1}{x^2} \Rightarrow f(4) = \dfrac{1}{4^2} = \dfrac{1}{16}$ and $f(-7) = \dfrac{1}{(-7)^2} = \dfrac{1}{49}$.

     (b)   All real numbers not equal to $0, (x \ne 0)$.

65. (a) $D \& R : (-\infty, \infty)$

     (b) $g(x) = 2x - 1 \Rightarrow g(-1) = 2(-1) - 1 = -2 - 1 = -3$ and $g(2) = 2(2) = 2(2) - 1 = 4 - 1 = 3$

     (c) $g(-1) = -3$ and $g(2) = 3$

67. (a) $D : (-\infty, \infty); R : (-\infty, 2]$

     (b) $g(x) = 2 - x^2 \Rightarrow g(-1) = 2 - (-1)^2 = 2 - 1 = 1$ and $g(2) = 2 - (2)^2 = 2 - 4 = -2$

     (c) $g(-1) = 1$ and $g(2) = -2$

69. (a) $D : [-2,2]; R : [-3,1]$

     (b) $g(x) = x^2 - 3 \Rightarrow g(-1) = (-1)^2 - 3 = 1 - 3 = -2$ and $g(2) = (2)^2 - 3 = 4 - 3 = 1$

(c) $g(-1) = -2$ and $g(2) = 1$

71. $D : [-3, 3]$

  $R : [0, 3]$

  $f(0) = 3$.

73. $D : (-\infty, \infty)$

  $R : (-\infty, 2]$

  $f(0) = 2$

75. $D : [-1, \infty)$

  $R : (-\infty, 2]$

  $f(0) = 0$

77. (a) $f(2) = 7$

  (b) $f = \{(1, 7), (2, 7), (3, 8)\}$

  (c) $D = \{1, 2, 3\}$; $R = \{7, 8\}$

79. Graph $f(x) = 0.25x^2$ in $[-4.7, 4.7, 1]$ by $[-3.1, 3.1, 1]$ by letting $Y_1 = 0.25X^\wedge 2$. See Figure 79.

  (a) From the graph, it appears that $f(2) = 1$.

  (b) $f(2) = 0.25(2)^2 = 0.25(4) = 1$

  (c) See Figure 79c.

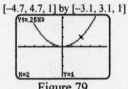
[-4.7, 4.7, 1] by [-3.1, 3.1, 1]

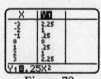

Figure 79   Figure 79c

81. Graph $f(x) = \sqrt{x + 2}$ in $[-4.7, 4.7, 1]$ by $[-3.1, 3.1, 1]$ by letting $Y_1 = \sqrt{(X + 2)}$. See Figure 81.

  (a) From the graph, it appears that $f(2) = 2$.

  (b) $f(2) = \sqrt{2 + 2} = \sqrt{4} = 2$

  (c) See Figure 81c.

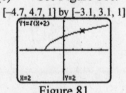
[-4.7, 4.7, 1] by [-3.1, 3.1, 1]

Figure 81   Figure 81c

83. Verbal: Square the input x.

  Graphical: Graph $Y_1 = X^\wedge 2$. See Figure 83.

  Numerical:

x	-2	-1	0	1	2
y	4	1	0	1	4

  $f(2) = 4$

85. Verbal: Multiply the input x by 2, add 1, and then take the absolute value.

  Graphical: Graph $Y_1 = abs(2X + 1)$. See Figure 85.

Numerical:

x	-2	-1	0	1	2
y	3	1	1	3	5

$f(2) = 5$

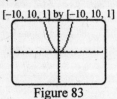 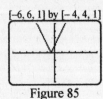

[-10, 10, 1] by [-10, 10, 1]   [-6, 6, 1] by [-4, 4, 1]

Figure 83   Figure 85

87.   Verbal: Subtract the input x from 5.

Graphical: $Y_1 = 5 - X$. See Figure 87.

Numerical:

x	-2	-1	0	1	2
y	7	6	5	4	3

$f(2) = 3$

89.   Verbal: Add 1 to the input x and then take the square root of the result.

Graphical: Graph $Y_1 = \sqrt{(X+1)}$ See Figure 89.

Numerical:

x	-2	-1	0	1	2
y	—	0	1	$\sqrt{2}$	$\sqrt{3}$

$f(2) = \sqrt{3}$

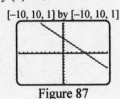 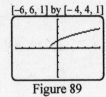

[-10, 10, 1] by [-10, 10, 1]   [-6, 6, 1] by [-4, 4, 1]

Figure 87   Figure 89

91.   It costs about $0.50 per mile.

Symbolic: $f(x) = 0.50x$.

Graphical: See Figure 91a.

Numerical: See Figure 91b.

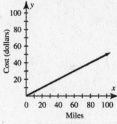

Figure 91a

Miles	1	2	3	4	5	6
Cost	0.50	1.00	1.50	2.00	2.50	3.00

Figure 91b

93. This is a graph of a function because every vertical line intersects the graph at most once. $D \& R : (-\infty, \infty)$

95. This is not a graph of a function because some vertical lines can intersect the graph twice. Because a vertical line can intersect the graph twice, two functions are necessary to create this graph.

97. This is a graph of a function because every vertical line intersects the graph at most once.

    The domain is $[-4, 4]$. The range is $[0, 4]$.

99. a) Yes

    b) Each real number has exactly one real cube root.

101. a) No.

    b) More than one student can have score x.

103. Yes, because the IDs are unique.

105. No. The ordered pairs $(1, 2)$ and $(1, 3)$ belong to the set S. The domain element 1 has more than one range element associated with it.

107. Yes. Each element in its domain is associated with exactly one range element.

109. No. The ordered pairs $(1, 10.5)$ and $(1, -0.5)$ belong to the set S. The domain element 1 has more than one range element associated with it.

111. No, for example, the ordered pairs $(1, -1)$ and $(1, 1)$ belong to the relation. The domain element 1 has more than one range element associated with it.

113. Yes. Each element in the domain of f is associated with exactly one range element.

115. No, for example, the ordered pairs $(0, \sqrt{70})$ and $(0, -\sqrt{70})$ belong to the relation. The domain element 0 has more than one range element associated with it.

117. Yes. Each element in the domain of f is associated with exactly one range element.

119. $g(x) = 12x \Rightarrow g(10) = 12(10) = 120$; there are 120 inches in 10 feet.

121. $g(x) = 0.25x \Rightarrow g(10) = 0.25(10) = 2.50$; there are 2.5 dollars in 10 quarters.

123. $g(x) = 60 \cdot 60 \cdot 24 \cdot x \Rightarrow g(x) = 86,400x \Rightarrow g(10) = 86,400(10) = 864,000$; there are 864,000 seconds in 10 days.

125. $f(x) = 3 - 4x^2 \Rightarrow f(a+2) = 3 - 4(a+2)^2 = -4a^2 - 16a - 13$

127. $f(x) = x^2 - x + 5 \Rightarrow f(a+h) = (a+h)^2 - (a+h) + 5 = a^2 + 2ah + h^2 - a - h + 5$

129. $f(x) = 2x^2 + 3 \Rightarrow f(a+h) - f(a) = \left(2(a+h)^2 + 3\right) - \left(2a^2 + 3\right) = 4ah + 2h^2$

131. a) $h(0) = 64(0) - 16(0)^2 = 0$ ft

   b) $h(4) = 64(4) - 16(4)^2 = 256 - 256 = 0$ ft

   c) Time to reach maximum height is half the time of flight, so it reaches maximum height in 2 s.
   $h(2) = 64(2) - 16(2)^2 = 128 - 64 = 64$ ft

   d) [0, 4]; This represents the total amount of time in which the ball is in the air (4 seconds).

   e) [0, 64]; The height of the golf ball varies from 0 to 64 ft during the 4-s time interval.

133. a) After 4 hours there would be 1500 gallons of water in the swimming pool.

   b) [0, 5]; The domain represents the 5-hour period during which the amount of the water in the pool was recorded.

   c) [500, 1500]; The range represents the amount of water in the pool during the 5-hour period.

135. (a) $\{(R, 37), (N, 30), (S, 17)\}$

   (b) $D = \{N, R, S\}$; $R = \{17, 30, 37\}$

137. $f(x) = 40x \Rightarrow f(5) = 40(5) = 200$; About 200 million tons of electronic waste will be piled up after 5 years.

139. $N(x) = 2200x$; $N(3) = 2200(3) = 6600$; in 3 years, the average person uses 6600 napkins.

141. Verbal: Multiply the input x by -5.8 to obtain the change in temperature.

   Symbolic: $f(x) = -5.8x$.

   Graphical: $Y_1 = -5.8X$. See Figure 141a.

   Numerical: Table $Y_1 = -5.8X$. See Figure 141b.

[0, 3, 1] by [−20, 20, 5]

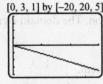
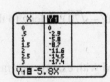

Figure 141a   Figure 141b

1.4: Types of Functions and Their Rates of Change

1. $f(x) = 5 - 2x \Rightarrow f(x) = -2x + 5; \ m = -2, \ b = 5$

3. $f(x) = -8x \Rightarrow f(x) = -8x + 0; \ m = -8, \ b = 0$

5. $m = \dfrac{5 - 6}{2 - 4} = \dfrac{-1}{-2} = \dfrac{1}{2} = 0.5$

7. $m = \dfrac{-2 - 4}{5 - (-1)} = \dfrac{-6}{6} = -1$

9. $m = \dfrac{-8 - (-8)}{7 - 12} = \dfrac{0}{-5} = 0$

11. $m = \dfrac{0.4 - (-0.1)}{-0.3 - 0.2} = \dfrac{0.5}{-0.5} = -1$

13. $m = \dfrac{7.6 - 9.2}{-0.3 - (-0.5)} = \dfrac{-1.6}{0.2} = -8$

15. $m = \dfrac{8 - 6}{-5 - (-5)} = \dfrac{2}{0} = \text{undefined}$

17. $m = \dfrac{\frac{7}{10} - \left(-\frac{3}{5}\right)}{-\frac{5}{6} - \frac{1}{3}} = \dfrac{\frac{13}{10}}{-\frac{7}{6}} = \dfrac{13}{10} \cdot \left(-\dfrac{6}{7}\right) = -\dfrac{39}{35} \approx -1.143$

19. Slope $= 2$; the graph rises 2 units for every unit increase in x.

21. Slope $= -\dfrac{3}{4}$; the graph falls $\dfrac{3}{4}$ unit for every unit increase in x, or equivalently, the graph falls 3 units for every 4-unit increase in x.

23. Slope $= -1$; the graph falls 1 unit for every unit increase in x.

25. (a) Buying no carpet should and does cost $0.

 (b) Slope $= \dfrac{100}{5} = 20$

 (c) The carpet costs $20 per square yard.

27. (a) $D(x) = 150 - 20x \Rightarrow D(5) = 150 - 20(5) = 50$. After 5 hours, the train is 50 miles from the station.

 (b) Slope equals -20. The train is traveling toward the station at 20 mph.

29. (a) $D(2) = 75(2) = 150$ miles

 (b) Slope $= 75$; the car is traveling away from the rest stop at 75 miles per hour.

31. $f(x) = -2x + 5$ is a linear function, but not a constant function, with a slope of $m = -2$. See Figure 31.

33. $f(x) = 1$ is a constant (and linear) function. See Figure 33.

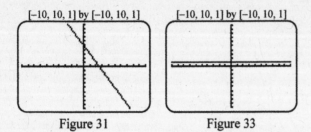

Figure 31   Figure 33

35.   From its graph, we see that $f(x) = |x+1|$ represents a nonlinear function. See Figure 35.

37.   From its graph, we see that $f(x) = x^2 - 1$ represents a nonlinear function. See Figure 37.

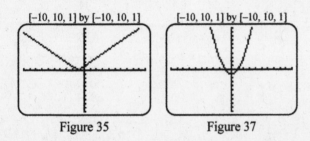

Figure 35   Figure 37

39.   Between each pair of points, the y-values increase 4 units for each unit increase in x. Therefore, the data is linear. The slope of the line passing through the data points is 4.

41.   The y-values do not increase by a constant amount for each 2-unit increase in x. The data is nonlinear.

43.   (a)   Slope $= \dfrac{\text{rise}}{\text{run}} = \dfrac{2}{1} = 2$; y-intercept: $(0, -1)$; x-intercept: $(0.5, 0)$

       (b) $f(x) = ax + b \Rightarrow f(x) = 2x - 1$

       (c)   0.5

45.   (a)   Slope $= \dfrac{\text{rise}}{\text{run}} = \dfrac{-1}{3} = -\dfrac{1}{3}$; y-intercept: $(0. 2)$; x-intercept: $(6, 0)$

       (b) $f(x) = ax + b \Rightarrow f(x) = -\dfrac{1}{3}x + 2$

       (c)   6

47. $f(x) = ax + b \Rightarrow f(x) = -\dfrac{3}{4}x + \dfrac{1}{3}$

49. $f(x) = ax + b \Rightarrow f(x) = 15x + 0$, or $f(x) = 15x$

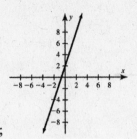

51.   Slope: 3, y-intercept: $(0,2)$;

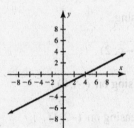

53.   Slope: $\dfrac{1}{2}$, y-intercept: $(0,-2)$;

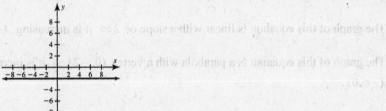

55.   Slope: 0, y-intercept: $(0,-2)$;

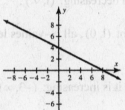

57.   Slope: $-\dfrac{1}{2}$, y-intercept: $(0,4)$;

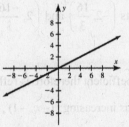

59.   Slope: $\dfrac{1}{2}$, y-intercept: $(0,0)$;

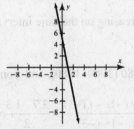

61.   Slope: -5, y-intercept: $(0,5)$;

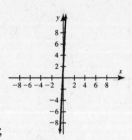

63.   Slope: 20, y-intercept: $(0,0)$;

65. f is decreasing on $(-\infty, \infty)$. It is never increasing.

67. f is increasing on $(2, \infty)$ and decreasing on $(-\infty, 2)$.

69. f is increasing on $(-2, -1)$, $(0, 2)$, and decreasing on $(-1, 0)$.

71. f is increasing on $(-\infty, -2)$, $(1, \infty)$ and decreasing on $(-2, 1)$.

73. f is increasing on $(-8, 0)$, $(8, \infty)$ and decreasing on $(-\infty, -8)$, $(0, 8)$.

75.   The graph of this equation is linear with a slope of $2 \Rightarrow$ it is increasing: $(-\infty, \infty)$, and decreasing: never.

77.   The graph of this equation is a parabola with a vertex $(0, -2) \Rightarrow$ it is increasing: $(0, \infty)$, and decreasing $(-\infty, 0)$.

79.   The graph of this equation is a parabola with a vertex $(1, 1)$, because of the negative coefficient before x^2 it opens downward $\Rightarrow$ it is increasing: $(-\infty, 1)$ and decreasing: $(1, \infty)$.

81.   The square root equation graph has a starting point $(1, 0)$, all x-values less than 1 are undefined $\Rightarrow$ it is increasing: $(1, \infty)$ and decreasing: never.

83.   The absolute value graph has a vertex $(-3, 0) \Rightarrow$ it is increasing: $(-3, \infty)$ and decreasing: $(-\infty, -3)$.

85.   The basic x^3 function is always increasing $\Rightarrow$ it is increasing: $(-\infty, \infty)$ and decreasing: never.

87.   The graph of this cubic equation has turning points $\left(-2, \dfrac{16}{3}\right)$ and $\left(2, \dfrac{-16}{3}\right) \Rightarrow$ it is increasing: $(-\infty, -2)$, $(2, \infty)$; and decreasing: $(-2, 2)$.

89.   The graph of this x^4 graph has a negative lead coefficient therefore is reflected through the x-axis has turning points $\left(-1, \dfrac{5}{12}\right)$, $(0, 0)$ and $\left(2, \dfrac{8}{3}\right) \Rightarrow$ it is increasing: $(-\infty, -1)$, $(0, 2)$; and decreasing: $(-1, 0)$, $(2, \infty)$.

91.   According to the graph, the water levels are increasing on the time intervals $(0, 2.4)$, $(8.7, 14.7)$ and $(21, 27)$.

93.   Energy consumption increased from 1960 to 1980 $(1960, 1980)$, and from 1990 to 2000 $(1990, 2000)$.

95.   The average rate of change from -3 to -1 is $\dfrac{f(-1) - f(-3)}{-1 - (-3)} = \dfrac{3.7 - 1.3}{2} = \dfrac{2.4}{2} = 1.2$.

The average rate of change from 1 to 3 is $\dfrac{f(3)-f(1)}{3-1}=\dfrac{1.3-3.7}{2}=-\dfrac{2.4}{2}=-1.2$.

97. (a) $f(x)=x^2 \Rightarrow f(1)=1^2=1$ and $f(2)=2^2=4 \Rightarrow (1,1),(2,4)$; using the slope formula for rate of change

we get $\dfrac{4-1}{2-1}=\dfrac{3}{1}=3$.

(b) See Figure 97.

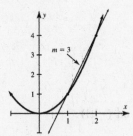

Figure 97

99. If $f(x)=7x-2$ then $\dfrac{f(4)-f(1)}{4-1}=7$. The slope of the graph is 7.

101. If $f(x)=\sqrt{2x-1}$ then $\dfrac{f(3)-f(1)}{3-1}\approx 0.62$. The slope of the line passing through the points is approximately 0.62.

103. (a) The average rate of change from 1900 to 1940 is calculated as $\dfrac{182-3}{1940-1900}=\dfrac{179}{40}=4.475$; from 1940 to

1980: $\dfrac{632-182}{1980-1940}=\dfrac{450}{40}=11.25$; from 1980 to 2010: $\dfrac{315-632}{2010-1980}=\dfrac{-317}{30}\approx -10.6$.

(b) The average rates of change in cigarette consumption in the time periods 1900 to 1940, 1940 to 1980, and 1980 to 2010 were 4.475, 11.25, and -10.06 , respectively.

105. See Figure 105.

107. See Figure 107. *Answers may vary.*

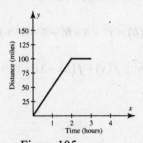

Figure 105

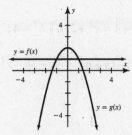

Figure 107

109. See Figure 109. *Answers may vary.*

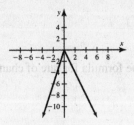

Figure 109

111. a) Zeroes are −2 and 2.

b) x-intercepts: $(-2,0)$, $(2,0)$; y-intercept: $(0,-2)$

c) $f(x)<0$ for $(-2,2)$, and $f(x)>0$ for $(-4,-2)$ and $(2,5)$.

d) $f(x)$ is increasing on $(1,5)$, decreasing on $(-4,-1)$, and unchanging on $(-1,1)$.

e) The average negative change is $\dfrac{-2-4}{-1-(-4)}=-2$, where the average positive change is $\dfrac{1-(-2)}{5-1}=\dfrac{3}{4}$, so the overall average change is negative.

113. a) Zeroes are $-4.5, -2$, 2 and 4.5.

b) x-intercepts: $(-4.5,0)$, $(-2,0)$, $(2,0)$, $(4.5,0)$; y-intercept: $(0,4)$.

c) $f(x)<0$ for $(-4.5,-2)$ and $(2,4.5)$, and $f(x)>0$ for $(-5,-4.5)$, $(-2,2)$, and $(2,4.5)$.

d) $f(x)$ is increasing on $(-3.5,0)$ and $(0,3.5)$, and is decreasing on $(-5,-3.5)$ and $(3.5,5)$.

e) The average negative changes are $\dfrac{-1-1}{-4-(-5)}=-2$, and $\dfrac{-1-4}{3-0}=-\dfrac{5}{3}$, whereas the average positive changes are $\dfrac{3.5-(-1)}{-1-(-3)}=\dfrac{4.5}{2}$ and $\dfrac{1-(-1)}{5-4}=2$, so the overall average change is positive.

115. $f(x)=2x^2 \Rightarrow f(3+4)=2(3+4)^2=2(7)^2=98$; $f(3)+f(4)=2(3)^2+2(4)^2=18+32=50$

117. $f(x)=x^2-4 \Rightarrow f(x+h)=(x+h)^2-4=x^2+2xh+h^2-4$; $f(x)+f(h)=x^2-4+h^2-4=x^2+h^2-8$

119. $f(x)=3x-x^2 \Rightarrow f(x+h)=3(x+h)-(x+h)^2=3x+3h-x^2-2xh-h^2$; $f(x)+f(h)=3x-x^2+3h-h^2$

121. a) $f(x)=3 \Rightarrow f(x+h)=3$

b) $f(x+h)-f(x)=3-3=0$

123. a) $f(x)=2x+1 \Rightarrow f(x+h)=2(x+h)+1=2x+2h+1$

b) $f(x+h)-f(x)=2x+2h+1-(2x+1)=2h$; $\dfrac{f(x+h)-f(x)}{h}=\dfrac{2h}{h}=2$

125. a) $f(x) = 4x - 3 \Rightarrow f(x+h) = 4(x+h) - 3 = 4x + 4h - 3$

   b) $f(x+h) - f(x) = 4x + 4h - 3 - (4x - 3) = 4h$; $\dfrac{f(x+h) - f(x)}{h} = \dfrac{4h}{h} = 4$

127. a) $f(x) = -6x^2 - x + 4 \Rightarrow f(x+h) = -6(x+h)^2 - (x+h) + 4 = -6x^2 - 12xh - 6h^2 - x - h + 4$

   b) $f(x+h) - f(x) = -6x^2 - 12xh - 6h^2 - x - h + 4 - (-6x^2 - x + 4) = -12xh - 6h^2 - h$;

   $$\dfrac{f(x+h) - f(x)}{h} = \dfrac{-12xh - 6h^2 - h}{h} = -12x - 1 - 6h$$

129. a) $f(x) = 1 - x^2 \Rightarrow f(x+h) = 1 - (x+h)^2 = 1 - x^2 - 2xh - h^2$

   b) $f(x+h) - f(x) = 1 - x^2 - 2xh - h^2 - (1 - x^2) = -2xh - h^2$; $\dfrac{f(x+h) - f(x)}{h} = \dfrac{-2xh - h^2}{h} = -2x - h$

131. a) $f(x) = \dfrac{1}{2x} \Rightarrow f(x+h) = \dfrac{1}{2(x+h)} = \dfrac{1}{2x + 2h}$

   b) $f(x+h) - f(x) = \dfrac{1}{2x + 2h} - \dfrac{1}{2x} = \dfrac{2x - 2x - 2h}{(2x + 2h)(2x)} = \dfrac{-h}{2x(x+h)}$;

   $$\dfrac{f(x+h) - f(x)}{h} = \dfrac{-h}{2x(x+h)h} = \dfrac{-1}{2x(x+h)}$$

133. a) $f(x) = 3x^2 + 1 \Rightarrow f(x+h) = 3(x+h)^2 + = 3x^2 + 6xh + 3h^2 + 1$

   b) $f(x+h) - f(x) = 3x^2 + 6xh + 3h^2 + 1 - (3x^2 + 1) = = 6xh + 3h^2$; $\dfrac{f(x+h) - f(x)}{h} = \dfrac{6xh + 3h^2}{h} = 6x + 3h$

135. a) $f(x) = -x^2 + 2x \Rightarrow f(x+h) = -(x+h)^2 + 2(x+h) = -x^2 - 2xh - h^2 + 2x + 2h$

   b) $f(x+h) - f(x) = -x^2 - 2xh - h^2 + 2x + 2h - (-x^2 + 2x) = -2xh - h^2 + 2h$;

   $$\dfrac{f(x+h) - f(x)}{h} = \dfrac{-2xh - h^2 + 2h}{h} = -2x + 2 - h$$

137. a) $f(x) = 2x^2 - x + 1 \Rightarrow f(x+h) = 2(x+h)^2 - (x+h) + 1 = 2x^2 + 4xh + 2h^2 - x - h + 1$

   b) $f(x+h) - f(x) = 2x^2 + 4xh + 2h^2 - x - h + 1 - (2x^2 - x + 1) = 4xh + 2h^2 - h$;

   $$\dfrac{f(x+h) - f(x)}{h} = \dfrac{4xh + 2h^2 - h}{h} = 4x - 1 + 2h$$

139. a) $f(x) = x^3 \Rightarrow f(x+h) = (x+h)^3 = x^3 + 3x^2h + 3xh^2 + h^3$

   b) $f(x+h) - f(x) = x^3 + 3x^2h + 3xh^2 + h^3 - (x^3) = 3x^2h + 3xh^2 + h^3$;

$$\frac{f(x+h)-f(x)}{h} = \frac{3x^2h+3xh^2+h^3}{h} = 3x^2+3xh+h^2$$

141.   a) $d(t) = 8t^2 \Rightarrow d(t+h) = 8(t+h)^2 = 8t^2 + 16th + 8h^2$

     b) $d(t+h) - d(t) = 8t^2 + 16th + 8h^2 - 8t^2 = 16th + 8h^2$; $\dfrac{d(t+h)-d(t)}{h} = \dfrac{16th+8h^2}{h} = 16t + 8h$

     c) $\dfrac{d(4+0.05)-d(4)}{0.05} = 64 + 0.40 = 64.4$; This quantity (64.4 ft/sec) represents the average speed of the car
between 4.00 s and 4.05 s.

Extended and Discovery Exercises for Section 1.4

1.   (a) $C(r) = 2\pi r \Rightarrow C(r+1) = 2\pi(r+1) = 2\pi r + 2\pi$; for every 1 inch increase in the radius, the circumference
increases by 2π inches. So the circumference increases at a constant rate of 2π inches per second.

     (b) No, because the area function, $A(r) = \pi r^2$, depends on the radius squared. The area function is not linear
and thus does not increase at a constant rate.

Checking Basic Concepts for Sections 1.3 and 1.4

1.   Symbolic: $f(x) = 5280x$

     Numerical: Use a table f starting at $x = 1$, incrementing by 1. See Figure 1a.

     Graphical: Graph $Y_1 = 5280X$ as shown in Figure 1b.

x	1	2	3	4	5
$f(x)$	5280	10,560	15,840	21,120	26,400

Figure 1a

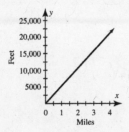

Figure 1b

3.   The slope is calculated as follows: $m = \dfrac{-5-4}{4-(-2)} = -\dfrac{9}{6} = -\dfrac{3}{2}$

If the graph of the linear function $f(x) = ax + b$, passes through the points $(-2, 4)$ and $(4, -5)$, its slope must be equal to $-\dfrac{3}{2}$; therefore, $a = -\dfrac{3}{2}$.

5.   (a) $(-\infty, 5]$

    (b) $[1, 6)$

7. $\dfrac{f(-1) - f(-3)}{-1 - (-3)} = \dfrac{4 - 18}{2} = -7$

Chapter 1 Review Exercises

1. -2 is an integer, rational number, and real number. $\dfrac{1}{2}$ is both a rational and a real number. 0 is an integer, rational number, and real number. 1.23 is both a rational and a real number. $\sqrt{7}$ is a real number. $\sqrt{16} = 4$ is a natural number, integer, rational number, and real number.

3. $1,891,000 = 1.891 \times 10^{6}$

5. $1.52 \times 10^{4} = 15,200$

7.   (a) $\sqrt[m]{1.2} + \pi^{3} \approx 32.07$

    (b) $\dfrac{3.2 + 5.7}{7.9 - 4.5} \approx 2.62$

    (c) $\sqrt{5^{2} + 2.1} \approx 5.21$

    (d) $1.2(6.3)^{2} + \dfrac{3.2}{\pi - 1} \approx 49.12$

9. $.4 - 3^{2} \cdot 5 = 4 - 9 \cdot 5 = 4 - 45 = -41$

11. $\dfrac{B - A}{A} \times 100 = \dfrac{\$120 - \$150}{\$150} \times 100 = -0.2 \times 100 = -20\%$

13.

-23	-5	8	19	24

    (a)   Maximum $= 24$; Minimum $= -23$

    (b)   Mean $= \dfrac{-23 + (-5) + 8 + 19 + 24}{5} = 4.6$; Median $= 8$

15.   (a) $S = \{(-15, -3), (-10, -1), (0, 1), (5, 3), (20, 5)\}$

    (b) $D = \{-15, -10, 0, 5, 20\}$ and $R = \{-3, -1, 1, 3, 5\}$

17.   The relation $\{(10, 13), (-12, 40), (-30, -23), (25, -22), (10, 20)\}$ is plotted in Figure 15. It is not a function since both $(10, 13)$ and $(10, 20)$ are contained in the set. Notice that these points are lined up vertically.

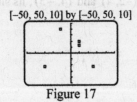

[−50, 50, 10] by [−50, 50, 10]

Figure 17

19. $d = \sqrt{(2-(-4))^2 + (-3-5)^2} = \sqrt{6^2 + (-8)^2} = \sqrt{36+64} = \sqrt{100} = 10$

21. $M = \left(\dfrac{24+(-20)}{2}, \dfrac{-16+13}{2}\right) = \left(\dfrac{4}{2}, \dfrac{-3}{2}\right) = \left(2, \dfrac{-3}{2}\right)$

23.   Center: $(1,-1)$; Radius: 2

25. $(x-2)^2 + (y-5)^2 = 17$

27.   See Figure 27.

29.   See Figure 29.

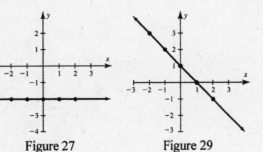

Figure 27   Figure 29

31.   See Figure 31.

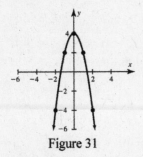

Figure 31

33.   See Figure 33.

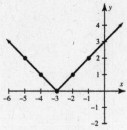

Figure 33

35. Symbolic: $f(x) = 16x$.

Numerical: Table f starting at $x = 0$, incrementing by 25. See Figure 35a.

Graphical: Graph $Y_1 = 16X$ in [0, 100, 10] by [0, 1800, 300]. See Figure 35b.

[0, 100, 10] by [0, 1800, 300]

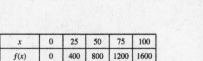

x	0	25	50	75	100
$f(x)$	0	400	800	1200	1600

Figure 35a   Figure 35b

37. (a) $f(x) = 5 \Rightarrow f(-3) = 5$ and $f(1.5) = 5$

    (b) $(-\infty, \infty)$

39. (a) $f(x) = x^2 - 3 \Rightarrow f(-10) = (-10)^2 - 3 = 97$ and $f(a+2) = (a+2)^2 - 3 = a^2 + 4a + 4 - 3 = a^2 + 4a + 1$

    (b) $(-\infty, \infty)$

41. (a) $f(-3) = \dfrac{1}{-3-4} = -\dfrac{1}{7}$; $f(a+1) = \dfrac{1}{a+1-4} = \dfrac{1}{a-3}$

    (b) $(-\infty, 4) \cup (4, \infty)$

43. No, for example, an input $x = 6$ produces outputs of $y = \pm 1$.

45. (a) Using the points (0, 6) and (2, 2), $m = \dfrac{2-6}{2-0} = \dfrac{-4}{2} = -2$; y-intercept: (0,6); x-intercept: (3,0).

    (b) $f(x) = -2x + 6$

    (c) The zeros of f are the same as the x-intercepts. That is $x = 3$.

47. Since any vertical line intersects the graph of f at most once, it is a function.

49. Yes, it is a function. Each input produces a single output.

51. $a = 0$, so the slope $= 0$.

53. $m = \dfrac{4-7}{3-(-1)} = -\dfrac{3}{4}$

55. $m = \dfrac{4-4}{-2-8} = \dfrac{0}{-10} = 0$

57. $f(x) = 8 - 3x$ represents a linear function.

59. $f(x) = |x+2|$ represents a nonlinear function.

61.   See Figure 61.

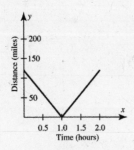

Figure 61

63.   Yes, $m = \dfrac{50-26}{-2-4} = \dfrac{24}{-6} = -4$. The best model is linear, but not constant, since the y-values decrease 8 units

for every 2-unit increase in x.

65. $\begin{aligned} & f(x+h) = 5(x+h)+1 = 5x+5h+1 \\ & \dfrac{f(x+h)-f(x)}{h} = \dfrac{5x+5h+1-(5x+1)}{h} = \dfrac{5h}{h} = 5 \end{aligned}$

67. $f(x) = 2x^2 \Rightarrow f(x+h) = 2(x+h)^2 = 2x^2 + 4xh + 2h^2$; $f(x) + f(h) = 2x^2 + 2h^2$

69.

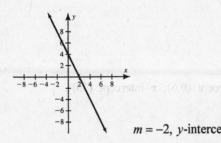

$m = -2$, y-intercept $= (0,4)$

71. $\dfrac{2.28 \times 10^8}{3 \times 10^5} = 760$ seconds $= 12\dfrac{2}{3}$ minutes

73.   (a)   Sketching a diagram of the pool and sidewalk (Not Shown) gives the following dimensions:

$l = 62$ ft. and $w = 37$ ft.   Thus, $P = 2(62) + 2(37) = 198$ ft.

(b) The area of the sidewalk would consist of the area of four 6×6 squares, two 50×6 rectangles, and two 25×6 rectangles. $A = 4(6 \cdot 6) + 2(50 \cdot 6) + 2(25 \cdot 6) = 4(36) + 2(300) + 2(150) = 1044$ ft^2

75. (a) Plot the points (0, 100), (1, 10), (2, 6), (3, 3), and (4, 2) and make a line graph. See Figure 75. The survival rates decrease rapidly at first. This means that a large number of eggs never develop into mature adults. See Figure 75a.

(b) Since any vertical line could intersect the graph at most once, this graph could represent a function.

(c) From 0 to 1, $\dfrac{10-100}{1-0} = -90$; from 1 to 2, $\dfrac{6-10}{2-1} = -4$; from 2 to 3, $\dfrac{3-6}{3-2} = -3$; from 3 to 4, $\dfrac{2-3}{4-3} = -1$; during the first year, the population of sparrows decreased, on average, by 90 birds. The other average rates of change can be interpreted similarly.

77. (a) See Figure 77a. f is nonlinear.

(b) $f(x) = 0.5x^2 + 50 \Rightarrow \dfrac{f(4) - f(1)}{4-1} = \dfrac{58 - 50.5}{3} = 2.5$

(c) The average rate of change in outside temperature from 1 P.M. to 4 P.M. was 2.5° F per hour. The slope of the line segment from (1, 50.5) to (4, 58) is 2.5. The temperature increased, on average, by 2.5° F per hour.

[−1, 5, 1] by [0, 110, 10]   [1, 5, 1] by [40, 70, 5]

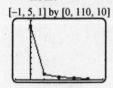   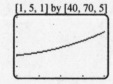

Figure 75   Figure 77a

Chapter 2: Linear Functions and Equations

2.1: Equations of Lines

1. Find slope: $m = \dfrac{-2-2}{3-1} = \dfrac{-4}{2} = -2$. Using $(x_1, y_1) = (1, 2)$ and point-slope form $y = m(x - x_1) + y_1$, we get

    $y = -2(x-1) + 2$. See Figure 1.

3. Find slope: $m = \dfrac{2-(-1)}{1-(-3)} = \dfrac{3}{4}$. Using $(x_1, y_1) = (-3, -1)$ and point-slope form $y = m(x - x_1) + y_1$, we get

    $y = \dfrac{3}{4}(x+3) - 1$. See Figure 3.

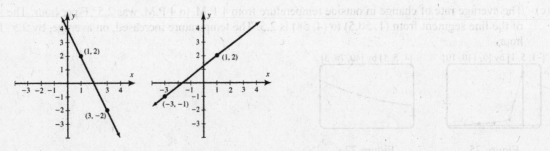

Figure 1   Figure 3

5. The point-slope form is given by $y = m(x - x_1) + y_1$. Thus, $m = -2.4$ and

    $(x_1, y_1) = (4, 5) \Rightarrow y = -2.4(x-4) + 5 \Rightarrow y = -2.4x + 9.6 + 5 \Rightarrow y = -2.4x + 14.6$ and $f(x) = -2.4x + 14.6$.

7. First find the slope between the points $(1, -2)$ and $(-9, 3)$: $m = \dfrac{3-(-2)}{-9-1} = -\dfrac{1}{2}$.

    $y = -\dfrac{1}{2}(x-1) - 2 \Rightarrow y = -\dfrac{1}{2}x + \dfrac{1}{2} - 2 \Rightarrow y = -\dfrac{1}{2}x - \dfrac{3}{2}$ and $f(x) = -\dfrac{1}{2}x - \dfrac{3}{2}$

9. $(4, 0)$, $(0, -3)$; $m = \dfrac{-3-0}{0-4} = \dfrac{3}{4}$. Thus, $y = \dfrac{3}{4}(x-4) + 0$ or $y = \dfrac{3}{4}x - 3$ and $f(x) = \dfrac{3}{4}x - 3$.

11. Using the points $(0, -1)$ and $(3, 1)$, we get $m = \dfrac{1-(-1)}{3-0} = \dfrac{2}{3}$ and $b = -1$; $y = mx + b \Rightarrow y = \dfrac{2}{3}x - 1$.

13. Using the points $(-2, 1.8)$ and $(1, 0)$, we get $m = \dfrac{0-1.8}{1-(-2)} = \dfrac{-1.8}{3} = -\dfrac{18}{30} = -\dfrac{3}{5}$; to find b, we use $(1, 0)$ in

    $y = mx + b$ and solve for b: $0 = -\dfrac{3}{5}(1) + b \Rightarrow b = \dfrac{3}{5}$; $y = -\dfrac{3}{5}x + \dfrac{3}{5}$.

15. c

17.   b

19.   e

21. $m = \dfrac{2-(-4)}{1-(-1)} = 3$; $y = 3(x+1)-4 = 3x+3-4 = 3x-1$

23. $m = \dfrac{-3-5}{1-4} = \dfrac{8}{3}$; $y = \dfrac{8}{3}(x-4)+5 = \dfrac{8}{3}x - \dfrac{32}{3} + 5 = \dfrac{8}{3}x - \dfrac{17}{3}$

25. $b = 5$ and $m = -7.8 \Rightarrow y = -7.8x+5$.

27.   The line passes through the points $(0, 45)$ and $(90, 0)$.

$m = \dfrac{0-45}{90-0} = -\dfrac{1}{2}$; $b = 45$ and $m = -\dfrac{1}{2} \Rightarrow y = -\dfrac{1}{2}x + 45$

29. $m = -3$ and $b = 5 \Rightarrow y = -3x+5$

31. $m = \dfrac{0-(-6)}{4-0} = \dfrac{6}{4} = \dfrac{3}{2}$ and $b = -6$; $y = mx+b \Rightarrow y = \dfrac{3}{2}x - 6$

33. $m = \dfrac{\frac{2}{3}-\frac{3}{4}}{\frac{1}{5}-\frac{1}{2}} = \dfrac{-\frac{1}{12}}{-\frac{3}{10}} = \dfrac{5}{18}$; using the point-slope form with $m = \dfrac{5}{18}$ and $\left(\dfrac{1}{2}, \dfrac{3}{4}\right)$, we get

$y = \dfrac{5}{18}\left(x - \dfrac{1}{2}\right) + \dfrac{3}{4} \Rightarrow y = \dfrac{5}{18}x - \dfrac{5}{36} + \dfrac{3}{4} \Rightarrow y = \dfrac{5}{18}x + \dfrac{11}{18}$.

35.   The line has a slope of 4 and passes through the point $(-4, -7)$; $y = 4(x+4)-7 \Rightarrow y = 4x+9$.

37.   The slope of the perpendicular line is equal to $\dfrac{3}{2}$ and the line passes through the point $(1980, 10)$;

$y = \dfrac{3}{2}(x-1980)+10 \Rightarrow y = \dfrac{3}{2}x - 2960$

39. $y = \dfrac{2}{3}x+3 \Rightarrow m = \dfrac{2}{3}$; the parallel line has slope $\dfrac{2}{3}$; since it passes through $(0, -2.1)$, the y-intercept

$= -2.1$; $y = mx+b \Rightarrow y = \dfrac{2}{3}x - 2.1$.

41. $y = -2x \Rightarrow m = -2$; the perpendicular line has slope $\dfrac{1}{2}$; since it passes through $(-2, 5)$, the equation is

$y = \dfrac{1}{2}(x+2)+5 = \dfrac{1}{2}x+1+5 = \dfrac{1}{2}x+6$.

43. $y = -x+4 \Rightarrow m = -1$; the perpendicular line has slope 1; since it passes through $(15, -5)$, the equation is
$y = 1(x-15)-5 = x-15-5 = x-20$.

45. $-3x + 4y = 12 \Rightarrow y = \frac{3}{4}x + 3 \Rightarrow m = \frac{3}{4}$; the parallel line has slope $\frac{3}{4}$; since it passes through $(-4, -6)$, the

equation is $y = \frac{3}{4}(x - (-4)) - 6 \Rightarrow y = \frac{3}{4}x + 3 - 6 \Rightarrow y = \frac{3}{4}x - 3$.

47. $m = \frac{1 - 3}{-3 - 1} = \frac{-2}{-4} = \frac{1}{2}$; a line parallel to this line also has slope $m = \frac{1}{2}$. Using $(x_1, y_1) = (5, 7)$, $m = \frac{1}{2}$, and

point-slope form $y = m(x - x_1) + y_1$, we get $y = \frac{1}{2}(x - 5) + 7 \Rightarrow y = \frac{1}{2}x + \frac{9}{2}$.

49. $m = \frac{\frac{2}{3} - \frac{1}{2}}{-3 - (-5)} = \frac{\frac{1}{6}}{2} = \frac{1}{12}$; a line perpendicular to this line has slope $m = -\frac{12}{1} = -12$. Using $(x_1, y_1) = (-2, 4)$,

$m = -12$, and point-slope form $y = m(x - x_1) + y_1$, we get

$y = -12(x + 2) + 4 \Rightarrow y = -12x - 24 + 4 \Rightarrow y = -12x - 20$.

51. $x = -5$. It is not possible to write as a linear function since a vertical line does not represent a function.

53. $y = 6$ and $f(x) = 6$.

55. Since the line $y = 15$ is horizontal, the perpendicular line through $(4, -9)$ is vertical and has equation $x = 4$.
    It is not possible to write as a linear function since a vertical line does not represent a function.

57. The line through $(19, 5.5)$ and parallel to $x = 4.5$ is also vertical and has equation $x = 19$. It is not possible to
    write as a linear function since a vertical line does not represent a function.

59. Let $4x - 5y = 20$.

    x-intercept: Substitute $y = 0$ and solve for x. $4x - 5(0) = 20 \Rightarrow 4x = 20 \Rightarrow x = 5$; x-intercept: $(5, 0)$

    y-intercept: Substitute $x = 0$ and solve for y. $4(0) - 5y = 20 \Rightarrow -5y = 20 \Rightarrow y = -4$; y-intercept: $(0, -4)$

    See Figure 59.

61. Let $x - y = 7$.

    x-intercept: Substitute $y = 0$ and solve for x. $x - 0 = 7 \Rightarrow x = 7$; x-intercept: $(7, 0)$

    y-intercept: Substitute $x = 0$ and solve for y. $0 - y = 7 \Rightarrow -y = 7 \Rightarrow y = -7$; y-intercept: $(0, -7)$

    See Figure 61.

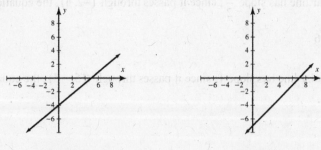

Figure 59   Figure 61

63.   Let $6x - 7y = -42$.

x-intercept: Substitute $y = 0$ and solve for x. $6x - 7(0) = -42 \Rightarrow 6x = -42 \Rightarrow x = -7$; x-intercept: $(-7, 0)$

y-intercept: Substitute $x = 0$ and solve for y. $6(0) - 7y = -42 \Rightarrow -7y = -42 \Rightarrow y = 6$; y-intercept: $(0, 6)$

See Figure 63.

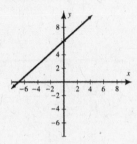

Figure 63

65.   Let $y - 3x = 7$.

x-intercept: Substitute $y = 0$ and solve for x. $0 - 3x = 7 \Rightarrow -3x = 7 \Rightarrow x = -\dfrac{7}{3}$; x-intercept: $\left(-\dfrac{7}{3}, 0\right)$

y-intercept: Substitute $x = 0$ and solve for y. $y - 3(0) = 7 \Rightarrow y - 0 = 7 \Rightarrow y = 7$; y-intercept: $(0, 7)$

See Figure 65.

67.   Let $0.2x + 0.4y = 0.8$.

x-intercept: Substitute $y = 0$ and solve for x. $0.2x + 0.4(0) = 0.8 \Rightarrow 0.2x = 0.8 \Rightarrow x = 4$; x-intercept: $(4, 0)$

y-intercept: Substitute $x = 0$ and solve for y. $0.2(0) + 0.4y = 0.8 \Rightarrow 0.4y = 0.8 \Rightarrow y = 2$; y-intercept: $(0, 2)$

See Figure 67.

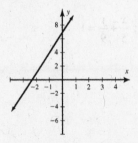

Figure 65

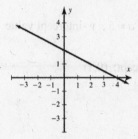

Figure 67

69.   Let $y = 8x - 5$.

x-intercept: Substitute $y = 0$ and solve for x. $0 = 8x - 5 \Rightarrow 5 = 8x \Rightarrow x = \dfrac{5}{8}$; x-intercept: $\left(\dfrac{5}{8}, 0\right)$

y-intercept: Substitute $x = 0$ and solve for y. $y = 8(0) - 5 \Rightarrow y = -5$; y-intercept: $(0, -5)$

See Figure 69.

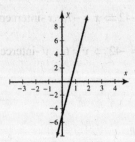

Figure 69

71. Let $\dfrac{x}{5} + \dfrac{y}{7} = 1$.

x-intercept: Substitute $y = 0$ and solve for x. $\dfrac{x}{5} + \dfrac{0}{7} = 1 \Rightarrow \dfrac{x}{5} = 1 \Rightarrow x = 5$; x-intercept: $(5, 0)$

y-intercept: Substitute $x = 0$ and solve for y. $\dfrac{0}{5} + \dfrac{y}{7} = 1 \Rightarrow \dfrac{y}{7} = 1 \Rightarrow y = 7$; y-intercept: $(0, 7)$

a and b represent the x- and y-intercepts, respectively.

73. Let $\dfrac{2x}{3} + \dfrac{4y}{5} = 1$.

x-intercept: Substitute $y = 0$ and solve for x. $\dfrac{2x}{3} + \dfrac{4(0)}{5} = 1 \Rightarrow \dfrac{2x}{3} = 1 \Rightarrow x = \dfrac{3}{2}$; x-intercept: $\left(\dfrac{3}{2}, 0\right)$

y-intercept: Substitute $x = 0$ and solve for y. $\dfrac{2(0)}{3} + \dfrac{4y}{5} = 1 \Rightarrow \dfrac{4y}{5} = 1 \Rightarrow y = \dfrac{5}{4}$; y-intercept: $\left(0, \dfrac{5}{4}\right)$

a and b represent the x- and y-intercepts, respectively.

75. $\dfrac{x}{a} + \dfrac{y}{b} = 1$; x-intercept value: $5 \Rightarrow a = 5$, y-intercept value: $9 \Rightarrow b = 9$; $\dfrac{x}{5} + \dfrac{y}{9} = 1$

77. (a) x-intercept: $(-3, 0)$; y-intercept: $(0, 1)$

    (b) $\dfrac{rise}{run} = \dfrac{1}{3} \Rightarrow m = \dfrac{1}{3}$

    (c)   zero: $x = -3$

    (d) f is positive on $(-3, \infty)$ and negative on $(-\infty, -3)$

    (e)   Increasing: $(-\infty, \infty)$; The function is increasing and negative on $(-\infty, -3)$.

    (f) $y = \dfrac{1}{3}x + 1$; Since m is positive, the linear function must be increasing.

79. (a) x -intercept: $(1,0)$; y -intercept: $(0,3)$

   (b) $\dfrac{rise}{run} = \dfrac{3}{-1} \Rightarrow m = -3$

   (c) zero: $x = 1$

   (d) f is negative on $(1,\infty)$ and positive on $(-\infty,1)$

   (e) Decreasing: $(-\infty,\infty)$; The function is decreasing and positive on $(-\infty,1)$.

   (f) $y = -3x + 3$ Since m is negative, the linear function must be decreasing.

81. (a) x -intercept: $(k,0)$; y -intercept: $(0,b)$

   (b) $\dfrac{rise}{run} = \dfrac{b}{-k} \Rightarrow m = \dfrac{b}{-k}$

   (c) zero: $x = k$

   (d) f is negative on (k,∞) and positive on $(-\infty,k)$

   (e) Decreasing: $(-\infty,\infty)$; The function is decreasing and positive on $(-\infty,k)$.

   (f) $y = \dfrac{-b}{k}x + b$ Since m is negative, the linear function must be decreasing.

83. (a) Since the point $(0,-3.2)$ is on the graph, the y -intercept value is -3.2. The data is exactly linear, so one can use any two points to determine the slope. Using the points $(0,-3.2)$ and $(1,-1.7)$,

   $m = \dfrac{-1.7-(-3.2)}{1-0} = 1.5$. The slope-intercept form of the line is $y = 1.5x - 3.2$.

   (b) When $x = -2.7$, $y = 1.5(-2.7) - 3.2 = -7.25$. This calculation involves interpolation.

   When $x = 6.3$, $y = 1.5(6.3) - 3.2 = 6.25$. This calculation involves extrapolation.

85. (a) Since the data is exactly linear, one can use any two points to determine the slope. Using the points $(5, 94.7)$ and $(23, 56.9)$, $m = \dfrac{56.9 - 94.7}{23 - 5} = -2.1$. The point-slope form of the line is $y = -2.1(x-5) + 94.7$ and the slope-intercept form of the line is $y = -2.1x + 105.2$.

   (b) When $x = -2.7$, $y = -2.1(-2.7) + 105.2 = 110.87$. This calculation involves extrapolation.

   When $x = 6.3$, $y = -2.1(6.3) + 105.2 = 91.97$. This calculation involves interpolation.

87. (a) Using the points $(2008, 3)$ and $(2011, 24)$, $m = \dfrac{24 - 3}{2011 - 2008} = \dfrac{21}{3} = 7$. The point slope form of the line is $f(x) = 7(x - 2008) + 3$. The function can also be written as $f(x) = 6.9x - 13,852.3$ (*answers may vary*). The function approximately models the given data.

   (b) $f(2007) = 7(2007 - 2008) + 3 = -7 + 3 = -4 \Rightarrow -4\%$

(c) The calculation involved extrapolation. The result was a negative so it is not possible. Numbers were decreasing but increased after 2011.

(d) $f(2015) = 7(2015) - 14,053 = 52 \Rightarrow 52\%$

(e) The calculation involved extrapolation. It is inaccurate because the percentage began decreasing after 2011.

89. (a) Find the slope: $m = \dfrac{32,000 - 21,000}{2015 - 2005} = \dfrac{11,000}{10} = 1100$. Using the first point $(2005, 21000)$ for (x_1, y_1) and $m = 1100$, we get $y = 1100(x - 2005) + 21,000 = 1100x - 2,184,500$. The cost of attending a private college or university is increasing by $1100 per year on average.

(b) $y = \dfrac{12,000}{7}(2007 - 2003) + 25,000 \Rightarrow y = 1100(2013) - 2184500 \Rightarrow y \approx \$29,800$.

91. (a) Water is leaving the tank because the amount of water in the tank is decreasing. After 3 minutes there are approximately 70 gallons of water in the tank.

(b) The x-intercept is $(10,0)$. This means that after 10 minutes the tank is empty. The y-intercept is $(0,100)$. This means that initially there are 100 gallons of water in the tank.

(c) To determine the equation of the line, we can use 2 points. The points $(0, 100)$ and $(10, 0)$ lie on the line. The slope of this line is $m = \dfrac{0 - 100}{10 - 0} = -10$. This slope means the water is being drained at a rate of 10 gallons per minute. Since the y-intercept is 100, the slope-intercept form of this line is given by $y = -10x + 100$.

(d) From the graph, when $y = 50$ the x-value appears to be 5. Symbolically, when $y = 50$ then $-10x + 100 = 50 \Rightarrow -10x = -50 \Rightarrow x = 5$. The x-coordinate is 5.

93. (a) First calculate the slope: $m = \dfrac{240 - 500}{2015 - 2007} = \dfrac{-260}{8} = -\dfrac{65}{2} = -32.5$, using the first point we have $y = -32.5(x - 2007) + 500 \Rightarrow y = -32.5x + 65,727.5$.

(b) The sales decreased, on average, by 32.5 million per year.

(c) $f(2011) = -32.5(2011) + 65727.5 = 370 \Rightarrow \$370$ million. The estimate is about $40 million higher than the true value of $330 million. The calculation involves interpolation.

95. (a) See Figure 95a.

(b) Using the points $(2006, 160)$ and $(2010, 425)$, $m = \dfrac{425 - 160}{2010 - 2006} = \dfrac{265}{4} = 66.25$. The point slope form of the line is $f(x) = 66.25(x - 2006) + 160$.

(c) See Figure 95c.

(d) Bankruptcies increased, on average, by 66,250 per year.

(e) $f(2014) = 66.25(2014 - 2006) + 160 = 690 \Rightarrow 690,000$; The calculation involves extrapolation.

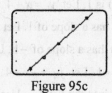

Figure 95a   Figure 95c

97. (a) Using the points (1970, 2000) and (2010, 1590), $m = \dfrac{1590 - 2000}{2010 - 1970} = \dfrac{-410}{40} = -10.25$. Since x

represents the number of years after 1970, we have a y-intercept of 2000, and the function is

$f(x) = -10.25x + 2000$.

(b) Hours worked decreased, on average, by 10.25 hours per year.

(c) Since 2014 is 44 years after 1970, we will let $x = 44$ and $f(44) = -10.25(44) + 2000 = 1549$. The

result is about 1.549 hours.

99. (a) Graph $Y_1 = X/1024 + 1$ in $[0, 3, 1]$ by $[-2, 2, 1]$ as in Figure 99. The line appears to be horizontal in

this viewing rectangle, however, we know that the graph of the line is not horizontal because its slope

is $\dfrac{1}{1024} \neq 0$.

(b) The resolution of most graphing calculator screens is not good enough to show the slight increase in

the y-values. Since the x-axis is 3 units long, this increase in y-values amounts to only

$\dfrac{1}{1024} \times 3 \approx 0.003$ units, which does not show up on the screen.

$[0, 3, 1]$ by $[-2, 2, 1]$

Figure 99

101. (a) From Figure 101a, one can see that the lines do not appear to be perpendicular. *(Answers may vary.)*

(b) The lines are graphed in the specified viewing rectangles and shown in Figures 101b-d, respectively. In

the windows $[-15, 15, 1]$ by $[-10, 10, 1]$ and $[-3, 3, 1]$ by $[-2, 2, 1]$ the lines appear to be

perpendicular.

(c) The lines appear perpendicular when the distance shown along the x-axis is approximately 1.5 times

the distance along the y-axis. For example, in window $[-12, 12, 1]$ by $[-8, 8, 1]$, the lines will appear

perpendicular. The distance along the x-axis is 24 while the distance along the y-axis is 16. Notice

that $1.5 \times 16 = 24$. This is called a "square window" and can be set automatically on some graphing

calculators.

$[-10, 10, 1]$ by $[-10, 10, 1]$ $[-15, 15, 1]$ by $[-10, 10, 1]$ $[-10, 10, 1]$ by $[-3, 3, 1]$ $[-3, 3, 1]$ by $[-2, 2, 1]$

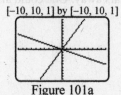   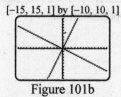   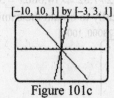   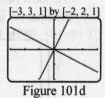

Figure 101a   Figure 101b   Figure 101c   Figure 101d

103. (i) The slope of the line connecting $(0, 0)$ and $(2, 2)$ is 1. Let $y_1 = x$.

(ii) A second line passing through $(0, 0)$ has a slope of -1. Let $y_2 = -x$.

(iii) A third line passing through $(1, 3)$ has a slope of 1. Let $y_3 = (x - 1) + 3 = x + 2$.

(iv) A fourth line passing through $(2, 2)$ has a slope of -1. Let $y_4 = -(x - 2) + 2 = -x + 4$.

105. (i) The slope of the line connecting $(-4, 0)$ and $(0, 4)$ is 1. Let $y_1 = x + 4$.

   (ii) A second line passing through $(4, 0)$ and $(0, -4)$ has a slope of 1. Let $y_2 = x - 4$.

   (iii) A third line passing through $(0, -4)$ and $(-4, 0)$ has a slope of -1. Let $y_3 = -x - 4$.

   (iv) A fourth line passing through $(0, 4)$ and $(4, 0)$ is -1. Let $y_4 = -x + 4$.

107. Enter the x-values into the list L_1 and the y-values into the list L_2. Use the statistical feature of your graphing calculator to find the correlation coefficient r and the regression equation. $r \neq -0.993$; $y \neq -0.789x + 0.526$ See Figure 107.

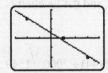

   Figure 107

109. (a) Enter the x-values into the list L_1 and the y-values into the list L_2 in the statistical feature of your graphing calculator; the scatterplot of the data indicates that the correlation coefficient will be positive (and very close to 1).

   (b) $y = ax + b$, where $a \approx 3.25$ and $b \approx -2.45$; $r \approx 0.9994$

   (c) $y \approx 3.25(2.4) - 2.45 = 5.35$

111. (a) Enter the x-values into the list L_1 and the y-values into the list L_2 in the statistical feature of your graphing calculator; the scatterplot of the data indicates that the correlation coefficient will be negative (and very close to -1).

   (b) $y = ax + b$, where $a \approx -3.8857$ and $b \approx 9.3254$; $r \approx -0.9996$

   (c) $y \approx 3.8857(2.4) + 9.3254 = -0.00028$. *Due to rounding answers may very slightly.*

113. Both percentages are increasing significantly.

115. Using technology, the regression is $f(x) \approx 10.8457x - 21795.3$;

   $f(2016) \approx 10.8457(2016) - 21795.3 \approx 69.63 \approx 70\%$

117. Using technology, the regression is $h(x) \approx 0.58401x + 0.667449$;

   $h(70) \approx 0.58401(70) + 0.667449 \approx 41.548 \approx 41.5\%$

119. (a) The data points $(50, 990)$, $(650, 9300)$, $(950, 15000)$ and $(1700, 25000)$ are plotted in Figure 119. The data appears to have a linear relationship.

   (b) Use the linear regression feature on your graphing calculator to find the values of a and b in the equation $y = ax + b$. In this instance, $a \approx 14.680$ and $b \approx 277.82$.

   (c) We must find the x-value when $y = 37,000$. This can be done by solving the equation $37,000 = 14.680x + 277.82 \Rightarrow 14.680x = 36,722.18 \Rightarrow x \approx 2500$ light years away. One could also solve the equation graphically to obtain the same approximation.

[−100, 1800, 100] by [−1000, 28000, 1000]

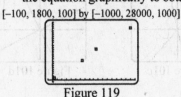

   Figure 119

2.2: Linear Equations

1. $ax + b = 0 \Rightarrow ax = -b \Rightarrow x = \dfrac{-b}{a}$. This shows that the equation $ax + b = 0$ has only one solution.

3. $4 - (5 - 4x) = 4 - 5 + 4x = -1 + 4x = 4x - 1$

5. The zero of f and the x-intercept of the graph of f are equal. The zero of f and the x-intercept of the graph of f are both found by finding the value of x when $y = 0$.

7. $3x - 1.5 = 7 \Rightarrow 3x - 1.5 - 7 = 0 \Rightarrow 3x - 8.5 = 0$; the equation is linear.

9. $2\sqrt{x} + 2 = 1$; since the equation cannot be written in the form $ax + b = 0$, it is nonlinear.

11. $7x - 5 = 3(x - 8) \Rightarrow 7x - 5 = 3x - 24 \Rightarrow 4x + 19 = 0$; the equation is linear.

13. $2x - 8 = 0 \Rightarrow 2x = 8 \Rightarrow x = 4$ Check: $2(4) - 8 = 0 \Rightarrow 8 - 8 = 0 \Rightarrow 0 = 0$

15. $-5x + 3 = 23 \Rightarrow -5x = 20 \Rightarrow x = -4$ Check: $-5(-4) + 3 = 23 \Rightarrow 20 + 3 = 23 \Rightarrow 23 = 23$

17. $6x = 4(x - 1) \Rightarrow 6x = 4x - 4 \Rightarrow 2x = -4 \Rightarrow x = -2$; Check: $6(-2) = 4((-2) - 1) \Rightarrow -12 = -12$

19. $-2 = 3 - 5x \Rightarrow -5 = -5x \Rightarrow 1 = x$; Check: $-2 = 3 - 5(1) \Rightarrow -2 = -2$

21. $4(z - 8) = z \Rightarrow 4z - 32 = z \Rightarrow 3z = 32 \Rightarrow z = \dfrac{32}{3}$ Check: $4\left(\dfrac{32}{3} - 8\right) = \dfrac{32}{3} \Rightarrow 4\left(\dfrac{8}{3}\right) = \dfrac{32}{3} \Rightarrow$

$\dfrac{32}{32} = \dfrac{32}{32}$

23. $-5(3 - 4t) = 65 \Rightarrow -15 + 20t = 65 \Rightarrow 20t = 80 \Rightarrow t = 4$ Check: $-5[3 - 4(4)] = 65 \Rightarrow$

$-5(3 - 16) = 65 \Rightarrow -5(-13) = 65 \Rightarrow 65 \Rightarrow 65$

25. $k + 8 = 5k - 4 \Rightarrow -4k = -12 \Rightarrow k = 3$ Check: $3 + 8 = 5(3) - 4 \Rightarrow 11 = 15 - 4 \Rightarrow 11 = 11$

27. $2(1 - 3x) + 1 = 3x \Rightarrow 2 - 6x + 1 = 3x \Rightarrow -6x + 3 = 3x \Rightarrow -9x = -3 \Rightarrow x = \dfrac{1}{3}$

Check: $2\left[1 - 3\left(\dfrac{1}{3}\right)\right] + 1 = 3\left(\dfrac{1}{3}\right) \Rightarrow 2(1 - 1) + 1 = 1 \Rightarrow 0 + 1 = 1 \Rightarrow 1 = 1$

29. $-5(3 - 2x) - (1 - x) = 4(x - 3) \Rightarrow -15 + 10x - 1 + x = 4x - 12 \Rightarrow 11x - 16 = 4x - 12 \Rightarrow$

$7x = 4 \Rightarrow x = \dfrac{4}{7}$ Check: $-5\left[3 - 2\left(\dfrac{4}{7}\right)\right] - \left(1 - \dfrac{4}{7}\right) = 4\left(\dfrac{4}{7} - 3\right) \Rightarrow$

$-5\left(\dfrac{13}{7}\right) - \dfrac{3}{7} = 4\left(-\dfrac{17}{7}\right) \Rightarrow -\dfrac{65}{7} - \dfrac{3}{7} = -\dfrac{68}{7} \Rightarrow -\dfrac{68}{7} = -\dfrac{68}{7}$

31. $-4(5x - 1) = 8 - (x + 2) \Rightarrow -20x + 4 = 8 - x - 2 \Rightarrow -20x + 4 = 6 - x \Rightarrow -19x = 2 \Rightarrow$

$x = -\dfrac{2}{19}$ Check: $-4\left[5\left(-\dfrac{2}{19}\right) - 1\right] = 8 - \left(-\dfrac{2}{19} + 2\right) \Rightarrow -4\left(-\dfrac{10}{19} - 1\right) = 8 + \dfrac{2}{19} - 2 \Rightarrow$

$\dfrac{40}{19} + 4 = 6 + \dfrac{2}{19} \Rightarrow \dfrac{116}{19} = \dfrac{116}{19}$

33. $\dfrac{2}{7}n + 2 = \dfrac{4}{7} \Rightarrow \dfrac{2}{7}n = -\dfrac{10}{7} \Rightarrow n = -5$ Check: $\dfrac{2}{7}(-5) + 2 = \dfrac{4}{7} \Rightarrow \dfrac{-10}{7} + \dfrac{14}{7} = \dfrac{4}{7} \Rightarrow \dfrac{4}{7} = \dfrac{4}{7}$

35. $\frac{1}{2}(d-3)-\frac{2}{3}(2d-5)=\frac{5}{12} \Rightarrow \frac{1}{2}d-\frac{3}{2}-\frac{4}{3}d+\frac{10}{3}=\frac{5}{12} \Rightarrow -\frac{5}{6}d+\frac{11}{6}=\frac{5}{12} \Rightarrow -\frac{5}{6}d=-\frac{17}{12} \Rightarrow d=\frac{17}{10}$

       Check: $\frac{1}{2}\left(\frac{17}{10}-3\right)-\frac{2}{3}\left[2\left(\frac{17}{10}\right)-5\right]=\frac{5}{12} \Rightarrow \frac{1}{2}\left(-\frac{13}{10}\right)-\frac{2}{3}\left(\frac{34}{10}-5\right)=\frac{5}{12} \Rightarrow$

       $\frac{1}{2}\left(-\frac{13}{10}\right)-\frac{2}{3}\left(-\frac{16}{10}\right)=\frac{5}{12} \Rightarrow -\frac{13}{20}+\frac{32}{30}=\frac{5}{12} \Rightarrow \frac{25}{60}=\frac{5}{12} \Rightarrow \frac{5}{12}=\frac{5}{12}$

37. $\frac{x-5}{3}+\frac{3-2x}{2}=\frac{5}{4}-2(1-x) \Rightarrow 4(x-5)+6(3-2x)=15-24(1-x) \Rightarrow$

       $4x-20+18-12x=15-24+24x \Rightarrow -8x-2=-9+24x \Rightarrow 7=32x \Rightarrow x=\frac{7}{32}$

       Check: $\frac{\frac{7}{32}-5}{3}+\frac{3-2\left(\frac{7}{32}\right)}{2}=\frac{5}{4}-2\left(1-\frac{7}{32}\right) \Rightarrow 4\left(\frac{7}{32}-5\right)+6\left(3-2\left(\frac{7}{32}\right)\right) \Rightarrow$

       $-\frac{153}{8}+\frac{123}{8}=15-\frac{75}{4} \Rightarrow -\frac{15}{4}=-\frac{15}{4}$

39. $0.1z-0.05=-0.07z \Rightarrow 0.17z=0.05 \Rightarrow z=\frac{0.05}{0.17} \Rightarrow z=\frac{5}{17}$ Check:

       $0.1\left(\frac{5}{17}\right)-0.05=-0.07\left(\frac{5}{17}\right) \Rightarrow \frac{5}{170}-\frac{5}{100}=-\frac{35}{1700} \Rightarrow \frac{50}{1700}-\frac{85}{1700}=-\frac{35}{1700} \Rightarrow$

       $-\frac{35}{1700}=-\frac{35}{1700}$

41. $0.15t+0.85(100-t)=0.45(100) \Rightarrow 0.15t+85-0.85t=45 \Rightarrow -0.7t=-40 \Rightarrow t=\frac{40}{0.7} \Rightarrow$

       $t=\frac{400}{7}$ Check: $0.15\left(\frac{400}{7}\right)+0.85\left(100-\frac{400}{7}\right)=0.45(100) \Rightarrow \frac{6000}{700}+85-\frac{34,000}{700}=45 \Rightarrow$

       $\frac{60}{7}+85-\frac{340}{7}=45 \Rightarrow 85-\frac{280}{7}=45 \Rightarrow 85-40=45 \Rightarrow 45=45$

43. (a) $5x-1=5x+4 \Rightarrow -1=4 \Rightarrow$ there is no solution.

    (b)   Since no x-value satisfies the equation, it is contradiction.

45. (a) $3(x-1)=5 \Rightarrow 3x-3=5 \Rightarrow 3x=8 \Rightarrow x=\frac{8}{3}$

    (b)   Since one x-value is a solution and other x-values are not, the equation is conditional.

47. (a) $0.5(x-2)+5=0.5x+4 \Rightarrow 0.5x-1+5=0.5x+4 \Rightarrow 0.5x+4=0.5x+4 \Rightarrow$ every x-value satisfies this
equation, so all real numbers is the solution.

(b) Since every x-value satisfies the equation, it is an identity.

49. (a) $\dfrac{t+1}{2}=\dfrac{3t-2}{6} \Rightarrow 6\left(\dfrac{t+1}{2}=\dfrac{3t-2}{6}\right) \Rightarrow 3t+3=3t-2 \Rightarrow 3=-2 \Rightarrow$ there is no solution.

(b) Since no x-value satisfies the equation, it is contradiction.

51. (a) $\dfrac{1-2x}{4}=\dfrac{3x-1.5}{-6} \Rightarrow -6(1-2x)=4(3x-1.5) \Rightarrow -6+12x=12x-6 \Rightarrow 0=0 \Rightarrow$ every x-value satisfies
this equation, so all real numbers is the solution.

(b) Since every x-value satisfies the equation, it is an identity.

53. In the graph, the lines intersect at $(3,-1)$. The solution is the x-value, 3.

55. (a) From the graph, when $f(x)$ or $y=-1$, $x=4$; the solution is the x-value, 4.

(b) From the graph, when $f(x)$ or $y=0$, $x=2$; the solution is the x-value, 2.

(c) From the graph, when $f(x)$ or $y=2$, $x=-2$; the solution is the x-value, -2.

57. (a) From the graph, when $f(x)$ or $y=-1$, $x=\dfrac{3}{2}$; the solution is the x-value, $\dfrac{3}{2}$.

(b) From the graph, when $f(x)$ or $y=0$, $x=1$; the solution is the x-value, 1.

(c) From the graph, when $f(x)$ or $y=2$, $x=0$; the solution is the x-value, 0.

59. Graph $Y_1=X+4$ and $Y_2=1-2X$. See Figure 59. The lines intersect at $x=-1$.
$x+4=1-2x \Rightarrow 3=-3x \Rightarrow x=-1$

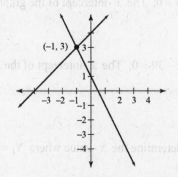

Figure 59

61. See Figure 61. The lines intersect at $x=2$. $x=4-x \Rightarrow 2x=4 \Rightarrow x=2$

63. Graph $Y_1=-X+4$ and $Y_2=3X$. See Figure 63. The lines intersect at $x=1$. $-x+4=3x \Rightarrow 4=4x \Rightarrow x=1$

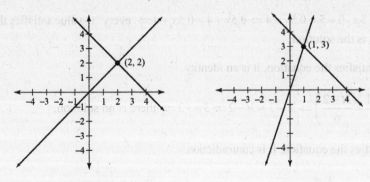

Figure 61   Figure 63

65.   Graph $Y_1 = 2(X-1)-2$ and $Y_2 = X$. See Figure 65. The lines intersect at $x = 4$.

$2(x-1)-2 = x \Rightarrow 2x-2-2 = x \Rightarrow 2x-4 = x \Rightarrow -4 = -x \Rightarrow x = 4$

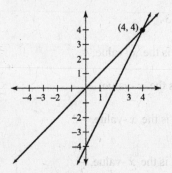

Figure 65

67. $h(x) = 2x - 4 = 0$; The x-intercept of the graph is at $(2,0) \Rightarrow x = 2$ is the solution.

69. $2x = -(3-x) \Rightarrow 2x = -3 + x \Rightarrow h(x) = x + 3 = 0$; The x-intercept of graph is at $(-3,0) \Rightarrow x = -3$ is the solution.

71. $-2(3-2x) = 4x - (1-x) \Rightarrow -6 + 4x = 4x - 1 + x \Rightarrow -5 = x \Rightarrow h(x) = x + 5 = 0$; The x-intercept of the graph is at $(-5,0) \Rightarrow x = -5$ is the solution.

73. $\dfrac{6-x}{7} = \dfrac{2x-3}{3} \Rightarrow 3(6-x) = 7(2x-3) \Rightarrow 18 - 3x = 14x - 21 \Rightarrow h(x) - 17x - 39 = 0$; The x-intercept of the

       graph is at $\left(\dfrac{39}{17}, 0\right) \approx (2.294, 0) \Rightarrow x \approx 2.294$ is a solution.

75.   One way to solve this equation is to make a table for $Y_1 = 2X - 7$ and determine the x-value where $Y_1 = -1$. See Figure 75. This occurs when $x = 3$, so the solution is 3.

77.   Make a table for $Y_1 = 2x - 7$ and determine the x-value where $Y_1 = -5$. See Figure 77. This occurs when $x = 1$, so the solution is 1.

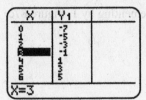

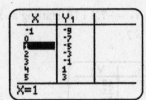

Figure 75   Figure 77

79.   Make a table for $Y_1 = \sqrt{(2)}(4X - 1) + \pi X$ and determine the x-value where $Y_1 = 0$. See Figure 79. This occurs when $x \neq 0.2$, so the solution is 0.2.

81.   Make a table for $Y_1 = 0.5 - 0.1(\sqrt{(2)} - 3X)$ and determine the x-value where $Y_1 = 0$. See Figure 81. This occurs when $x \neq -1.2$, so the solution is -1.2.

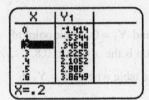

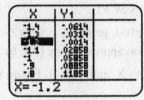

Figure 79   Figure 81

83.   (a) $5 - (x + 1) = 3 \Rightarrow 5 - x - 1 = 3 \Rightarrow 4 - x = 3 \Rightarrow x = 1$

      (b)   Using the intersection of graphs method, graph $Y_1 = 5 - (X + 1)$ and $Y_2 = 3$. Their point of intersection is shown in Figure 83b as $(1, 3)$. The solution is the x-value, 1.

      (c)   Make a table for $Y_1 = 5 - (X + 1)$ and $Y_2 = 3$. Figure 83c shows a table where $Y_1 = Y_2$ at $x = 1$.

$[-10, 10, 1]$ by $[-10, 10, 1]$

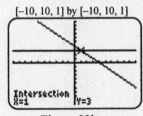

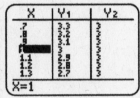

Figure 83b   Figure 83c

85.   (a) $x - 3 = 2x + 1 \Rightarrow -x = 4 \Rightarrow x = -4$

      (b)   Using the intersection-of-graphs method, graph $Y_1 = X - 3$ and $Y_2 = 2X + 1$. Their point of intersection is shown in Figure 85b as $(-4, -7)$. The solution is the x-value, -4.

      (c)   Make a table for $Y_1 = X - 3$ and $Y_2 = 2X + 1$, starting at $x = -7$, incrementing by 1. Figure 85c shows a table where $Y_1 = Y_2$ at $x = -4$.

$[-12, 8, 1]$ by $[-12, 8, 1]$

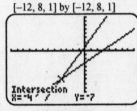

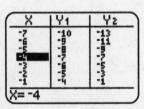

Figure 85b   Figure 85c

87.   (a) $6x - 8 = -7x + 18 \Rightarrow 13x = 26 \Rightarrow x = 2$

      (b)   Using the intersection-of-graphs method, graph $Y_1 = 6X - 8$ and $Y_2 = -7X + 18$. Their point of intersection is shown in Figure 87b as $(2, 4)$. The solution is the x-value, 2.

(c)   Make a table for $Y_1 = 6X - 8$ and $Y_2 = -7X + 18$ starting at $x = 0$, incrementing by 1. Figure 87c shows a table where $Y_1 = Y_2$ at $x = 2$.

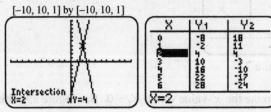

Figure 87b   Figure 87c   c

89.   (a) $\sqrt{3}(2 - \pi x) + x = 0 \Rightarrow 2\sqrt{3} - \sqrt{3}\pi x + x = 0 \Rightarrow -\sqrt{3}\pi x + x = -2\sqrt{3} \Rightarrow$

$$x(-\sqrt{3}\pi + 1) = -2\sqrt{3} \Rightarrow x = \frac{-2\sqrt{3}}{(-\sqrt{3}\pi + 1)} \approx 0.8.$$

(b)   Using the intersection of graphs method, graph $Y_1 = \sqrt{(3)}(2 - \pi X) + X$ and $Y_2 = 0$. Their point of intersection is shown in Figure 89b as approximately $(0.8, 0)$. The solution is the x-value, 0.8.

(c)   Make a table for $Y_1 = \sqrt{(3)}(2 - \pi X) + X$ and $Y_2 = 0$. Figure 89c shows a table where $Y_1 = Y_2$ at $x \approx 0.8$.

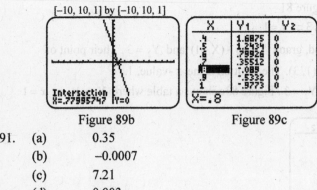

Figure 89b   Figure 89c

91.   (a)   0.35
       (b)   −0.0007
       (c)   7.21
       (d)   0.003
93.   (a)   −0.055
       (b)   −0.0154
       (c)   1.2
       (d)   0.0015
95.   (a)   37%
       (b)   −9.5%
       (c)   190%
       (d)   35%
97.   (a)   −12.1%
       (b)   140%
       (c)   320%
       (d)   −25%

99. $A = LW \Rightarrow W = \dfrac{A}{L}$

101. $P = 2L + 2W \Rightarrow P - 2W = 2L \Rightarrow L = \dfrac{P - 2W}{2} \Rightarrow L = \dfrac{1}{2}P - W$

103. $S = 2LW + 2WH + 2LH \Rightarrow S - 2LW = 2WH + 2LH \Rightarrow S - 2LW = (2W + 2L)H \Rightarrow H = \dfrac{S - 2LW}{2W + 2L}$

105. $3x + 2y = 8 \Rightarrow 2y = 8 - 3x \Rightarrow y = 4 - \dfrac{3}{2}x$

107. $y = 3(x-2) + x \Rightarrow y = 3x - 6 + x \Rightarrow y = 4x - 6 \Rightarrow 4x = y + 6 \Rightarrow x = \dfrac{1}{4}y + \dfrac{3}{2}$

109. (a) $2x + y = 8 \Rightarrow y = 8 - 2x \Rightarrow y = -2x + 8$

    (b) $f(x) = -2x + 8$

110. (a) $3x - y = 5 \Rightarrow y = 3x - 5$

    (b) $f(x) = 3x - 5$

111. (a) $2x - 4y = -1 \Rightarrow -4y = -1 - 2x \Rightarrow y = \dfrac{1}{4} + \dfrac{1}{2}x$

    (b) $f(x) = \dfrac{1}{4} + \dfrac{1}{2}x$

113. (a) $-9x + 8y = 9 \Rightarrow 8y = 9 + 9x \Rightarrow y = \dfrac{9}{8} + \dfrac{9}{8}x$

    (b) $f(x) = \dfrac{9}{8} + \dfrac{9}{8}x$

115. $f(x) = 62400$ and $f(x) = 1500(x - 2011) + 50,000$

    $\Rightarrow 62400 = 1550(x - 2011) + 50000 \Rightarrow 62400 = 1550x - 3117050 + 50000 \Rightarrow 3129450 = 1550x \Rightarrow x = 2019$.
The per capita income might be $62400 in 2019.

117. Using the intersection of graphs method, graph $Y_1 = 51.6(X - 1985) + 9.1$ and $Y_2 = -31.9(X - 1985) + 167.7$.
Their approximate point of intersection is (1987, 107). In approximately 1987 the sales of LP records and compact discs were equal.

119. $P = 2w + 2l \Rightarrow 180 = 2w + 2(w + 18) \Rightarrow 180 = 4w + 36 \Rightarrow 4w = 144 \Rightarrow w = 36$ and
$w + 18 = 54$; the window is 36 inches by 54 inches.

121. To calculate the sale price subtract 25% of the regular price from the regular price.
$f(x) = x - 0.25x \Rightarrow f(x) = 0.75x$. An item which normally costs $56.24 will be on sale for
$f(56.24) = 0.75(56.24) = \$42.18$.

123. (a)   The number of skin cancer cases is given by $0.043x$.

    (b) $73,000 = 0.043x \Rightarrow \dfrac{73,000}{0.043} = x \Rightarrow x \approx 1,697,674 \approx 1,698,000$

125. (a)   It would take a little less time than the faster gardener, who can rake the lawn alone in 3 hours. It would take both gardeners about 2 hours working together. *Answers may vary.*

    (b)   Let $x =$ time to rake the lawn working together. In 1 hour the first gardener can rake $\dfrac{1}{3}$ of the

lawn, whereas the second gardener can rake $\dfrac{1}{5}$ of the lawn; in x hours both gardeners working together can

rake $\dfrac{x}{3} + \dfrac{x}{5}$ of the lawn; $\dfrac{x}{3} + \dfrac{x}{5} = 1 \Rightarrow 5x + 3x = 15 \Rightarrow 8x = 15 \Rightarrow x = \dfrac{15}{8} = 1.875$ hours.

127. Let $x =$ the speed of the airplane; Then traveling with the wind $\Rightarrow 3(x + 30)$ and traveling against the wind
$\Rightarrow 4(x - 30)$; For the same trip: $4(x - 30) = 3(x + 30) \Rightarrow 4x - 120 = 3x + 90 \Rightarrow x = 210$; Therefore, the speed of the airplane without any wind is 210 mi/h.

129. Let $t =$ time spent traveling at 55 mph and $6 - t =$ time spent traveling at 70 mi/h. Using $d = rt$, we get
$d = 55t + 70(6 - t) \Rightarrow 372 = 55t + 420 - 70t \Rightarrow -48 = -15t \Rightarrow t = 3.2$ and $6 - t = 2.8$; the car traveled 3.2 hours at 55 mi/h and 2.8 hours at 70 mi/h.

131. Let $t =$ time traveled by car at 55 mi/h and $t + \frac{1}{2} =$ time traveled by runner at 10 mi/h ; since $d = rt$ and the

distance is the same for both runner and driver, we get $55t = 10\left(t + \frac{1}{2}\right) \Rightarrow 55t = 10t + 5 \Rightarrow 45t = 5 \Rightarrow t = \frac{1}{9}$; it

takes the driver $\frac{1}{9}$ hour or $6\frac{2}{3}$ minutes to catch the runner.

133. This problem can be solved using similar triangles or a proportion. Let x be the height of the tree; then,

$\frac{5}{4} = \frac{x}{33} \Rightarrow x = \frac{5 \times 33}{4} = 41.25$. The height of the tree is 41.25 feet.

135. Use similar triangles to find the radius of the cone when the water is 7 feet deep: $\frac{r}{3.5} = \frac{7}{11} \Rightarrow r \approx \frac{49}{22}$ ft . Use

$V = \frac{1}{3}\pi r^2 h$ to find the volume of the water in the cone at $h = 7$ ft.: $V = \frac{1}{3}\pi\left(\frac{49}{22}\right)^2 (7) \approx 36.4 \text{ft}^3$.

137. Let $x =$ the amount of pure water to be added.

Then $x + 0 = 4\%, x + 2 = 3\% \Rightarrow 0.03(x + 2) = 0.04(2) \Rightarrow 0.03x + .06 = .08 \Rightarrow x = \frac{2}{3}$; $\frac{2}{3}L$ of pure water should

be added.

139. $d = r_1 t_1 + r_2 t_2 \Rightarrow 12 = 12x + 8(1.25 - x) \Rightarrow 12 = 12x + 10 - 8x \Rightarrow 2 = 4x \Rightarrow x = \frac{1}{2} \Rightarrow$ the runner runs for $\frac{1}{2}$ an

hour, or 30 minutes, at a speed of 12 mi/h, and runs for 45 minutes, at a speed of 8 mi/h.

141. Let $x =$ amount of pure water to be added and $x + 5 =$ final amount of the 15% solution . Since pure water is

0% sulfuric acid, we get $0\%x + 40\%(5) = 15\%(x + 5) \Rightarrow 0.40(5) = 0.15(x + 5) \Rightarrow$

$40(5) = 15(x + 5) \Rightarrow 200 = 15x + 75 \Rightarrow 15x = 125 \Rightarrow x = \frac{125}{15} \approx 8.333$; about 8.33 liters of pure water should be

added.

143. (a)   The linear function S must fit the coordinates (2011, 192) and (2014, 249).

$m = \frac{249 - 192}{2014 - 2011} = \frac{57}{3} = 19; S = 19(x - 2014) + 249 \Rightarrow S = 19x - 38,017$

(b)   The slope shows that the sales increased, on average, by $19 billion per year.

(c) $230 = 19x - 38,017 \Rightarrow 38,247 = 19x \Rightarrow x = 2013$

145. $C = \frac{5}{9}(F - 32)$ and $F = C \Rightarrow F = \frac{5}{9}(F - 32) \Rightarrow F = \frac{5}{9}F - \frac{160}{9} \Rightarrow \frac{4}{9}F = -\frac{160}{9} \Rightarrow F = -40$;

$-40° F$ is equivalent to $-40°C$.

147. (a)   It is reasonable to expect that f is linear because if the number of gallons of gas doubles so should the

amount of oil. Five gallons of gasoline requires five times the oil that one gallon of gasoline would. The

increase in oil is always equal to 0.16 pint for each additional gallon of gasoline. Oil is mixed at a

constant rate, so a linear function describes this amount.

(b) $f(3) = 0.16(3) = 0.48$; 0.48 pint of oil should be added to 3 gallons of gasoline to get the correct

mixture.

(c) $0.16x = 2 \Rightarrow x = 12.5$; 12.5 gallons of gasoline should be mixed with 2 pints of oil.

149. Linear regression gives the model:

$y = 0.36x - 0.21$. $y = 2.99 \Rightarrow 2.99 = 0.36x - 0.21 \Rightarrow 3.2 = 0.36x \Rightarrow x \approx 8.89$

Extended and Discovery Exercises for Section 2.2

1. (a) Yes; since multiplication distributes over addition, doubling the lengths gives double the sum of the lengths.

   (b) No; If the length and width are doubled, the product of the length and width is multiplied by 4.

3. (a) $(100 \text{ ft}^2)(140 \,\mu g/\text{ft}^2) = 14,000 \,\mu g$; $f(x) = 14,000x$.

   (b) $(800 \text{ ft}^3)(33 \,\mu g/\text{ft}^3) = 26,400 \,\mu g$; $f(x) = 14,000x \Rightarrow 26,400 = 14,000x \Rightarrow x \approx 1.9$; it takes about 1.9 hours for the concentrations to reach $33 \,\mu g/\text{ft}^3$.

5. Let d_1 = the distance covered by person 1 and d_2 = the distance covered by person 2.

   Then $\dfrac{d_1}{4} + \dfrac{15-d_1}{28} = \dfrac{d_2}{28} + \dfrac{15-d_2}{4} = \dfrac{d_2}{28} + \dfrac{d_2-d_1}{28} + \dfrac{15-d_1}{28} \Rightarrow$

   $\dfrac{7d_1+15-d_1}{28} = \dfrac{d_2+7(15-d_2)}{28} = \dfrac{d_2+d_2-d_1+15-d_1}{28}$. Then the following equations hold:

   $d_1 + d_2 = 15$, $6d_1 + 15 = 2d_2 - 2d_1 + 15$, $7d_1 + 15 - d_1 = d_2 + 105 - 7d_2$. Then it follows

   $d_1 + d_2 = 15$ and $d_2 = 4d_1 \Rightarrow d_1 + 4d_1 = 15 \Rightarrow 5d_1 = 15 \Rightarrow d_1 = 3$. Therefore, each person walked 3 miles.

Checking Basic Concepts for Sections 2.1 and 2.2

1. The slope of the line passing through $(-3, 4)$ and $(5, -2)$ is $m = \dfrac{-2-4}{5-(-3)} = -\dfrac{3}{4}$. Using the point-slope form

   of a line results in $y = -\dfrac{3}{4}(x+3)+4$ or $y = -\dfrac{3}{4}x+\dfrac{7}{4}$. The line $y = -\dfrac{3}{4}x$ is parallel to $y = -\dfrac{3}{4}x+\dfrac{7}{4}$ and

   $y = \dfrac{4}{3}x$ is perpendicular. *Answers may vary.*

3. $y = -\dfrac{3}{2}x+\dfrac{1}{2}$

5. $5x+2 = 2x-3 \Rightarrow 3x+2 = -3 \Rightarrow 3x = -5 \Rightarrow x = -\dfrac{5}{3}$

   Check: $5\left(-\dfrac{5}{3}\right)+2 = 2\left(-\dfrac{5}{3}\right)-3 \Rightarrow -\dfrac{25}{3}+2 = -\dfrac{10}{3}-3 \Rightarrow -\dfrac{19}{3} = \dfrac{19}{3}$

7. All methods result in the same solution: $4(x-2) = 2(5-x)-3 \Rightarrow 4x-8 = 10-2x-3 \Rightarrow 6x = 15 \Rightarrow x = 2.5$

2.3: Linear Inequalities

1. $(-\infty, 2)$

3. $[-1, \infty)$

5. $[1, 8)$

7. $(-\infty, 1]$

9. $x \neq 5 \Rightarrow (-\infty,5) \cup (5,\infty)$

11. $2x+6 \geq 10 \Rightarrow 2x \geq 4 \Rightarrow x \geq 2$;$[2, \infty)$; set-builder notation is $\{x \mid x \geq 2\}$.

13. $-2(x-10)+1>0 \Rightarrow -2x+21>0 \Rightarrow -2x>-21 \Rightarrow x<10.5$; $(-\infty, 10.5)$; set-builder notation is $\{x \mid x<10.5\}$.

15. $2x-1 \geq 4-x \Rightarrow 3x \geq 5 \Rightarrow x \geq \frac{5}{3}$; $\left[\frac{5}{3}, \infty\right)$; set-builder notation is $\left\{x \mid x \geq \frac{5}{3}\right\}$.

17. $-2x<-(x+1) \Rightarrow -2x<-x-1 \Rightarrow -x<-1 \Rightarrow x>1$; $(1, \infty)$; set-builder notation is $\{x \mid x>1\}$.

19. $\frac{t+2}{3} \geq 5 \Rightarrow t+2 \geq 15 \Rightarrow t \geq 13$; $[13, \infty)$; set-builder notation the interval is $\{t \mid t \geq 13\}$.

21. $4x-1<\frac{3-x}{-3}=7 \Rightarrow -12x+3>3-x \Rightarrow -11x>0 \Rightarrow x<0$; $(-\infty, 0)$; set-builder notation the interval is $\{x \mid x<0\}$

23. $-3(z-4) \geq 2(1-2z) \Rightarrow -3z+12 \geq 2-4z \Rightarrow z \geq -10$; $[-10, \infty)$; set-builder notation the interval is $\{z \mid z \geq -10\}$.

25. $\frac{1-x}{4}<\frac{2x-2}{3} \Rightarrow 3(1-x)<4(2x-2) \Rightarrow 3-3x<8x-8 \Rightarrow -11x<-11 \Rightarrow x>1$; $(1, \infty)$; set-builder notation the interval is $\{x \mid x>1\}$.

27. $2x-3>\frac{1}{2}(x+1) \Rightarrow 2x-3>\frac{1}{2}x+\frac{1}{2} \Rightarrow \frac{3}{2}x>\frac{7}{2} \Rightarrow x>\frac{7}{3}$; $\left(\frac{7}{3}, \infty\right)$; set-builder notation the interval is $\left\{x \mid x>\frac{7}{3}\right\}$.

29. $5<4t-1 \leq 11 \Rightarrow 6<4t \leq 12 \Rightarrow \frac{3}{2}<t \leq 3$; $\left(\frac{3}{2}, 3\right]$; set-builder notation the interval is $\left\{t \mid \frac{3}{2}<t \leq 3\right\}$.

31. $3 \leq 4-x \leq 20 \Rightarrow -1 \leq -x \leq 16 \Rightarrow 1 \geq x \geq -16$; $[-16, 1]$; set-builder notation the interval is $\{x \mid -16 \leq x \leq 1\}$.

33. $-7 \leq \frac{1-4x}{7}<12 \Rightarrow -49 \leq 1-4x<84 \Rightarrow -50 \leq -4x<83 \Rightarrow 12.5 \geq x>-20.75$; $(-20.75, 12.5]$; set-builder notation the interval is $\{x \mid -20.75<x \leq 12.5\}$.

35. $5>2(x+4)-5>-5 \Rightarrow 5>2x+8-5>-5 \Rightarrow 2>2x>-8 \Rightarrow 1>x>-4$; $(-4, 1)$; set-builder notation the interval is $\{x \mid -4<x<1\}$.

37. $3 \leq \frac{1}{2}x+\frac{3}{4} \leq 6 \Rightarrow 12 \leq 2x+3 \leq 24 \Rightarrow 9 \leq 2x \leq 21 \Rightarrow \frac{9}{2} \leq x \leq \frac{21}{2}$; $\left[\frac{9}{2}, \frac{21}{2}\right]$; set-builder notation the interval is $\left\{x \mid \frac{9}{2} \leq x \leq \frac{21}{2}\right\}$.

39. $5x-2(x+3) \geq 4-3x \Rightarrow 5x-2x-6 \geq 4-3x \Rightarrow 6x \geq 10 \Rightarrow x \geq \frac{5}{3}$; $\left[\frac{5}{3}, \infty\right)$; set-builder notation the interval is $\left\{x \mid x \geq \frac{5}{3}\right\}$.

41. $\frac{1}{2} \leq \frac{1-2t}{3}<\frac{2}{3} \Rightarrow \frac{3}{2} \leq 1-2t<2 \Rightarrow \frac{1}{2} \leq -2t<1 \Rightarrow -\frac{1}{4} \geq t>-\frac{1}{2} \Rightarrow -\frac{1}{2}<t \leq -\frac{1}{4}$; $\left(-\frac{1}{2}, -\frac{1}{4}\right]$; set-builder notation the interval is $\left\{t \mid -\frac{1}{2}<t \leq -\frac{1}{4}\right\}$.

43. $\frac{1}{2}z+\frac{2}{3}(3-z)-\frac{5}{4}z \geq \frac{3}{4}(z-2)+z \Rightarrow \frac{1}{2}z+2-\frac{2}{3}z-\frac{5}{4}z \geq \frac{3}{4}z-\frac{3}{2}+z \Rightarrow$

$-\dfrac{17}{12}z+2\geq\dfrac{7}{4}z-\dfrac{3}{2}\Rightarrow-\dfrac{38}{12}z\geq-\dfrac{7}{2}\Rightarrow z\leq\dfrac{21}{19};\ \left(-\infty,\dfrac{21}{19}\right];$ set-builder notation the interval is $\left\{z\mid z\leq\dfrac{21}{19}\right\}.$

45. Graph $y_1=x+2$ and $y_2=2x$. See Figure 45. $y_1\geq y_2$ when the graph of y_1 is above the graph of y_2, which is left of the intersection point $(2,4)$ and includes point $(2,4)\Rightarrow(-\infty,2]$.

47. Graph $y_1=\dfrac{2}{3}x-2$ and $y_2=-\dfrac{4}{3}x+4$. See Figure 47. $y_1>y_2$ when the graph of y_1 is above the graph of y_2, which is right of the intersection point $(3,0)$ and does not include point $(3,0)\Rightarrow(3,\infty)$.

[−10, 10, 1] by [−10, 10, 1]   [−10, 10, 1] by [−10, 10, 1]

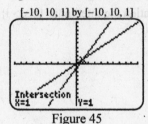   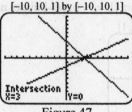

Figure 45   Figure 47

49. Graph $y_1=-1$, $y_2=2x-1$, and $y_3=3$. See Figure 49. $y_1\leq y_2\leq y_3$ when the graph of y_2 is in between the graphs of y_1 and y_3, which is in between the intersection points $(0,-1)$ and $(2,3)$ and it does include each point $\Rightarrow[0,2]$.

51. Graph $y_1=-3$, $y_2=x-2$, and $y_3=2$. See Figure 51. $y_1<y_2\leq y_3$ when the graph of y_2 is in between the graphs of y_1 and y_3, which is in between the intersection points $(-1,-3)$ and $(4,2)$ and it does not include the point $(-1,-3)$ but does include $(4,2)\Rightarrow(-1,4]$.

[−10, 10, 1] by [−10, 10, 1]   [−10, 10, 1] by [−10, 10, 1]

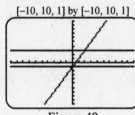   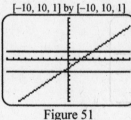

Figure 49   Figure 51

53. (a) $y=\dfrac{3}{2}x-3$, then $ax+b=0$ gives us $\dfrac{3}{2}x-3=0\Rightarrow\dfrac{3}{2}x=3\Rightarrow x=2$

(b) $ax+b<0$ gives us $\dfrac{3}{2}x-3<0\Rightarrow x<2\Rightarrow(-\infty,2)$ or in set builder notation, $\{x\mid x<2\}$.

(c) $ax+b\geq0$ gives us $\dfrac{3}{2}x-3\geq0\Rightarrow x\geq2\Rightarrow[2,\infty)$ or in set builder notation, $\{x\mid x\geq2\}$.

55. (a) $y=-x-2$, then $ax+b=0$ gives us $-x-2=0\Rightarrow-x=2\Rightarrow x=-2$

(b) $ax+b<0$ gives us $-x-2<0\Rightarrow x>-2\Rightarrow(-2,\infty)$ or in set builder notation, $\{x\mid x>-2\}$.

(c) $ax+b\geq0$ gives us $-x-2\geq0\Rightarrow x\leq-2\Rightarrow(-\infty,-2]$ or in set builder notation, $\{x\mid x\leq-2\}$.

57. $x - 3 \le \frac{1}{2}x - 2 \Rightarrow x - 3 - \frac{1}{2}x + 2 \le 0 \Rightarrow \frac{1}{2}x - 1 \le 0$. Figure 57 shows the graph of $y_1 = \frac{1}{2}x - 1$. The solution set

for $y_1 \le 0$ occurs when the graph is on or below the x -axis, or when $x \le 2$. The solution set is $(-\infty, 2]$.

Solving symbolically, $x - 3 \le \frac{1}{2}x - 2 \Rightarrow \frac{1}{2}x \le 1 \Rightarrow x \le 2 \Rightarrow (-\infty, 2]$. In set-builder notation the interval is

$\{x | x \le 2\}$.

59. $2 - x < 3x - 2 \Rightarrow 2 - x - 2x + 2 < 0 \Rightarrow -4x + 4 < 0$. Figure 59 shows the graph of $y_1 = -4x + 4$. The solution

set for $y_1 < 0$ occurs when the graph is below the x -axis, or when $x > 1$. The solution set is $(1, \infty)$. Solving

symbolically, $2 - x < 3x - 2 \Rightarrow -4x < -4 \Rightarrow x > 1 \Rightarrow (1, \infty)$. In set-builder notation the interval is $\{x | x > 1\}$.

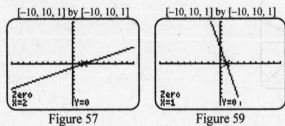

[-10, 10, 1] by [-10, 10, 1]   [-10, 10, 1] by [-10, 10, 1]

Figure 57   Figure 59

61.   Graph $Y_1 = 5X - 4$ and $Y_2 = 10$ The graphs intersect at the point $(2.8, 10)$. The graph of Y_1 is above the

graph of Y_2 for x -values to the right of this intersection point or where $x > 2.8$, $\{x | x > 2.8\}$. See Figure 61.

63.   Graph $Y_1 = -2(X - 1990) + 55$ and $Y_2 = 60$ The graphs intersect at the point $(1987.5, 60)$. The graph of Y_1 is

above the graph of Y_2 for x -values to the left of this intersection point, so $y_1 \ge y_2$ when $x \le 1987.5$,

$\{x | x \le 1987.5\}$. See Figure 63.

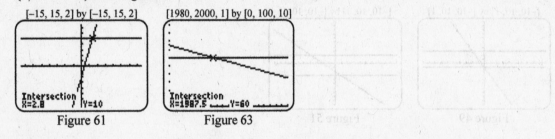

[-15, 15, 2] by [-15, 15, 2]   [1980, 2000, 1] by [0, 100, 10]

Figure 61   Figure 63

65.   Graph $Y_1 = \sqrt{(5)}(X - 1.2) - \sqrt{(3)}X$ and $Y_2 = 5(X + 1.1)$ The graphs intersect near the point $(-1.820, -3.601)$.

The graph of Y_1 is below the graph of Y_2 for x -values to the right of this intersection point or when $x > k$,

where $k \approx -1.82$, $\{x | x > -1.82\}$. See Figure 65.

67.   Graph $Y_1 = 3$, $Y_2 = 5X - 17$ and $Y_3 = 15$, as shown in Figure 67. The graphs intersect at the points $(4, 3)$

and $(6.4, 15)$. The solutions to $Y_1 \le Y_2 < Y_3$ are the x -values between 4 and 6.4, including $4 \Rightarrow [4, 6.4)$. In

set-builder notation the interval is $\{x | 4 \le x < 6.4\}$.

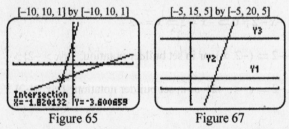

[-10, 10, 1] by [-10, 10, 1]   [-5, 15, 5] by [-5, 20, 5]

Figure 65   Figure 67

69.   Graph $Y_1 = 1.5$, $Y_2 = 9.1 - 0.5X$ and $Y_3 = 6.8$, as shown in Figure 69. The graphs intersect at the points (4.6, 6.8) and (15.2, 1.5). The solutions to $Y_1 \le Y_2 \le Y_3$ are the x-values between 4.6 and 15.2 (inclusive) or $4.6 \le x \le 15.2 \Rightarrow [4.6, 15.2]$. In set builder notation the interval is $\{x | 4.6 \le x \le 15.2\}$.

71.   Graph $Y_1 = X - 4$, $Y_2 = 2X - 5$ and $Y_3 = 6$ as shown in Figure 71. The graph of y_2 intersects the graphs of y_1 and y_3 at $(1, -3)$ and $(5.5, 6)$. The solutions to $Y_1 < Y_2 < Y_3$ are the x-values between 1 and 5.5 or $1 < x < 5.5 \Rightarrow (1, 5.5)$. In set-builder notation the interval is $\{x | 1 < x < 5.5\}$.

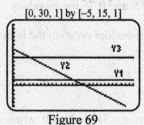   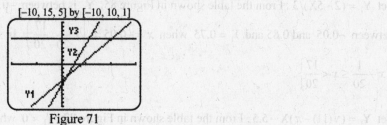

[0, 30, 1] by [−5, 15, 1]   [−10, 15, 5] by [−10, 10, 1]

          Figure 69   Figure 71

73.   (a) $x = -1$

       (b) $(-\infty, -1)$

       (c) $[-1, \infty)$

75.   (a) $x = 3$

       (b) $(3, \infty)$

       (c) $(-\infty, 3]$

77.   (a)   The graphs intersect at the point (8, 7). Therefore, $g(x) = f(x)$ is satisfied when $x = 8$. The solution is 8.

       (b) $g(x) > f(x)$ whenever the y-values on the graph of g are above the y-values on the graph of f. This occurs to the left of the point of intersection. Therefore the x-values that satisfy this inequality are $x < 8$. In set-builder notation the interval is $\{x | x < 8\}$.

79.   From the table,

       $Y_1 = 0$ when $x = 4$. $Y_1 > 0$ when $x < 4 \Rightarrow \{x | x < 4\}$; $Y_1 \le 0$ when $x \ge 4 \Rightarrow \{x | x \ge 4\}$.

81.   Let $Y_1 = -4X - 6$. From the table shown in Figure 81, $Y_1 = 0$ when $x = -1.5$ or $-\dfrac{3}{2}$. $Y_1 > 0$ when $x < -\dfrac{3}{2}$.

       The solution is $\left(-\infty, -\dfrac{3}{2}\right)$. In set-builder notation the interval is $\left\{x \middle| x < -\dfrac{3}{2}\right\}$.

83.   Let $Y_1 = 3X - 2$. From the table shown in Figure 83, Y_1 is between 10 and 4 (inclusive) for x-values between 1 and 4 (inclusive) $\Rightarrow [1, 4]$. In set-builder notation the interval is $\{x | 1 \le x \le 4\}$.

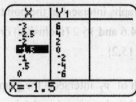

Figure 81   Figure 83

85. Let $Y_1 = (2 - 5X)/3$. From the table shown in Figure 85, Y_1 is between -0.75 and 0.75 for x-values

between -0.05 and 0.85 and $Y_1 = 0.75$ when $x = -0.05 \Rightarrow \left[-\dfrac{1}{20}, \dfrac{17}{20} \right)$. In set-builder notation the interval is

$\left\{ x \middle| -\dfrac{1}{20} \le x < \dfrac{17}{20} \right\}$.

87. Let $Y_1 = (\sqrt{(11)} - \pi)X - 5.5$. From the table shown in Figure 87, $Y_1 \approx 0$ when $x = 31.4$. $Y_1 \le 0$ when

$x \le 31.4$; $(-\infty, 31.4]$. In set-builder notation the interval is $\{x | x \le 31.4\}$.

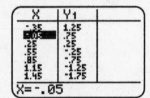   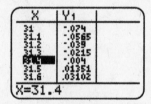

Figure 85   Figure 87

89. Symbolically: $2x - 8 > 5 \Rightarrow 2x > 13 \Rightarrow x > \dfrac{13}{2}$. The solution set is $\left(\dfrac{13}{2}, \infty \right)$. In set-builder notation the

interval is $\left\{ x \middle| x > \dfrac{13}{2} \right\}$.

91. Graphically: Let $Y_1 = \pi X - 5.12$ and $Y_2 = \sqrt{(2)}X - 5.7(X - 1.1.)$ Graph Y_1 and Y_2 as shown in Figure 91.
     The graphs intersect near $(1.534, -0.302)$. The graph of Y_1 is below Y_2 for $x < 1.534$, so $Y_1 \le Y_2$ when
     $x \le 1.534$. The solution set is $(-\infty, 1.534]$.

[−10, 10, 1] by [−10, 10, 1]

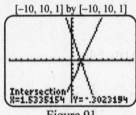

Figure 91

93. (a)   Car A is traveling faster since it passes Car B. Its graph has the greater slope.

     (b)   The cars are the same distance from St. Louis when their graphs intersect. This point of
                intersection occurs at $(2.5, 225)$. The cars are both 225 miles from St. Louis after 2.5 hours.

     (c)   Car B is ahead of Car A when $0 \le x < 2.5$.

95. (a)   Since the air temperature cools at $19°F$ for 1 mile increase in altitude we have a slope of -19 .
                The ground level air temperature is $65°F$, and we will have the function $T(x) = 65 - 19x$.

(b)   Since the dew point decreases by $5.8°F$ for 1 mile increase in altitude we have a slope of -5.8.

            The ground level dew point is $50°F$ and we will have the function $D(x) = 50 - 5.8x$.

(c) $65 - 19x > 50 - 5.8x \Rightarrow 15 > 13.2x \Rightarrow x < 1.14$. The result is below approximately 1.14 miles.

(d)   See Figure 95.

[0, 3, 1] by [0, 70, 10]

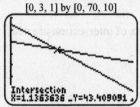

Figure 95

97. (a)   Revenue increased, on average, by \$0.86 billion per year.

    (b) $0.86x + 1.2 > 3 \Rightarrow 0.86x > 1.8 \Rightarrow x > 2.09$, the revenue is expected to be more than \$3 billion from 2012 to 2015.

99. (a)   The linear function U must fit the following points (0, 100) and (7, 1000). Using

$$m = \frac{1000 - 100}{7 - 0} = \frac{900}{7} \approx 128.57; b = 100; \text{ so } U(x) = 128.57x + 100.$$

    (b) $128.57x + 100 \geq 485 \Rightarrow 128.57x \geq 385 \Rightarrow x \geq 3;$, since x represents the number of years after 2008, the result is 2011 or later.

101. (a)   The linear function V must fit the following points (2006, 33) and (2011, 71). Using

$$m = \frac{71 - 33}{2011 - 2006} = \frac{38}{5} = 7.6 \text{ and the point (2011, 71) we have}$$

            $71 = 7.6(2011) + b \Rightarrow b = -15,212.6$ and $V(x) = 7.6x - 15,212.6$.

    (b) $40 \leq 7.6x - 15,212.6 \leq 55 \Rightarrow 15,252.6 \leq 7.6x \leq 15,267.6 \Rightarrow 2007 \leq x \leq 2009$

103. $(2011, 0.6), (2016, 2.4) \Rightarrow m = \frac{2.4 - 0.6}{2016 - 2011} = \frac{1.8}{5} = 0.36$; Using point-slope form,

$y = 0.36(x - 2016) + 2.4 \Rightarrow y = 0.36x - 723.36;$ So $y = 0.36x - 723.36 \Rightarrow 3.12 = 0.36x - 723.36 \Rightarrow x \approx 2018$ and $y = 0.36x - 723.36 \Rightarrow 4.2 = 0.36x - 723.36 \Rightarrow x \approx 2021$; The music sales might be between \$3.12 billion and \$4.2 billion sometime between 2018 and 2021.

105.   The graph of linear function will intersect the points (90, 6.5) and (129, 5.5).

$$m = \frac{6.5 - 5.5}{90 - 129} = -\frac{1}{39} \Rightarrow f(x) = -\frac{1}{39}(x - 129) + 5.5 \Rightarrow 5.75 < -\frac{1}{39}(x - 129) + 5.5 < 6 \Rightarrow$$

$$0.25 < -\frac{1}{39}(x - 129) < 0.5 \Rightarrow -9.75 > x - 129 > 79.5 \Rightarrow 119.25 > x > 109.5$$

So, about day 110 (April 19th) to day 119 (April 28th).

107. $r = \frac{C}{2\pi}$ and $1.99 \leq r \leq 2.01 \Rightarrow 1.99 \leq \frac{C}{2\pi} \leq 2.01 \Rightarrow 3.98\pi \leq C \leq 4.02\pi$

109. (a) $m = \dfrac{4.5 - (-1.5)}{2 - 0} = \dfrac{6}{2} = 3$ and y-intercept $= -1.5$; $f(x) = 3x - 1.5$ models the data.

(b) $f(x) > 2.25 \Rightarrow 3x - 1.5 > 2.25 \Rightarrow 3x > 3.75 \Rightarrow x > 1.25$

111. (a)   Using the linear regression function on the calculator the function P is found to be
$P(x) = 0.658x - 1290.76$.

(b)   Let Y_1 be $0.658X - 1290.76$, $Y_2 = 18$, and $Y_3 = 28.5$. The points of intersection are near
(1989, 18) and (2005, 28.5).

(c)   The answer was a result of interpolation.

Extended and Discovery Exercises for Section 2.3

1. $a < b \Rightarrow 2a < a + b < 2b \Rightarrow a < \dfrac{a+b}{2} < b$

2.4: More Modeling with Functions

1.   (a) $f(x) = \dfrac{x}{16}$

(b) $f(x) = 10x$

(c) $f(x) = 0.06x + 6.50$

(d) $f(x) = 500$

3.   Using the points (0, 3) and (4, 1) we have $m = \dfrac{1-3}{4-0} = \dfrac{-2}{4} = -\dfrac{1}{2}$ and a y-intercept of 3. Therefore, the

equation of the line is slope intercept form is $y = -\dfrac{1}{2}x + 3$.

5.   Using the points (3, 11) and (1, 7) we have $m = \dfrac{11-7}{3-1} = \dfrac{4}{2} = 2$. Substituting the point (1, 7) into the equation

$y = 2x + b \Rightarrow 7 = 2(2) + b \Rightarrow b = 5$. Therefore, the equation of the line is slope intercept form is $y = 2x + 5$.

7.   The height of the Empire State Building is constant; the graph that has no rate of change is d.

9.   As time increases the distance to the finish line decreases; the graph that shows this decline in distance as time increases is c.

11    $B(t) = 1.2t + 27$; t represents years after 2010; $D = \{t|0 \le t \le 4\}$

13. $T(t) = 0.08t + 0.6$; t represents months after January; $D = \{t|t = 0,1,2,...11\}$

15. $V(t) = \dfrac{180 - 580}{40}t + 580 = -10t + 580$; t represents minutes; $D = \{0 \le t \le 40\}$

17. $P(t) = 21.5 + 0.6t$; t represents years after 1900; $D = \{0 \le t \le 120\}$

19.   (a) $W(t) = -10t + 300$

      (b) $W(7) = -10(7) + 300 = 230$ gallons

      (c)   See Figure 19. x -intercept: $(30,0)$, after 30 minutes the tank is empty; y -intercept: $(0,300)$, the tank initially contains 300 gallons of water.

      (d) $D = \{t|0 \le t \le 30\}$

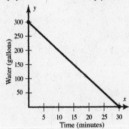

      Figure 19

21.   (a) $f(x) = 0.044x + 1.2$

      (b)   Since 2014 corresponds to $x = 0$, 2018 corresponds to $x = 4$; $f(x) = 0.044(4) + 1.2 = 1.376$, which means that about 1,376,000 people may be infected by 2018.

23.   (a) $f(x) = 0.25x + 0.5$

      (b) $f(2.5) = 0.25(2.5) + 0.5 = 1.125$ inches

25.   (a) $(5, 84)$, $(10, 169) \Rightarrow$ slope $= \dfrac{169 - 84}{10 - 5} = 17$

                     $(10, 169)$, $(15, 255) \Rightarrow$ slope $= \dfrac{255 - 169}{15 - 10} = 17.2$

                     $(15, 255)$, $(20, 338) \Rightarrow$ slope $= \dfrac{338 - 255}{20 - 15} = 16.6$

      (b) $f(x) = 17x$

      (c)   See Figure 25. The slope indicates that the number of miles traveled per gallon is 17.

      (d) $f(30) = 17(30) = 510$ miles . This indicates that the vehicle traveled 510 miles on 30 gallons of gasoline.

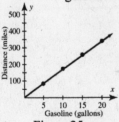

      Figure 25

27.   (a)   The maximum speed limit is 55 mph and the minimum is 30 mph.

      (b)   The speed limit is 55 for $0 \le x < 4$, $8 \le x < 12$, and $16 \le x < 20$. This is $4 + 4 + 4 = 12$ miles .

      (c) $f(4) = 40$, $f(12) = 30$, and $f(18) = 55$.

    (d)   The graph is discontinuous when $x = 4, 6, 8, 12$, and 16. The speed limit changes at each discontinuity.

29.   (a) $f(1.5) = 30; f(4) = 10$

    (b) $m_1 = 20$ indicates that the car is moving away from home at 20 mph; $m_2 = -30$ indicates that the car is moving toward home at 30 mph; $m_3 = 0$ indicates that the car is not moving; $m_4 = -10$ indicates that the car is moving toward home at 10 mph.

    (c)   The driver starts at home and drives away from home at 20 mph for 2 hours. The driver then travels toward home at 30 mph for 1 hour. Then the car does not move for 1 hour. Finally, the driver returns home in 1 hour at 10 mph.

    (d)   Increasing: $(0,2)$; Decreasing: $(2,3) \cup (4,5)$; Constant: $(3,4)$

31.   (a) $P(1.5) = 0.97; P(3) = 1.14;$ It costs \$0.97 to mail something that weighs 1.5oz and \$1.14 to mail something that weighs 3oz.

    (b)   See Figure 31. Domain: $(0,5]$.

    (c) P is discontinuous at $x = 1, 2, 3, 4$.

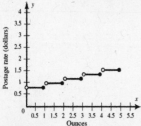

Figure 31

33.   The graphs start at points $f(x) = x^2 - 4$ and $f(x) = -x + 5$ are both shown in graph b.

35.   The graphs start at points $f(x) = -4$ and $f(x) = 4$ are both shown in graph d.

37.   From the graph, $P(2007) = 43\%$, $P(2009) = 10\%$,

39. $P(x) = \begin{cases} 5.5(x - 2003) + 21, & 2003 \le x \le 2007 \\ -16.5(x - 2007) + 43, & 2007 \le x \le 2009 \quad \textit{Answers may vary.} \\ 5.3(x - 2009) + 10, & 2009 \le x \le 2015 \end{cases}$

41.   (a) $D:[-5,5]$

    (b) $f(-2) = 2, f(0) = 0 + 3 = 3, f(3) = 3 + 3 = 6$

    (c)   See Figure 41.

    (d) f is continuous.

43.   (a) $D:[-2,2]$

    (b) $f(-2) = 2(-2) = -4, f(0) = 2(0) = 0, f(3)$ is undefined

    (c)   See Figure 43.

    (d) f is continuous.

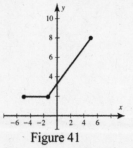

Figure 41

Figure 43

45.   (a) $D:[-4,3)$

    (b) $f(-2) = -2 - 1 = -3, f(0) = 0 - 1 = -1, f(3)$ is undefined

    (c)   See Figure 45.

(d) f is not continuous.

47.   (a) $D:[-1,2]$

      (b) $f(-2)$ is undefined, $f(0) = 3(0) = 0$, $f(3)$ is undefined

      (c)   See Figure 47.

      (d) f is not continuous.

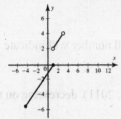

Figure 45

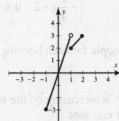

Figure 47

49.   (a) $D:[-3,3]$

      (b) $f(-2) = -2$, $f(0) = 1$, $f(3) = 2 - 3 = -1$

      (c)   See Figure 49.

      (d) f is not continuous.

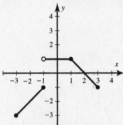

Figure 49

51. $f(-4) = -\dfrac{1}{2}(-4) + 1 = 3$, $(-4,3)$; $f(-2) = -\dfrac{1}{2}(-2) + 1 = 2$, $(-2,2)$; graph a segment from $(-4,3)$ to

      $(-2,2)$, use a closed dot for each point. See Figure 51.

      $f(-2) = 1 - 2(-2) = 5$, $(-2,5)$; $f(1) = 1 - 2(1) = -1$, $(1,-1)$; graph a segment from $(-2,5)$ to $(1,-1)$, use an

      open dot at $(-2,5)$ and a closed dot for $(1,-1)$. See Figure 51.

      $f(1) = \dfrac{2}{3}(1) + \dfrac{4}{3} = 2$, $(1,2)$; $f(4) = \dfrac{2}{3}(4) + \dfrac{4}{3} = 4$, $(4,4)$; graph a segment from $(1,2)$ to $(4,4)$, use an open dot

      at $(1,2)$ and a closed dot for $(4,4)$. See Figure 51.

53.   (a) $f(-3) = 3(-3) - 1 = -10$, $f(1) = 4$, $f(2) = 4$, and $f(5) = 6 - 5 = 1$

      (b)   The function f is constant with a value of 4 on the interval $[1,3]$.

      (c)   See Figure 53. f is not continuous.

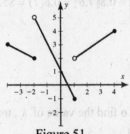

Figure 51

Figure 53

55.   We must determine two equations for the lines that represent the two line segments. Given the slope and y -

      intercept we can write the slope intercept form of the line. The first line segment has a slope of $\dfrac{3}{4}$ and

y-intercept of 3, so $f(x) = \frac{3}{4}x + 3$. The second line segment has a slope of $-\frac{2}{3}$ and a y-intercept of 2, so

$f(x) = -\frac{2}{3}x + 2$. The piecewise function is $f(x) = \begin{cases} \frac{3}{4}x + 3 & -4 \leq x < 0 \\ -\frac{2}{3}x + 2 & 0 \leq x \leq 3 \end{cases}$

57. (a)   See Figure 57.
    (b) $R(2009) = 700$, there were 700 people for each housing start. A small number will indicate a strong housing market.
    (c) R is continuous on its domain.
    (d)   From the graph we can see that R is increasing on the interval (2005, 2011), decreasing on the interval (2000, 2005), and is never constant.
    (e)   The piecewise function is $R(x) = \begin{cases} -9(x - 2000) + 225 & 2000 \leq x \leq 2005 \\ 130(x - 2005) + 180 & 2005 < x \leq 2009. \\ 13.5(x - 2009) + 700 & 2009 < x \leq 2011 \end{cases}$

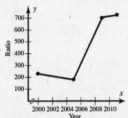

Figure 57

59. (a)   Graph $Y_1 = \text{int}(2X - 1)$ as shown in Figure 59.
    (b) $f(-3.1) = [\![2(-3.1) - 1]\!] = [\![-7.2]\!] = -8$ and $f(1.7) = [\![2(1.7) - 1]\!] = [\![2.4]\!] = 2$

61. (a)   Graph $Y_1 = (\text{int}(X)) + 1$ as shown in Figure 61.
    (b) $f(-3.1) = [\![-3.1]\!] + 1 = (-4) + 1 = -3$ and $f(1.7) = [\![1.7]\!] + 1 = (1) + 1 = 2$

Figure 59   Figure 61

63. (a) $f(x) = 0.8 \left[\!\left[\frac{x}{2} \right]\!\right]$ for $6 \leq x \leq 18$
    (b)   Graph $Y_1 = 0.8(\text{int}(X/2))$ as shown in Figure 63b.
    (c) $f(8.5) = 0.8 \left[\!\left[\frac{8.5}{2} \right]\!\right] = 0.8[\![4.25]\!] = 0.8(4) = \$3.20$; $f(15.2) = 0.8 \left[\!\left[\frac{15.2}{2} \right]\!\right] = 0.8[\![7.6]\!] = 0.8(7) = \$5.60$

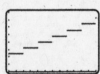

Figure 63b

65. Since y is directly proportional to x, the variation equation $y = kx$ must hold. To find the value of k, use the value $y = 7$ when $x = 14$. Solve the equation $7 = k(14) \Rightarrow k = \frac{1}{2}$. Then $y = \frac{1}{2}(5) = \frac{5}{2} = 2.5$.

67. Since y is directly proportional to x, the variation equation $y = kx$ must hold. To find the value of k, use the value $y = \dfrac{3}{2}$ when $x = \dfrac{2}{3}$. Solve the equation $\dfrac{3}{2} = k\left(\dfrac{2}{3}\right) \Rightarrow k = \dfrac{9}{4}$. Then $y = \dfrac{9}{4}\left(\dfrac{1}{2}\right) = \dfrac{9}{8}$.

69. Since y is directly proportional to x, the variation equation $y = kx$ must hold. To find the value of k use the value $y = 7.5$ when $x = 3$ from the table. Solve the equation $7.5 = k(3) \Rightarrow k = 2.5$. The variation equation is $y = 2.5x$ and hence $y = 2.5(8) = 20$ when $x = 8$.

71. Since y is directly proportional to x, the variation equation $y = kx$ must hold. To find the value of k use the value $y = 1.50$ when $x = 25$ from the table. Solve the equation $1.50 = k(25) \Rightarrow k = 0.06$. The variation equation is $y = 0.06x$ and hence when $y = 5.10$, $x = \dfrac{5.1}{0.06} = 85$.

73. Let y represent the cost of tuition and x represent the number of credits taken. Since the cost of tuition is directly proportional to the number of credits taken, the variation equation $y = kx$ must hold. If cost $y = \$720.50$ when the number of credits $x = 11$, we find the constant of proportionality k by solving $720.50 = k(11) \Rightarrow k = 65.50$. The variation equation is $y = 65.50x$. Therefore, the cost of taking 16 credits is $y = 65.50(16) = \$1048$.

75. (a)   Since the points $(0, 0)$ and $(300, 3)$ lie on the graph of $y = kx$, the slope of the graph is
$$\dfrac{3-0}{300-0} = 0.01 \text{ and } y = 0.01x, \text{ so } k = 0.01.$$
     (b) $y = 0.01(110) = 1.1$ millimeters.

77. (a)   Using $F = kx \Rightarrow 15 = k(8) \Rightarrow k = \dfrac{15}{8}$

     (b)   The variation equation is $y = \dfrac{15}{8}x$; $25 = \dfrac{15}{8}(x) \Rightarrow x = 13\dfrac{1}{3}$ inches.

79.   (a)   For $(150, 26)$, $\dfrac{F}{x} = \dfrac{26}{150} \approx 0.173$; for $(180, 31)$, $\dfrac{F}{x} = \dfrac{31}{180} \approx 0.172$; for $(210, 36)$,
$\dfrac{F}{x} = \dfrac{36}{210} \approx 0.171$; for $(320, 54)$, $\dfrac{F}{x} = \dfrac{54}{320} \approx 0.169$; the ratios give the force needed to push 1 lb box.
     (b)   From the table it appears that approximately 0.17 lb of force is needed to push a 1 lb cargo box $\Rightarrow k = 0.17$.
     (c)   See Figure 79.
     (d) $F \approx 0.17(275) \Rightarrow F = 46.75$ lbs of force.

Figure 79

81. (a)   Linear regression gives the model: $S(x) \approx 3.974x - 14,479$. *Answers may vary.*
     (b) $S(x) = 6 \Rightarrow 6 = 3.974x - 14.479 \Rightarrow 20.479 = 3.974x \Rightarrow x \approx 5.15$; The circumference of a finger with ring size 6 is approximately 5.15 cm.

83. (a)   Linear regression gives the model: $f(x) \approx 0.15878(x) - 315.69$ (Answers may vary)
     (b) $f(2009) \approx 0.15878(2009) - 315.69 \Rightarrow f(2009) \approx 3.3$, or \$3.3 million. The estimate was found using interpolation.

Extended and Discovery Exercises for Section 2.4

1. Let x = number of fish in the sample and y = number of tagged fish. Then $y = kx$, where k represents the proportion of fish tagged. From the data point $(94, 13)$, get $13 = k(94) \Rightarrow k \approx 0.138298$. Letting the sample represent the entire number of fish, we get $y = 0.138298x \Rightarrow 85 = 0.138298x \Rightarrow x \approx 615$.

3. *Answers may vary.*

Checking Basic Concepts for Sections 2.3 and 2.4

1. $2(x-4) > 1-x \Rightarrow 2x-8 > 1-x \Rightarrow 3x > 9 \Rightarrow x > 3; \{x \mid x > 3\}$

3. (a) $-3(2-x) - \dfrac{1}{2}x - \dfrac{3}{2} = 0$ when $x = 3$; symbolically,

   $-3(2-x) - \dfrac{1}{2}x - \dfrac{3}{2} = 0 \Rightarrow -6 + 3x - \dfrac{1}{2}x - \dfrac{3}{2} = 0 \Rightarrow \dfrac{5}{2}x - \dfrac{15}{2} = 0 \Rightarrow \dfrac{5}{2}x = \dfrac{15}{2} \Rightarrow x = 3$

   (b) $-3(2-x) - \dfrac{1}{2}x - \dfrac{3}{2} > 0$ when $x > 3$; symbolically,

   $-3(2-x) - \dfrac{1}{2}x - \dfrac{3}{2} > 0 \Rightarrow -6 + 3x - \dfrac{1}{2}x - \dfrac{3}{2} > 0 \Rightarrow \dfrac{5}{2}x - \dfrac{15}{2} > 0 \Rightarrow \dfrac{5}{2}x > \dfrac{15}{2} \Rightarrow x > 3 \Rightarrow (3, \infty)$. In set-builder notation the interval is $\{x \mid x > 3\}$.

   (c) $-3(2-x) - \dfrac{1}{2}x - \dfrac{3}{2} \leq 0$ when $x \leq 3$; symbolically,

   $-3(2-x) - \dfrac{1}{2}x - \dfrac{3}{2} \leq 0 \Rightarrow -6 + 3x - \dfrac{1}{2}x - \dfrac{3}{2} \leq 0 \Rightarrow \dfrac{5}{2}x - \dfrac{15}{2} \leq 0 \Rightarrow \dfrac{5}{2}x \leq \dfrac{15}{2} \Rightarrow x \leq 3 \Rightarrow (-\infty, 3]$. In set-builder notation the interval is $\{x \mid x \leq 3\}$.

5. Since the car is initially 50 miles south of home and driving south at 60 mph, the y-intercept is 50 and $m = 60$; $f(t) = 60t + 50$, where t is in hours.

2.5: Absolute Value Equations and Inequalities

1. $|x| = 3 \Rightarrow x = 3$ or $x = -3$

3. $|x| > 3 \Rightarrow x > 3$ or $x < -3; (-\infty, -3) \cup (3, \infty)$

5. The graph of $y = |ax + b|$ is V-shaped with the vertex on the x-axis.

7. $\sqrt{36a^2} = |6a|$ since 36 and a^2 are always positive values.

9. (a) $x + 1 = 0 \Rightarrow x = -1 \Rightarrow$ the vertex is $(-1, 0)$. Find any other point, $x = 0 \Rightarrow (0, 1)$; graph the absolute value graph with the vertex $(-1, 0)$, point $(0, 1)$ and its reflection through $x = -1$.
   See Figure 9.
   (b) $y = |x + 1|$ is increasing on $x > -1$ or $(-1, \infty)$ and decreasing on $x < -1$ or $(-\infty, -1)$.

11. (a) $2x - 3 = 0 \Rightarrow 2x = 3 \Rightarrow x = \dfrac{3}{2} \Rightarrow$ the vertex is $\left(\dfrac{3}{2}, 0\right)$. Find another point such as $(0, 3)$; graph the absolute value function. See Figure 11.

(b) $y = |2x - 3|$ is increasing on $x > \dfrac{3}{2}$ or $\left(\dfrac{3}{2}, \infty\right)$ and decreasing on $x < \dfrac{3}{2}$ or $\left(-\infty, \dfrac{3}{2}\right)$.

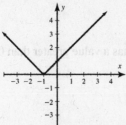

Figure 9

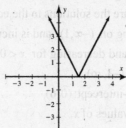

Figure 11

13.   (a)   The graph of $y_1 = 2x$ is shown in Figure 13a.

(b)   The graph of $y = |2x|$ is similar to the graph of $y = 2x$ except that it is reflected across the x-axis whenever $2x < 0$. The graph of $y_1 = |2x|$ is shown in Figure 13b.

(c)   The x-intercept occurs when $2x = 0$ or when $x = 0$. The x-intercept is 0.

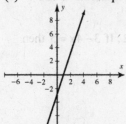

Figure 13a

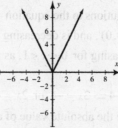

Figure 13b

15.   (a)   The graph of $y_1 = 3x - 3$ is shown in Figure 15a.

(b)   The graph of $y = |3x - 3|$ is similar to the graph of $y = 3x - 3$ except that it is reflected across the x-axis whenever $3x - 3 < 0$ or $x < 1$. The graph of $y_1 = |3x - 3|$ is shown in Figure 15b.

(c)   The x-intercept occurs when $3x - 3 = 0$ or when $x = 1$. The x-intercept is located at 1.

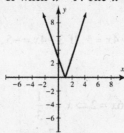

Figure 15a

Figure 15b

17.   (a)   The graph of $y_1 = 6 - 2x$ is shown in Figure 17a.

(b)   The graph of $y = |6 - 2x|$ is similar to the graph of $y = 6 - 2x$ except that it is reflected across the x-axis whenever $6 - 2x < 0$ or $x > 3$. The graph of $y_1 = |6 - 2x|$ is shown in Figure 17b.

(c)   The x-intercept occurs when $6 - 2x = 0$ or when $x = 3$. The x-intercept is located at 3.

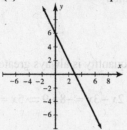

Figure 17a

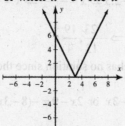

Figure 17b

19. (a) x-intercepts: $(0,0)$, $(2,0)$; y-intercept: $(0,0)$

    (b) $f(x)>0$ for $x<0$ and $x>2$, $f(x)<0$ for $0<x<2$, and $f(0)=0$, $f(2)=0$.

    (c)   The x-intercepts are the solutions to the equation. $f(x)=0$.

    (d) $f(x)$ is decreasing on $(-\infty,1)$, and is increasing on $(1,\infty)$.

    (e) $f(x)$ is positive and decreasing for $x<0$, as it has a negative slope but has a value greater than 0.

    (f) $\boldsymbol{D}:(-\infty,\infty)$; $\boldsymbol{R}:(-1,\infty)$

21. (a)   no x-intercepts; y-intercept: $(0,6)$

    (b) $f(x)>0$ for all values of x.

    (c)   There are no x-intercepts, indicating no real solutions for $f(x)$.

    (d) $f(x)$ is decreasing on $(-\infty,2)$, and is increasing on $(2,\infty)$.

    (e) $f(x)$ is positive and decreasing for $x<2$, as it has a negative slope but has a value greater than 0.

    (f) $\boldsymbol{D}:(-\infty,\infty)$; $\boldsymbol{R}:(2,\infty)$

23. (a) x-intercepts: $(-1,0)$, $(1,0)$; y-intercept: $(0,3)$

    (b) $f(x)>0$ for $-1<x<1$, $f(x)<0$ for $x<-1$ and $x>1$, and $f(-1)=0$, $f(1)=0$.

    (c)   The x-intercepts are the solutions to the equation $f(x)=0$.

    (d) $f(x)$ is increasing on $(-\infty,0)$, and is decreasing on $(0,\infty)$.

    (e) $f(x)$ is positive and decreasing for $0<x<1$, as it has a negative slope but has a value greater than 0.

    (f) $\boldsymbol{D}:(-\infty,\infty)$; $\boldsymbol{R}:(-\infty,3)$

25. $|-2x|=4 \Rightarrow -2x=4$ or $-2x=-4 \Rightarrow 2x=-4$ or $2x=4 \Rightarrow x=-2$ or 2.

27. $|-5x|=-2$ has no solutions since the absolute value of any quantity is always greater than or equal to 0.

29. $|2x-7|=15 \Rightarrow 2x-7=-15$ or $2x-7=15 \Rightarrow x=-4$ or 11.

31. $|-9-4x|=1 \Rightarrow -9-4x=-1$ or $-9-4x=1 \Rightarrow x=-\dfrac{5}{2}$ or -2

33. $|5x-7|=2 \Rightarrow 5x-7=-2$ or $5x-7=2$. If $5x-7=-2$, then $5x=5 \Rightarrow x=1$; If $5x-7=2$, then

       $5x=9 \Rightarrow x=\dfrac{9}{5}$; $1,\dfrac{9}{5}$.

35. $|3-4x|=5 \Rightarrow 3-4x=-5$ or $3-4x=5$. If $3-4x=-5$, then $-4x=-8 \Rightarrow x=2$; If $3-4x=5$ then

       $-4x=2 \Rightarrow x=-\dfrac{1}{2}$; $-\dfrac{1}{2},2$.

37. $|-6x-2|=0 \Rightarrow -6x-2=0 \Rightarrow -6x=2 \Rightarrow x=-\dfrac{1}{3}$.

39. $|7-16x|=0 \Rightarrow 7-16x=0 \Rightarrow 7=16x \Rightarrow x=\dfrac{7}{16}$.

41. $|17x-6|=-3$ has no solutions since the absolute value of any quantity is always greater than or equal to 0.

43. $|1.2x-1.7|-5=-1 \Rightarrow |1.2x-1.7|=4$, then $1.2x-1.7=-4$ or $1.2x-1.7=4$.

       If $1.2x-1.7=-4$ then, $1.2x=-2.3 \Rightarrow x=-\dfrac{2.3}{1.2} \Rightarrow x=-\dfrac{23}{12}$; if $1.2x-1.7=4$ then

       $1.2x=5.7 \Rightarrow x=\dfrac{5.7}{1.2} \Rightarrow x=\dfrac{19}{4}$; $\Rightarrow -\dfrac{23}{12},\dfrac{19}{4}$.

45. $|4x-5|+3=2 \Rightarrow |4x-5|=-1$ has no solution since the absolute value of any quantity is always greater than or equal to 0.

47. $|2x-9|=|8-3x| \Rightarrow 2x-9=8-3x$ or $2x-9=-(8-3x) \Rightarrow 2x+3x=8+9$ or $2x-3x=-8+9 \Rightarrow 5x=17$ or

       $-x=1 \Rightarrow x=\dfrac{17}{5},-1$

49. $\left|\dfrac{3}{4}x-\dfrac{1}{4}\right|=\left|\dfrac{3}{4}-\dfrac{1}{4}x\right|\Rightarrow \dfrac{3}{4}x-\dfrac{1}{4}=\dfrac{3}{4}-\dfrac{1}{4}x$ or $\dfrac{3}{4}x-\dfrac{1}{4}=-\left(\dfrac{3}{4}-\dfrac{1}{4}x\right)\Rightarrow \dfrac{3}{4}x+\dfrac{1}{4}x=\dfrac{3}{4}+\dfrac{1}{4}$ or

$\dfrac{3}{4}x-\dfrac{1}{4}x=-\dfrac{3}{4}+\dfrac{1}{4}\Rightarrow x=1$ or $\dfrac{1}{2}x=-\dfrac{1}{2}\Rightarrow x=-1,1$

51. $|15x-5|=|35-5x|\Rightarrow 15x-5=35-5x$ or $15x-5=-(35-5x)\Rightarrow 15x-5=35-5x\Rightarrow 20x=40\Rightarrow x=2$ or

$15x-5=-(35-5x)\Rightarrow 15x-5=-35+5x\Rightarrow 10x=-30\Rightarrow x=-3\Rightarrow x=-3$ or 2

53. (a) $f(x)=g(x)$ when $x=-1$ or 7.

(b) $f(x)<g(x)$ between these x-values or when $-1<x<7$; $(-1,7)$

(c) $f(x)>g(x)$ outside of these x-values or when $x<-1$ or $x>7$; $(-\infty,-1)\cup(7,\infty)$

55. (a) $f(x)=g(x)$ when $x=-1$ or 6.

(b) $f(x)<g(x)$ between these x-values or when $-1<x<6$; $(-1,6)$

(c) $f(x)>g(x)$ outside of these x-values or when $x<-1$ or $x>6$; $(-\infty,-1)\cup(6,\infty)$

57. (a) $f(x)=g(x)$ when $x=4$.

(b) $f(x)<g(x)$ has no solutions.

(c) $f(x)>g(x)$ outside of these x-values or when $|x|>4$; $(-\infty,4)\cup(4,\infty)$

59. (a) $|2x-3|=1\Rightarrow 2x-3=1$ or $2x-3=-1$. If $2x-3=1$, then $2x=4\Rightarrow x=2$; If $2x-3=-1$ then

$2x=2\Rightarrow x=1$; $x=1$ or $x=2$

(b) $|2x-3|<1\Rightarrow -1<2x-3<1\Rightarrow 2<2x<4\Rightarrow 1<x<2$; $(1,2)$

(c) $|2x-3|>1\Rightarrow 2x-3>1$ or $2x-3<-1$. If $2x-3>1$, then $2x>4\Rightarrow x>2$. If $2x-3<-1$, then

$2x<2\Rightarrow x<1$. $x<1$ or $x>2$ or $(-\infty,1)\cup(2,\infty)$

61. (a) $|4-2x|=6\Rightarrow 4-2x=-6$ or $4-2x=6$. If $4-2x=6$ then $-2x=2\Rightarrow x=-1$. Or

$4-2x=-6\Rightarrow -2x=-10$ then. $x=-1$ or $x=5$.

(b) $|4-2x|\le 6\Rightarrow 4-2x\le 1$ or 5, $[-1,5]$, or $-1\le x\le 5$

(c) $|4-2x|\ge 6\Rightarrow x\le -1$ or $x\ge 5$; $(-\infty,-1]\cup[5,\infty)$.

63. (a) Graph $Y_1=\text{abs}(2X-5)$ and $Y_2=10$ See Figures 63a and 63b. The solutions are $-\dfrac{5}{2}$ and $\dfrac{15}{2}$.

(b) See Figure 63c. The solutions are -2.5 and 7.5.

(c) $|2x-5|=10\Rightarrow 2x-5=-10$ or $2x-5=10\Rightarrow x=-\dfrac{5}{2}$ or $\dfrac{15}{2}$

From each method, the solution to $|2x-5|<10$ lies between -2.5 and 7.5, exclusively: $-\dfrac{5}{2}<x<\dfrac{15}{2}$ or

$\left(-\dfrac{5}{2},\dfrac{15}{2}\right)$.

[-10, 10, 1] by [-5, 15, 1]

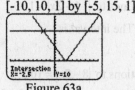
Figure 63a

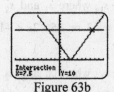
Figure 63b

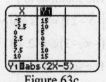
Figure 63c

65. (a) Graph $Y_1=\text{abs}(6-3X)$ and $Y_2=3$ See Figures 65a and 65b. The solutions are 1 and 3.

(b) See Figure 65c. The solutions are 1 and 3.

(c) $|6-3x|=3\Rightarrow 6-3x=-3$ or $5-3x=3\Rightarrow x=1,3$.

From each method, the solution to $|6-3x|>3$ lies outside of 1 and 3 exclusively: $x<1$ or $x>3$ or

$(-\infty,1)\cup(3,\infty)$.

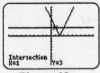

Figure 65a

Figure 65b

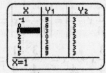

Figure 65c

67. $|2.1x - 0.7| = 2.4 \Rightarrow 2.1x - 0.7 = -2.4$ or $2.1x - 0.7 = 2.4 \Rightarrow x = -\dfrac{17}{21}$ or $\dfrac{31}{21}$ The solution to

$|2.1x - 0.7| \geq 2.4$ lies outside of $-\dfrac{17}{21}$ and $\dfrac{31}{21}$, inclusively: $x \leq -\dfrac{17}{21}$ or $x \geq \dfrac{31}{21}$ or $\left(-\infty, -\dfrac{17}{21}\right] \cup \left[\dfrac{31}{21}, \infty\right)$.

69. $|3x| + 5 = 6 \Rightarrow |3x| = 1 \Rightarrow 3x = -1$ or $3x = 1 \Rightarrow x = -\dfrac{1}{3}$ or $\dfrac{1}{3}$.

The solution to $|3x| + 5 > 6$ lies outside of $-\dfrac{1}{3}$ and $\dfrac{1}{3}$, exclusively: $x < -\dfrac{1}{3}, x > \dfrac{1}{3}, \left(-\infty, -\dfrac{1}{3}\right) \cup \left(\dfrac{1}{3}, \infty\right)$.

71. $\left|\dfrac{2}{3}x - \dfrac{1}{2}\right| = -\dfrac{1}{4}$ has no solutions since the absolute value of any quantity is always greater than or equal to 0.

There are no solutions to $\left|\dfrac{2}{3}x - \dfrac{1}{2}\right| \leq -\dfrac{1}{4}$.

73.   The solutions to $|3x - 1| < 8$ satisfy $s_1 < x < s_2$ where s_1 and s_2, are the solutions to $|3x - 1| = 8$.

$|3x - 1| = 8$ is equivalent to $3x - 1 = -8 \Rightarrow x = -\dfrac{7}{3}$ or $3x - 1 = 8 \Rightarrow x = 3$. The interval is $\left(-\dfrac{7}{3}, 3\right)$.

75.   The solutions to $|7 - 4x| \leq 11$ satisfy $s_1 \leq x \leq s_2$, where s_1 and s_2 are the solutions to $|7 - 4x| = 11$.

$|7 - 4x| = 11$ is equivalent to $7 - 4x = -11 \Rightarrow x = \dfrac{9}{2}$ or $7 - 4x = 11 \Rightarrow x = -1$. The interval is $\left[-1, \dfrac{9}{2}\right]$.

77.   The solutions to $\left|-\dfrac{1}{2}x\right| < 3$ satisfy $s_1 < x < s_2$, where s_1 and s_2 are the solutions to $\left|-\dfrac{1}{2}x\right| = 3$. $\left|-\dfrac{1}{2}x\right| = 3$ is

equivalent to $-\dfrac{1}{2}x = 3 \Rightarrow x = -6$ or $-\dfrac{1}{2}x = -3 \Rightarrow x = 6$. The interval is $(-6, 6)$.

79.   The solutions to $\left|\dfrac{1}{3}x\right| \geq 9$ satisfy $s_2 \geq x \geq s_1$, where s_1 and s_2 are the solutions to $\left|\dfrac{1}{3}x\right| = 9$. $\left|\dfrac{1}{3}x\right| = 9$ is

equivalent to $-\dfrac{1}{3}x = 9 \Rightarrow x = -27$ or $-\dfrac{1}{3}x = -9 \Rightarrow x = 27$. The interval is $(-\infty, -27] \cup [27, \infty)$.

81.   The solutions to $|3x - 5| < 7$ satisfy $s_1 < x < s_2$, where s_1 and s_2 are the solutions to $|3x - 5| = 7$. $|3x - 5| = 7$

is equivalent to $3x - 5 = -7 \Rightarrow x = -\dfrac{2}{3}$ or $3x - 5 = 7 \Rightarrow x = 4$. The interval is $\left(-\dfrac{2}{3}, 4\right)$.

83.   The solutions to $|-5 - 2x| > 1$ satisfy $s_2 > x > s_1$, where s_1 and s_2 are the solutions to $|-5 - 2x| = 1$.

$|-5 - 2x| = 1$ is equivalent to $-5 - 2x = 1 \Rightarrow x = -3$ or $-5 - 2x = -1 \Rightarrow x = -2$. The interval is
$(-\infty, -3) \cup (-2, \infty)$.

85.   The solutions to $|8 - 2x| \geq 10$ satisfy $s_2 \geq x \geq s_1$, where s_1 and s_2 are the solutions to $|8 - 2x| = 10$.

$|8 - 2x| = 10$ is equivalent to $8 - 2x = 10 \Rightarrow x = -1$ or $8 - 2x = -10 \Rightarrow x = 9$. The interval is $(-\infty, -1] \cup [9, \infty)$.

87.   There are no solutions to $|4x + 1| \leq -1$, because the absolute value of any quantity must always be greater than
or equal to 0.

89.   The solutions to $|x + 1| - 1 \geq -2$ satisfy $s_2 \geq x \geq s_1$, where s_1 and s_2 are the solutions to $|x + 1| - 1 = -2$.

$|x + 1| - 1 = -2 \Rightarrow |x + 1| = -1$. But, this equality has no solutions, so the only possible solutions to the
inequality are those that satisfy $|x + 1| \geq 0$, that is, all values of x. The interval is therefore $(-\infty, \infty)$.

91. The solutions to $|0.5x-0.75|<2$ satisfy $s_1 < x < s_2$, where s_1 and s_2 are the solutions to $|0.5x-0.75|=2$. $|0.5x-0.75|=2$ is equivalent to $0.5x-0.75=-2 \Rightarrow x=-\dfrac{5}{2}$ or $0.5x-0.75=2 \Rightarrow x=\dfrac{11}{2}$. The interval is $\left(-\dfrac{5}{2}, \dfrac{11}{2}\right)$.

93. The solutions to $|2x-3|>1$ satisfy $x < s_1$ or $x > s_2$, where s_1 and s_2 are the solutions to $|2x-3|=1$. $|2x-3|=1$ is equivalent to $2x-3=-1 \Rightarrow x=1$ or $2x-3=1 \Rightarrow x=2$. The solution set is $(-\infty, 1) \cup (2, \infty)$.

95. The solutions to $|-3x+8| \geq 3$ satisfy $x \leq s_1$ or $x \geq s_2$, where s_1 and s_2 are the solutions to $|-3x+8|=3$. $|-3x+8|=3$ is equivalent to $-3x+8=-3 \Rightarrow x=\dfrac{11}{3}$ or $-3x+8=3 \Rightarrow x=\dfrac{5}{3}$. The solution set is $\left(-\infty, \dfrac{5}{3}\right] \cup \left[\dfrac{11}{3}, \infty\right)$.

97. The solutions to $|0.25x-1|>3$ satisfy $x < s_1$ or $x > s_2$, where s_1 and s_2 are the solutions to $|0.25x-1|=3$. $|0.25x-1|=3$ is equivalent to $0.25x-1=-3 \Rightarrow x=-8$ or $0.25x-1=3 \Rightarrow x=16$. The solution set is $(-\infty, -8) \cup (16, \infty)$.

99. $|-6|=6$

101. Since the inputs of absolute values can be positive or negative, the domain of $|f(x)|$ is also $[-2, 4]$.

103. Since all solutions or the range of absolute values must be non-negative, all negative solutions will change to positive solutions; therefore, if the range of $f(x)$ is $(-\infty, 0]$, the range of $|f(x)|$ is $[0, \infty)$.

105. (a) The domain is not affected, so $\boldsymbol{D}:[-2,3]$;

$\boldsymbol{R}:(-5,8]$, so $|f(x)| \Rightarrow \boldsymbol{R}:[0,8]; -|f(x)| \Rightarrow \boldsymbol{R}:[-8,0]; -|f(x)|+2 \Rightarrow \boldsymbol{R}:[-8+2,0+2]=[-6,2]$

(b) The domain is not affected, so $\boldsymbol{D}:[-2,3]$;

$\boldsymbol{R}:(-5,8]$, so $|f(x)| \Rightarrow \boldsymbol{R}:[0,8]; -|f(x)| \Rightarrow \boldsymbol{R}:[-8,0]; -|f(x)|-3 \Rightarrow \boldsymbol{R}:[-8-3,0-3]=[-11,-3]$

(c) $|f(x-1)| \Rightarrow \boldsymbol{D}:[-2+1,3+1]=[-1,4]$

$\boldsymbol{R}:(-5,8]$, so $|f(x-1)| \Rightarrow \boldsymbol{R}:[0,8]; |f(x-1)|+1 \Rightarrow \boldsymbol{R}:[0+1,8+1]=[1,9]$

107. For $f(x)$, $\boldsymbol{D}:[-2,3]$; $\boldsymbol{R}:[-2,3]$; For $|f(x)|$, $\boldsymbol{D}:[-2,3]$; $\boldsymbol{R}:[0,3]$.

109. $|T-(-10)| \leq 5$; $-15 \leq T \leq -5$ in degrees Fahrenheit

111. $|S-57.5|=17.5 \Rightarrow S-57.5=17.5$ or $S-57.5=-17.5$ If $S-57.5=17.5$, then $S=75$. If $S-57.5=-17.5$, then $S=40$. Therefore, the maximum speed limit is 75 mph and the minimum speed limit is 40 mph.

113. (a) $0 \leq 80-19x \leq 32 \Rightarrow -80 \leq -19x \leq -48 \Rightarrow \dfrac{80}{19} \geq x \geq \dfrac{48}{19} \Rightarrow \dfrac{48}{19} \leq x \leq \dfrac{80}{19}$. The air temperature is between $0°F$ and $32°F$ when the altitudes are between $\dfrac{48}{19}$ and $\dfrac{80}{19}$ miles inclusively.

(b) The air temperature is between $0°F$ and $32°F$ inclusively when the altitude is within $\dfrac{16}{19}$ mile of $\dfrac{64}{19}$ miles. $\left|x-\dfrac{64}{19}\right| \leq \dfrac{16}{19}$.

115. (a) $|T-43|=24 \Rightarrow T-43=-24$ or $T-43=24 \Rightarrow T=19$ or 67. The average monthly temperature range is $19°F \leq T \leq 67°F$.

(b) The monthly average temperatures in Marquette vary between a low of $19°F$ and a high of $67°F$. The monthly averages are always within $24°$ of $43°F$.

117. (a) $|T-50|=22 \Rightarrow T-50=-22$ or $T-50=22 \Rightarrow T=28$ or 72. The average monthly temperature range is $28°F \leq T \leq 72°F$.

(b) The monthly average temperatures in Boston vary between a low of $28°F$ and a high of $72°F$.

The monthly averages are always within 22° of 50°F.

119. (a) $|T - 61.5| = 12.5 \Rightarrow T - 61.5 = -12.5$ or $T - 61.5 = 12.5 \Rightarrow T = 49$ or 74 The average monthly
        temperature range is $49°F \leq T \leq 74°F$.

     (b) The monthly average temperatures in Buenos Aires vary between a low of 49°F and a high of 74°F.
        The monthly averages are always within 12.5° of 61.5°F.

121. (a) $|T - 10.5| < 0.05$

     (b) $|T - 10.5| < 0.05 \Rightarrow -0.05 < T - 10.5 < 0.05 \Rightarrow 10.45 < T < 10.55$, the actual thickness must be greater than
        10.45 mm and less than 10.55 mm.

123. (a) $|D - 2.118| \leq 0.07$

     (b) $|D - 2.118| \leq 0.07 \Rightarrow -0.07 \leq D - 2.118 \leq 0.07 \Rightarrow 2.111 \leq T \leq 2.125$, the diameter must be greater than or
        equal to 2.111 inches and less than or equal to 2.125 inches.

125. $\left|\dfrac{Q-A}{A}\right| \leq 0.02 \Rightarrow \left|\dfrac{Q-35}{35}\right| \leq 0.02$, so $-0.02 \leq \dfrac{Q-35}{35} \leq 0.02 \Rightarrow -0.7 \leq Q - 35 \leq 0.7 \Rightarrow 34.3 \leq Q \leq 35.7$

Extended and Discovery Exercises for Section 2.5

1. The distance between points x and c on a number line can be shown by $|x - c|$. This distance is given to be
    less than some positive value δ. Then $|x - c| < \delta$.

Checking Basic Concepts for Section 2.5

1. $\sqrt{4x^2} = |2x|$

3. (a) Graphically: Graph $Y_1 = abs(2X - 1)$ and $Y_2 = 5$. Their graphs intersect at the points $(-2, 5)$ and $(3,5)$.
        The solutions are $-2, 3$. See Figures 3a & 3b.
        The solutions are $-2, 3$. See Figure 3c.
        Symbolically: $|2x - 1| = 5 \Rightarrow 2x - 1 = 5$ or $2x - 1 = -5 \Rightarrow 2x = 6$ or $2x = -4 \Rightarrow x = 3, -2$. The solutions are
        $-2, 3$.

    (b) The solutions to $|2x - 1| \leq 5$ lie between $x = -2$ and $x = 3$, inclusively. Thus, $-2 \leq x \leq 3$ or $[-2, 3]$. The
        solutions to $|2x - 1| > 5$ lie left of $x = -2$ or right of $x = 3$. Thus, $x < -2$ or $x > 3$ or $(-\infty, -2) \cup (3, \infty)$.

Figure 3a

Figure 3b

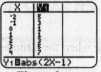
Figure 3c

5. $|x + 1| = |2x| \Rightarrow x + 1 = 2x$ or $x + 1 = -2x$. If $x + 1 = 2x$, then $x = 1$;

    $x + 1 = -2x \Rightarrow 1 = -3x \Rightarrow x = -\dfrac{1}{3}; \ -\dfrac{1}{3}, 1$

Chapter 2 Review Exercises

1. The slope of the line is $m = \dfrac{5-4}{2-(-3)} = \dfrac{1}{5}$, and then we substitute the slope and the point $(-3, 4)$ into the

    equation $y = m(x - x_1) + y_1$ to find $y = \dfrac{1}{5}(x - (-3)) + 4 \Rightarrow y = \dfrac{1}{5}(x + 3) + 4$.

3. Substituting the slope and the given point into the equation $y = m(x - x_1) + y_1$, we have

    $y = -\dfrac{7}{5}(x + 5) + 6 \Rightarrow y = -\dfrac{7}{5}x - 7 + 6 \Rightarrow y = -\dfrac{7}{5}x - 1$. The function is $f(x) = -\dfrac{7}{5}x - 1$.

5. The slope of the line is $m = \dfrac{-2-0}{0-5} = \dfrac{-2}{-5} = \dfrac{2}{5}$, and then we substitute the slope and the point $(5, 0)$ into the

    equation $y = m(x - x_1) + y_1$ to find $y = \dfrac{2}{5}(x - 5) + 0 \Rightarrow y = \dfrac{2}{5}x - 2$. The function is $f(x) = \dfrac{2}{5}x - 2$.

7. Using point-slope form $y = m(x - x_1) + y_1$, we get $y = -2(x + 2) + 3 \Rightarrow f(x) = -2x - 1$

9. $y = 7(x + 3) + 9 \Rightarrow y = 7x + 21 + 9 \Rightarrow y = 7x + 30$

11. Let $m = -3$. Then, $y = -3(x - 1) - 1 \Rightarrow y = -3x + 2$.

13. The line segment has slope $m = \dfrac{0-3}{6-0} = -\dfrac{1}{2}$; the parallel line has slope $m = -\dfrac{1}{2}$;

    $y = -\dfrac{1}{2}(x - 1) - 7 \Rightarrow y = -\dfrac{1}{2}x - \dfrac{13}{2}$

15. The line is vertical passing through $(6, -7)$, so the equation is $x = 6$.

17. The line is horizontal passing through $(1, 3)$, so the equation is $y = 3$.

19. The equation of the vertical line with x-intercept 2.7 is $x = 2.7$.

21. For x-intercept: $y = 0 \Rightarrow 5x - 4(0) = 20 \Rightarrow 5x = 20 \Rightarrow x = 4$; for y-intercept:

    $x = 0 \Rightarrow 5(0) - 4y = 20 \Rightarrow -4y = 20 \Rightarrow y = -5$; use $(4, 0)$ and $(0, -5)$ to graph the equation. See Figure 21.

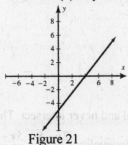

Figure 21

23. Graphical: Graph $Y_1 = 5X - 25$ and $Y_2 = 10$. Their graphs intersect at $(7, 10)$ as shown in Figure 23. The
    solution is $x = 7$.

    Symbolic: $5x - 25 = 10 \Rightarrow 5x = 35 \Rightarrow x = \dfrac{35}{5} = 7$

    [-15, 15, 5] by [-15, 15, 5]

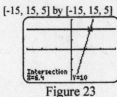

Figure 23

25. Graphical: Graph $Y_1 = -2(3X - 7) + X$ and $Y_2 = 2X - 1$. Their graphs intersect near $(2.143, 3.286)$ as shown in Figure 25. The solution is approximately 2.143.

   Symbolic: $-2(3x - 7) + x = 2x - 1 \Rightarrow -6x + 14 + x = 2x - 1 \Rightarrow -7x = -15 \Rightarrow x = \dfrac{15}{7} \approx 2.143$

27. Graphical: Graph $Y_1 = \pi X + 1$ and $Y_2 = 6$. Their graphs intersect near $(1.592, 6)$ as shown in Figure 27. The solution is approximately 1.592.

   Symbolic: $\pi x + 1 = 6 \Rightarrow \pi x = 5 \Rightarrow x = \dfrac{5}{\pi} \approx 1.592$

Figure 25   Figure 27

29. Let $Y_1 = 3.1X - 0.2 - 2(X - 1.7)$ and approximate where $Y_1 = 0$. From Figure 29 this occurs when $x \approx -2.9$.

Figure 29

31. (a) $4(6 - x) = -4x + 24 \Rightarrow 24 - 4x = -4x + 24 \Rightarrow 0 = 0 \Rightarrow$ all real numbers are solutions.

   (b) Because all real numbers are solutions, the equation is an identity.

33. (a) $5 - 2(4 - 3x) + x = 4(x - 3) \Rightarrow 5 - 8 + 6x + x = 4x - 12 \Rightarrow 7x - 3 = 4x - 12 \Rightarrow 3x = -9 \Rightarrow x = -3$

   (b) Because there are finitely many solutions, the equation is conditional.

35. $(-3, \infty)$

37. $\left[-2, \dfrac{3}{4} \right)$

39. Graphical: Graph $Y_1 = 3X - 4$ and $Y_2 = 2 + X$. Their graphs intersect at $(3, 5)$. The graph of Y_1 is below the graph of Y_2 to the left of the point of intersection. Thus, $3x - 4 \leq 2 + x$ holds when $x \leq 3$ or $(-\infty, 3]$. See Figure 39. Symbolic: $3x - 4 \leq 2 + x \Rightarrow 2x \leq 6 \Rightarrow x \leq 3$ or $(-\infty, 3]$. In set-builder notation, the interval is $\{ x \mid x \leq 3 \}$.

Figure 39

41. Graphical: Graph $Y_1 = (2X - 5)/2$ and $Y_2 = (5X + 1)/5$. Their graphs are parallel and never intersect. The graph of Y_1 is always below the graph of Y_2, so $Y_1 < Y_2$ for all values of x; the inequality $\dfrac{2x - 5}{2} < \dfrac{5x + 1}{5}$ holds when $-\infty < x < \infty$, or $(-\infty, \infty)$. See Figure 41. In set-builder notation the interval is $\{ x \mid -\infty < x < \infty \}$.

43. Graphical: Graph $Y_1 = -2$, $Y_2 = 5 - 2X$ and $Y_3 = 7$. See Figure 43. Their graphs intersect at the points $(-1, 7)$ and $(3.5, -2)$. The graph of Y_2 is between the graphs of Y_1 and Y_3 when $-1 < x \leq 3.5$. In interval notation the solution is $(-1, 3.5]$.

   Symbolic: $-2 \leq 5 - 2x < 7 \Rightarrow -7 \leq -2x < 2 \Rightarrow \dfrac{7}{2} \geq x > -1 \Rightarrow -1 < x \leq \dfrac{7}{2}$ or $\left(-1, \dfrac{7}{2} \right]$

   In set-builder notation, the interval is $\left\{ x \mid -1 < x \leq \dfrac{7}{2} \right\}$.

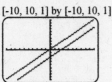
Figure 41

Figure 43

45.   Graph $Y_1 = 2x$ and $Y_2 = x - 1$. See Figure 45. The lines intersect at $(-1, -2)$. $Y_1 > Y_2$ when the graph of Y_1 is above the graph of Y_2; this happens when $x > -1 \Rightarrow (-1, \infty)$. In set-builder notation the interval is $\{x \mid x > -1\}$.

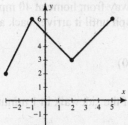
Figure 44

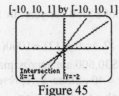
Figure 45

Figure 46

47.   (a)   The graphs intersect at $(2, 1)$. The solution to $f(x) = g(x)$ is 2.

       (b)   The graph of f is below the graph of g to the right of $(2, 1)$. Thus, $f(x) < g(x)$ when $x > 2$ or on $(2, \infty)$.

       (c)   The graph of f is above the graph of g to the left of $(2, 1)$. Thus, $f(x) > g(x)$ when $x < 2$ or on $(-\infty, 2)$.

49.   (a) $f(-2) = 8 + 2(-2) = 4$; $f(-1) = 8 + 2(-1) = 6$; $f(2) = 5 - 2 = 3$; $f(3) = 3 + 1 = 4$.

       (b)   The graph of f is shown in Figure 49. It is essentially a piecewise line graph with the points $(-3, 2)$, $(-1, 6)$, $(2, 3)$, and $(5, 6)$. Since there are no breaks in the graph, f is continuous.

       (c)   From the graph we can see that there are two x-values where $f(x) = 3$. They occur when $8 + 2x = 3 \Rightarrow x = -2.5$ and when $5 - x = 3 \Rightarrow x = 2$. The solutions are $x = -2.5$ or 2.

Figure 49

51. $|2x - 5| - 1 = 8 \Rightarrow |2x - 5| = 9 \Rightarrow 2x - 5 = -9$ or $2x - 5 = 9$; $2x - 5 = -9 \Rightarrow 2x = -4 \Rightarrow x = -2$; $2x - 5 = 9 \Rightarrow 2x = 14 \Rightarrow x = 7 \Rightarrow -2, 7$

53. $|6 - 4x| = -2$ has no solutions since the absolute value of any quantity is always greater than or equal to 0.

55. $|x| = 3 \Rightarrow x = \pm 3$. The solutions to $|x| > 3$ lie to the left of -3 and to the right of 3. That is, $x < -3$ or $x > 3$. This can be supported by graphing $Y_1 = |x|$ and $Y_2 = 3$ and determining where the graph of Y_1 is above the graph of Y_2. To support this result numerically, table $Y_1 = \text{abs}(X)$ starting at -9 and incrementing by 3.

57. $|3x - 7| = 10 \Rightarrow 3x - 7 = 10$ or $3x - 7 = -10 \Rightarrow x = \dfrac{17}{3}$ or $x = -1$. The solutions to $|3x - 7| > 10$ lie to the left of -1 or to the right of $x = \dfrac{17}{3}$; that is, $x < -1$ or $x > \dfrac{17}{3}$. This can be supported by graphing $Y_1 = |3x - 7|$ and $Y_2 = 10$ and determining where the graph of Y_1 is above the graph of Y_2. To support this result numerically, table $Y_1 = \text{abs}(3X - 7)$ starting at -3 and incrementing by $\dfrac{1}{3}$.

59.   The solutions to $|3 - 2x| < 9$ satisfy $s_1 < x < s_2$ where s_1 and s_2, are the solutions to $|3 - 2x| = 9$.

$|3-2x|=9$ is equivalent to $3-2x=-9 \Rightarrow x=6$ or $3-2x=9 \Rightarrow x=-3$.

The solutions are $-3 < x < 6$ or $(-3, 6)$.

61. The solutions to $\left|\dfrac{1}{3}x-\dfrac{1}{6}\right| \geq -1$ are satisfied by all real numbers, or $(-\infty, \infty)$.

63. (a) Using the points $(1980, 17{,}700)$ and $(2010, 49{,}500)$, $m=\dfrac{49{,}500-17{,}700}{2010-1980}=\dfrac{31{,}800}{30}=1060$. Since x

   represents the number of years after 1980, we have a y-intercept of 17,700, and the function is

   $f(x)=1060x+17{,}700$.

   (b) The median income increased, on average, by about \$1060 per year. In 1980 median income was \$17,700.

   (c) Since 1992 is 12 years after 1980, we will let $x=12$ and $f(12)=1060(12)+17{,}700=30{,}420$. The calculated result of \$30,420 and the true value of \$30,600 are approximately equal.

   (d) $60{,}000=1060x+17{,}700 \Rightarrow 42{,}300=1060x \Rightarrow x \approx 40$, Since x represents the number of years after 1980 the result is about 2020. The calculation involved extrapolation.

65. (a) Using the points $(2010, 524)$ and $(2020, 949)$, $m=\dfrac{524-949}{2010-2020}=\dfrac{-425}{-10}=42.5$. Since x represents the

   number of years after 2010, we have a y-intercept of 524, and the function is $f(x)=42.5x+524$.

   (b) The spending increased, on average, by about \$42.5 billion per year. In 2010 spending was \$524 billion.

   (c) Since 2016 is 6 years after 2010, we will let $x=6$ and $f(6)=42.5(6)+524=779$. The result is \$779 billion. The calculation involved interpolation.

   (d) $694 \leq 42.5x+524 \leq 864 \Rightarrow 170 \leq 42.5x \leq 340 \Rightarrow 4 \leq x \leq 8$, since x represents the number of years after 2010 the result is about from 2014 to 2018.

67. Since the graph is piecewise linear, the slope each line segment represents a constant speed. Initially, the car is home. After 1 hour it is 30 miles from home and has traveled at a constant speed of 30 mph. After 2 hours it is 50 miles away. During the second hour the car travels 20 mph. During the third hour the car travels toward home at 30 mph until it is 20 miles away. During the fourth hour the car travels away from home at 40 mph until it is 60 miles away from home. The last hour the car travels 60 miles at 60 mph until it arrives back at home.

69. The midpoint is computed by $\left(\dfrac{2012+2016}{2}, \dfrac{167{,}933+143{,}247}{2}\right)=(2014, 155590)$.

71. Let $x=$ time it takes for both working together; the first worker can shovel $\dfrac{1}{50}$ of the sidewalk in 1 minute,

   and the second worker can shovel $\dfrac{1}{30}$ of the sidewalk in 1 minute; for the entire job, we get the equation

   $\dfrac{x}{50}+\dfrac{x}{30}=1 \Rightarrow 3x+5x=150 \Rightarrow 8x=150 \Rightarrow x=18.75$; it takes the two workers 18.75 minutes to shovel the sidewalk together.

73. Let $t=$ time spent jogging at 7 mph; then $1.8-t=$ time spent jogging at 8 mph; since $d=rt$ and the total distance jogged is 13.5 miles, we get the equation

   $7t+8(1.8-t)=13.5 \Rightarrow 7t+14.4-8t=13.5 \Rightarrow -t=-0.9 \Rightarrow t=0.9$ and $1.8-t=0.9$; the runner jogged 0.9 hour at 7 mph and 0.9 hour at 8 mph.

75. (a) Begin by selecting any two points to determine the equation of the line. For example, if we use

   $(-1, 4.2)$ and $(2, 0.6)$, then $m=\dfrac{4.2-0.6}{-1-2}=\dfrac{3.6}{-3}=-1.2$.

   $y-y_1=m(x-x_1) \Rightarrow y-0.6=-1.2(x-2) \Rightarrow y-0.6=-1.2x+2.4 \Rightarrow y=-1.2x+3$.

   (b) When $x=-1.5$, then $y=-1.2(-1.5)+3=4.8$. This involves interpolation.

   When $x=3.5$, then $y=-1.2(3.5)+3=-1.2$. This involves extrapolation.

(c) $1.3 = -1.2x + 3 \Rightarrow -1.7 = -1.2x \Rightarrow x = \dfrac{17}{12}$.

77.   The tank is initially empty. When $0 \le x \le 3$, the slope is 5. The inlet pipe is open; the outlet pipe is closed. When $3 < x \le 5$, the slope is 2. Both pipes are open. When $5 < x \le 8$, the slope is 0. Both pipes are closed. When $8 < x \le 10$, the slope is -3. The inlet pipe is closed; the outlet pipe is open.

79.   Let x represent the distance above the ground and let y represent the temperature. Since the ground temperature is 25°C, the point $(0, 25)$ is on the graph of the function which models the situation. Since the rate of change is a constant -6°C per kilometer, the model is linear with a slope of $m = -6$. Therefore, the equation of the linear model is $y = -6x + 25$.

Graphically: Graph $Y_1 = 15$, $Y_2 = -6x + 25$, and $Y_3 = 5$ in [0, 4, 1] by [0, 30, 5] See Figure 79. The intersection points are $\left(1\dfrac{2}{3}, 15\right)$ and $\left(3\dfrac{1}{3}, 5\right)$. The distance above the ground is between $1\dfrac{2}{3}$ km and $3\dfrac{1}{3}$ km.

Symbolically: Solve $5 \le -6x + 25 \le 15 \Rightarrow -20 \le -6x \le -10 \Rightarrow \dfrac{20}{6} \ge x \ge \dfrac{10}{6} \Rightarrow 1\dfrac{2}{3} \le x \le 3\dfrac{1}{3}$.

The solution interval is the same for either method, $\left[1\dfrac{2}{3}, 3\dfrac{1}{3}\right]$. The distance above the ground is between $1\dfrac{2}{3}$ km and $3\dfrac{1}{3}$ km.

[0, 4, 1] by [0, 30, 5]

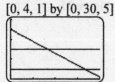

Figure 79

81. $\left|\dfrac{C - A}{A}\right| \le 0.003 \Rightarrow -0.003 \le \dfrac{C - 52.3}{52.3} \le 0.003 \Rightarrow -0.1569 \le C - 52.3 \le 0.1569 \Rightarrow$

$52.1431 \le C \le 52.4569 \Rightarrow$ between 52.1431 and 52.4569 ft.

Chapters 1-2 Cumulative Review Exercises

1.   Move the decimal point five places to the left; $123{,}000 = 1.23 \times 10^5$
     Move the decimal point three places to the right; $0.005 = 5.1 \times 10^{-3}$

3. $\dfrac{4 + \sqrt{2}}{4 - \sqrt{2}} \approx 2.09$

5.   The standard equation of a circle must fit the form $(x - h)^2 + (y - k)^2 = r^2$, where (h, k) is the center and the radius r. The equation of the circle with center $(-2, 3)$ and radius 7 is $(x + 2)^2 + (y - 3)^2 = 49$.

7. $d = \sqrt{[2 - (-3)]^2 + ((-3) - 5)^2} = \sqrt{25 + 64} = \sqrt{89}$

9.   (a) $D = $ all real numbers $\Rightarrow \{x \mid -\infty < x < \infty\} \Rightarrow (-\infty, \infty)$; $R = \{y \mid y \ge -2\} \Rightarrow [-2, \infty)$; $f(-1) = -1$

     (b) $D = \{x \mid -3 \le x \le 3\} \Rightarrow [-3, 3]$; $R = \{y \mid -3 \le y \le 2\} \Rightarrow [-3, 2]$; $f(-1) = -\dfrac{1}{2}$

11. (a) $f(2) = 5(2) - 3 = 7$; $f(a - 1) = 5(a - 1) - 3 = 5a - 5 - 3 = 5a - 8$

     (b)   The domain of f includes all real numbers $\Rightarrow D : (-\infty, \infty)$

13. No, this is not a graph of a function because some vertical lines intersect the graph twice.

15. $f(1) = (1)^2 - 2(1) + 1 = 1 - 2 + 1 = 0 \Rightarrow (1,0)$; $f(2) = (2)^2 - 2(2) + 1 = 4 - 4 + 1 = 1 \Rightarrow (2,1)$. The slope

$m = \dfrac{1-0}{2-1} = \dfrac{1}{1} = 1$, so the average rate of change is 1.

17.   (a) $m = \dfrac{2}{3}$; y-intercept: $(0,-2)$, x-intercept: $(3,0)$

(b) $f(x) = mx + b \Rightarrow f(x) = \dfrac{2}{3}x - 2$

(c)   3

19. $m = \dfrac{\frac{1}{2} - (-5)}{-3 - 1} = \dfrac{\frac{11}{2}}{-4} = -\dfrac{11}{8}$; using $(1, -5)$ and point-slope form:

$y = -\dfrac{11}{8}(x-1) - 5 \Rightarrow y = -\dfrac{11}{8}x + \dfrac{11}{8} - 5 \Rightarrow y = -\dfrac{11}{8}x - \dfrac{29}{8}$

21.   All lines parallel to the y-axis have undefined slope $\Rightarrow y$ changes but x remains constant $\Rightarrow x = -1$.

23.   For the points $(2.4, 5.6)$ and $(3.9, 8.6)$ we get $m = \dfrac{8.6 - 5.6}{3.9 - 2.4} = \dfrac{3}{1.5} = 2$. A line parallel to this has the same

slope. Using point-slope form: $y = 2(x+3) + 5 \Rightarrow y = 2x + 11$.

25.   For $-2x + 3y = 6$: x-intercept, then $y = 0 \Rightarrow -2x + 3(0) = 6 \Rightarrow -2x = 6 \Rightarrow x = -3$; x-intercept: $(-3,0)$ y-

intercept, then $x = 0 \Rightarrow -2(0) + 3y = 6 \Rightarrow 3y = 6 \Rightarrow y = 2$; y-intercept: $(0,2)$. See Figure 25.

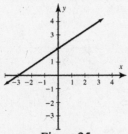

Figure 25

27. $4x - 5 = 1 - 2x \Rightarrow 6x = 6 \Rightarrow x = 1$; 1

29. $\dfrac{2}{3}(x-2) - \dfrac{4}{5}x = \dfrac{4}{15} + x \Rightarrow \dfrac{2}{3}x - \dfrac{4}{3} - \dfrac{4}{5}x = \dfrac{4}{15} + x \Rightarrow 15\left(\dfrac{2}{3}x - \dfrac{4}{3} - \dfrac{4}{5}x = \dfrac{4}{15} + x\right) \Rightarrow$

$10x - 20 - 12x = 4 + 15x \Rightarrow -17x = 24 \Rightarrow x = -\dfrac{24}{17}$; $-\dfrac{24}{17}$

31.   Graph $Y_1 = X + 1$ and $Y_2 = 2X - 2$. See Figure 31a. The lines intersect at point $(3, 4) \Rightarrow x = 3$. Make a table

of $Y_1 = X + 1$ and $Y_2 = 2X - 2$ for x values from 0 to 5. See Figure 31b.

Both equations have y-value 4 at $x = 3$.

[-10, 10, 1] by [-10, 10, 1]

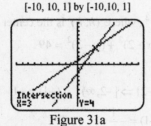

Figure 31a   Figure 31b

33. $(-\infty, 5)$

35. $(-\infty, -2) \cup (2, \infty)$

37. $-3(1 - 2x) + x \le 4 - (x + 2) \Rightarrow -3 + 6x + x \le 4 - x - 2 \Rightarrow 7x - 3 \le -x + 2 \Rightarrow 8x \le 5 \Rightarrow$

$x \le \dfrac{5}{8} \Rightarrow \left(-\infty, \dfrac{5}{8}\right]$. In set builder notation the interval is $\left\{x \mid x \le \dfrac{5}{8}\right\}$.

39. (a) 2

  (b) The graph of $f(x)$ is above the graph of $g(x)$ to the left of $x = 2 \Rightarrow x < 2$

  (c) The graph of $f(x)$ intersects or is below the graph of $g(x)$ to the right of $x = 2 \Rightarrow f(x) \le g(x)$ when $x \ge 2$.

41. $|d+1| = 5 \Rightarrow d+1 = 5$ or $d+1 = -5$. If $d+1 = -5$, $d = -6$; if $d+1 = 5$, $d = 4 \Rightarrow -6, 4$.

43. $|2t| - 4 = 10 \Rightarrow |2t| = 14 \Rightarrow 2t = 14$ or $2t = -14$. If $2t = -14$, $t = -\dfrac{14}{2} \Rightarrow t = -7$; if $2t = 14$, $t = 7 \Rightarrow -7, 7$.

45. The solutions to $|2t-5| \le 5$ satisfy $s_1 \le t \le s_2$ where s_1 and s_2, are the solutions to $|2t-5| = 5$. $|2t-5| = 5$ is equivalent to $2t-5 = -5 \Rightarrow t = 0$ or $2t-5 = 5 \Rightarrow t = 5$. The interval is $[0, 5]$. In set-builder notation the interval is $\{t \mid 0 \le t \le 5\}$.

47. (a) $C(1500) = 500(1500) + 20,000 = 770,000$; it costs \$770,000 to manufacture 1500 computers.

  (b) 500; each additional computer costs \$500 to manufacture and fixed costs are \$20,000.

49. (a) $T(2) = 70 + \dfrac{3}{2}(2)^2 = 70 + 6 = 76$; $T(4) = 70 + \dfrac{3}{2}(4)^2 = 70 + 24 = 94$ Using (2, 76) and (4, 94):

  $m = \dfrac{94 - 76}{4 - 2} = \dfrac{18}{2} = 9°\text{F}$ F increase per hour.

  (b) On average the temperature increased by 9°F per hour over this 2-hour period.

51. Let t = time for the two to mow the lawn together. Then the first person mows $\dfrac{1}{5}t$ of the lawn and the second

  person mows $\dfrac{1}{12}t$ of the lawn $\Rightarrow \dfrac{1}{5}t + \dfrac{1}{12}t = 1 \Rightarrow \dfrac{12}{60}t + \dfrac{5}{60}t = 1 \Rightarrow \dfrac{17}{60}t = 1 \Rightarrow t = \dfrac{60}{17} = 3.53$ hours.

53. (a) Using (2001, 56) and (2010, 84), $m = \dfrac{84 - 56}{2010 - 2001} = \dfrac{28}{9}$; $f(x) = \dfrac{28}{9}(x - 2001) + 56$ or

  $f(x) = \dfrac{28}{9}(x - 2010) + 84$.

  (b) $f(2019) = \dfrac{28}{9}(2019 - 2010) + 84 = 28 + 84 = 112$ pounds.

Chapter 3: Quadratic Functions and Equations

3.1: Quadratic Functions and Models

1. $f(x) = 1 - 2x + 3x^2$ is quadratic; $a = 3$; $f(-2) = 1 - 2(-2) + 3(-2)^2 = 17$.

3. $f(x) = \dfrac{1}{x^2 - 1}$ is neither linear nor quadratic.

5. $f(x) = \dfrac{1}{2} - \dfrac{3}{10}x$ is linear.

7.   (a) $a > 0$
     (b)   vertex: $(1, 0)$
     (c)   axis of symmetry: $x = 1$
     (d) f is increasing for $x > 1$ and decreasing for $x < 1$.
     (e) $D = (-\infty, \infty); R = [0, \infty)$

9.   (a) $a < 0$
     (b)   vertex: $(-3, -2)$
     (c)   axis of symmetry: $x = -3$
     (d) f is increasing for $x < -3$ and decreasing for $x > -3$.
     (e) $D = (-\infty, \infty); R = (-\infty, -2]$

11. The graph of g is narrower than the graph of f.

13. The graph of g is wider than the graph of f and opens downward rather than upward.

15. $f(x) = -3(x-1)^2 + 2 \Rightarrow$ vertex : $(1, 2)$; leading coefficient: -3; $f(x) = -3x^2 + 6x - 1$

17. $f(x) = 5 - 2(x-4)^2 \Rightarrow$ vertex : $(4, 5)$; leading coefficient: -2; $f(x) = -2x^2 + 16x - 27$

19. $f(x) = \dfrac{3}{4}(x+5)^2 - \dfrac{7}{4} \Rightarrow$ vertex : $\left(-5, -\dfrac{7}{4}\right)$; leading coefficient: $\dfrac{3}{4}$; $f(x) = \dfrac{3}{4}x^2 + \dfrac{15}{2}x + 17$

21. The vertex of the parabola is $(-1, 0)$, so $f(x) = a(x-h)^2 + k \Rightarrow f(x) = a(x+1)^2$; since $(0, 1)$ is a point on the parabola, $f(0) = a(0+1)^2 \Rightarrow 1 = a(1)^2 \Rightarrow 1 = a \Rightarrow a = 1$; $f(x) = (x+1)^2$.

23. The vertex of the parabola is $(0, 0)$, so $f(x) = a(x-h)^2 + k \Rightarrow f(x) = a(x)^2$; since $(1, 2)$ is a point on the parabola, $f(1) = a(1)^2 \Rightarrow 2 = a \Rightarrow a = 2$; $f(x) = 2x^2$.

25. The vertex of the parabola is $(-1, 1)$, so $f(x) = a(x-(-1))^2 + 1 \Rightarrow f(x) = a(x+1)^2 + 1$; since $(0, 2)$ is a point on the parabola, $f(0) = a(0+1)^2 + 1 \Rightarrow 2 = a(1)^2 + 1 \Rightarrow 2 = a + 1 \Rightarrow a = 1$; $f(x) = (x+1)^2 + 1$.

27. The vertex of the parabola is $(2, -2)$, so $f(x) = a(x-h)^2 + k \Rightarrow f(x) = a(x-2)^2 - 2$; since $(0, 2)$ is a point on the parabola, $f(0) = a(0-2)^2 - 2 \Rightarrow 2 = a(-2)^2 - 2 \Rightarrow 4 = 4a \Rightarrow a = 1$; $f(x) = (x-2)^2 - 2$.

29. The vertex of the parabola is $(2, -3)$, so $f(x) = a(x-h)^2 + k \Rightarrow f(x) = a(x-2)^2 - 3$; since $(0, -1)$ is a point on the parabola, $f(0) = a(0-2)^2 - 3 \Rightarrow -1 = a(4) - 3 \Rightarrow -1 = 4a - 3 \Rightarrow 2 = 4a \Rightarrow a = \dfrac{1}{2}$; $f(x) = \dfrac{1}{2}(x-2)^2 - 3$.

31. The vertex of the parabola is $(-1, 3)$, so $f(x) = a(x-h)^2 + k \Rightarrow f(x) = a(x+1)^2 + 3$; since $(0, 1)$ is a point on the parabola, $f(0) = a(0+1)^2 + 3 \Rightarrow 1 = a(1) + 3 \Rightarrow 1 = a + 3 \Rightarrow a = -2$; $f(x) = -2(x+1)^2 + 3$.

33. The vertex of the parabola is $(2, 6)$, so $f(x) = a(x-h)^2 + k \Rightarrow f(x) = a(x-2)^2 + 6$; since $(0, -6)$ is a point on the parabola, $f(0) = a(0-2)^2 + 6 \Rightarrow -6 = 4a + 6 \Rightarrow -12 = 4a \Rightarrow a = -3$; $f(x) = -3(x-2)^2 + 6$.

35. $f(x) = x^2 - 3x \Rightarrow x = -\dfrac{b}{2a} = -\dfrac{(-3)}{2(1)} = \dfrac{3}{2}$ and $f\left(\dfrac{3}{2}\right) = \left(\dfrac{3}{2}\right)^2 - 3\left(\dfrac{3}{2}\right) = -\dfrac{9}{4} \Rightarrow$ vertex: $\left(\dfrac{3}{2}, -\dfrac{9}{4}\right)$; since

    $a = 1, f(x) = 1\left(x - \dfrac{3}{2}\right)^2 - \dfrac{9}{4}$, or $f(x) = \left(x - \dfrac{3}{2}\right)^2 - \dfrac{9}{4}$

37. $f(x) = 2x^2 - 5x + 3 \Rightarrow x = -\dfrac{b}{2a} = -\dfrac{(-5)}{2(2)} = \dfrac{5}{4}$ and $f\left(\dfrac{5}{4}\right) = 2\left(\dfrac{5}{4}\right)^2 - 5\left(\dfrac{5}{4}\right) + 3 = -\dfrac{1}{8} \Rightarrow$ vertex: $\left(\dfrac{5}{4}, -\dfrac{1}{8}\right)$;

    since $a = 2, f(x) = 2\left(x - \dfrac{5}{4}\right)^2 - \dfrac{1}{8}$

39. $f(x) = 2x^2 - 8x - 1 \Rightarrow x = -\dfrac{b}{2a} = -\dfrac{(-8)}{2(2)} = 2$ and $f(2) = 2(2)^2 - 8(2) - 1 = -9 \Rightarrow$ vertex: $(2, -9)$; since

    $a = 2, f(x) = 2(x-2)^2 - 9$

41. $f(x) = 2 - 6x - 3x^2 \Rightarrow x = -\dfrac{b}{2a} = -\dfrac{(-6)}{2(-3)} = -1$ and $f(-1) = 2 - 6(-1) - 3(-1)^2 = 5 \Rightarrow$ vertex: $(-1, 5)$; since

    $a = -3, f(x) = -3(x-(-1))^2 + 5$, or $f(x) = -3(x+1)^2 + 5$

43. $f(x) = x^2 + 4x - 5 \Rightarrow x = -\dfrac{b}{2a} = -\dfrac{4}{2(1)} = -2$ and $f(-2) = (-2)^2 + 4(-2) - 5 = -9 \Rightarrow$ vertex: $(-2, -9)$; since

    $a = 1, f(x) = 1(x-(-2))^2 - 9$, or $f(x) = (x+2)^2 - 9$

45. $f(x) = \dfrac{1}{3}x^2 + x + 1 \Rightarrow x = -\dfrac{b}{2a} = -\dfrac{1}{2(\frac{1}{3})} = -\dfrac{1}{\frac{2}{3}} = -\dfrac{3}{2}$ and $f\left(-\dfrac{3}{2}\right) = \dfrac{1}{3}\left(-\dfrac{3}{2}\right)^2 - \dfrac{3}{2} + 1 = \dfrac{1}{4} \Rightarrow$ vertex: $\left(-\dfrac{3}{2}, \dfrac{1}{4}\right)$;

    since $a = \dfrac{1}{3}, f(x) = \dfrac{1}{3}\left(x + \dfrac{3}{2}\right)^2 + \dfrac{1}{4}$

47. (a) It may be helpful to write f in standard form as follows: $f(x) = -x^2 + 0x + 6$.

    To find the vertex symbolically, use the vertex formula with $a = -1$ and $b = 0$.

    $x = -\dfrac{b}{2a} = -\dfrac{0}{2(-1)} = \dfrac{0}{2} = 0$. The x-coordinate of the vertex is 0.

    $y = f\left(-\dfrac{b}{2a}\right) = f(0) = 6 - (0)^2 = 6$. The y-coordinate of the vertex is 6. Thus, the vertex is $(0, 6)$.

    (b) f is increasing on $x \leq 0$ or $(-\infty, 0]$ and decreasing on $x \geq 0$ or $[0, \infty)$.

49. (a) It may be helpful to write f in standard form as follows: $f(x) = x^2 - 6x + 0$.

    To find the vertex symbolically, use the vertex formula with $a = 1$ and $b = -6$.

    $x = -\dfrac{b}{2a} = -\dfrac{(-6)}{2(1)} = \dfrac{6}{2} = 3$. The x-coordinate of the vertex is 3.

    $y = f\left(-\dfrac{b}{2a}\right) = f(3) = (3)^2 - 6(3) = -9$. The y-coordinate of the vertex is -9. Thus, the vertex is $(3, -9)$.

(b) f is increasing on $x \geq 3$ or $[3, \infty)$ and decreasing on $x \leq 3$ or $(-\infty, 3]$.

51.   (a)   The function is already in standard form: $f(x) = 2x^2 - 4x + 1$.

To find the vertex symbolically, use the vertex formula with $a = 2$ and $b = -4$.

$x = -\dfrac{b}{2a} = -\dfrac{(-4)}{2(2)} = \dfrac{4}{4} = 1$. The x-coordinate of the vertex is 1.

$y = f\left(-\dfrac{b}{2a}\right) = f(1) = 2(1)^2 - 4(1) + 1 = -1$. The y-coordinate of the vertex is -1 Thus, the vertex is $(1, -1)$.

(b) f is increasing on $x \geq 1$ or $[1, \infty)$ and decreasing on $x \leq 1$ or $(-\infty, 1]$.

53.   (a)   It may be helpful to write f in standard form as follows: $f(x) = \dfrac{1}{2}x^2 + 0x + 10$.

To find the vertex symbolically, use the vertex formula with $a = \dfrac{1}{2}$ and $b = 0$.

$x = -\dfrac{b}{2a} = -\dfrac{0}{2(\frac{1}{2})} = 0$. The x-coordinate of the vertex is 0.

$y = f\left(-\dfrac{b}{2a}\right) = f(0) = \dfrac{1}{2}(0)^2 + 0(0) + 10 = 10$. The y-coordinate of the vertex is 10. Thus, the vertex is $(0, 10)$.

(b) f is increasing on $x \geq 0$ or $[0, \infty)$ and decreasing on $x \leq 0$ or $(-\infty, 0]$.

55.   (a)   The function is already in standard form: $f(x) = -\dfrac{3}{4}x^2 + \dfrac{1}{2}x - 3$.

To find the vertex symbolically, use the vertex formula with $a = -\dfrac{3}{4}$ and $b = \dfrac{1}{2}$.

$x = -\dfrac{b}{2a} = -\dfrac{\frac{1}{2}}{2(-\frac{3}{4})} = \dfrac{1}{3}$. The x-coordinate of the vertex is $\dfrac{1}{3}$.

$y = f\left(-\dfrac{b}{2a}\right) = f\left(\dfrac{1}{3}\right) = -\dfrac{3}{4}\left(\dfrac{1}{3}\right)^2 + \dfrac{1}{2}\left(\dfrac{1}{3}\right) - 3 = -\dfrac{35}{12}$. The y-coordinate of the vertex is $-\dfrac{35}{12}$. Thus, the vertex is $\left(\dfrac{1}{3}, -\dfrac{35}{12}\right)$.

(b) f is increasing on $x \leq \dfrac{1}{3}$ or $\left(-\infty, \dfrac{1}{3}\right]$ and decreasing on $x \geq \dfrac{1}{3}$ or $\left[\dfrac{1}{3}, \infty\right)$.

57.   (a)   It may be helpful to write f in standard form as follows: $f(x) = -6x^2 - 12x + 1.5$.

To find the vertex symbolically, use the vertex formula with $a = -6$ and $b = -12$.

$x = -\dfrac{b}{2a} = -\dfrac{(-12)}{2(-6)} = -\dfrac{12}{12} = -1$. The x-coordinate of the vertex is -1.

$y = f\left(-\dfrac{b}{2a}\right) = f(-1) = 1.5 - 12(-1) - 6(-1)^2 = 7.5$. The y-coordinate of the vertex is 7.5. Thus, the vertex is $(-1, 7.5)$.

(b) f is increasing on $x \leq -1$ or $(-\infty, -1]$ and decreasing on $x \geq -1$ or $[-1, \infty)$.

59. Because $a > 0$ the graph of f is a parabola opening upward. The vertex is the lowest point on the graph, with an x-coordinate of $x = \dfrac{-4}{2(1)} = -2$. The corresponding y-coordinate is $f(-2) = (-2)^2 + 4(-2) - 2 = -6$. Thus the vertex is $-2, -6$ and the minimum y-value is -6.

61. Because $a > 0$ he graph of f is a parabola opening upward. The vertex is the lowest point on the graph, with an x-coordinate of $x = \dfrac{4}{2(3)} = \dfrac{2}{3}$. The corresponding y-coordinate is $f\left(\dfrac{2}{3}\right) = 3\left(\dfrac{2}{3}\right)^2 - 4\left(\dfrac{2}{3}\right) + 2 = \dfrac{2}{3}$. Thus the vertex is $\left(\dfrac{2}{3}, \dfrac{2}{3}\right)$ and the minimum y-value is $\dfrac{2}{3}$.

63. Because $a > 0$ the graph of f is a parabola opening upward. The vertex is the lowest point on the graph, with an x-coordinate of $x = \dfrac{-3}{2(1)} = -\dfrac{3}{2}$. The corresponding y-coordinate is $f\left(-\dfrac{3}{2}\right) = \left(-\dfrac{3}{2}\right)^2 + 3\left(-\dfrac{3}{2}\right) + 5 = \dfrac{11}{4}$. Thus the vertex is $\left(-\dfrac{3}{2}, \dfrac{11}{4}\right)$ and the minimum y-value is $\dfrac{11}{4}$.

65. Because $a > 0$ the graph of f is a parabola opening upward. The vertex is the lowest point on the graph, with an x-coordinate of $x = \dfrac{-3}{2(-1)} = \dfrac{3}{2}$. The corresponding y-coordinate is $f\left(\dfrac{3}{2}\right) = -\left(\dfrac{3}{2}\right)^2 + 3\left(\dfrac{3}{2}\right) - 2 = \dfrac{1}{4}$. Thus the vertex is $\left(\dfrac{3}{2}, \dfrac{1}{4}\right)$ and the maximum y-value is $\dfrac{1}{4}$.

67. Because $a < 0$ the graph of f is a parabola opening downward. The vertex is the highest point on the graph, with an x-coordinate of $x = \dfrac{-5}{2(-1)} = \dfrac{5}{2}$. The corresponding y-coordinate is $f\left(\dfrac{5}{2}\right) = 5\left(\dfrac{5}{2}\right) - \left(\dfrac{5}{2}\right)^2 = \dfrac{25}{4}$. Thus the vertex is $\left(\dfrac{5}{2}, \dfrac{25}{4}\right)$ and the maximum y-value is $\dfrac{25}{4}$.

69. $0+3$Because $a < 0$ the graph of f is a parabola opening downward. The vertex is the highest point on the graph, with an x-coordinate of $x = \dfrac{-2}{2(-3)} = \dfrac{1}{3}$. The corresponding y-coordinate is $f\left(\dfrac{1}{3}\right) = 2\left(\dfrac{1}{3}\right) - 3\left(\dfrac{1}{3}\right)^2 = \dfrac{1}{3}$. and the maximum y-value is $\dfrac{1}{3}$.

71. The graph of $f(x) = x^2$ is shown in Figure 71.

73. The graph of $f(x) = \dfrac{1}{4}x^2$ is shown in Figure 73.

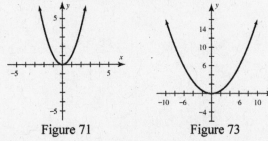

Figure 71   Figure 73

75. The graph of $f(x) = 1 - x^2$ is shown in Figure 75.

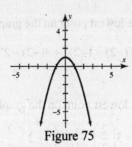

Figure 75

77.   The graph of $f(x) = \frac{1}{2}x^2 - 2$ is shown in Figure 77.

79.   The graph of $f(x) = -\frac{1}{2}x^2$ is shown in Figure 79.

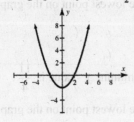   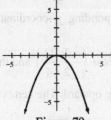

Figure 77   Figure 79

81.   The graph of $f(x) = x^2 - 3$ is shown in Figure 81.

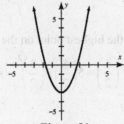

Figure 81

83.   The graph of $f(x) = (x-2)^2 + 1$ is shown in Figure 83.

85.   The graph of $f(x) = -3(x+1)^2 + 3$ is shown in Figure 85.

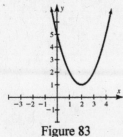   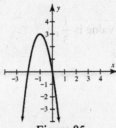

Figure 83   Figure 85

87.   The graph of $f(x) = 9x - x^2$ is shown in Figure 87.

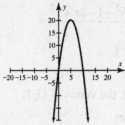

Figure 87

89.   The graph of $f(x) = x^2 - 2x - 2$ is shown in Figure 89.

91.   The graph of $f(x) = -x^2 + 4x - 2$ is shown in Figure 91.

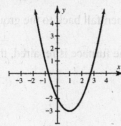

Figure 89

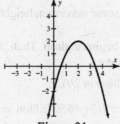

Figure 91

93.   The graph of $f(x) = 2x^2 - 4x - 1$ is shown in Figure 93.

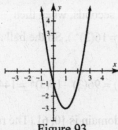

Figure 93

95.   The graph of $f(x) = -3x^2 - 6x + 1$ is shown in Figure 95.

97.   The graph of $f(x) = -\dfrac{1}{2}x^2 + x + 1$ is shown in Figure 97.

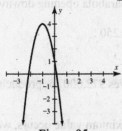

Figure 95

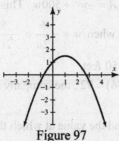

Figure 97

99.   The average rate of change from 1 to 3 for $f(x) = -3x^2 + 5x$ is $\dfrac{f(3) - f(1)}{3 - 1} = \dfrac{-14}{2} = -7$.

101. $\dfrac{f(x+h) - f(x)}{h} = \dfrac{3(x+h)^2 - 2(x+h) - [3x^2 - 2x]}{h} =$

$\dfrac{3x^2 + 6xh + 3h^2 - 2x - 2h - 3x^2 + 2x}{h} = \dfrac{6xh + 3h^2 - 2h}{h} = 6x + 3h - 2$

103. $\dfrac{f(x+h)-f(x)}{h} = \dfrac{1+2(x+h)-(x+h)^2-[1+2x-x^2]}{h} = \dfrac{1+2x+2h-x^2-2xh-h^2-1-2x+x^2}{h} =$

$\dfrac{2h-2xh-h^2}{h} = 2-2x-h$

105. Since the axis of symmetry is $x=3$ and the graph passes through the point $(3, 1)$, the vertex is $(3, 1)$.

$f(x) = a(x-h)^2 + k \Rightarrow f(x) = a(x-3)^2 + 1$. Using the point $(1, 9)$ to find a gives

$9 = a(1-3)^2 + 1 \Rightarrow a = 2; f(x) = 2(x-3)^2 + 1$.

107. $f(x) = a(x-h)^2 + k \Rightarrow f(x) = a(x+1)^2 - 5$. Using the point $(0, -7)$ to find a gives

$-7 = a(0+1)^2 - 5 \Rightarrow a = -2; f(x) = -2(x+1)^2 - 5$.

109. A stone thrown from ground level would first rise to some maximum height and then fall back to the ground. This is represented in figure d.

111. When the furnace first fails to work, the temperature begins to drop. Then, after the furnace is repaired, the temperature begins to rise. This is represented in figure a.

113. (a) $I(0) = 0.6(0)^2 + 5.2(0) + 29 = 29 \Rightarrow I(0) = \$29$ billion in 2015;

$I(3) = 0.6(3)^2 + 5.2(3) + 29 = 5.4 + 15.6 + 29 = 50 \Rightarrow I(3) = \$50$ billion in 2018;

$I(5) = 0.6(5)^2 + 5.2(5) + 29 = 15 + 15.6 + 29 = 70 \Rightarrow I(5) = \$70$ billion in 2020

(b) These values agree exactly with those in the graph in the chapter opener.

115. (a) $h(4) = 96(4) - 16(4)^2 = 128$ ft The maximum height of the ball is reached at 3 seconds, when then quadratic term is equal to the height of the ball ($h(3) = 96(3) - 16(3)^2 = 144$ ft $= 16(3)^2$). So the ball at 4 seconds is moving downward.

(b) By the vertex formula, $t = -\dfrac{b}{2a} = -\dfrac{96}{2(-16)} = 3$ s ; the maximum height is $h(3) = 96(3) - 16(3)^2 = 144$ ft.

(c) The ball lands after 6 seconds ($h(6) = 96(6) - 16(6)^2 = 576 - 576 = 0$), so the domain is $[0, 6]$. The range is $[0, 144]$.

117. Perimeter of fence $= 2l + 2w = 1000 \Rightarrow l = \dfrac{1000-2w}{2} \Rightarrow l = 500 - w$.

If $A = lw$ then $A = (500 - w)w \Rightarrow A = 500w - w^2 \Rightarrow A = -w^2 + 500w$. This is a parabola opening downward

and by the vertex formula, the maximum area occurs when $w = -\dfrac{b}{2a} = -\dfrac{500}{2(-1)} = 250$.

The dimensions that maximize area are 250 feet by 250 feet.

119. (a) $R(x) = x(40 - 2x) \Rightarrow R(2) = 2(40 - 2(2)) = 2(36) = 72$; the company receives \$72,000 for producing 2000 CD players.

(b) $R(x) = 40x - 2x^2$ is a quadratic function; to find the value at which the maximum value occurs, we

need to find the vertex: $x = -\dfrac{b}{2a} = -\dfrac{40}{2(-2)} = 10$ and $R(10) = 40(10) - 2(10)^2 = 200 \Rightarrow (10, 200)$ is the

vertex, thus, the company needs to produce 10,000 CD players to maximize its revenue.

(c)   Since $R(10) = 200$, the maximum revenue for the company is \$200,000.

121. (a)   Because $a > 0$ the graph of f is a parabola opening upward.

(b)   The vertex is the lowest point on the graph, with an x-coordinate of $x = \dfrac{0.0145}{2(0.00000093)} \approx 7796$. The

corresponding y-coordinate is $f(7796) = 0.00000093(7796)^2 - 0.0145(7796) + 60 \approx 3.5$. Thus the vertex is approximately (7796, 3.5). The minimum average cost per copy is 3.5 cents when 7796 copies are made.

123. (a) $s(1) = -16(1)^2 + 44(1) + 4 = 32$; the baseball is 32 feet high after 1 second.

   (b)   For $s(t) = -16t^2 + 44t + 4$, the vertex formula gives $t = -\dfrac{b}{2a} = -\dfrac{44}{2(-16)} = 1.375$ and

   $f(1.375) = -16(1.375)^2 + 44(1.375) + 4 = 34.25$; the maximum height 34.25 feet. See Figure 123.

   (c) $D : [0, 2.84]$ (approx); $R : [0, 34.25]$.

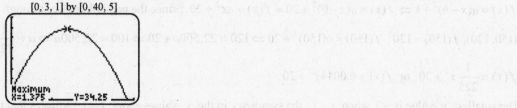

[0, 3, 1] by [0, 40, 5]

Figure 123

125. (a)   The initial velocity $v_0 = -66$ and the initial height is $h_0 = 120$, so

   $s(t) = -16t^2 + v_0 t + h_0 \Rightarrow s(t) = -16t^2 - 66t + 120$.

   (b)   When the stone hits the water, $s(t) = 0$;

   $0 = -16t^2 - 66t + 120 \Rightarrow 0 = -2(8t^2 + 33t - 60) \Rightarrow 0 = 8t^2 + 33t - 60$; using the quadratic formula or graphing the parabola, we find that $s(t) = 0$ when $t \approx 1.37$ seconds, so the stone hits the water within the first 2 seconds.

127. Using the wall of the barn for one side gives us $W + L + W = 160$ for the three sides $\Rightarrow L = 160 - 2W$. If $A = LW$ then $A = (160 - 2W)W \Rightarrow A = 160W - 2W^2$. Since this is a parabola opening downward, maximum

   area occurs when $W = -\dfrac{b}{2a} \Rightarrow W = \dfrac{-160}{2(-2)} \Rightarrow W = 40$. Then $L = 160 - 2(40) = 80$. Thus, the dimensions that

   yield maximum area are 40 feet by 80 feet.

129. (a)   When $g = 32$, $v_0 = 88$, and $h_0 = 25$, then $f(x) = -\dfrac{1}{2}(32)x^2 + 88x + 25 = -16x^2 + 88x + 25$. Graph

   $Y_1 = -16X^2 + 88X + 25$. By using the calculator, the maximum height is found to be approximately 146 feet. The maximum height of 146 ft occurs when $x = 2.75$ seconds. See Figure 129.

   (b)   To find the maximum height symbolically, use the vertex formula with $a = -16$ and $b = 88$.

   $x = -\dfrac{b}{2a} = -\dfrac{88}{2(-16)} = \dfrac{88}{32} = 2.75$. The maximum height occurs at $x = 2.75$ seconds.

   $y = f\left(\dfrac{b}{2a}\right) = f(2.75) = -16(2.75)^2 + 88(2.75) + 25 = 146$. The maximum height is 146 feet.

131. (a)   When $g = 13$, $v_0 = 88$, and $h_0 = 25$, then $f(x) = -\dfrac{1}{2}(13)x^2 + 88x + 25 = -6.5x^2 + 88x + 25$. Graph

   $Y_1 = -6.5X^2 + 88X + 25$. By using the calculator, the maximum height is found to be approximately 323 feet. The maximum height of about 323 ft occurs when $x \approx 6.77$ seconds. See Figure 131.

   (b)   To find the maximum height symbolically, use the vertex formula with $a = -6.5$ and $b = 88$.

   $x = -\dfrac{b}{2a} = -\dfrac{88}{2(-6.5)} = \dfrac{88}{13} \approx 6.77$. The maximum height occurs at $x \approx 6.77$ seconds.

   $y = f\left(-\dfrac{b}{2a}\right) = f(6.77) = -6.5(6.77)^2 + 88(6.77) + 25 \approx 323$. The maximum height is about 323 feet.

[0, 6, 1] by [0, 160, 10]   [0, 15, 5] by [0, 400, 100]

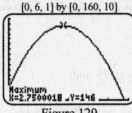   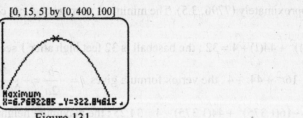

Maximum
X=2.7500018 Y=146

Maximum
X=6.7692285 Y=322.84615

Figure 129   Figure 131

133. The cables drawn as a parabola can be placed in the coordinate plane. Use $(0, 20)$ as the vertex.

$f(x) = a(x-h)^2 + k \Rightarrow f(x) = a(x-0)^2 + 20 = f(x) = ax^2 + 20$; since the parabola passes through

$(150, 120)$, $f(150) = 120$; $f(150) = a(150)^2 + 20 \Rightarrow 120 = 22,500a + 20 \Rightarrow 100 = 22,500a \Rightarrow a = \dfrac{1}{225}$;

$f(x) = \dfrac{1}{225}x^2 + 20$, or $f(x) \approx 0.0044x^2 + 20$.

135. The smallest y -value is -3 when $x = 1$; the symmetry in the y -values about $x = 1$ indicates that the axis of

symmetry is $x = 1$ and the vertex is $(1, -3)$, so $f(x) = a(x-1)^2 - 3$. Since $(0, -1)$ is a data point, $f(0) = -1$;

$f(0) = a(0-1)^2 - 3 \Rightarrow -1 = a - 3 \Rightarrow a = 2$; the function $f(x) = 2(x-1)^2 - 3$ models the data exactly.

137. (a) Since the minimum occurs at $t = 4$, let $(4, 90)$ be the vertex and write $H(t) = a(t-4)^2 + 90$. Use the

   data point $(0, 122)$ to find

   $a: H(0) = a(0-4)^2 + 90 \Rightarrow 122 = a(0-4)^2 + 90 \Rightarrow 122 = 16a + 90 \Rightarrow 16a = 32 \Rightarrow a = 2$;

   $H(t) = 2(t-4)^2 + 90$; Domain of $D = \{t \,|\, 0 \le t \le 4\}$; $[0,4]$.

   (b) $H(1.5) = 2(1.5 - 4)^2 + 90 \Rightarrow H(1.5) = 2(-2.5)^2 + 90 \Rightarrow H(1.5) = 12.5 + 90 \Rightarrow$

   $H(1.5) = 102.5$ beats per minute .

139. (a) Sales increase until they peak at 365 million in 2011. They then decrease to nearly the 2007 levels by
       2015.
   (b) No, the data do not follow a linear pattern. A quadratic function may be more suitable. This is because the
       scatter plot suggests that the data are more parabolic than linear.
   (c) Downward. The data increase, then decrease, so the vertex is the highest point.
   (d) The vertex of the parabola is approximately $(2011, 365)$, so

       $S(x) = a(x-h)^2 + k \Rightarrow S(x) = a(x-2011)^2 + 365$; for $(2015, 285)$,

       $S(2015) = a(2015-2011)^2 + 365 \Rightarrow 285 = a(4)^2 + 365 \Rightarrow -80 = 16a \Rightarrow a = -5$;

       $S(x) = -5(x-2011)^2 + 365$ (value for a may vary with choice of data point)

   (e) $(2011, 365)$; This pair of values, which indicated the amount and year of maximum sales, corresponded to
       the vertex of a downward-opening parabola.
   (f) 320 million. The answer involves interpolation between the given values. Other reasonable values would
       include 310 million and 315 million.

141. Enter the data into your calculator. See Figure 141a. Select quadratic regression from the STAT menu. See
    Figure 141b. In Figure 141c, the modeling function is given (approximately) by

    $f(x) = 3.125x^2 + 2.05x - 0.9$. $f(3.5) = 3.125(3.5)^2 + 2.05(3.5) - 0.9 \approx 44.56$

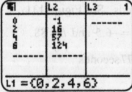   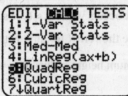   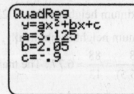

L1	L2	L3	1
0	-1	------	
2	16		
4	57		
6	124		
------	------		
L1 = {0, 2, 4, 6}			

EDIT CALC TESTS
1:1-Var Stats
2:2-Var Stats
3:Med-Med
4:LinReg(ax+b)
5:QuadReg
6:CubicReg
7↓QuartReg

QuadReg
y=ax²+bx+c
a=3.125
b=2.05
c=-.9

Figure 141a   Figure 141b   Figure 141c

143. (a) From the quadratic regression function on the calculator, the model for the data is
$f(x) \approx 0.10573214x^2 - 425.8552x + 428,801.98$. See Figure 143.

(b) $f(2022) \approx 0.10573214(2022)^2 - 425.8552(2022) + 428,801.98 = 6.928$, or about \$6.9 billion.

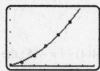

Figure 143

145. (a) The number of sales increased, but not at a constant rate.
(b) No, the data do not increase at a constant rate.
(c) A quadratic or other nonlinear function might model the data satisfactorily.
(d) Upward, if the right side of the parabola is used. This is because the data increase steadily, as if from the right of the vertex.
(e) From the quadratic regression function on the calculator, the model for the data is
$S(x) \approx 0.45x^2 - 1814.93x + 1,829,984.04$.

(f) $S(2015) \approx 0.45(2015)^2 - 1814.93(2015) + 1,829,984.04 \approx 1.34$, or about 1.34 million. This is too large because of extrapolation with a parabola near a point corresponding to the vertex.

Extended and Discovery Exercises for Section 3.1

1. (a)

x	1	2	3	4	5
$f(x)$	-2	1	6	13	22

(b) For $(1, -2)$ and $(2, 1)$, $m = \dfrac{1-(-2)}{2-1} = 3$; for $(2, 1)$ and $(3, 6)$, $m = \dfrac{6-1}{3-2} = 5$;

for $(3, 6)$ and $(4, 13)$, $m = \dfrac{13-6}{4-3} = 7$; for $(4, 13)$ and, $(5, 22)$, $m = \dfrac{22-13}{5-4} = 9 \Rightarrow 3, 5, 7, 9$.

(c) $f(x+h) = (x+h)^2 - 3 = x^2 + 2xh + h^2 - 3$; the difference quotient

$= \dfrac{f(x+h)-f(x)}{h} = \dfrac{x^2 + 2xh + h^2 - 3 - (x^2 - 3)}{h} = \dfrac{2xh + h^2}{h} = 2x + h$; when $h = 1$, the difference

quotient is $2x + 1$.

(d) $x = 1, 2(1)+1 = 3$; $x = 2, 2(2)+1 = 5$; $x = 3, 2(3)+1 = 7$; $x = 4, 2(4)+1 = 9$; the results are the same.

3. (a)

x	1	2	3	4	5
$f(x)$	0	-3	-10	-21	-36

(b) For $(1, 0)$ and $(2, -3)$, $m = \dfrac{-3-0}{2-1} = -3$; for $(2, -3)$ and $(3, -10)$, $m = \dfrac{-10-(-3)}{3-2} = -7$; for $(3, -10)$

and $(4, -21)$, $m = \dfrac{-21-(-10)}{4-3} = -11$; for $(4, -21)$ and $(5, -36)$,

$m = \dfrac{-36-(-21)}{5-4} = -15 \Rightarrow -3, -7, -11, -15$.

(c) $f(x) = -2x^2 + 3x - 1, f(x+h) = -2(x+h)^2 + 3(x+h) - 1 = -2x^2 - 4xh - 2h^2 + 3x + 3h - 1$; the

difference quotient $= \dfrac{f(x+h) - f(x)}{h} = \dfrac{-2x^2 - 4xh - 2h^2 + 3x + 3h - 1 - (-2x^2 + 3x - 1)}{h} =$

$\dfrac{-4xh - 2h^2 + 3h}{h} = -4x - 2h + 3$; when $h = 1$, the difference quotient is $-4x + 1$.

(d) $x = 1, -4(1) + 1 = -3$; $x = 2, -4(2) + 1 = -7$; $x = 3, -4(3) + 1 = -11$; $x = 4, -4(4) + 1 = -15$; the results are the same.

3.2: Quadratic Equations and Problem Solving

1. $x^2 = 4 \Rightarrow x = \pm 2$. Check: $(-2)^2 = 4$; $(2)^2 = 4$

3. $x^2 + 11 = 0 \Rightarrow x^2 = -11$ Since $\sqrt{-11}$ is not real, there are no real solutions to the equation.

5. $2x^2 - 32 = 0 \Rightarrow x^2 = 16 \Rightarrow x = \pm 4$. Check: $2(-4)^2 - 32 = 0$; $2(4)^2 - 32 = 0$

7. $\dfrac{1}{2}x^2 - 5 = 0 \Rightarrow x^2 = 10 \Rightarrow x = \pm\sqrt{10}$. Check: $\dfrac{1}{2}(-\sqrt{10})^2 - 5 = \dfrac{10}{2} - 5 = 0$; $\dfrac{1}{2}(\sqrt{10})^2 - 5 = \dfrac{10}{2} - 5 = 0$

9. $x^2 + 3x + 2 = 0$. Factoring this we get: $(x+2)(x+1) = 0$. Then $x + 2 = 0 \Rightarrow x = -2$ or $x + 1 = 0 \Rightarrow x = -1$. Therefore $x = -2, -1$. Check: $(-2)^2 + 3(-2) + 2 = 4 - 6 + 2 = 0$; $(-1)^2 + 3(-1) + 2 = 1 - 3 + 2 = 0$

11. $x^2 - 6x + 8 = 0$. Factoring this we get: $(x-2)(x-4) = 0$. Then $x - 2 = 0 \Rightarrow x = 2$ or $x - 4 = 0 \Rightarrow x = 4$. Therefore $x = 2, 4$. Check: $(2)^2 - 6(2) + 8 = 4 - 12 + 8 = 0$; $(4)^2 - 6(4) + 8 = 16 - 24 + 8 = 0$

13. $x^2 + x - 11 = 1 \Rightarrow x^2 + x - 12 = 0$. Factoring this we get: $(x+4)(x-3) = 0$. Then $x + 4 = 0 \Rightarrow x = -4$ or $x - 3 = 0 \Rightarrow x = 3$. Therefore $x = -4, 3$.
Check: $(-4)^2 + (-4) - 11 = 1 \Rightarrow 16 - 4 - 11 = 1 \Rightarrow 12 - 11 = 1 \Rightarrow 1 = 1$;
$(3)^2 + (3) - 11 = 1 \Rightarrow 9 + 3 - 11 = 1 \Rightarrow 12 - 11 = 1 \Rightarrow 1 = 1$

15. $t^2 = 2t \Rightarrow t^2 - 2t = 0$. Factoring this we get: $t(t-2) = 0$. Then $t = 0$ or $t - 2 = 0 \Rightarrow t = 2$. Therefore $t = 0, 2$.
Check: $0^2 = 2(0) \Rightarrow 0 = 0; 2^2 = 2(2) \Rightarrow 4 = 4$

17. $3x^2 - 7x = 0$. Factoring we get $x(3x - 7) = 0$. Then $x = 0$ or $3x - 7 = 0 \Rightarrow x = \dfrac{7}{3}$. Therefore $x = 0, \dfrac{7}{3}$. Check:

$3(0)^2 - 7(0) = 0 \Rightarrow 0 = 0$; $3\left(\dfrac{7}{3}\right)^2 - 7\left(\dfrac{7}{3}\right) = 0 \Rightarrow \dfrac{49}{3} - \dfrac{49}{3} = 0 \Rightarrow 0 = 0$

19. $2z^2 = 13z + 15 \Rightarrow 2z^2 - 13z - 15 = 0$. Factoring we get $(2z - 15)(z + 1) = 0$. Then $2z - 15 = 0 \Rightarrow z = \dfrac{15}{2}$ or

$z + 1 = 0 \Rightarrow z = -1$. Check: $2\left(\dfrac{15}{2}\right)^2 = 13\left(\dfrac{15}{2}\right) + 15 \Rightarrow \dfrac{225}{2} = \dfrac{195}{2} + \dfrac{30}{2} \Rightarrow \dfrac{225}{2} = \dfrac{225}{2}$;

$2(-1)^2 = 13(-1) + 15 \Rightarrow 2 = -13 + 15 \Rightarrow 2 = 2$.

21. $x^2 + 6x + 9 = 0$ Factoring this we get $(x + 3)^2 = 0 \Rightarrow x + 3 = 0 \Rightarrow x = -3$. Therefore, $x = -3$. Check:
$(-3)^2 + 6(-3) + 9 = 0 \Rightarrow 9 - 18 + 0 = 0 \Rightarrow 0 = 0$.

23. $4x^2 + 1 = 4x \Rightarrow 4x^2 - 4x + 1 = 0$ Factoring this we get $(2x-1)^2 = 0 \Rightarrow 2x - 1 = 0 \Rightarrow 2x - 1 = 0 \Rightarrow x = \frac{1}{2}$.

Therefore, $x = \frac{1}{2}$. Check: $4\left(\frac{1}{2}\right)^2 + 1 = 4\left(\frac{1}{2}\right) \Rightarrow 1 + 1 = 2 \Rightarrow 2$

25. $x(3x + 14) = 5 \Rightarrow 3x^2 + 14x - 5 = 0$. Factoring this we get: $(3x - 1)(x + 5) = 0$. Then

$3x - 1 = 0 \Rightarrow 3x = 1 \Rightarrow x = \frac{1}{3}$ or $x + 5 = 0 \Rightarrow x = -5$. Therefore $x = -5, \frac{1}{3}$. Check:

$-5(3(-5) + 14) = 5 \Rightarrow -5(-1) = 5 \Rightarrow 5 = 5$; $\frac{1}{3}\left(3\left(\frac{1}{3}\right) + 14\right) = 5 \Rightarrow \frac{1}{3}(15) = 5 \Rightarrow 5 = 5$

27. $6x^2 + \frac{5}{2} = 8x \Rightarrow 6x^2 - 8x + \frac{5}{2} = 0 \Rightarrow 12x^2 - 16x + 5 = 0$. Factoring this we get: $(2x - 1)(6x - 5) = 0$.

Then $2x - 1 = 0 \Rightarrow 2x = 1 \Rightarrow x = \frac{1}{2}$ or $6x - 5 = 0 \Rightarrow 6x = 5 \Rightarrow x = \frac{5}{6}$. Therefore $x = \frac{1}{2}, \frac{5}{6}$.

Check: $6\left(\frac{1}{2}\right)^2 + \frac{5}{2} = 8\left(\frac{1}{2}\right) \Rightarrow \frac{3}{2} + \frac{5}{2} = 4 \Rightarrow 4 = 4$;

$6\left(\frac{5}{6}\right)^2 + \frac{5}{2} = 8\left(\frac{5}{6}\right) \Rightarrow \frac{25}{6} + \frac{15}{6} = \frac{40}{6} \Rightarrow \frac{40}{6} = \frac{40}{6}$

29. $(t + 3)^2 = 5 \Rightarrow t^2 + 6t + 4 = 0$. Using the quadratic formula:

$t = \frac{-b \pm \sqrt{b^2 - 4ac}}{2a} \Rightarrow t = \frac{-6 \pm \sqrt{6^2 - 4(1)(4)}}{2(1)} \Rightarrow t = \frac{-6 \pm \sqrt{20}}{2} \Rightarrow t = \frac{-6 \pm 2\sqrt{5}}{2} \Rightarrow t = -3 \pm \sqrt{5}$

31. $4x^2 - 13 = 0 \Rightarrow 4x^2 + 0x - 13 = 0$. Using the quadratic formula:

$x = \frac{-b \pm \sqrt{b^2 - 4ac}}{2a} \Rightarrow x = \frac{0 \pm \sqrt{0^2 - 4(4)(-13)}}{2(4)} \Rightarrow x = \frac{\pm\sqrt{208}}{8} \Rightarrow x = \frac{\pm 4\sqrt{13}}{8} \Rightarrow x = \frac{\pm\sqrt{13}}{2}$

33. $2(x - 1)^2 + 4 = 0 \Rightarrow 2(x^2 - 2x + 1) + 4 = 0 \Rightarrow 2x^2 - 4x + 6 = 0$. Using the quadratic formula:

$x = \frac{-b \pm \sqrt{b^2 - 4ac}}{2a} \Rightarrow x = \frac{-(-4) \pm \sqrt{(-4)^2 - 4(2)(6)}}{2(2)} \Rightarrow x = \frac{4 \pm \sqrt{-32}}{4} = \frac{4 \pm 4\sqrt{-2}}{4} = 1 \pm \sqrt{-2}$. Since $\sqrt{-2}$ is

not real, there is no real solution to the equation.

35. $\frac{1}{2}x^2 - 3x + \frac{1}{2} = 0 \Rightarrow x^2 - 6x + 1 = 0$. Using the quadratic formula:

$x = \frac{-b \pm \sqrt{b^2 - 4ac}}{2a} \Rightarrow x = \frac{-(-6) \pm \sqrt{(-6)^2 - 4(1)(1)}}{2(1)} \Rightarrow x = \frac{6 \pm \sqrt{32}}{2} \Rightarrow x = \frac{6 \pm 4\sqrt{2}}{2} \Rightarrow x = 3 \pm 2\sqrt{2}$

37. $-3z^2 - 2z + 4 = 0$. Using the quadratic formula:

$x = \frac{-b \pm \sqrt{b^2 - 4ac}}{2a} \Rightarrow x = \frac{-(-2) \pm \sqrt{(-2)^2 - 4(-3)(4)}}{2(-3)} \Rightarrow x = \frac{2 \pm \sqrt{52}}{-6} \Rightarrow x = \frac{-2 \pm 2\sqrt{13}}{6} \Rightarrow x = \frac{-1 \pm \sqrt{13}}{3}$

39. $25k^2 + 1 = 10k \Rightarrow 25k^2 - 10k + 1 = 0$. Using the quadratic formula:

$$x = \frac{-b \pm \sqrt{b^2 - 4ac}}{2a} \Rightarrow x = \frac{-(-10) \pm \sqrt{(-10)^2 - 4(25)(1)}}{2(25)} \Rightarrow x = \frac{10 \pm \sqrt{0}}{50} \Rightarrow x = \frac{1}{5}$$

41. $-0.3x^2 + 0.1x = -0.02 \Rightarrow -30x^2 + 10x + 2 = 0$. Using the quadratic formula:

$$x = \frac{-b \pm \sqrt{b^2 - 4ac}}{2a} \Rightarrow x = \frac{-10 \pm \sqrt{10^2 - 4(-30)(2)}}{2(-30)} \Rightarrow x = \frac{-10 \pm \sqrt{340}}{-60} \Rightarrow x = \frac{10 \pm 2\sqrt{85}}{60} \Rightarrow x = \frac{5 \pm \sqrt{85}}{30}$$

43. $2x(x + 2) = (x - 1)(x + 2) \Rightarrow 2x^2 + 4x = x^2 + x - 2 \Rightarrow x^2 + 3x + 2 = 0$. Factoring we get $(x + 2)(x + 1) = 0$. Then $x + 2 = 0 \Rightarrow x = -2$ or $x + 1 = 0 \Rightarrow x = -1$.

45. For the x-intercepts, $y = 0 \Rightarrow 6x^2 + 13x - 5 = 0$. Factoring this we get $(2x + 5)(3x - 1) = 0$. Then

$2x + 5 = 0 \Rightarrow 2x = -5 \Rightarrow x = -\frac{5}{2}$ or $3x - 1 = 0 \Rightarrow 3x = 1 \Rightarrow x = \frac{1}{3}$. Therefore $x - \text{int} \rightarrow \left(\frac{5}{2}, 0\right), \left(\frac{1}{3}, 0\right)$. For the

y-intercept, $x = 0 \Rightarrow y = 6(0)^2 + 13(0) - 5 \Rightarrow y - \text{int} \rightarrow (0, -5)$

47. For the x-intercept, $y = 0 \Rightarrow -4x^2 + 12x - 9 = 0$. Using the quadratic formula:

$$x = \frac{-12 \pm \sqrt{12^2 - 4(-4)(-9)}}{2(-4)} \Rightarrow x = \frac{-12 \pm \sqrt{144 - 144}}{-8} \Rightarrow x = \frac{-12 \pm \sqrt{0}}{-8} \Rightarrow x - \text{int} \rightarrow \left(\frac{3}{2}, 0\right). \text{ For the } y\text{-}$$

intercept, $x = 0 \Rightarrow y = -4(0)^2 + 12(0) - 9 \Rightarrow y - \text{int} \rightarrow (0, -9)$

49. For the x-intercepts, $y = 0 \Rightarrow -3x^2 + 11x - 6 = 0$. Using the quadratic formula:

$$x = \frac{-11 \pm \sqrt{11^2 - 4(-3)(-6)}}{2(-3)} \Rightarrow x = \frac{-11 \pm \sqrt{121 - 72}}{-6} \Rightarrow x = \frac{-11 \pm \sqrt{49}}{-6} \Rightarrow x = \frac{11 \pm 7}{6} \Rightarrow x - \text{int} \rightarrow \left(\frac{2}{3}, 0\right), (3, 0).$$

For the y-intercept, $x = 0 \Rightarrow y = -3(0)^2 + 11(0) - 6 \Rightarrow y = -6$

51. (a) Graphical: Graph $Y_1 = X^2 + 2X$ and locate the x-intercepts. See Figure 51a and Figure 51b; the solutions to the equation are -2 and 0.

   (b) Numerical: Table $Y_1 = X^2 + 2X$ starting at $x = -4$, incrementing by 1. $Y_1 = 0$ when $x = -2$ or 0. See Figure 51c.

   (c) Symbolic: $x^2 + 2x = 0 \Rightarrow x(x + 2) = 0$. Then $x = 0$ or $x + 2 = 0 \Rightarrow x = -2$. Therefore $x = -2, 0$.

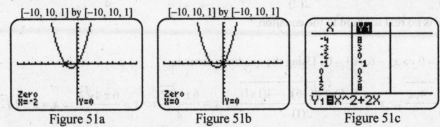

Figure 51a   Figure 51b   Figure 51c

53. (a) Graphical: Graph $Y_1 = X^2 - X - 6$ and locate the x-intercepts. See Figures 53a and 53b; the solutions to the equation are -2 and 3.

   (b) Numerical: Table $Y_1 = X^2 - X - 6$ starting at $x = -3$, incrementing by 1. $Y_1 = 0$ when $x = -2$ or 3. See Figure 53c.

   (c) Symbolic: $x^2 - x - 6 = 0 \Rightarrow (x - 3)(x + 2) = 0 \Rightarrow x = 3$ or -2.

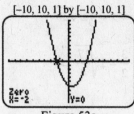
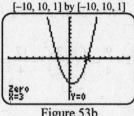
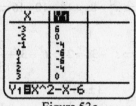

[−10, 10, 1] by [−10, 10, 1]   [−10, 10, 1] by [−10, 10, 1]

Figure 53a   Figure 53b   Figure 53c

55. (a) Graphical: Graph $Y_1 = 2X^2 - 6$ and locate the x-intercepts. See Figures 55a and 55b; the solutions to the equation are $x \approx \pm 1.7$.

    (b) Numerical: Table $Y_1 = 2X^2 - 6$ starting at $x = -2\sqrt{3}$, incrementing by $\sqrt{3}$. $Y_1 \approx 0$ when $x \approx \pm 1.7$. See Figure 55c.

    (c) Symbolic: $2x^2 = 6 \Rightarrow x^2 = 3 \Rightarrow x = \pm\sqrt{3} \approx \pm 1.7$.

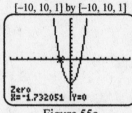
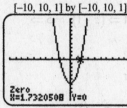
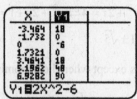

[−10, 10, 1] by [−10, 10, 1]   [−10, 10, 1] by [−10, 10, 1]

Figure 55a   Figure 55b   Figure 55c

57. $20x^2 + 11x = 3 \Rightarrow 20x^2 + 11x - 3 = 0$. Graph $Y_1 = 20X^2 + 11X - 3$ and locate the x-intercepts. See Figures 57a and 57b; the solutions to the equation are $x = -0.75$ or 0.2.

59. $2.5x^2 = 4.75x - 2.1 \Rightarrow 2.5x^2 - 4.75x + 2.1 = 0$. Graph $Y_1 = 2.5X^2 - 4.75X + 2.1$ and locate the x-intercepts. See Figures 59a and 59b; the solutions to the equation are $x = 0.7$ or 1.2.

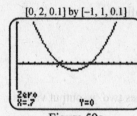

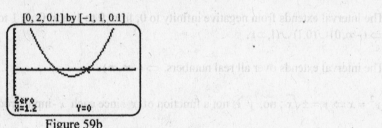

[0, 2, 0.1] by [−1, 1, 0.1]   [0, 2, 0.1] by [−1, 1, 0.1]

Figure 59a   Figure 59b

61. $x^2 + 4x - 6 = 0 \Rightarrow x^2 + 4x = 6 \Rightarrow x^2 + 4x + 4 = 6 + 4 \Rightarrow (x+2)^2 = 10 \Rightarrow$
     $x + 2 = \pm\sqrt{10} \Rightarrow x = -2 \pm \sqrt{10}$

63. $x^2 + 5x = 4 \Rightarrow x^2 + 5x + \dfrac{25}{4} = 4 + \dfrac{25}{4} \Rightarrow \left(x + \dfrac{5}{2}\right)^2 = \dfrac{41}{4} \Rightarrow$
     $x + \dfrac{5}{2} = \pm\sqrt{\dfrac{41}{4}} \Rightarrow x = -\dfrac{5}{2} \pm \dfrac{\sqrt{41}}{2} \Rightarrow x = -\dfrac{5}{2} \pm \dfrac{1}{2}\sqrt{41}$

65. $3x^2 - 6x = 2 \Rightarrow x^2 - 2x = \dfrac{2}{3} \Rightarrow x^2 - 2x + 1 = \dfrac{2}{3} + 1 \Rightarrow (x-1)^2 = \dfrac{5}{3} \Rightarrow$
     $x - 1 = \pm\sqrt{\dfrac{5}{3}} \Rightarrow x = 1 \pm \sqrt{\dfrac{5}{3}} \Rightarrow 1 \pm \dfrac{\sqrt{15}}{3}$

67. $x^2 - 8x = 10 \Rightarrow x^2 - 8x + 16 = 10 + 16 \Rightarrow (x-4)^2 = 26 \Rightarrow x - 4 = \pm\sqrt{26} \Rightarrow x = 4 \pm \sqrt{26}$

69. $\dfrac{1}{2}t^2 - \dfrac{3}{2}t = 1 \Rightarrow t^2 - 3t = 2 \Rightarrow t^2 - 3t + \dfrac{9}{4} = 2 + \dfrac{9}{4} \Rightarrow \left(t - \dfrac{3}{2}\right)^2 = \dfrac{17}{4} \Rightarrow t - \dfrac{3}{2} = \pm\sqrt{\dfrac{17}{4}} \Rightarrow$

    $t = \dfrac{3}{2} \pm \dfrac{\sqrt{17}}{2} \Rightarrow t = \dfrac{3 \pm \sqrt{17}}{2}$

71. $-2z^2 + 3z + 1 = 0 \Rightarrow z^2 - \dfrac{3}{2}z = \dfrac{1}{2} \Rightarrow z^2 - \dfrac{3}{2}z + \dfrac{9}{16} = \dfrac{1}{2} + \dfrac{9}{16} \Rightarrow \left(z - \dfrac{3}{4}\right)^2 = \dfrac{17}{16} \Rightarrow$

    $z - \dfrac{3}{4} = \pm\sqrt{\dfrac{17}{16}} \Rightarrow z = \dfrac{3}{4} \pm \dfrac{\sqrt{7}}{4} \Rightarrow z = \dfrac{3 \pm \sqrt{17}}{4}$

73. $-z^2 - 2z + 4 = 0 \Rightarrow z^2 + 2z = 4 \Rightarrow z^2 + 2z + 1 = 4 + 1 \Rightarrow (z+1)^2 = 5 \Rightarrow$

    $z + 1 = \pm\sqrt{5} \Rightarrow z = -1 \pm \sqrt{5}$

75. $D =$ all real numbers except when the denominator

    $x^2 - 5 = 0 \Rightarrow x^2 = 5 \Rightarrow x = \pm\sqrt{5} \Rightarrow D = \{x \mid x \neq \sqrt{5}, x \neq -\sqrt{5}\}$

77. $D =$ all real numbers except when the denominator

    $t^2 - t - 2 = 0 \Rightarrow (t-2)(t+1) = 0 \Rightarrow t = -1, 2 \Rightarrow D = \{t \mid t \neq -1, t \neq 2\}$

79. The interval extends from negative infinity to -1, from -1 to 1, and from 1 to infinity,
    $\Rightarrow (-\infty, -1) \cup (-1, 1) \cup (1, \infty)$.

81. The interval extends from negative infinity to 0, from 0 to 1, and from 1 to infinity,
    $\Rightarrow (-\infty, 0) \cup (0, 1) \cup (1, \infty)$.

83. The interval extends over all real numbers, $\Rightarrow (-\infty, \infty)$.

85. $y^2 = x \Rightarrow y = \pm\sqrt{x}$; no, y is not a function of x since each x-input produces two y-output values.

87. $2 - 4y^2 = x \Rightarrow y^2 = \dfrac{2-x}{4} \Rightarrow y = \pm\sqrt{\dfrac{2-x}{4}} = \pm\dfrac{1}{2}\sqrt{2-x}$; no, y is not a function of x since for each x-input

    less than 2 it produces two y-output values.

89. $4x^2 + 3y = \dfrac{y+1}{3} \Rightarrow 3y - \dfrac{y+1}{3} = -4x^2 \Rightarrow 9y - y - 1 = -12x^2 \Rightarrow 8y - 1 = -12x^2 \Rightarrow$

    $8y = -12x^2 + 1 \Rightarrow y = \dfrac{-12x^2 + 1}{8}$; yes, y is a function of x since each x-input produces only one y-output.

91. $3y = \dfrac{2x - y}{3} \Rightarrow 9y = 2x - y \Rightarrow 10y = 2x \Rightarrow y = \dfrac{x}{5}$; yes, y is a function of x since each x-input produces

    only one y-output.

93. $x^2 + (y-3)^2 = 9 \Rightarrow (y-3)^2 = 9 - x^2 \Rightarrow y - 3 = \pm\sqrt{9 - x^2} \Rightarrow y = 3 \pm \sqrt{9 - x^2}$; no, y is not a function of x since some x-inputs produce two y-outputs.

95. $3x^2 + 4y^2 = 12 \Rightarrow 4y^2 = 12 - 3x^2 \Rightarrow 2y = \pm\sqrt{12 - 3x^2} \Rightarrow y = \pm\dfrac{\sqrt{12 - 3x^2}}{2}$; no, y is not a function of x because some x-inputs produce two y-outputs.

97. $V = \dfrac{1}{3}\pi r^2 h$ for $r \Rightarrow \dfrac{3V}{\pi h} = r^2 \Rightarrow r = \pm\sqrt{\dfrac{3V}{\pi h}}$

99. $K = \dfrac{1}{2}mv^2$ for $v \Rightarrow \dfrac{2K}{m} = v^2 \Rightarrow v = \pm\sqrt{\dfrac{2K}{m}}$

101. $a^2 + b^2 = c^2$ for $b \Rightarrow b^2 = c^2 - a^2 \Rightarrow b = \pm\sqrt{c^2 - a^2}$

103. $s = -16t^2 + 100t$ for $t \Rightarrow -s = 16t^2 - 100t \Rightarrow \dfrac{-s}{16} = t^2 - \dfrac{25}{4}t \Rightarrow \dfrac{-s}{16} + \dfrac{625}{64} = t^2 - \dfrac{25}{4}t + \dfrac{625}{64} \Rightarrow$

$\left(t - \dfrac{25}{8}\right)^2 = \dfrac{625 - 4s}{64} \Rightarrow t - \dfrac{25}{8} = \pm\sqrt{\dfrac{625 - 4s}{64}} \Rightarrow t = \dfrac{25}{8} \pm \sqrt{\dfrac{625 - 4s}{64}} \Rightarrow t = \dfrac{25 \pm \sqrt{625 - 4s}}{8}$

105. (a) $3x^2 = 12 \Rightarrow 3x^2 - 12 = 0$

   (b) $b^2 - 4ac = 0^2 - 4(3)(-12) = 144 > 0$. There are two real solutions.

   (c) $3x^2 = 12 \Rightarrow x^2 = 4 \Rightarrow x = \pm 2$

107. (a) $x^2 - 2x = -1 \Rightarrow x^2 - 2x + 1 = 0$

   (b) $b^2 - 4ac = (-2)^2 - 4(1)(1) = 0$. There is one real solution.

   (c) $x^2 - 2x + 1 = 0 \Rightarrow (x-1)^2 = 0 \Rightarrow x = 1$

109. (a) $4x = x^2 \Rightarrow x^2 - 4x = 0$

   (b) $b^2 - 4ac = (-4)^2 - 4(1)(0) = 16 > 0$. There are two real solutions.

   (c) $x^2 - 4x = 0 \Rightarrow x(x-4) = 0 \Rightarrow x = 0$ or $x - 4 = 0 \Rightarrow x = 4$. Therefore, $x = 0, 4$.

111. (a) $x^2 + 1 = x \Rightarrow x^2 - x + 1 = 0$

   (b) $b^2 - 4ac = (-1)^2 - 4(1)(1) = -3 < 0$. There are no real solutions.

   (c) There are no real solutions.

113. (a) $2x^2 + 3x = 12 - 2x \Rightarrow 2x^2 + 5x - 12 = 0$.

   (b) $b^2 - 4ac = (5)^2 - 4(2)(-12) = 121 > 0$. There are two real solutions.

   (c) $2x^2 + 5x - 12 = 0 \Rightarrow (2x-3)(x+4) = 0 \Rightarrow 2x = 3 \Rightarrow x = 1.5$ or $x = -4$. So, $x = 1.5, -4$.

115. (a) $9x(x-4) = -36 \Rightarrow 9x^2 - 36x + 36 = 0$

   (b) $b^2 - 4ac = (-36)^2 - 4(9)(36) = 0$. There is one real solution.

   (c) $9x^2 - 36x + 36 = 0 \Rightarrow x^2 - 4x + 4 = 0 \Rightarrow (x-2)^2 = 0 \Rightarrow x = 2$

117. (a) $x\left(\dfrac{1}{2}x+1\right)=-\dfrac{13}{2}\Rightarrow\dfrac{1}{2}x^2+x+\dfrac{13}{2}=0$

(b) $b^2-4ac=(1)^2-4\left(\dfrac{1}{2}\right)\left(\dfrac{13}{2}\right)=-12<0$. There are no real solutions.

(c) There are no real solutions.

119. (a) $3x^2=1-x\Rightarrow 3x^2+x-1=0$

(b) $b^2-4ac=1^2-4(3)(-1)=13>0$. There are two real solutions.

(c) $x=\dfrac{-1\pm\sqrt{1^2-4(3)(-1)}}{2(3)}=\dfrac{-1\pm\sqrt{13}}{2(3)}=\dfrac{-1\pm\sqrt{13}}{6}$

121. (a) Since the parabola opens upward, $a>0$.

(b) Since the zeros of f are -6 and 2, the solutions to $ax^2+bx+c=0$ are also -6 and 2.

(c) Since there are two real solutions, the discriminant is positive.

123. (a) Since the parabola opens upward, $a>0$.

(b) Since the only zero of f is -4, the solution is also -4.

(c) Since there is one real solution, the discriminant is equal to zero.

125. (a) Since the parabola opens downward, $a<0$.

(b) Since f has no real zeros, there are no solutions to the quadratic equation.

(c) Since there are no real solutions, the discriminant is negative.

127. (a) The zeros of f are at 0 and 2. Because there are two real solutions, the discriminant is positive.

(b) x-intercepts: $(0,0)$, $(2,0)$; y-intercept: $(0,0)$

(c) $f(x)<0$ for $(0,2)$, and $f(x)>0$ for $(-\infty,0),(2,\infty)$.

(d) $f(x)$ is increasing on $(1,\infty)$, and decreasing on $(-\infty,1)$.

(e) The function f decreases from -1 to 1, so its average rate of change is negative.

(f) The average rate of change of f over the given interval is $\dfrac{-1-3}{1-(-1)}=\dfrac{-4}{2}=-2$.

129. The area of the square must equal the perimeter, or $A(x)=x^2=P(x)=4x\Rightarrow x=4$. The square's dimensions are 4 by 4.

131. The smaller square's sides are given by x, whereas the larger square's sides, which are 4 inches longer, are given by $x+4$. $x^2+(x+4)^2=80\Rightarrow 2x^2+8x+16=80\Rightarrow 2x^2+8x-64=0\Rightarrow x^2+4x-32=0$. From the quadratic formula, $x=\dfrac{-4\pm\sqrt{16-4(1)(-32)}}{2(1)}=\dfrac{-4\pm\sqrt{144}}{2}=-2\pm6\Rightarrow x=-8,4$. As the negative root is not physically possible, the only possible answer is $x=4$, so that the smaller square's dimensions are 4 inches by 4 inches, and the larger square's dimensions are 8 inches by 8 inches.

133. The height when it hits the ground will be

$$0 \Rightarrow 75 - 16t^2 = 0 \Rightarrow -16t^2 = -75 \Rightarrow t^2 = \frac{75}{16} \Rightarrow t = \sqrt{\frac{75}{16}} \Rightarrow t = \frac{\sqrt{75}}{4} \Rightarrow t \approx 2.2 \text{ seconds}.$$

135. $55 = 16x^2 + 7x + 32 \Rightarrow 16x^2 + 7x - 23 = 0$ Factoring this we get $(x-1)(16x+23) = 0 \Rightarrow x - 1 = 0 \Rightarrow x = 1$ or

$16x + 23 = 0 \Rightarrow x = -\dfrac{23}{16}$. Since x represents the number of years after 2008 the result cannot be negative and

the only answer is $x = 1$. Therefore, the year is 2009.

137. Graphical: First, we must determine a formula for the area. Let x represent the height of the computer screen.
Then, $x + 2$ is the width. The area of the screen is height times width, computed $A(x) = x(x+2)$. We must
solve the quadratic equation $x(x+2) = 143$ or $x^2 + 2x - 143 = 0$. Graph $Y_1 = x^2 + 2x - 143$ and determine
any zeros. Figure 137a shows that the equation has two zeros, one negative and one positive. The positive
zero is located at $x = 11$. The height is 11 inches and the width is $11 + 2 = 13$ inches.

Symbolic: The quadratic equation $x^2 + 2x - 143 = 0$ can be solved by factoring.
$x^2 + 2x - 143 = 0 \Rightarrow (x+13)(x-11) = 0 \Rightarrow x = -13, 11$. The positive answer gives a height of 11 inches. It
follows that the width is 13 inches. The numerical solution is shown in Figure 137b. Yes, the symbolic,
graphical and numerical answers agree.

139. Let $x = $ width of the metal sheet in inches and $x + 10 = $ length of the metal sheet in inches. Make a sketch to
find expressions for the dimensions of the box. See Figure 139. The width of the box is $x - 8$ inches, the
length of the box is $x + 2$ inches, and the height of the box is 4 inches; the volume of the box, which is given
as 476 cubic inches, is determined by the length times the width times the height:
$4(x-8)(x+2) = 476 \Rightarrow (x-8)(x+2) = 119 \Rightarrow$

$x^2 - 6x - 16 = 119 \Rightarrow x^2 - 6x - 135 = 0 \Rightarrow (x+9)(x-15) = 0 \Rightarrow x = -9$ or $x = 15$. Since width cannot be
negative, $x = 15$ inches; thus, the dimensions of the metal sheet is 15 inches by 25 inches.

[−15, 15, 2] by [−15, 15, 2]

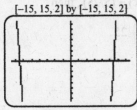

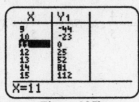

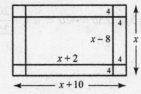

Figure 137a   Figure 137b   Figure 139

141. $V = \pi r^2 h$, $V = 28$ cubic inches, and $h = 4$ inches $\Rightarrow 28 = \pi r^2(4) \Rightarrow r^2 = \dfrac{28}{4\pi} \Rightarrow r = \sqrt{\dfrac{28}{4\pi}} \approx 1.49$; the radius
of the cylinder is approximately 1.49 inches.

143. Since the diameter of the semicircle is x, the radius is $\dfrac{x}{2}$; the area of the semicircle

$= \dfrac{1}{2}\pi\left(\dfrac{x}{2}\right)^2 = \dfrac{1}{2}\pi\left(\dfrac{x^2}{4}\right) = \dfrac{1}{8}\pi x^2$. The area of the square $= x^2$; thus, the total area of the window, which is 463

square inches, is $x^2 + \dfrac{1}{8}\pi x^2$;

$463 = x^2 + \dfrac{1}{8}\pi x^2 \Rightarrow 463 = \left(1 + \dfrac{\pi}{8}\right)x^2 \Rightarrow x^2 = \dfrac{463}{(1+\frac{\pi}{8})} \Rightarrow x = \sqrt{\dfrac{463}{1+\frac{\pi}{8}}} \approx 18.23$ inches.

145. Let x be the number of shirts ordered. Revenue equals the number of shirts sold times the price of each shirt. If x shirts are sold, then the price in dollars of each shirt is $20 - 0.10(x-1)$. The revenue $R(x)$ is given by $R(x) = x(20 - 0.10(x-1))$. We must solve the equation

$989 = x(20 - 0.10(x-1)) \Rightarrow -0.10x^2 + 20.1x - 989 = 0$. Let $y_1 = -0.10x^2 + 20.1x - 989$ and use the graphing calculator to find the x-intercept. Because the discount applies to orders up to 100 shirts, we find that $x = 86$ shirts. See Figure 145.

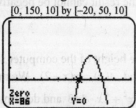

[0, 150, 10] by [−20, 50, 10]

Figure 145

147. (a) Since $s(0) = 32$, $s(t) = -16t^2 + v_0 t + h_0 \Rightarrow 32 = -16(0)^2 + v_0(0) + h_0 \Rightarrow h_0 = 32$ and so

$s(t) = -16t^2 + v_0 t + 32$; since $s(1) = 176, 176 = -16(1)^2 + v_0(1) + 32 \Rightarrow v_0 = 160$; thus,

$s(t) = -16t^2 + 160t + 32$ models the data.

(b) When the projectile strikes the ground, $s(t) = 0$;

$0 = -16t^2 + 160t + 32 \Rightarrow 0 = t^2 - 10t - 2 \Rightarrow t = \dfrac{10 \pm \sqrt{10^2 - 4(1)(-2)}}{2} \approx 10.2$ or -0.2; since t cannot be

negative, $t \approx 10.2$; the projectile strikes the ground after about 10.2 seconds.

149. If $L(t)$ is linear, then the function is of the form $L(t) = mt + b$. $L(0) = 5 = m(0) + b = b$, and

$L(5) = 25 = m(5) + 5 \Rightarrow 5m = 20 \Rightarrow m = 4$, so $L(t) = 4t + 5$.

151. If $S(t)$ is quadratic, then the function is of the form $S(t) = a(t-h)^2 + k$. If $(h, k) = (5, 25)$, then

$S(0) = a(0-5)^2 + 25 = 5 \Rightarrow 25a = -20 \Rightarrow a = -\dfrac{20}{25} = -\dfrac{4}{5}$, so $S(t) = -\dfrac{4}{5}(t-5)^2 + 25$.

153.

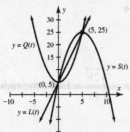

Figure 153

155. (a) The vertex of (2009, 1200) substituted into the function will be written as $B(x) = a(x - 2009)^2 + 1200$. We now us the point (2015, 840) to substitute into the equation and solve for a as follows,

$840 = a(2015 - 2009)^2 + 1200 \Rightarrow 840 = 36a + 1200 \Rightarrow -360 = 36a \Rightarrow a = -10$. Thus

$B(x) = -10(x - 2009)^2 + 1200$ can be used to model the budget for pedestrian and bicycle programs in selected years.

(b) $710 = -10(x - 2009)^2 + 1200 \Rightarrow -490 = -10(x - 2009)^2 \Rightarrow 49 = (x - 2009)^2$

$\Rightarrow \pm 7 = x - 2009 \Rightarrow x = 2002, 2016$

157. (a) Use the Quadratic regression function on the calculator to find the function
$$I(x) = -2.277x^2 + 11.71x + 40.4 .$$

(b) $28 = -2.27x^2 + 11.71x + 40.4 \Rightarrow -2.277x^2 + 11.71x + 12.4 = 0$, using the quadratic formula:

$$\frac{-11.71 \pm \sqrt{(11.71)^2 - 4(-2.277)(12.4)}}{2(-2.277)} \Rightarrow x \approx -1 \text{ and } 6 . \text{ Since } x = 0 \text{ corresponds to 2006 the results are}$$

2005 and 2012.

159. (a) $E(15) = 1.4, 1987 + 15 = 2002$; in 2002 there were 1.4 million Wal-Mart employees.

(b) Use the quadratic regression function on your graphing calculator to find
$f(x) = 0.00474x^2 + 0.00554x + 0.205$. *Answers may vary.*

(c) See Figure 159c.

(d) Let $Y_1 = (0.00474)X^2 + 0.00554X + 0.205$ and $Y_2 = 3$. Use the CALC function on your calculator to find the intersection points of the Y_1 and Y_2. See Figure 159d. When $x \approx 24$, $f(x) = 3$. Therefore, in the year 2011 the number of employees may reach 3 million. *Answers may vary.*

[0, 25, 5] by [0, 3.6, 0.2]

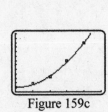   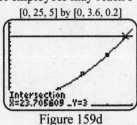

Figure 159c   Figure 159d

Extended and Discovery Exercises for Section 3.2

1. $b^2 - 4ac = 14^2 - 4(8)(-15) = 676 = 26^2$. Since 676 is a perfect square the equation can be solved by factoring.

$8x^2 + 14x - 15 = 0 \Rightarrow (4x - 3)(2x + 5) = 0$. Then $4x - 3 = 0 \Rightarrow x = \dfrac{3}{4}$ or $2x + 5 = 0 \Rightarrow x = -\dfrac{5}{2}$.

3. $b^2 - 4ac = (-3)^2 - 4(5)(-3) = 69$. Since 69 is not a perfect square the equation cannot be solved by factoring.

$x = \dfrac{-b \pm \sqrt{b^2 - 4ac}}{2a} \Rightarrow x = \dfrac{3 \pm \sqrt{(-3)^2 - 4(5)(-3)}}{2(5)} \Rightarrow x = \dfrac{3 \pm \sqrt{69}}{10}$.

5. (a) Follow the steps for completing the square in this equation:

$$ax^2 + bx + c = 0 \Rightarrow x^2 + \frac{b}{a}x + \frac{c}{a} = 0 \Rightarrow x^2 + \frac{b}{a}x = -\frac{c}{a}$$

(b) Add $\left(\dfrac{b}{2a}\right)^2 = \dfrac{b^2}{4a^2}$ to both sides to obtain $x^2 + \dfrac{b}{a}x + \dfrac{b^2}{4a^2} = -\dfrac{c}{a} + \dfrac{b^2}{4a^2}$ or $\left(x + \dfrac{b}{2a}\right)^2 = \dfrac{b^2 - 4ac}{4a^2}$.

(c) In order to derive the quadratic formula we must solve this second equation for x.

$$\left(x+\frac{b}{2a}\right)^2 = \frac{b^2-4ac}{4a^2} \Rightarrow \left(x+\frac{b}{2a}\right) = \pm\sqrt{\frac{b^2-4ac}{4a^2}} \Rightarrow \quad \{\text{Square root property}\}$$

$$x = -\frac{b}{2a}\pm\sqrt{\frac{b^2-4ac}{4a^2}} \Rightarrow x = \frac{b}{2a}\pm\frac{\sqrt{b^2-4ac}}{|2a|} \Rightarrow \quad \{\sqrt{4a^2}=|2a|\}$$

$$\qquad\qquad\qquad\qquad\qquad\qquad\qquad\qquad \{\text{Since there is a } \pm \text{ in front of the}$$
$$\text{expression, the absolute value does}$$

$$x = -\frac{b}{2a}\pm\sqrt{\frac{b^2-4ac}{4a^2}} \Rightarrow x = \frac{-b\pm\sqrt{b^2-4ac}}{2a} \Rightarrow \quad \text{not matter.}\}$$

7. $x^2 = k \Rightarrow \sqrt{x^2} = \sqrt{k} \Rightarrow |x| = \sqrt{k} \Rightarrow x = \pm\sqrt{k}$

Checking Basic Concepts for Sections 3.1 and 3.2

1.   Vertex: $(1,-4)$; axis of symmetry: $x=1$; x-intercepts: $(-1,0),(3,0)$. See Figure 1.

        Figure 1

3.   Since 3 is the smallest y-value and there is symmetry of the y-values around $x=-1$, the vertex is $(-1,3)$;
      $f(x) = a(x-h)^2 + k \Rightarrow f(x) = a(x+1)^2 + 3$; since $(0,5)$ is a data point, $f(0)=5$; $5 = a(0+1)^2 + 3 \Rightarrow a = 2$;
      $f(x) = 2(x+1)^2 + 3$ models the data exactly.

5.   Find the vertex:
      $x = -\dfrac{b}{2a} \Rightarrow x = -\dfrac{4}{2(1)} \Rightarrow x = -2$. Since $f(-2) = (-2)2 + 4(-2) - 3 \Rightarrow f(-2) = -7$, $(-2,-7)$ is the vertex. The

      leading coefficient is $a=1$. Converting to $f(x) = a(x-h)^2 + k$, we get
      $f(x) = 1(x-(-2))^2 + (-7) \Rightarrow f(x) = (x+2)^2 - 7$. The minimum value is at the vertex or -7.

7.   Let x = width of rectangle, then $x+4$ = length of rectangle; since area = 165 square inches,
      $x(x+4) = 165 \Rightarrow x^2 + 4x - 165 = 0 \Rightarrow (x-11)(x+15) = 0 \Rightarrow x = 11, -15$; since $x = -15$ has no physical
      meaning for the width, the width of the rectangle is 11 inches and the length is $11+4 = 15$ inches .

3.3: Complex Numbers

1. $\sqrt{-4} = i\sqrt{4} = 2i$

3. $\sqrt{-100} = i\sqrt{100} = 10i$

5. $\sqrt{-23} = i\sqrt{23}$

7. $\sqrt{-12} = i\sqrt{12} = i\sqrt{4}\sqrt{3} = 2i\sqrt{3}$

9. $\sqrt{-54} = \sqrt{(9)(6)(-1)} = \sqrt{9(-1)}\sqrt{6} = 3i\sqrt{6}$

11. $\dfrac{4 \pm \sqrt{-16}}{2} = \dfrac{4 \pm \sqrt{(16)(-1)}}{2} = \dfrac{4 \pm 4i}{2} = 2 \pm 2i$

13. $\dfrac{-6 \pm \sqrt{-72}}{3} = \dfrac{-6 \pm \sqrt{(36)(-1)}\sqrt{2}}{3} = \dfrac{-6 \pm 6i\sqrt{2}}{3} = -2 \pm 2i\sqrt{2}$

15. $\sqrt{-5} \cdot \sqrt{-5} = \sqrt{5}\sqrt{-1} \cdot \sqrt{5}\sqrt{-1} = \sqrt{5}i \cdot \sqrt{5}i = \sqrt{25}i^2 = 5(-1) = -5.$

17. $\sqrt{-18} \cdot \sqrt{-2} = (\sqrt{9}\sqrt{2}\sqrt{-1})(\sqrt{2}\sqrt{-1}) = (3\sqrt{2}i)(\sqrt{2}i) = 3\sqrt{4}i^2 = (3)(2)(-1) = -6$

19. $\sqrt{-3} \cdot \sqrt{-6} = (\sqrt{3}\sqrt{-1})(\sqrt{6}\sqrt{-1}) = (\sqrt{3}i)(\sqrt{6}i) = \sqrt{18}i^2 = \sqrt{9}\sqrt{2}(-1) = -3\sqrt{2}$

21. $3i + 5i = (3+5)i = 8i$

23. $(3+i) + (-5-2i) = (3+(-5)) + (1-2)i = -2-i$

25. $2i - (-5+23i) = (-5) + (2-23)i = 5-21i$

27. $3 - (4-6i) = (3-4) + i = -1+6i$

29. $(2)(2+4i) = 4+8i$

31. $(1+i)(2-3i) = (1)(2) + (1)(-3i) + (i)(2) + (i)(-3i) = 2-i-3i^2 = 2-i-3(-1) = 5-i$

33. $(-3+2i)(-2+i) = (-3)(-2) + (-3)(i) + (2i)(-2) + (2i)(i) = 6-7i+2i^2 = 6-7i+2(-1) = 4-7i$

35. $(-2+3i)^2 = (-2+3i)(-2+3i) = (-2)(-2) + (-2)(3i) + (-2)(3i) + (3i)(3i) =$
    $4 + (-12i) + (9i^2) = 4-12i+(-9) = -5-12i$

37. $2i(1-i)^2 = 2i(1-i)(1-i) = 2i[(1)(1) + (1)(-i) + (1)(-i) + (-i)(-i)] =$
    $2i[1-2i+(-1)] = 2i(-2i) = -4i^2 = 4$

39. $\dfrac{1}{1+i} = \dfrac{1}{1+i} \cdot \dfrac{1-i}{1-i} = \dfrac{1-i}{(1+i)(1-i)} = \dfrac{1-i}{1-i^2} = \dfrac{1-i}{2} = \dfrac{1}{2} - \dfrac{1}{2}i$

41. $\dfrac{4+i}{5-i} = \dfrac{4+i}{5-i} \cdot \dfrac{5+i}{5+i} = \dfrac{(4+i)(5+i)}{(5-i)(5+i)} = \dfrac{20+9i+i^2}{25-i^2} = \dfrac{19+9i}{26} = \dfrac{19}{26} + \dfrac{9}{26}i$

43. $\dfrac{2i}{10-5i} = \dfrac{2i}{10-5i} \cdot \dfrac{10+5i}{10+5i} = \dfrac{20i+10i^2}{(10-5i)(10+5i)} = \dfrac{-10+20i}{100-25i^2} = \dfrac{-10+20i}{125} = -\dfrac{2}{25} + \dfrac{4}{25}i$

45. $\dfrac{3}{-i} \cdot \dfrac{i}{i} = \dfrac{3i}{-i^2} = \dfrac{3i}{-i^2} = \dfrac{3i}{-(-1)} = 3i$

47. $\dfrac{-2+i}{(1+i)^2} = \dfrac{-2+i}{(1+i)(1+i)} = \dfrac{-2+i}{(1)(1)+(1)(i)+(1)(i)+(i)(i)} = \dfrac{-2+i}{1+2i+i^2} = \dfrac{-2+i}{1+2i-1} =$
    $\dfrac{-2+i}{2i} \cdot \dfrac{i}{i} = \dfrac{-2i+i^2}{2i^2} = \dfrac{-2i-1}{2(-1)} = \dfrac{-2i-1}{-2} = \dfrac{1}{2}+i$

49. $(23-5.6i) + (-41.5+93i) = -18.5+87.4i$

51. $(17.1-6i) - (8.4+0.7i) = 8.7-6.7i$

53. $(-12.6-5.7i)(5.1-9.3i) = -117.27+88.11i$

55. $\dfrac{17-135i}{18+142i} \approx -0.921 - 0.236i$

57. $i^{50} = (i^4)^{12} \cdot i^2 = (1)^{12} \cdot (-1) = -1$

59. $i^{32} = (i^4)^7 \cdot i^3 = (1)^7 \cdot (-i) = -i$

61. $i^{12} = (i^4)^3 = (1)^3 = 1$

63. $i^{57} = (i^4)^{14} \cdot i = (1)^{14} \cdot (i) = i$

65. $x^2 + 5 = 0 \Rightarrow x^2 = -5 \Rightarrow x = \pm\sqrt{-5} = \pm i\sqrt{5}$

67. $5x^2 + 1 = 3x^2 \Rightarrow 2x^2 = -1 \Rightarrow x^2 = -\dfrac{1}{2} \Rightarrow x = \pm i\sqrt{\dfrac{1}{2}}$

69. $3x = 5x^2 + 1 \Rightarrow 5x^2 - 3x + 1 = 0$; use the quadratic formula with $a = 5$, $b = -3$, and $c = 1$.

$x = \dfrac{-(-3) \pm \sqrt{(-3)^2 - 4(5)(1)}}{2(5)} = \dfrac{3 \pm \sqrt{9 - 20}}{10} = \dfrac{3 \pm \sqrt{-11}}{10} = \dfrac{3 \pm i\sqrt{11}}{10} = \dfrac{3}{10} \pm \dfrac{i\sqrt{11}}{10}$

71. $x(x - 4) = -5 \Rightarrow x^2 - 4x = -5 \Rightarrow x^2 - 4x + 5 = 0$; use the quadratic formula to solve

$x^2 - 4x + 5 = 0$ with $a = 1$, $b = -4$ and $c = 5$.

$x = \dfrac{-(-4) \pm \sqrt{(-4)^2 - 4(1)(5)}}{2(1)} = \dfrac{4 \pm \sqrt{16 - 20}}{2} = \dfrac{4 \pm \sqrt{-4}}{2} = \dfrac{4 \pm 2i}{2} = 2 \pm i$

73. Use the quadratic formula to solve $x^2 - 3x + 5 = 0$ with $a = 1$, $b = -3$, and $c = 5$.

$x = \dfrac{-(-3) \pm \sqrt{(-3)^2 - 4(1)(5)}}{2(1)} = \dfrac{3 \pm \sqrt{9 - 20}}{2} = \dfrac{3 \pm \sqrt{-11}}{2} = \dfrac{3 \pm i\sqrt{11}}{2} = \dfrac{3}{2} \pm \dfrac{i\sqrt{11}}{2}$

75. Use the quadratic formula to solve $x^2 + 2x + 4 = 0$ with $a = 1$, $b = 2$, and $c = 4$.

$x = \dfrac{-2 \pm \sqrt{2^2 - 4(1)(4)}}{2(1)} = \dfrac{-2 \pm \sqrt{4 - 16}}{2} = \dfrac{-2 \pm \sqrt{-12}}{2} = \dfrac{-2 \pm 2i\sqrt{3}}{2} = -1 \pm i\sqrt{3}$

77. $3x^2 - 4x = x^2 - 3 \Rightarrow 2x^2 - 4x + 3 = 0$. Using the quadratic formula we get:

$\dfrac{4 \pm \sqrt{16 - 4(2)(3)}}{2(2)} = \dfrac{4 \pm \sqrt{-8}}{4} = \dfrac{4 \pm 2\sqrt{-2}}{4} = 1 + \dfrac{\sqrt{-2}}{2} = 1 \pm \dfrac{i\sqrt{2}}{2}$

79. $2x(x - 2) = x - 4 \Rightarrow 2x^2 - 4x = x - 4 \Rightarrow 2x^2 - 5x + 4 = 0$.

Using the quadratic formula we get: $\dfrac{5 \pm \sqrt{(5)^2 - 4(2)(4)}}{2(2)} = \dfrac{5 \pm \sqrt{-7}}{4} = \dfrac{5}{4} \pm \dfrac{i\sqrt{7}}{4}$

81. $3x(3 - x) - 8 = x(x - 2) \Rightarrow 9x - 3x^2 - 8 = x^2 - 2x \Rightarrow -4x^2 + 11x - 8 = 0$.

Using the quadratic formula we get: $\dfrac{-11 \pm \sqrt{(11)^2 - 4(-4)(-8)}}{2(-4)} = \dfrac{-11 \pm \sqrt{-7}}{-8} = \dfrac{11}{8} \pm \dfrac{i\sqrt{7}}{8}$

83. (a) The graph of $y = 2x^2 - x - 3$ intersects the x-axis twice, so there are two real zeros.

   (b) $x = \dfrac{-b \pm \sqrt{b^2 - 4ac}}{2a} = \dfrac{1 \pm \sqrt{(-1)^2 - 4(2)(-3)}}{2(2)} = \dfrac{1 \pm 5}{4} = \dfrac{3}{2}, -1$

85. (a) The graph of $f(x) = x^2 + x + 2$ does not intersect the x-axis. Both of its zeros are imaginary.

   (b) $x = \dfrac{-b \pm \sqrt{b^2 - 4ac}}{2a} = \dfrac{-1 \pm \sqrt{1^2 - 4(1)(2)}}{2(1)} = \dfrac{-1 \pm \sqrt{-7}}{2} = -\dfrac{1}{2} \pm \dfrac{i\sqrt{7}}{2}$

87. (a) The graph of $y = -x^2 - 2$ does not intersect the x-axis. Both of its zeros are imaginary.

   (b) $x = \dfrac{-b \pm \sqrt{b^2 - 4ac}}{2a} = \dfrac{0 \pm \sqrt{0^2 - 4(-1)(-2)}}{2(-1)} = \dfrac{\pm\sqrt{-8}}{-2} = \dfrac{\pm 2i\sqrt{2}}{-2} = \pm i\sqrt{2}$

89. $Z = \dfrac{V}{I} \Rightarrow Z = \dfrac{50+98i}{8+5i} \Rightarrow Z = \dfrac{50+98i}{8+5i} \cdot \dfrac{8-5i}{8-5i} \Rightarrow Z = \dfrac{400-250i+784i-490i^2}{64-25i^2} \Rightarrow$

$Z = \dfrac{400+534i+490}{64+25} \Rightarrow Z = \dfrac{890+534i}{89} \Rightarrow Z = 10+6i$

91. $Z = \dfrac{V}{I} \Rightarrow 3-4i = \dfrac{V}{1+2i} \Rightarrow V = (3-4i)(1+2i) \Rightarrow V = 3+6i-4i-8i^2 \Rightarrow$

$V = 3+2i+8 \Rightarrow V = 11+2i$

93. $Z = \dfrac{V}{I} \Rightarrow 22-5i = \dfrac{27+17i}{I} \Rightarrow (22-5i)V = 27+17i \Rightarrow V = \dfrac{27+17i}{22-5i} \Rightarrow V = \dfrac{27+17i}{22-5i} \cdot \dfrac{22+5i}{22+5i} \Rightarrow$

$V = \dfrac{594+135i+374i+85i^2}{484-25i^2} \Rightarrow V = \dfrac{594+509i-85}{484+25} \Rightarrow V = \dfrac{509+509i}{509} \Rightarrow V = 1+i$

Extended and Discovery Exercise for Section 3.3

1. (a) $i^1 = i$, $i^2 = -1$, $i^3 = -i$, $i^4 = 1$, $i^5 = i$, $i^6 = -1$, $i^7 = -i$, $i^8 = 1$, and so on.

(b)   Divide n by 4. If the remainder is r, then $i^n = i^r$, where $i^0 = 1$, $i^1 = i$, $i^2 = -1$ and $i^3 = -i$.

3.4: Quadratic Inequalities

1. (a)   To solve the inequality $x^2 - 1 > 0$, we will first solve the quadratic equation by factoring. $x^2 - 1 = 0 \Rightarrow (x-1)(x+1) = 0 \Rightarrow x = \pm1$ These two boundary numbers separate the number line into three intervals $(-\infty, -1)$, $(-1, 1)$ and $(1, \infty)$. We choose the test values $x = -2, 0$, and 2. The expression $x^2 - 1$ is positive (or greater than zero) when x is in the interval $(-\infty, -1) \cup (1, \infty)$. See table 1 below.

(b)   To solve the inequality $x^2 - 1 < 0$, we will first solve the quadratic equation by factoring. $x^2 - 1 = 0 \Rightarrow (x-1)(x+1) = 0 \Rightarrow x = \pm1$ These two boundary numbers separate the number line into three intervals $(-\infty, -1), (-1, 1)$, and $(1, \infty)$. We choose the test values $x = -2, 0$, and 2. The expression $x^2 - 1$ is negative (or less than zero) when x is in the interval $(-1, 1)$. See Table 1 below.

Table 1

Interval	Test Value	Positive (+) or Negative (−) Result
$-\infty, -1$	-2	$+$
$-1, 1$	0	$-$
$1, \infty$	2	$+$

3. (a)   To solve the inequality $x^2 - 16 \le 0$, we will first solve the quadratic equation by factoring. $x^2 - 16 = 0 \Rightarrow (x-4)(x+4) = 0 \Rightarrow x = \pm4$ These two boundary numbers separate the number line into three intervals $(-\infty, -4]$, $[-4, 4]$, and $[4, \infty)$. We choose the test values $x = -5, 0$, and 5. the expression $x^2 - 16$ is negative (or less than or equal to zero) when x is in the interval $[-4, 4]$. See Table 3 below.

(b)   To solve the inequality $x^2 - 16 \ge 0$, we will first solve the quadratic equation by factoring.

$x^2 - 16 = 0 \Rightarrow (x-4)(x+4) = 0 \Rightarrow x = \pm 4$ These two boundary numbers separate the number line into three intervals $(-\infty, -4]$, $[-4, 4]$, and $[4, \infty)$. We choose test values $x = -5, 0$, and 5. The expression $x^2 - 16$ is positive (or greater than or equal to zero) when x is in the interval $(-\infty, -4] \cup [4, \infty)$. See Table 3 below.

Table 3

Interval	Test Value	Positive (+) or Negative (−) Result
$-\infty, -4$	-5	+
$-4, 4$	0	−
$4, \infty$	5	+

5. (a) To solve the inequality $4 - x^2 > 0$, we will first solve the quadratic equation by factoring.
    $4 - x^2 = 0 \Rightarrow (2-x)(2+x) = 0 \Rightarrow x = \pm 2$ These two boundary numbers separate the number line into three intervals $(-\infty, -2)$, $(-2, 2)$ and $(2, \infty)$. We choose test values $x = -3, 0$, and 3. The expression $4 - x^2$ is positive (or greater than zero) when x is in the interval $(-2, 2)$. See Table 5 below.

    (b) To solve the inequality $4 - x^2 < 0$, we will first solve the quadratic equation by factoring.
    $4 - x^2 = 0 \Rightarrow (2-x)(2+x) = 0 \Rightarrow x = \pm 2$ These two boundary numbers separate the number line into three intervals $(-\infty, -2)$, $(-2, 2)$, and $(2, \infty)$. We choose test values $x = -3, 0$, and 3. The expression $4 - x^2$ is negative (or less than zero) when x is in the interval $(-\infty, -2) \cup (2, \infty)$. See Table 5 below.

Table 5

Interval	Test Value	Positive (+) or Negative (−) Result
$-\infty, -2$	-3	−
$-2, 2$	0	+
$2, \infty$	3	−

7. (a) $x^2 - x - 12 = 0 \Rightarrow (x-4)(x+3) = 0 \Rightarrow x = -3, 4$.

    (b) Using test intervals, $x^2 - x - 12 < 0$ on $(-3, 4)$ or $\{x \mid -3 < x < 4\}$.

    (c) Using test intervals, $x^2 - x - 12 > 0$ on $(-\infty, -3) \cup (4, \infty)$ or $\{x \mid x < -3 \text{ or } x > 4\}$

9. (a) $k^2 - 4 = 0 \Rightarrow k^2 = 4 \Rightarrow k = \pm\sqrt{4} = \pm 2$.

    (b) Using test intervals, $k^2 - 4 \le 0$ on $[-2, 2]$ or $\{k \mid -2 \le k \le 2\}$.

    (c) Using test intervals, $k^2 - 4 \ge 0$ on $(-\infty, -2] \cup [2, \infty)$ or $\{k \mid k \le -2 \text{ or } k \ge 2\}$.

11. (a) $3x^2 + 8x = 0 \Rightarrow x(3x+8) = 0 \Rightarrow x = -\frac{8}{3}, 0$. Using $x = -1$, which is between $-\frac{8}{3}$ and 0, produces
    $3(-1)^2 + 8(-1) = -5 \Rightarrow$ less than 0.

    (b) Using test intervals, $3x^2 + 8x \le 0$ on $\left[-\frac{8}{3}, 0\right]$ or $\left\{x \mid -\frac{8}{3} \le x \le 0\right\}$.

    (c) Using test intervals, $3x^2 + 8x \ge 0$ on $\left(-\infty, -\frac{8}{3}\right] \cup [0, \infty)$ or $\left\{x \mid x \le -\frac{8}{3} \text{ or } x \ge 0\right\}$.

13. (a) $x^2 - 3x + 2 = 0 \Rightarrow (x-1)(x-2) = 0 \Rightarrow x = 1, 2$.

    (b) Using test intervals, $x^2 - 3x + 2 < 0$ on $(1, 2)$ or $\{x \mid 1 < x < 2\}$

(c)   Using test intervals, $x^2 - 3x + 2 > 0$ on $(-\infty, 1) \cup (2, \infty)$ or $\{x \mid x < 1 \text{ or } x > 2\}$

15. (a) $-x^2 + x + 6 = 0 \Rightarrow (-x-2)(x-3) = 0 \Rightarrow x = -2, 3$.

     (b)   Using test intervals, $-x^2 + x + 6 < 0$ on $(-\infty, -2) \cup (3, \infty)$ or $\{x \mid x < -2 \text{ or } x > 3\}$

     (c)   Using test intervals, $-x^2 + x + 6 > 0$ on $(-2, 3)$ or $\{x \mid -2 < x < 3\}$

17. (a) $-4x^2 + 12x - 9 = 0 \Rightarrow (2x-3)(-2x+3) = 0 \Rightarrow x = \dfrac{3}{2}$.

     (b)   Using test intervals, $-4x^2 + 12x - 9 < 0$ on $x \neq \dfrac{3}{2}$ on $\left(-\infty, \dfrac{3}{2}\right) \cup \left(\dfrac{3}{2}, \infty\right)$ or $\left\{x \mid x < \dfrac{3}{2} \text{ or } x > \dfrac{3}{2}\right\}$

     (c)   Using test intervals, $-4x^2 + 12x - 9 > 0$ has no solution

19. (a) $12z^2 - 23z + 10 = 0 \Rightarrow (3z-2)(4z-5) = 0 \Rightarrow z = \dfrac{2}{3}, \dfrac{5}{4}$.

     (b)   Using test intervals, $12z^2 - 23z + 10 \leq 0$ on $\left[\dfrac{2}{3}, \dfrac{5}{4}\right]$ or $\left\{z \mid \dfrac{2}{3} \leq z \leq \dfrac{5}{4}\right\}$

     (c)   Using test intervals, $12z^2 - 23z + 10 \geq 0$ on $\left(-\infty, \dfrac{2}{3}\right] \cup \left[\dfrac{5}{4}, \infty\right)$ or $\left\{z \mid z \leq \dfrac{2}{3} \text{ or } z \geq \dfrac{5}{4}\right\}$

21. (a) $x^2 + 2x - 1 = 0$, using the quadratic formula we get: $x = \dfrac{-2 \pm \sqrt{(2)^2 - 4(1)(-1)}}{2(1)} \Rightarrow$

              $x = \dfrac{-2 \pm \sqrt{8}}{2} \Rightarrow x = \dfrac{-2 \pm 2\sqrt{2}}{2} \Rightarrow x = -1 \pm \sqrt{2}$.

     (b)   Using test intervals, $x^2 + 2x - 1 < 0$ on $(-1-\sqrt{2}, -1+\sqrt{2})$ or $\{x \mid -1-\sqrt{2} < x < -1+\sqrt{2}\}$

     (c)   Using test intervals, $x^2 + 2x - 1 > 0$ on $(-\infty, -1-\sqrt{2}) \cup (-1+\sqrt{2}, \infty)$ or $\{x \mid x < -1-\sqrt{2} \text{ or } x > -1+\sqrt{2}\}$

23. (a)   The inequality $f(x) < 0$ is satisfied when the graph of f is below the x-axis. This occurs when $-3 < x < 2$.

     (b)   The inequality $f(x) \geq 0$ is satisfied when the graph of f is above the x-axis or intersects it. This occurs when $x \leq -3$ or $x \geq 2$.

25. (a)   The inequality $f(x) \leq 0$ is satisfied when the graph of f is below the x-axis or intersects it. This occurs only when $x = -2$.

     (b)   The inequality $f(x) > 0$ is satisfied when the graph of f is above the x-axis. This occurs when $x \neq -2$.

27. (a)   The inequality $f(x) > 0$ is satisfied when the graph of f is above the x-axis. Since the graph of f is always below the x-axis, this inequality has no solutions.

     (b)   The inequality $f(x) < 0$ is satisfied when the graph of f is below the x-axis. This occurs for all real values of x.

29. (a)   The x-intercepts are the solutions to $f(x) = 0$: $0 = 2x^2 + 6x + \dfrac{5}{2} \Rightarrow 4x^2 + 12x + 5 = 0$.

              $x = \dfrac{-12 \pm \sqrt{(12)^2 - 4(4)(5)}}{2(4)} \Rightarrow x = \dfrac{-12 \pm \sqrt{64}}{8} \Rightarrow x = \dfrac{-12 \pm 8}{8} \Rightarrow x = \dfrac{-4}{8} \Rightarrow$

              $x = -\dfrac{1}{2}$ or $x = -\dfrac{20}{8} \Rightarrow x = -2\dfrac{1}{2}, x = -2\dfrac{1}{2}, -\dfrac{1}{2}$

(b) The inequality $f(x)<0$ is satisfied when the graph of f is above the x-axis. This occurs in the interval $\left(-\dfrac{5}{2},-\dfrac{1}{2}\right)$ or $\left\{x\middle|-\dfrac{5}{2}<x<-\dfrac{1}{2}\right\}$.

(c) The inequality $f(x)>0$ is satisfied when the graph of f is above the x-axis. This occurs in the interval $\left(-\infty,-\dfrac{5}{2}\right)\cup\left(-\dfrac{1}{2},\infty\right)$ or $\left\{x\middle|x<-\dfrac{5}{2}\text{ or }x>-\dfrac{1}{2}\right\}$.

31. (a) The x-intercepts are the solutions to $f(x)=0$: $0=-5x^2+2x+7$.

$x=\dfrac{-2\pm\sqrt{(2)^2-4(-5)(7)}}{2(-5)}\Rightarrow x=\dfrac{-2\pm\sqrt{144}}{-10}\Rightarrow x=\dfrac{-2\pm12}{-10}\Rightarrow x=-1,\dfrac{7}{5}$

(b) The inequality $f(x)<0$ is satisfied when the graph of f is above the x-axis. This occurs in the interval $(-\infty,-1)\cup\left(\dfrac{7}{5},\infty\right)$ or $\left\{x\middle|x<-1\text{ or }x>\dfrac{7}{5}\right\}$

(c) The inequality $f(x)>0$ is satisfied when the graph of f is above the x-axis. This occurs in the interval $\left(-1,\dfrac{7}{5}\right)$ or $\left\{x\middle|-1<x<\dfrac{7}{5}\right\}$

33. (a) $f(x)>0$ when $x<-1$ or $x>1$
    (b) $f(x)\le0$ when $-1\le x\le1$

35. (a) $f(x)>0$ when $-6<x<-2$
    (b) $f(x)\le0$ when $x\le-6$ or $x\ge-2$

37. Start by solving $x^2+2x-35=0\Rightarrow(x-5)(x+7)=0\Rightarrow x=-7$ or 5; thus, $x^2+2x-35\le0$ when $-7\le x\le5$.

39. Start by solving $x^2+6x-16=0\Rightarrow(x+8)(x-2)=0\Rightarrow x=-8$ or 2; thus, $x^2+6x-16\ge0$ when $x\le-8$ or $x\ge2$.

41. Start by solving $-x^2-x+20=0\Rightarrow(-x+4)(x+5)=0\Rightarrow x=-5$ or 4; thus, $-x^2-x+20>0$ when $-5<x<4$.

43. Start by solving $2x^2+9x-5=0\Rightarrow(2x-1)(x+5)=0\Rightarrow x=-5$ or $\dfrac{1}{2}$; thus, $2x^2+9x-5\ge0$ when $x\le-5$ or $x\ge\dfrac{1}{2}$.

45. Start by solving $2x^2+5x+2=0\Rightarrow(2x+1)(x+2)=0\Rightarrow x=-2$ or -0.5; thus, $2x^2+5x+2\le0$ when $-2\le x\le-0.5$.

47. Start by solving $x^2+x-6=0\Rightarrow(x+3)(x-2)=0\Rightarrow x=-3$ or 2; thus, $x^2+x-6>0$ when $x<-3$ or $x>2$.

49. Start by solving the equation $x^2=4\Rightarrow x=\pm2$. Next write the quadratic inequality with $a>0$. $x^2\le4\Rightarrow x^2-4\le0$. The graph of $y=x^2-4$ is a parabola opening up. It will be less than or equal to 0 between (and including) the solutions to the equation. The solutions are $-2\le x\le2$.

51. Since $x(x-4)\ge4\Rightarrow x^2-4x+4\ge0\Rightarrow(x-2)^2\ge0$, the solutions are all real numbers.

53. Start by solving the equation $-x^2+x+6=0\Rightarrow x^2-x-6=0\Rightarrow(x+2)(x-3)=0\Rightarrow x=-2,3$. The graph of $y=-x^2+x+6$ is a parabola opening downward. It will be below the x-axis or intersect it outside of the solutions of the equation. The solutions are $x\le-2$ or $x\ge3$.

55.   Start by solving the equation $6x^2 - x = 1 \Rightarrow 6x^2 - x - 1 = 0 \Rightarrow (3x+1)(2x-1) = 0 \Rightarrow x = -\frac{1}{3}, \frac{1}{2}$. The graph of

$y = 6x^2 - x - 1$ is a parabola opening up. It will be below the x-axis between the solutions of the equation.

The solutions to $6x^2 - x < 1$ are $-\frac{1}{3} < x < \frac{1}{2}$.

57.   Start by solving the equation $(x+4)(x-10) = 0 \Rightarrow x = -4, 10$. The graph of $y = (x+4)(x-10) = x^2 - 6x - 40$
is a parabola opening up. It will intersect or be below the x-axis between (and including) the solutions of the
equation. The solutions to $(x+4)(x-10) \le 0$ are $-4 \le x \le 10$.

59.   Since $2x^2 + 4x + 3 > 0$ for all values of x, there are no solutions to $2x^2 + 4x + 3 < 0$.

61.   Start by solving the equation $9x^2 - 12x + 4 = 0 \Rightarrow (3x-2)(3x-2) = 0 \Rightarrow x = \frac{2}{3}$. The graph of

$y = 9x^2 - 12x + 4 = (3x-2)^2$ is a parabola opening up and intersecting the x-axis at $x = \frac{2}{3}$. It will always be

above the x-axis except at $x = \frac{2}{3}$. The solutions to $9x^2 + 4 > 12x$ are all real numbers except $\frac{2}{3} \Rightarrow x \ne \frac{2}{3}$.

63.   Start by solving the equation $x^2 = x \Rightarrow x^2 - x = 0 \Rightarrow x(x-1) = 0 \Rightarrow x = 0, 1$. The graph of $y = x^2 - x$ is a
parabola opening up. It will be above or intersect the x-axis outside (and including) the solutions of the
equation. The solutions to $x^2 \ge x$ are $x \le 0$ or $x \ge 1$.

65.   Start by solving the equation $x(x-1) - 6 = 0 \Rightarrow x^2 - x - 6 = 0 \Rightarrow (x+2)(x-3) = 0 \Rightarrow x = -2, 3$. The graph of
$y = x^2 - x - 6$ is a parabola opening up. It will be above or intersect the x-axis outside (and including) the
solutions of the equation. The solutions to $x(x-1) \ge 6$ are $x \le -2$ or $x \ge 3$.

67.   Start by solving the equation $x^2 - 5 = 0 \Rightarrow x^2 = 5 \Rightarrow x = \pm\sqrt{5}$. The graph of $y = x^2 - 5$ is a parabola opening
up. It will intersect or be below the x-axis between (and including) the solutions of the equation. The
solutions are $-\sqrt{5} \le x \le \sqrt{5}$.

69.   Start by solving

$$7x^2 + 515.2 = 179.8x \Rightarrow 7x^2 - 179.8x + 515.2 = 0 \Rightarrow \frac{-b \pm \sqrt{b^2 - 4ac}}{2a} \Rightarrow \frac{-(-179.8) \pm \sqrt{(-179.8)^2 - 4(7)(515.2)}}{2(7)} \Rightarrow$$

$\frac{179.8 \pm \sqrt{17902.44}}{14} = \frac{179.8 \pm 133.8}{14} \Rightarrow x = \frac{23}{7}, 22.4$. The graph of $7x^2 - 179.8x + 515.2 = y$ is a parabola

opening upward. It will intersect or be above the x-axis on the ends of the solutions, but not in-between. The

solutions are $x \le \frac{23}{7}$ or $x \ge 22.4$.

71.   For $x^2 - 9x + 14 \le 0$, first solve $x^2 - 9x + 14 = 0 \Rightarrow (x-7)(x-2) = 0 \Rightarrow x = 2, 7$. These boundary numbers
give us the disjoint intervals: $(-\infty, 2), (2, 7), (7, \infty)$. See Figure 71. $x^2 - 9x + 14 \le 0$ when $2 \le x \le 7$.

Interval	Test Value x	$x^2 - 9x + 14$	Positive or Negative?
$(-\infty, 2)$	0	14	Positive
$(2, 7)$	4	-6	Negative
$(7, \infty)$	10	24	Positive

Figure 71

73. For $x^2 \geq 3x + 10 \Rightarrow x^2 - 3x - 10 \geq 0$, first solve $x^2 - 3x - 10 = 0 \Rightarrow (x-5)(x+2) = 0 \Rightarrow x = -2, 5$. These boundary numbers give us the disjoint intervals: $(-\infty, -2), (-2, 5), (5, \infty)$. See Figure 73. $x^2 - 3x - 10 \geq 0$ when $x \leq -2$ or $x \geq 5$.

Interval	Test Value x	$x^2 - 3x - 10$	Positive or Negative?
$(-\infty, -2)$	-3	8	Positive
$(-2, 5)$	0	-10	Negative
$(5, \infty)$	6	8	Positive

Figure 73

75. For $\dfrac{1}{8}x^2 + x + 2 \geq 0 \Rightarrow x^2 + 8x + 16 \geq 0$, first solve $x^2 + 8x + 16 = 0 \Rightarrow (x+4)(x+4) = 0 \Rightarrow x = -4$. This boundary gives us the disjoint intervals: $(-\infty, -4), (-4, \infty)$. See Figure 75. $\dfrac{1}{8}x^2 + x + 2 \geq 0$ for all real numbers.

Interval	Test Value x	$1/8x^2 + x + 2$	Positive or Negative?
$(-\infty, -4)$	-8	2	Positive
$(-4, \infty)$	0	2	Positive

Figure 75

77. For $x^2 > 3 - 4x \Rightarrow x^2 + 4x - 3 > 0$, first solve $x^2 + 4x - 3 = 0$. Using the quadratic formula we get:

$$x = \frac{-4 \pm \sqrt{(4)^2 - 4(1)(-3)}}{2(1)} \Rightarrow x = \frac{-4 \pm \sqrt{28}}{2} \Rightarrow x = -2 + \sqrt{7} \text{ or } x = -2 - \sqrt{7}.$$ These boundary numbers give us the disjoint intervals: $(-\infty, -2-\sqrt{7}), (-2-\sqrt{7}, -2+\sqrt{7}), (-2+\sqrt{7}, \infty)$. See Figure 77. $x^2 + 4x - 3 > 0$ when $x < -2 - \sqrt{7}$ or $x > -2 + \sqrt{7}$.

Interval	Test Value x	$x^2 + 4x - 3$	Positive or Negative?
$(-\infty, -4.6)$	-5	2	Positive
$(-4.6, 0.6)$	0	-3	Negative
$(0.6, \infty)$	1	2	Positive

Figure 77

79. The area of the square is given as follows: $A = x^2 \leq 289 \Rightarrow x^2 - 289 \leq 0 \Rightarrow (x+17)(x-17) \leq 0$. The parabola of $y = x^2 - 289$ opens upward and the x-intercepts are $x = -17$ and $x = 17$. Therefore, $x^2 - 289 \leq 0$ when x is $17 \leq x \leq 17$. Given that the width cannot be negative or 0, the only possible values for the width are given by the interval $0 < x \leq 17$ feet.

81. (a) $s(t) = -16t^2 + 80t$. Except for the initial value of $t = 0$, the height of the ball is zero at

$$s(t) = 0 = -16t^2 + 80t \Rightarrow t \geq \frac{80}{16} = 5,$$ so the ball struck the ground at or after 5 seconds.

(b) $s(t) = 64 = -16t^2 + 80t \Rightarrow -16t^2 + 80t - 64 = 0 \Rightarrow t^2 - 5t + 4 = 0 \Rightarrow (t-4)(t-1) = 0$, so $x = 1$ or $x = 4$. The baseball is above 52 feet in the interval $1 \leq t \leq 4$ seconds.

83.   Graph $Y_1 = (1/9)X^2 + (11/3)X$, $Y_2 = 300$ in [30, 40, 1] by [0, 400, 100] (Not Shown). The intersection is (38.018, 300). Thus, the safe stopping speed is to be no more than 38 mph. The speed limit might be 35 mph.

85.   To find the possible values of the radius of the can, solve the following inequality:

$24\pi \le \pi(r^2)6 \le 54\pi \Rightarrow 24 \le 6r^2 \le 54 \Rightarrow 4 \le r^2 \le 9 \Rightarrow 2 \le r \le 3$. The possible values of r range between 2 and 3 inches.

87.   (a) $f(0) = \dfrac{4}{5}(0-10)^2 + 80 = 160$ and $f(2) = \dfrac{4}{5}(2-10)^2 + 80 = 131.2$. Initially when the person stops exercising the heart rate is 160 beats per minute, and after 2 minutes the heart rate has dropped to about 131 beats per minute.

      (b)   Graph $Y_1 = 80, Y_2 = \dfrac{4}{5}(x-10)^2 + 80$ and $Y_3 = 100$ (Not shown). To find the points of intersection, use the calc function on the calculator. The person's heart rate is between 80 and 100 when the graph of Y_2 is between the graphs of Y_1 and Y_3. This occurs between approximately 5 minutes and 10 minutes after the person stops exercising, so $5 \le x \le 10$.

89.   Graph $Y_1 = 2375X^2 + 5134X - 5020, Y_2 = 90,000$ and $Y_3 = 200,000$ in [−1, 11, 1] by [5000, 210000, 10000] (Not shown). The intersection points are approximately (4.9977492, 90000) and (8.044127, 200,000). From the graphs, we see that $90,000 \le Y_1 \le 200,000$ when $4.998 \le X \le 8.044$. Since 1984 corresponds to $x = 0$, the number of AIDS deaths was from 90,000 to 200,000 for the years from 1989 to 1992.

91.   To find when the sales were between 50 and 55 million iPods, graph the function. $Y_1 = 50, Y_2 = -2.277x^2 + 11.71x + 40.4, Y_3 = 55$ in the same window. The intersection points are approximately (1.02, 50), (2.12, 55), and (4.11, 50). Since x represents the number of years 2006, the result is from 2007 to 2008 and 2009 to 2010. See Figures 91c and 91d.

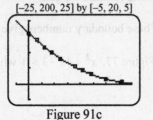
Figure 91c

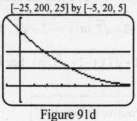
Figure 91d

Checking Basic Concepts for Sections 3.3 and 3.4

1.   (a) $\sqrt{-25} = \sqrt{25}\sqrt{-1} = 5i$

     (b) $\sqrt{-3} \cdot \sqrt{-18} = (\sqrt{3}\sqrt{-1})(\sqrt{9}\sqrt{2}\sqrt{-1}) = (i\sqrt{3})(3i\sqrt{2}) = 3i^2\sqrt{6} = -3\sqrt{6}$

     (c) $\dfrac{7+\sqrt{-98}}{14} = \dfrac{7\pm\sqrt{49}\sqrt{2}\sqrt{-1}}{14} = \dfrac{7\pm 7i\sqrt{2}}{14} = \dfrac{1}{2}\pm\dfrac{i\sqrt{2}}{2}$

3.   (a)   The x-intercepts are −3 and 0. The graph of $y = f(x)$ is below the x-axis between these values, so $f(x) \le 0$ on the interval $[-3, 0]$. The graph of $y = f(x)$ is above the x-axis outside of these values, so $f(x) > 0$ on the intervals $(-\infty, -3) \cup (0, \infty)$ or $\{x \mid x < -3 \text{ or } x > 0\}$.

     (b)   The x-intercept is −2. The graph of $y = f(x)$ is always below the x-axis for all real numbers except at $x = -2$. That is, $f(x) \le 0$ on $(-\infty, \infty)$. The graph of $y = f(x)$ is never above the x-axis, so there are no solutions to $f(x) > 0$.

5.   (a)   Start by solving the equation $x^2 - 36 = 0 \Rightarrow x^2 = 36 \Rightarrow x = \pm 6$. The graph of $y = x^2 - 36$ is a parabola opening up and y is positive when x is greater than 6 or less than negative 6. The solution is $(-\infty, -6] \cup [6, \infty)$ or $\{x \mid x \le -6 \text{ or } x \ge 6\}$.

   (b)   Start by solving the equation $4x^2 + 9 = 9x \Rightarrow 4x^2 - 9x + 9 = 0$. Using the quadratic formula we get:
   $$x = \frac{9 \pm \sqrt{81 - 4(4)(9)}}{2(4)} \Rightarrow x = \frac{9 \pm \sqrt{-63}}{8} \Rightarrow \text{ no real solutions} \Rightarrow \text{ no intersection with the } x\text{-axis. This is}$$
   a parabola opening up completely above the x-axis $\Rightarrow$ the solutions to $4x^2 + 9 > 9x$ are all real numbers, or $(-\infty, \infty)$.

   (c)   Start by solving the equation $2x(x-1) = 2 \Rightarrow 2x^2 - 2x = 2 \Rightarrow 2x^2 - 2x - 2 = 0$. Using the quadratic
   formula we get: $x = \dfrac{2 \pm \sqrt{4 - 4(2)(-2)}}{2(2)} \Rightarrow x = \dfrac{2 \pm \sqrt{20}}{4} \Rightarrow \dfrac{1 \pm \sqrt{5}}{2}$. The graph of $y = 2x^2 - 2x - 2$ is a
   parabola opening up. It will be below the x-axis between the solutions of the equation. The solutions to
   $2x(x-1) \le 2$ is $\left[\dfrac{1 - \sqrt{5}}{2}, \dfrac{1 + \sqrt{5}}{2}\right]$ or $\left\{x \mid \dfrac{1 - \sqrt{5}}{2} \le x \le \dfrac{1 + \sqrt{5}}{2}\right\}$.

3.5: Transformations of Graphs

1.   The vertex is $(-2, 0)$. The parabola $y = x^2$ has been shifted 2 units to the left. The equation is $y = (x+2)^2$.

3.   The endpoint is $(-3, 0)$. The curve $y = \sqrt{x}$ has been shifted 3 units to the left. The equation is $y = \sqrt{x+3}$.

5.   The vertex is $(-2, -1)$. The graph of $y = |x|$ has been shifted 2 units to the left and 1 unit down.
   The equation is $y = |x+2| - 1$.

7.   The endpoint is $(-2, -3)$. The curve $y = \sqrt{x}$ has been shifted 2 units to the left and 3 units down.
   The equation is $y = \sqrt{x+2} - 3$.

9. $y = (x-h)^2 + k$ describes a transformation of the parent quadratic function, which for positive h and k shifts the curve to the right and upward, as depicted by figure c.

11. $y = \sqrt{x+h} - k$ describes a transformation of the parent square root function, which for positive h and k shifts the curve to the left and downward, as depicted by figure a.

13. To shift the graph of $f(x) = x^2$ to the right 2 units, replace x with $(x-2)$ in the formula for $f(x)$.
   This results in $y = f(x-2) = (x-2)^2$. To shift the graph of this new equation down 3 units, subtract 3 from the formula to obtain $y = f(x-2) - 3 = (x-2)^2 - 3$. See Figure 13.

15. To shift the graph of $f(x) = x^2 - 4x + 1$ to the left 6 units, replace x with $(x+6)$ in the formula for $f(x)$.
   This results in $y = f(x+6) = (x+6)^2 - 4(x+6) + 1$. To shift the graph of this new equation up 4 units, add 4 to the formula to obtain $y = f(x+6) + 4 = (x+6)^2 - 4(x+6) + 1 + 4 = (x+6)^2 - 4(x+6) + 5$. See Figure 15.

17. To shift the graph of $f(x) = \dfrac{1}{2}x^2 + 2x - 1$ to the left 3 units, replace x with $(x+3)$ in the formula for $f(x)$.
   The result is $y = f(x+3) = \dfrac{1}{2}(x+3)^2 + 2(x+3) - 1$. To shift the graph of this new equation down 2 units,

subtract 2 to obtain $y = f(x+3) - 2 = \frac{1}{2}(x+3)^2 + 2(x+3) - 1 - 2 \Rightarrow y = \frac{1}{2}(x+3)^2 + 2(x+3) - 3$. See Figure 17.

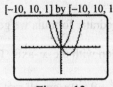

[−10, 10, 1] by [−10, 10, 1]

Figure 13

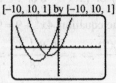

[−10, 10, 1] by [−10, 10, 1]

Figure 15

[−10, 10, 1] by [−10, 10, 1]

Figure 17

19. (a) $g(x) = 3(x+3)^2 + 2(x+3) - 5$

    (b) $g(x) = 3x^2 + 2x - 9$

21. (a) $g(x) = 2(x-2)^2 + 4$

    (b) $g(x) = 2(x+8)^2 - 5$

23. (a) $g(x) = 3(x-2000)^2 - 3(x-2000) + 72$

    (b) $g(x) = 3(x+300)^2 - 3(x+300) - 28$

25. (a) $g(x) = -\sqrt{x-4}$

    (b) $g(x) = \sqrt{-x+2}$

27. $(x-3)^2 + (y+4)^2 = 4$; Center: $(3, -4)$; Radius $= 2$

29. $(x+5)^2 + (y-3)^2 = 5$; Center: $(-5, 3)$; Radius $= \sqrt{5}$

31. (a) See Figure 31a.
    (b) See Figure 31b.
    (c) See Figure 31c.

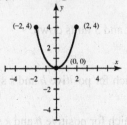

Figure 31a

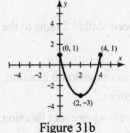

Figure 31b

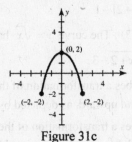

Figure 31c

33. (a) See Figure 33a.
    (b) See Figure 33b.
    (c) See Figure 33c.

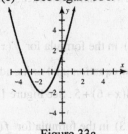

Figure 33a

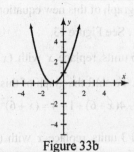

Figure 33b

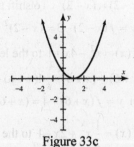

Figure 33c

35. (a) See Figure 35a.
    (b) See Figure 35b.
    (c) See Figure 35c.

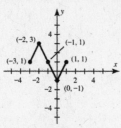

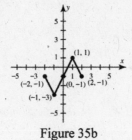

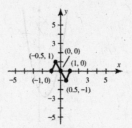

Figure 35a

Figure 35b

Figure 35c

37. (a)   See Figure 37a.
     (b)   See Figure 37b.
     (c)   See Figure 37c.

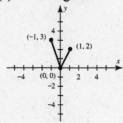

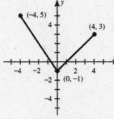

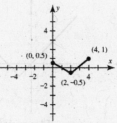

Figure 37a

Figure 37b

Figure 37c

39. The graph of $y = -f(x)$ is a reflection of the graph of $y = f(x)$ across the x-axis. Therefore the new equation is $y = -\sqrt{x} - 1$. See Figure 39a and Figure 39b.

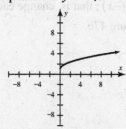

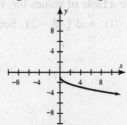

Figure 39a

Figure 39b

41. The graph of $y = f(-x)$ is a reflection of the graph of $y = f(x)$ across the x-axis. Therefore the new equation is $y = x^2 + x$. See Figure 41a and Figure 41b.

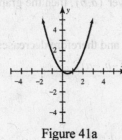

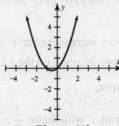

Figure 41a

Figure 41b

43. To reflect this function across the x-axis, graph $y = -f(x) = -(x^2 - 2x - 3) = -x^2 + 2x + 3$. See Figure 43a.

To reflect this function across the y-axis, graph $y = f(-x) = (-x)^2 - 2(-x) - 3 = x^2 + 2x - 3$. See Figure 43b.

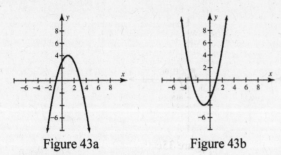

Figure 43a   Figure 43b

45. To reflect this function across the x-axis, graph $y = -f(x) = -(|x+1|-1) = -|x+1|+1$. See Figure 45a.

To reflect this function across the y-axis, graph $y = f(-x) = |-x+1|-1 = |-x+1|-1$. See Figure 45b.

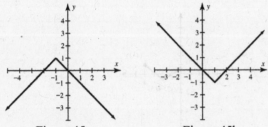

Figure 45a   Figure 45b

47. (a) To reflect this line graph across the x-axis, make a table of values for $y = -f(x)$; that is, change each (x, y) to $(x, -y)$. Plot the points $(-3, -2)$, $(-1, -3)$, $(1, 1)$ and $(2, 2)$. See Figure 47a.

(b) To reflect this line graph across the y-axis, make a table of values for $y = f(-x)$; that is, change each (x, y) to $(-x, y)$. Plot the points $(3, 2)$, $(1, 3)$, $(-1, -1)$, and $(-2, -2)$. See Figure 47b.

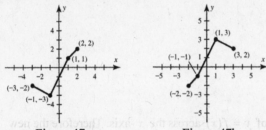

Figure 47a   Figure 47b

49. Because the function $-f(x)$ has negative values over (a, b), if $f(x)$ increases over (a, b), then the graph of $y = -f(x)$ must decrease on the interval (a, b).

51. Because the function $f(x)$ increases over (a, b), $f(-x)$ increases over $(-a, -b)$, and therefore decreases over $(-b, -a)$. The graph of $y = -f(-x)$ must therefore increase on the interval $(-b, -a)$.

53. Shift the graph of $y = x^2$ right 3 units and upward 1 unit.

55. Shift the graph of $y = x^2$ left 1 unit and vertically shrink it with factor $\dfrac{1}{4}$.

57. Reflect the graph of $y = \sqrt{x}$ across the x-axis and shift it left 5 units.

59. Reflect the graph of $y = \sqrt{x}$ across the y-axis and vertically stretch it with factor 2.

61. Reflect the graph of $y = |x|$ across the y-axis and shift it left 1 unit.

63. Shift the graph of $y = x^2$ down 3 units. The vertex is located at $(0, -3)$. See Figure 63.

65. Shift the graph of $y = x^2$ to the right 5 units and up 3 units. The vertex is located at $(5, 3)$. See Figure 65.

67. Reflect the graph of $y = \sqrt{x}$ across the x-axis. See Figure 67.

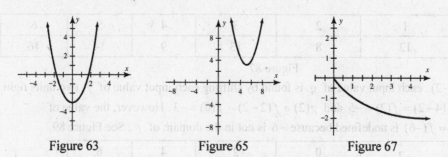

Figure 63   Figure 65   Figure 67

69. Reflect the graph of $y = x^2$ across the x-axis and shift the graph 4 units up. See Figure 69.

71. Shift the graph of $y = |x|$ down 4 units. See Figure 71.

73. Shift the graph of $y = \sqrt{x}$ to the right 3 units and up 2 units. See Figure 73.

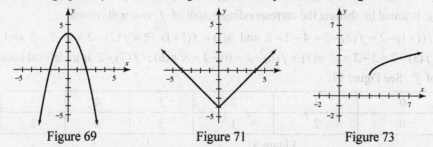

Figure 69   Figure 71   Figure 73

75. The graph is like $f(x) = |x|$ but narrower by a factor of 2, vertex $(0, 0)$. See Figure 75.

77. Reflect the graph of $f(x) = \sqrt{x}$ across the x-axis and then shift up 1 unit, the vertex is $(0, 1)$. See Figure 77.

79. Shift the graph of $f(x) = \sqrt{x}$ left 1 unit, then reflect this graph across the y-axis and then reflect this graph across the x-axis. See Figure 79.

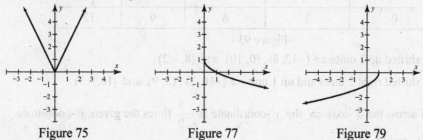

Figure 75   Figure 77   Figure 79

81. Reflect the graph of $f(x) = \sqrt{x}$ across the y-axis, then shift left 1 unit. See Figure 81.

83. Shift the graph of $f(x) = x^3$ right 1 unit. See Figure 83.

85. Reflect the graph of $f(x) = x^3$ across the x-axis. See Figure 85.

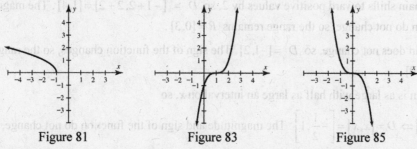

Figure 81   Figure 83   Figure 85

87. Since $g(x) = f(x) + 7$, the output for $g(x)$ can be determined by adding 7 to the output of $f(x)$. See Figure 87.

x	1	2	3	4	5	6
$g(x)$	12	8	13	9	14	16

Figure 87

89. Since $g(x) = f(x-2)$, each input value of g is found by shifting each input value of f two units right. For example, $g(4) = f(4-2) = f(2) = -5$ and $g(2) = f(2-2) = f(0) = -3$. However, the value of $g(-4) = f(-4-2) = f(-6)$ is undefined because -6 is not in the domain of f. See Figure 89.

x	-2	0	2	4	6
$g(x)$	5	2	-3	-5	-9

Figure 89

91. Since $g(x) = f(x+1) - 2$, each input value of g is found by shifting each input value of f one unit left. Then each output of g is found by shifting the corresponding output of f two units down.

For example, $g(1) = f(1+1) - 2 = f(2) - 2 = 4 - 2 = 2$ and $g(1) = f(1+1) - 2 = f(2) - 2 = 4 - 2 = 2$ and $g(2) = f(2+1) - 2 = f(3) - 2 = 3 - 2 = 1$. $g(5) = f(6) - 2 = 10 - 2 = 8$; $g(6) = f(7) - 2$ is undefined because 7 is not in the domain of f. See Figure 91.

x	0	1	2	3	4	5
$g(x)$	0	2	1	5	6	8

Figure 91

93. Since $g(x) = f(-x) + 1$, each input value of g is the opposite of each input value of f. Then each output of g is found by shifting the corresponding output of f one unit up. For example, $g(-2) = f(-(-2)) + 1 = f(2) + 1 = -1 + 1 = 0$ and $g(-1) = f(-(-1)) + 1 = f(1) + 1 = 2 + 1 = 3$. See Figure 93.

x	-2	-1	0	1	2
$g(x)$	0	3	6	9	12

Figure 93

95. The graph of $f(x)$ is shifted up 2 units $\Rightarrow (-12, 8)$, $(0, 10)$, and $(8, -2)$.

97. The graph of $f(x)$ is shifted right 2 units and up 1 unit $\Rightarrow (-10, 7)$, $(2, 9)$, and $(10, -3)$.

99. The graph is reflected across the x-axis $\Rightarrow$ the y-coordinate is $-\dfrac{1}{2}$ times the given y-coordinate $\Rightarrow (-12, -3)$, $(0, -4)$, and $(8, 2)$.

101. The graph is reflected across the y-axis $\Rightarrow$ the x-coordinate is the reciprocal of -2 or $-\dfrac{1}{2}$ times the given x-coordinate $\Rightarrow (6, 6)$, $(0, 8)$, and $(-4, -4)$.

103. For $f(x-2)$, the domain shifts toward positive values by 2, so $D = [-1+2, 2+2] = [1, 4]$. The magnitude and sign of the function do not change, so the range remains $R = [0, 3]$.

105. For $-f(x)$, the domain does not change, so $D = [-1, 2]$. The sign of the function changes, so the range is $R = [-1 \cdot 0, -1 \cdot 3] = [-3, 0]$.

107. For $f(2x)$, the domain is as large with half as large an interval on x, so

$$D = [-1, 2] = [2x, 2x] \Rightarrow D = [x, x] = \left[-\frac{1}{2}, 1\right].$$ The magnitude and sign of the function do not change, so the range remains $R = [0, 3]$.

109. For $f(-x)$, the domain values change sign, so $D = [-1 \cdot (-1), -1 \cdot 2] = [-2, 1]$. The magnitude and sign of the function do not change, so the range remains $R = [0, 3]$.

111. For $f(x) = -(x+1)^2 - 5$, the domain is not restricted, and so consists of all real numbers, whereas the range extends from negative infinity to and including a maximum of -5. ($D : (-\infty, \infty)$; $R : (-\infty, -5]$).

113. For $f(x) = \sqrt{-x-4} - 2$, the domain extends from negative infinity to, and including, -4, whereas the range extends from, and includes, -2 to infinity. ($D : (-\infty, -4]$; $R : [-2, \infty)$).

115. Let $(2009, 20.7)$ be the vertex (h, k). Translate the graph of $y = x^2$ to the right 2009 units and up 20.7 units to find $f(x) = a(x - 2009)^2 + 20.7$. Choose the point $(2014, 169)$ to determine the value for a as follows: $169 = a(2014 - 2009)^2 + 20.7 \Rightarrow 148.3 = 25a \Rightarrow a \approx 6$. The function is $f(x) = 6(x - 2009)^2 + 20.7$.

117. Let $(2008, 22)$ be the vertex (h, k). Translate the graph of $y = x^2$ to the right 2008 units up 22 units to find $f(x) = a(x - 2008)^2 + 22$. Choose the point $(2014, 66)$ to determine the value for a as follows: $66 = a(2014 - 2008)^2 + 22 \Rightarrow 44 = 36a \Rightarrow a \approx 1.1$. The function is $f(x) = 1.1(x - 2008)^2 + 22$.

119. We must determine a function g such that $g(1991) = P(1), g(1992) = P(2)\ldots$. Thus, the relationship $g(x) = P(x - 1990)$ holds for $x = 1990, 1991, 1992\ldots$. It follows that the representation for $g(x)$ is $g(x) = 0.00075(x - 1990)^2 + 0.17(x - 1990) + 44$.

121. In 15 seconds, the plane has moved $15(0.2) = 3$ kilometers to the left. To show this movement, translate the graph of the mountain 3 kilometers (units) to the right; $y = -0.4(x - 3)^2 + 4$. See Figure 121.

123. (a) In 4 hours, the cold front has moved $4(40) = 160$ miles, which is $\frac{160}{100} = 1.6$ units on the graph. To show this movement, shift the graph 1.6 units down; $y = \frac{1}{20}x^2 - 1.6$. See Figure 123a.

(b) The graph of the cold front has shifted $\frac{250}{100} = 2.5$ units down and $\frac{210}{100} = 2.1$ units to the right. To show this movement, the new equation should be $y = \frac{1}{20}(x - 2.1)^2 - 2.5$. The location of Columbus, Ohio is at $(5.5, -0.8)$ relative to the location of Des Moines at $(0, 0)$. Since the parabola representing the cold front at midnight has past the point $(5.5, -0.8)$, the front reaches Columbus by midnight. See Figure 123b.

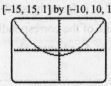 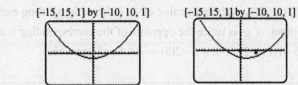

[−4, 4, 1] by [0, 6, 1]   [−15, 15, 1] by [−10, 10, 1]   [−15, 15, 1] by [−10, 10, 1]

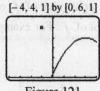

Figure 121   Figure 123a   Figure 123b

125. (a)

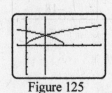

Figure 125

(b) The reflection is of $y = f(x)$ across the line $x = 2$.

127. (a)

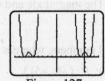

Figure 127

(b) The reflection is of $y = f(x)$ across the line $x = -6$.

Checking Basic Concepts for Section 3.5

1. (a) The graph of $y = (x+4)^2$ is the graph of $f(x) = x^2$ shifted 4 units to the left.

    (b) The graph of $y = x^2 - 3$ is the graph of $f(x) = x^2$ shifted 3 units down.

    (c) The graph of $y = (x-5)^2 + 3$ is the graph of $f(x) = x^2$ shifted 5 units to the right and 3 units up.

3. (a) To shift $y = x^2$ to the right 3 units replace x with $(x-3)$. To shift the graph of the new function down 4 units subtract 4 from y. The new equation is $y = (x-3)^2 - 4$.

    (b) To reflect $y = x^2$ about the y-axis, replace $f(x)$ with $-f(x)$. The new equation is $y = -x^2$.

    (c) To shift $y = x^2$ to the left 6 units replace x with $(x+6)$. To reflect the graph of the new function about the y-axis, replace $f(x)$ with $f(-x)$. The new equation is $y = (-x+6)^2$.

    (d) To reflect the graph of $f(x) = x^2$ about the y-axis, replace x with $-x$. This results in $y = f(-x) = (-x)^2$. To shift the new equation to the left 6 units, replace x with $(x+6)$. This results in $f(-(x+6)) = (-(x+6))^2$, or $y = (x+6)^2$.

5. (a) Since $g(x) = f(x-2) + 3$, each input value of g is found by shifting each input value of f two units right. Then each output of g is found by shifting the corresponding output of f three units up. For example, $g(-2) = f(-2-2) + 3 = f(-4) + 3 = 1 + 3 = 4$. See Figure 5a.

x	−2	0	2	4	6
$g(x)$	4	6	9	11	12

Figure 5a

    (b) Since $h(x) = -2(f(x+1))$, each input value of g is found by shifting each input value of f one unit left. Then each output of g is twice the opposite of the corresponding output of f. For example, $h(-3) = -2(f(-3+1)) = -2(f(-2)) = -2(3) = -6$. See Figure 5b.

x	−5	−3	−1	1	3
$h(x)$	−2	−6	−12	−16	−18

Figure 5b

Chapter 3 Review Exercises

1. (a) The parabola opens downward, so $a < 0$.

    (b) Vertex: (2, 4)

(c)   Axis of symmetry: $x = 2$.

(d) f is increasing for $x < 2$ and decreasing for $x > 2$.

3. $f(x) = -2(x-5)^2 + 1 \Rightarrow f(x) = -2(x^2 - 10x + 25) + 1 \Rightarrow f(x) = -2x^2 + 20x - 49$ The leading coefficient is -2.

5.   The graph of $y = x^2$ has been shifted 1 unit left, reflected across the x-axis, and shifted 2 units up $\Rightarrow f(x) = -(x+1)^2 + 2$.

7. $f(x) = x^2 + 6x - 1 \Rightarrow f(x) = x^2 + 6x + 9 - 9 - 1 \Rightarrow f(x) = (x^2 + 6x + 9) - 10 \Rightarrow$
     $f(x) = (x+3)^2 - 10$; the vertex is $(-3, -10)$.

9. $-\dfrac{b}{2a} = -\dfrac{2}{2(-3)} = \dfrac{1}{3}$ and $f\left(\dfrac{1}{3}\right) = -3\left(\dfrac{1}{3}\right)^2 + 2\left(\dfrac{1}{3}\right) - 4 = -\dfrac{11}{3}$; the vertex is $\left(\dfrac{1}{3}, -\dfrac{11}{3}\right)$.

11. Shift the graph of $f(x) = x^2$ up 3 units, reflect it across the x-axis, and make it narrower by the scale factor 3. The vertex is located at $(0, 3)$. See Figure 11.

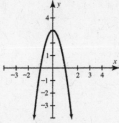

Figure 11

13. Shift the graph of $f(x) = |x|$ left 3 units, and reflect it across the x-axis. The vertex is located at $(-3, 0)$. See Figure 13.

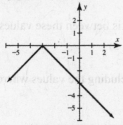

Figure 13

15. Average rate of change $= \dfrac{f(b) - f(a)}{b - a} \Rightarrow \dfrac{f(4) - f(2)}{4 - 2} = \dfrac{-63 - (-5)}{2} = \dfrac{-58}{2} = -29$

17. $x^2 - x - 20 = 0 \Rightarrow (x-5)(x+4) = 0 \Rightarrow x = -4, 5$

19. $4z^2 - 7 = 0 \Rightarrow 4z^2 = 7 \Rightarrow z^2 = \dfrac{7}{4} \Rightarrow z = \pm\sqrt{\dfrac{7}{4}} \Rightarrow z = \pm\dfrac{\sqrt{7}}{2}$

21. $-2t^2 - 3t + 14 = 0 \Rightarrow (2t+7)(-t+2) = 0 \Rightarrow 2t+7 = 0 \Rightarrow 2t = -7 \Rightarrow t = -\frac{7}{2}$ or

$-t + 2 = 0 \Rightarrow -t = -2 \Rightarrow t = 2 \Rightarrow t = -\frac{7}{2}, 2$

23. $0.1x^2 - 0.3x = 1 \Rightarrow x^2 - 3x - 10 = 0 \Rightarrow (x-5)(x+2) = 0 \Rightarrow x = -2, 5$

25. $x^2 + 2x = 5 \Rightarrow x^2 + 2x + 1 = 5 + 1 \Rightarrow (x+1)^2 = 6 \Rightarrow x + 1 = \pm\sqrt{6} \Rightarrow x = -1 \pm \sqrt{6}$

27. $2z^2 - 6z - 1 = 0 \Rightarrow z^2 - 3z = \frac{1}{2} \Rightarrow z^2 - 3z + \frac{9}{4} = \frac{1}{2} + \frac{9}{4} \Rightarrow \left(z - \frac{3}{2}\right)^2 = \frac{11}{4} \Rightarrow$

$z - \frac{3}{2} = \pm\sqrt{\frac{11}{4}} \Rightarrow z = \frac{3}{2} \pm \frac{\sqrt{11}}{2} \Rightarrow z = \frac{3 \pm \sqrt{11}}{2}$

29. $2x^2 - 3y^2 = 6 \Rightarrow -3y^2 = -2x^2 + 6 \Rightarrow y^2 = \frac{2x^2 - 6}{3} \Rightarrow y = \pm\sqrt{\frac{2x^2 - 6}{3}}$

Since there are two output values for each input, y is not a function of x.

31. (a) $\sqrt{-16} = \sqrt{16}\sqrt{-1} = 4i$

(b) $\sqrt{-48} = \sqrt{16}\sqrt{-1}\sqrt{3} = 4i\sqrt{3}$

(c) $\sqrt{-5} \cdot \sqrt{-15} = i\sqrt{5} \cdot i\sqrt{5}\sqrt{3} = i^2(5)\sqrt{3} = 5(-1)\sqrt{3} = -5\sqrt{3}$

33. (a) The x-intercepts of f are $\left(-\frac{5}{2}, 0\right)$ and $\left(\frac{1}{2}, 0\right)$.

(b) The complex zeros of f are $-\frac{5}{2}$ and $\frac{1}{2}$.

35. $4x^2 + 9 = 0 \Rightarrow 4x^2 = -9 \Rightarrow x^2 = -\frac{9}{4} \Rightarrow x = \pm\sqrt{-\frac{9}{4}} = \pm\frac{3}{2}i$

37. (a) The graph of $y = f(x)$ has x-intercepts at $x = -3, 2$. It is above the x-axis between these values, which are solutions to $f(x) = 0$, so $f(x) > 0$ when $-3 < x < 2$ or on $(-3, 2)$.

(b) The graph of $y = f(x)$ is below or intersects the x-axis outside of and including the values where $f(x) = 0$, so $f(x) \le 0$ when $x \le -3$ or $x \ge 2$ or on $(-\infty, -3] \cup [2, \infty)$.

39. Start by solving $x^2 - 3x + 2 = 0 \Rightarrow (x-1)(x-2) = 0 \Rightarrow x = 1$ or 2; thus, $x^2 - 3x + 2 \le 0$ on $[1, 2]$ or $\{x | 1 \le x \le 2\}$. Note that a graph of $y = x^2 - 3x + 2$ intersects or is below the x-axis for values of x between the x-intercepts.

41. Start by solving $n(n-2) = 15 \Rightarrow n^2 - 2n - 15 = 0 \Rightarrow (n-5)(n+3) = 0 \Rightarrow n = -3$ or 5; thus $n(n-2) \ge 15$ on $(-\infty, -3] \cup [5, \infty)$ or $\{x | x \le -3 \text{ or } x \ge 5\}$. Note that a graph of $y = n^2 - 2n - 15$ intersects or is above the n-axis for values of n outside of the n-intercepts.

43. Because $a < 0$, the graph is a parabola opening downward. The discriminant $b^2 - 4ac = 0$ indicates that there is only one real solution, and the vertex is the only zero, on the x-axis. An example graph is shown in Figure 43.

45.   Because $a < 0$, the graph is a parabola opening downward. The discriminant $b^2 - 4ac < 0$ indicates that there are no real solutions, and the vertex is below the x-axis. An example graph is shown in Figure 45.

47.   Because $a > 0$, the graph is a parabola opening upward. The discriminant $b^2 - 4ac > 0$ indicates that there are two real solutions, so the vertex is below the x-axis. An example graph is shown in Figure 47.

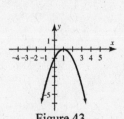
Figure 43

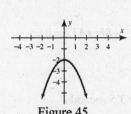
Figure 45

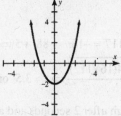
Figure 47

49. $y = -f(x) = -(2x^2 - 3x + 1) = -2x^2 + 3x - 1$. See Figure 49a.

      $y = f(-x) = 2(-x)^2 - 3(-x) + 1 = 2x^2 + 3x + 1$. See Figure 49b.

51.   Shift the graph of $y = x^2$ four units down. See Figure 51.

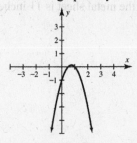
Figure 49a

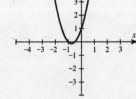
Figure 49b

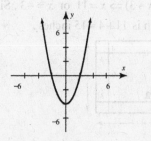
Figure 51

53.   Reflect the graph of $y = x^2$ across the x-axis, shift it two units right and three units up, then stretch it vertically. See Figure 53.

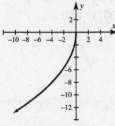
Figure 52

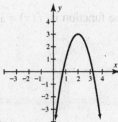
Figure 53

55.   Let $W + L + W = 44 \Rightarrow L = 44 - 2W$. If $A = L \cdot W$ then $A = (44 - 2W)(W) \Rightarrow A = 44W - 2W^2$ or

      $A = -2W^2 + 44W$. Now use the vertex formula: $W = -\dfrac{44}{2(-2)} \Rightarrow W = 11$ and $L = 44 - 2(11) \Rightarrow L = 22$. So the

      dimensions are 11 feet by 22 feet.

57.   (a) $h(t) = -16t^2 + 88t + 5 \Rightarrow h(0) = -16(0)^2 + 88(0) + 5 \Rightarrow h(0) = 5$

      The stone was 5 feet above the ground when it was released.

(b) $h(2) = -16(2)^2 + 88(2) + 5 = 117$; the stone was 117 feet high after 2 seconds.

(c) $t = -\dfrac{b}{2a} = -\dfrac{88}{2(-16)} = 2.75$ and $h(2.75) = -16(2.75)^2 + 88(2.75) + 5 = 126$; the maximum height of the

stone was 126 feet.

(d)
$$h(t) = 117 \Rightarrow 117 = -16t^2 + 88t + 5 \Rightarrow 16t^2 - 88t + 112 = 0 \Rightarrow$$

$$t = \frac{88 \pm \sqrt{(-88)^2 - 4(16)(112)}}{2(16)} \Rightarrow t = 3.5 \text{ or } t = 2$$

The stone was 117 feet high after 2 seconds and after 3.5 seconds.

(e) $D : [0, 5.56]$ (approx); $R : [0, 146]$

59.   Use the sketch shown in Figure 59 to determine the dimensions of the box; let $x =$ width of the metal sheet.
The dimensions of the box are 3 inches by $(x - 6)$ inches by $(x - 2)$ inches, and the volume is 135 cubic
inches.

$135 = 3(x - 6)(x - 2) \Rightarrow 135 = 3x^2 - 24x + 36 \Rightarrow 0 = 3x^2 - 24x - 99 \Rightarrow 0 = x^2 - 8x - 33 \Rightarrow$
$0 = (x - 11)(x + 3) \Rightarrow x = 11$ or $x = -3$. Since $x = -3$ is impossible, the width of the metal sheet is 11 inches
and the length is $11 + 4 = 15$ inches.

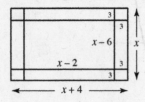

Figure 59

61.   Let $(2007, 60)$ be the vertex (h, k). Translate the graph of $y = x^2$ to the right 2006 units and up 60 units to
find $f(x) = a(x - 2007)^2 + 60$. Choose the point $(2009, 180)$ to determine the value for a as follows:
$180 = a(2009 - 2007)^2 + 60 \Rightarrow 120 = 4a \Rightarrow a = 30$. The function is $f(x) = 30(x - 2007)^2 + 60$.

Chapter 4: More Nonlinear Functions and Equations

4.1: More Nonlinear Functions and Their Graphs

1. $f(x) = 2x^3 - x + 5$ is a polynomial; degree: 3; leading coefficient: 2.

3. $f(x) = \sqrt{x}$ is not a polynomial.

5. $f(x) = 1 - 4x - 5x^4$ is a polynomial; degree: 4; leading coefficient: -5.

7. $g(t) = \dfrac{1}{t^2 + 3t - 1}$ is not a polynomial.

9. $g(t) = 22$ is a polynomial; degree: 0; leading coefficient: 22.

11. (a) A local maximum of approximately 5.5 occurs when $x \approx -2$. A local minimum of approximately -5.5 occurs when $x \approx 2$. *Answers may vary Slightly.*

    (b) There are no absolute extrema.

13. (a) Local maxima of approximately 17 and 27 occur when $x \approx -3$ and 2, respectively. Local minima of approximately -10 and 24 occur when $x \approx -1$ and 3, respectively. *Answers may vary slightly.*

    (b) There are no absolute extrema.

15. (a) Local maxima of approximately 0.5 and 2.8 occur when $x \approx 1$ and -2, respectively. A local minimum of approximately 0 occur when $x = 0$. *Answers may vary slightly.*

    (b) The absolute maximum is 2.8 when $x \approx -2$. There is no absolute minimum.

17. (a) A Local maximum of 0 occurs when $x = 0$. A local minimum of $-1,000$ occurs when $x = \pm 8$.

    (b) There is no absolute maximum. The absolute minimum of $-1,000$ occurs when $x = \pm 8$.

19. (a) Each local minimum is -1. Each local maximum is 1.

    (b) The absolute minimum is -1 and the absolute maximum is 1.

21. (a) A local maximum of 4 occurs when $x \approx \pm 1.4$. A local minimum of 0 occurs when $x = 0$. *Answers may vary slightly.*

    (b) The absolute maximum of 4 occurs when $x \approx \pm 1.4$. There is no absolute minimum.

23. (a) A local maximum of 1 occurs when $x = 0$. The local minima of -1 and -2 occur when $x = -1$ and $x = 1$, respectively.

    (b) The absolute minimum is of -2 occurs when $x = 1$. There is no absolute maximum.

25. (a) A local minimum of approximately -3.2 occurs when $x \approx 0.5$. There are no local maxima. *Answers may vary slightly.*

    (b) The absolute minimum is approximately -3.2. The absolute maximum is 3 and occurs when $x = -2$.

27. (a) Local minima are approximately -0.5 and -2. Local maxima are approximately 0.5 and 2.

    (b) The absolute minimum is -2 and the absolute maximum is 2.

29. (a) There is no local maximum. A local minimum of -2 occurs when $x \approx 1$.

    (b) There is no absolute maximum. An absolute minimum of -2 occurs when $x \approx 1$.

31. (a) A local maximum of 0 occurs when $x = 1$. The local minima of -2 and -1 occur when $x = -1$ and $x = 2$, respectively.

    (b) The absolute minimum of -2 occurs when $x = -1$. An absolute maximum of 3 occurs when $x = 3$.

33. (a) The graph of g is linear $\Rightarrow$ no local extrema.

    (b) The graph of g is linear $\Rightarrow$ no absolute extrema.

35. (a)   The graph of g is a parabola opening up with a vertex $(0, 1) \Rightarrow$ it has a local minimum: 1; and no local maximum.

     (b)   The graph of g is a parabola opening up with a vertex $(0, 1) \Rightarrow$ it has an absolute minimum: 1; and no absolute maximum.

37. (a)   The graph of g is a parabola opening down with a vertex $(-3, 4) \Rightarrow$ it has a local maximum: 4; and no local minimum.

     (b)   The graph of g is a parabola opening down with a vertex $(-3, 4) \Rightarrow$ it has an absolute maximum: 4; and no absolute minimum.

39. (a)   The graph of g is a parabola opening up with a vertex $\left(\frac{3}{4}, -\frac{1}{8}\right) \Rightarrow$ it has a local minimum: $-\frac{1}{8}$; and no local maximum.

     (b)   The graph of g is a parabola opening up with a vertex $\left(\frac{3}{4}, -\frac{1}{8}\right) \Rightarrow$ it has an absolute minimum: $-\frac{1}{8}$; and no absolute maximum.

41. (a)   The absolute value graph opening up has a vertex $(-3, 0) \Rightarrow$ it has a local minimum: 0; and no local maximum.

     (b)   The absolute value graph opening up has a vertex $(-3, 0) \Rightarrow$ it has an absolute minimum: 0; and no absolute maximum.

43. (a)   The graph of g is the basic $\sqrt[3]{x}$ graph and has no local extrema.

     (b)   The graph of g is the basic $\sqrt[3]{x}$ graph and has no absolute extrema.

45. (a)   The cubic function graph of g has turning points $(-1, -2)$ and $(1, 2) \Rightarrow$ it has a local minimum: -2; and a local maximum: 2.

     (b)   The cubic function graph of g has turning points $(-1, -2)$ and $(1, 2)$ but then continues on to $-\infty$ and $\infty \Rightarrow$ it has no absolute extrema.

47. (a)   A graph of $f(x) = -3x^4 + 8x^3 + 6x^2 - 24x$ is shown in Figure 47. Local maxima of 19 and -8 occur at $x \approx -1$ and $x \approx 2$ respectively. A local minima of -13 occurs when $x \approx 1$.

     (b)   The absolute maximum of 19 occurs at $x \approx -1$, and there is no absolute minimum.

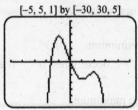

[−5, 5, 1] by [−30, 30, 5]

Figure 47

49. (a)   A graph of $f(x) = x^4 - 2x^2 + 1$ is shown in Figure 49. A local maximum of 1 occurs at $x = 0$. A local minimum of 0 occurs when $x \approx \pm 1$.

     (b)   There is no absolute maximum. An absolute minimum of 0 occurs when $x \approx \pm 1$.

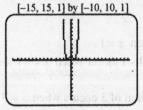

[−15, 15, 1] by [−10, 10, 1]

Figure 49

51. (a)   A graph of $f(x) = \dfrac{8}{1+x^2}$ is shown in Figure 51. A local maximum of 8 occurs at $x = 0$. There are no
           local minima.

     (b)   An absolute maximum of 8 occurs at $x = 0$. There is no absolute minimum.

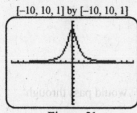

[-10, 10, 1] by [-10, 10, 1]

Figure 51

53. From the graph we can see that the function f is neither even or odd.

55. From the graph we can see that the function is even. Since the point $(1, -1)$ is on the graph of f, the point
     $(-1, -1)$ must also be located on the graph.

57. From the graph, we can see that the function is odd. Since the point $(1.5, -2)$ is on the graph of f, the point
     $(-15, 2)$ must also be located on the graph.

59. The function f is a monomial with only an odd power $\Rightarrow$ it is odd.

61. The function f is a polynomial with an odd power and a constant term, which is an even power $\Rightarrow$ it is
     neither.

63. The function f is a polynomial with an even power and a constant term, which is also even $\Rightarrow$ it is even.

65. The function f is a polynomial with all even powers and a constant term, which is also even $\Rightarrow$ it is even.

67. The function f is a polynomial with all odd powers $\Rightarrow$ it is odd.

69. The function f is a polynomial with both an odd powered term and an even powered term $\Rightarrow$ it is neither.

71. If $f(x) = \sqrt[3]{x^2}$, then $f(-x) = \sqrt[3]{(-x)^2} = \sqrt[3]{x^2}$, which is equal to $f(x) \Rightarrow$ it is even.

73. If $f(x) = \sqrt{1-x^2}$, then $f(-x) = \sqrt{1-(-x)^2} = \sqrt{1-x^2}$, which is equal to $f(x) \Rightarrow$ it is even.

75. If $f(x) = \dfrac{1}{1+x^2}$, then $f(-x) = \dfrac{1}{1+(-x)^2} = \dfrac{1}{1+x^2}$, which is equal to $f(x) \Rightarrow$ it is even.

77. If $f(x) = |x+2|$, then $f(-x) = |-x+2|$, and $-f(x) = -|x+2|$. Since $f(x) \neq f(-x)$ and $f(-x) \neq -f(x)$.
     It is neither.

79. Notice that $f(-100) = 56$ and $f(100) = -56$. In general, since $f(-x) = -f(x)$ holds for each x in the table,
     f is odd.

81. Since f is even, the condition $f(-x) = f(x)$ must hold. For example, since $f(-3) = 21$, $f(3)$ must also
     equal 21. The value assigned to $f(0)$ makes no difference. See Figure 81.

x	-3	-2	-1	0	1	2	3
$f(x)$	21	-12	-25	1	-25	-12	21

Figure 81

83. Since f is an odd function and $(-5, -6)$ and $(-3, 4)$ are on the graph of f then $(5, 6)$ and $(3, -4)$ are on
     the graph. Thus, $f(5) = 6$ and $f(3) = -4$.

85. Sketch any linear function that passes through the origin will work. For example, the graph of $y = x$ is shown
     in Figure 85.

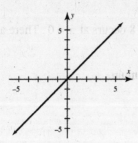

Figure 85

87.   No. If (2, 5) is on the graph of an odd function f, then so is $(-2, -5)$. Since f would pass through $(-3, -4)$ and then $(-2, -5)$, it could not always be increasing.

89.   Plot points $(-2, -3)$ and $(2, -1)$ and make them minima. See Figure 89.

91.   Plot point (2, 3) and make it a maxima. See Figure 91. Yes, it could be quadratic, but it does not have to be quadratic.

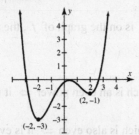

Figure 89

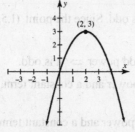

Figure 91

93.   Shift the graph of $y = f(x) = 4x - \dfrac{x^3}{3}$ to the left 1 unit to get the graph of $y = f(x+1) = 4(x+1) - \dfrac{(x+1)^3}{3}$.
       See Figure 93.

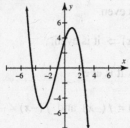

Figure 93

95.   Sketch the graph of $y = f(x) = 4x - \dfrac{x^3}{3}$ vertically by multiplying each y-value by 2, that is, each (x, y)

       point becomes $(x, 2y)$, to get the graph of $y = 2f(x) = 2\left(4x - \dfrac{x^3}{3}\right) = 8x - \dfrac{2}{3}x^3$. See Figure 95.

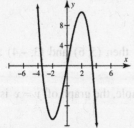

Figure 95

97.   The first equation graph is transformed 1 unit left $\Rightarrow y = f(x+1)-2$ increases on $(0, 3)$. The second equation graph is transformed across the x-axis and transformed 2 units right $\Rightarrow y = -f(x-2)$ decreases on $(3, 6)$.

99.   Polynomial P is odd, since it is a polynomial with only an odd power. Polynomial Q is odd, since it is a polynomial with all odd powers. If $P = 4x^3$ and $Q = 2x + x^5$, then $PQ = 4x^3(2x+x^5) = 8x^4 + 4x^8$. Polynomial PQ is even, since it is a polynomial with all even powers.

101. *(Examples will vary.)* The product of two odd polynomials will be an even polynomial.

103. Polynomial P is odd, since it is a polynomial with only an odd power. Polynomial Q is even, since it is a polynomial with all even powers. If $P = x^3$ and $Q = x^2 + 2x^4$, then $PQ = x^3(x^2+2x^4) = x^5 + 2x^7$. Polynomial PQ is odd, since it is a polynomial with all odd powers.

105. (a)   An absolute maximum of approximately 84°F occurs when $x \approx 5$. An absolute minimum of approximately 63°F occurs when $x \approx 1.6$. These are the high and low temperatures between 1 p.m. and 6 p.m. *Answers may vary slightly.*

      (b)   Local maxima of approximately 78°F and 84°F occur when $x \approx 2.9$ and 5, respectively. Local minima of approximately 63°F and 72°F occur when $x \approx 1.6$ and 3.8. *Answers may vary slightly.*

      (c)   The temperature is increasing on the intervals $1.6 < x < 2.9$ and $3.8 < x < 5$. *Answers may vary slightly.*

107. (a) $F(1) = 0.0484(1) - 1.504(1)^3 + 17.7(1)^2 + 53 = 0.0484 - 1.504 + 17.7 + 53 \approx 69$

            $F(12) = 0.0484(12)^3 - 1.504(12)^2 + 17.7(12) + 53 \approx 132$

            In January 2009, Facebook had about 69 million unique monthly users. In December 2009 this number increased to 132 million.

      (b)   See Figure 107. One can see that F does not have any local extrema for the given interval.

      (c)   According to the graph one can see that F is increasing on the interval $(1, 12)$.

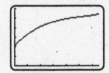

Figure 107

109. (a)   One might expect an absolute maximum to occur in January since it is the coldest month and requires the heating. An absolute minimum might occur in July since it is the warmest month.

      (b)   The graph f is shown in both Figures 109a and 109b. One can see from the graph that there is one local maximum and one local minimum where $1 \leq x \leq 12$. The peak is near the point $(1.46, 140.06)$ and the valley is located near $(7.12, 14.78)$. This means that the peak heating expense of $140 occurs in January $(x \approx 1)$. The minimum heating cost is during July $(x \approx 7)$ when it is roughly $15.

[1, 12, 1] by [0, 150, 10]

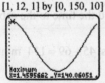

[1, 12, 1] by [0, 150, 10]

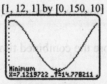

Figure 109a   Figure 109b

111. (a)   Since the graph of f is symmetric with respect to the y-axis, it is a graphical representation of an even function.

      (b)   If f represents the average temperature in the month x, $f(1) = f(-1)$. Thus, the average monthly temperature in August is also 83°F.

      (c)   The average monthly temperatures in March and November are equal, since $f(-4) = f(4)$.

      (d)   In Austin the average monthly temperature is symmetric about July. In July the highest average monthly temperature occurs. In January the lowest temperature occurs. The pairs June-August, May-

September, April-October, March-November, and February-December have approximately the same average temperatures.

Extended and Discovery Exercise for Section 4.1

1. The semicircle with radius 3 has the equation $y = \sqrt{9-x^2}$. (Recall that the equation for a circle with center

    (0, 0) and radius 3 is $x^2 + y^2 = 9 \Rightarrow y = \pm\sqrt{9-x^2}$. So any point (x, y) on the semicircle would have

    coordinates $(x, \sqrt{9-x^2})$. Since the rectangle is symmetric about the y-axis, the length of the rectangle

    $= 2x$, as shown in Figure 1a. The width of the rectangle $= y$.

    The area of the rectangle is: $A = lw = 2x(y) = 2x(\sqrt{9-x^2})$. To find the maximum area, graph the function

    $A(x) = 2x\sqrt{9-x^2}$ on a graphing calculator, and then find the maximum value. See Figure 1b. The

    maximum area of 9 occurs when $x \approx 2.12$. Thus, the length of the rectangle is $2x \approx 2(2.12) \approx 4.24$,

    and the height is $y = \sqrt{9-x^2} \approx \sqrt{9-(2.12)^2} \approx 2.12$.

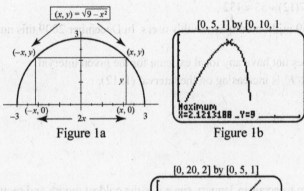

Figure 1a   Figure 1b

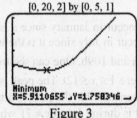

Figure 3

3. (a) Since $r \cdot t = D$ the time to shore is $4 \cdot t = 3 \Rightarrow t = \frac{3}{4}$ hour or 45 minutes; and the time to the cabin is

        $7 \cdot t = 8 \Rightarrow t = \frac{8}{7}$ hours or about 69 minutes. Therefore the combined time is $45 + 69 = 114$ minutes or 1

        hour and 54 minutes.

    (b) Using Pythagorean theorem; the distance to the cabin is

        $3^2 + 8^2 = d^2 \Rightarrow 9 + 64 = d^2 \Rightarrow d^2 = 73 \Rightarrow d \approx 8.5$. Now the time is

        $4 \cdot t = 8.5 \Rightarrow t = \frac{8.5}{4} \Rightarrow t \approx 128$ minutes or 2 hours and 8 minutes.

    (c) Let $x =$ the distance jogged, then the distance rowed by using Pythagorean theorem is;

        $3^2 + (8-x)^2 = d^2$ or $d = \sqrt{x^2 - 16x + 73} \Rightarrow$ the time needed to reach the cabin if $t = \frac{d}{r}$ is

$t = \dfrac{d(\text{jogged})}{7} + \dfrac{d(\text{rowed})}{4}$ or $t = \dfrac{x}{7} + \dfrac{\sqrt{x^2 - 16x + 73}}{4}$. Graph this equation to find the minimum time.

See Figure 3. The minimum time is $t \approx 1.76$ hours or about 1 hour 46 minutes.

4.2: Polynomial Functions and Models

1. (a) Turning points occur near $(1.6, 3.6)$, $(3, 1.2)$, and $(4.4, 3.6)$.

    (b) The turning points represent the times and distances where the runner turned around and jogged the other direction. In this example, after 1.6 minutes the runner is 360 feet from the starting line. The runner turns and jogs toward the starting line. After 3 minutes the runner is 120 feet from the starting line, turns, and jogs away from the starting line. After 4.4 minutes the runner is again 360 feet from the starting line. The runner turns and jogs back to the starting line.

3. (a) It has no turning points and one x-intercept of $(0.5, 0)$.

    (b) The leading coefficient is positive, since the slope of the line is positive.

    (c) Since the graph appears to be a line that is not horizontal, its minimum degree is 1.

5. (a) It has three turning points and three x-intercepts of $(-6, 0)$, $(-1, 0)$, and $(6, 0)$.

    (b) Since the graph goes down for large values of x, the leading coefficient is negative.

    (c) Since the graph has three turning points, its minimum degree is 4.

7. (a) It has four turning points and five x-intercepts of $(-3, 0)$, $(-1, 0)$, $(0, 0)$, $(1, 0)$, and $(2, 0)$.

    (b) Since the graph goes up for large values of x, the leading coefficient is positive.

    (c) Since the graph has four turning points, its minimum degree is 5.

9. (a) It has two turning points and one x-intercept of $(-3, 0)$.

    (b) Since the graph goes down to the left and up to the right, the leading coefficient is positive.

    (c) The graph has two turning points. The minimum degree of f is 3.

11. (a) It has one turning point and two x-intercepts of $(-1, 0)$ and $(2, 0)$.

    (b) Since the graph is a parabola opening up, the leading coefficient is positive.

    (c) The graph has one turning point. The minimum degree of f is 2.

13. (a) It has four turning points and three x-intercepts of $(-1, 0)$, $(0, 0)$, and $(1, 0)$.

    (b) Since the graph goes up to the left and down to the right, the leading coefficient is negative.

    (c) The graph has four turning points. The minimum degree of f is 5.

15. (a) Graph d.

    (b) There is one turning point at $(1, 0)$.

    (c) It has one x-intercept: $(1, 0)$.

    (d) The only local minimum is 0.

    (e) The absolute minimum is 0. There is no absolute maximum.

17. (a) Graph b.

    (b) The turning points are at $(-3, 27)$ and $(1, -5)$.

    (c) The three x-intercepts are at approximately $(-4.9, 0)$ and $(1.9, 0)$, and exactly $(0, 0)$.

    (d) The local minimum is -5, and the local maximum is 27.

    (e) There are no absolute extrema.

19. (a) Graph a.

    (b) The turning points are at $(-2, 16)$, $(0, 0)$, and $(2, 16)$.

    (c) The three x-intercepts are at approximately $(-2.8, 0)$ and $(2.8, 0)$, and exactly $(0, 0)$.

    (d)   The local minimum is 0, and the local maximum is 16.

    (e)   The absolute maximum is 16. There is no absolute minimum.

21.   (a)   The graph of $f(x) = \frac{1}{9}x^3 - 3x$ is shown in Figure 21.

    (b)   There are two turning points located at $(-3, 6)$ and $(3, -6)$.

    (c)   At $x = -3$ there is a local maximum of 6 and at $x = 3$ there is a local minimum of -6.

23.   (a)   The graph of $f(x) = 0.025x^4 - 0.45x^2 - 3$ is shown in Figure 23.

    (b)   There are three turning points located at $(-3, -7.025)$, $(0, -5)$, and $(3, -7.025)$.

    (c)   At $x = \pm 3$ there is a local minimum of -7.025. At $x = 0$ there is a local maximum of -5.

[−10, 10, 1] by [−10, 10, 1]   [−10, 10, 1] by [−10, 10, 1]

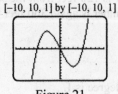   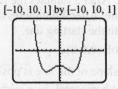

      Figure 21   Figure 23

25.   (a)   The graph of $f(x) = 1 - 2x + 3x^2$ is shown in Figure 25.

    (b)   There is one turning point located at $\left(\frac{1}{3}, \frac{2}{3}\right) \approx (0.333, 0.667)$.

    (c)   At $x = \frac{1}{3}$ there is a local minimum of $\frac{2}{3} \approx 0.667$.

[−10, 10, 1] by [−10, 10, 1]

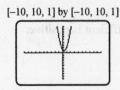

    Figure 25

27.   (a)   The graph of $f(x) = \frac{1}{3}x^3 + \frac{1}{2}x^2 - 2x$ is shown in Figure 27.

    (b)   The turning points are at $\left(-2, \frac{10}{3}\right) \approx (-2, 3.333)$ and $\left(1, -\frac{7}{6}\right) \approx (1, -1.167)$.

    (c)   There is a local minimum of $-\frac{7}{6} \approx -1.167$, and there is a local maximum of $\frac{10}{3} \approx 3.333$.

[−10, 10, 1] by [−10, 10, 1]

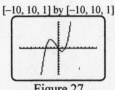

    Figure 27

29.   (a)   The degree is 1 and the leading coefficient is -2.

    (b)   Since the degree of f is 1 and the lead coefficient is negative the graph is a line with a negative slope $\Rightarrow$ it goes up on the left end and down on the right.

31.   (a)   The degree is 2 and the leading coefficient is 1.

    (b)   Since the degree of f is 2 and the lead coefficient is positive the graph is a parabola that opens up on each side.

33. (a) The degree is 3 and the leading coefficient is -2.

   (b) Since the degree of f is odd and the lead coefficient is negative its graph will go up on the left and down on the right.

35. (a) The degree is 1 and the leading coefficient is 5.

   (b) Since the degree of f is 1 and the leading coefficient is positive, the graph is a line with a positive slope $\Rightarrow$ it does down on the left end and up on the right.

37. (a) The degree is 2 and the leading coefficient is -5.

   (b) Since the degree of f is even and the leading coefficient is negative, the graph will go down on both sides.

39. (a) The degree is 3 and the leading coefficient is 2.

   (b) Since the degree of f is odd and the leading coefficient is positive, the graph will go down on the left and up on the right.

41. (a) The degree is 4 and the leading coefficient is -1.

   (b) Since the degree of f is even and the leading coefficient is negative, the graph will go down on both sides.

43. (a) The degree is 3 and the leading coefficient is -1.

   (b) Since the degree of f is odd and the leading coefficient is negative, its graph will go up on the left and down on the right.

45. (a) The degree is 5 and the leading coefficient is 0.1.

   (b) Since the degree of f is odd and the leading coefficient is positive, its graph will go down on the left and up on the right.

47. (a) The degree is 2 and the leading coefficient is $-\dfrac{1}{2}$.

   (b) Since the degree of f is even and the leading coefficient is negative, its graph will go down on both sides.

49. (a) A line graph of the data is shown in Figure 49.

   (b) The data appears to decrease, increase, and decrease. The line graph crosses the x-axis three times. Neither a linear or quadratic function could do this. Thus, f must be degree 3.

   (c) Since f rises to the left and falls to the right $a < 0$.

   (d) Using the cubic regression function on the calculator we find $f(x) = -x^3 + 8x$.

51. (a) A line graph of the data is shown in Figure 51.

   (b) The data decreases and then increases. This could not be a linear function. The data appears to be parabolic and the line graph crosses the x-axis twice. Since all zeros are real and between -3 and 3, f must be degree 2 with two real zeros.

   (c) Since f is opening upward $a > 0$.

   (d) Using the quadratic regression function on the calculator we find $f(x) = x^2 - 2x - 1$.

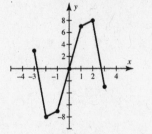

Figure 49

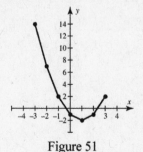

Figure 51

53. (a)   See Figure 53.

     (b)   The data increases then decreases then increases and the decreases. The graph crosses the x-axis 4
            times. Therefore, f is degree 4.

     (c)   Since both ends of f go down $a < 0$.

     (d)   Using the quartic regression function on the calculator we find $f(x) = -x^4 + 3x^2 - 1$.

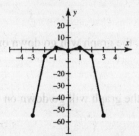

Figure 53

55. One possible graph is shown in Figure 55.

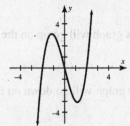

Figure 55

57. One possible graph is shown in Figure 57.

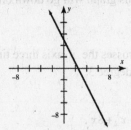

Figure 57

59. No such graph is possible.

61. No such graph is possible.

63. Start by plotting the points $(-1, 2)$ and $\left(1, \dfrac{2}{3}\right)$. Then sketch a cubic polynomial having these two turning

     points. One possible graph is shown in Figure 63.

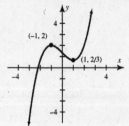

Figure 63

65. Since a quadratic polynomial has only one turning point, $(-1, 2)$ must be the vertex. Plot the points $(-1, 2)$, $(-3, 4)$, and $(1, 4)$. Sketch a parabola passing through these points with a vertex of $(-1, 2)$. The parabola must open up. See Figure 65.

Figure 65

67. The graphs of $f(x) = 2x^4$, $g(x) = 2x^4 - 5x^2 + 1$, and $h(x) = 2x^4 + 3x^2 - x - 2$ are shown in Figures 67a-c. As the viewing rectangle increases in size, the graphs begin to look alike. Each formula contains the term $2x^4$. This term determines the end behavior of the graph for large values of $|x|$. The other terms are relatively small in absolute value by comparison when $|x| \geq 100$. End behavior is determined by the leading term.

$[-4, 4, 1]$ by $[-4, 4, 1]$ $[-10, 10, 1]$ by $[-100, 100, 10]$ $[-100, 100, 10]$ by $[-10^6, 10^6, 10^5]$

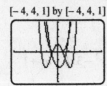 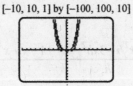 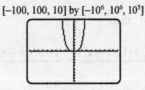

Figure 67a   Figure 67b   Figure 67c

69. For $f(x) = x$ we get $(0, 0)$ and $\left(\frac{1}{2}, \frac{1}{2}\right) \Rightarrow$ the average rate of change is $\frac{\frac{1}{2} - 0}{\frac{1}{2} - 0} = \frac{\frac{1}{2}}{\frac{1}{2}} = 1$

For $g(x) = x^2$ we get $(0, 0)$ and $\left(\frac{1}{2}, \frac{1}{4}\right) \Rightarrow$ the average rate of change is $\frac{\frac{1}{4} - 0}{\frac{1}{2} - 0} = \frac{\frac{1}{4}}{\frac{1}{2}} = 0.5$

For $h(x) = x^3$ we get $(0, 0)$ and $\left(\frac{1}{2}, \frac{1}{8}\right) \Rightarrow$ the average rate of change is $\frac{\frac{1}{8} - 0}{\frac{1}{2} - 0} = \frac{\frac{1}{8}}{\frac{1}{2}} = 0.25$ It decreases with higher degree.

71. (a) For $[1.9, 2.1]$ we get: $(1.9, 6.859)$ and $(2.1, 9.261) \Rightarrow$ the average rate of change is

$$\frac{9.261 - 6.859}{2.1 - 1.9} = 12.01$$

(b) For $[1.99, 2.01]$ we get: $(1.99, 7.880599)$ and $(2.01, 8.120601) \Rightarrow$ the average rate of change is

$$\frac{8.120601 - 7.880599}{2.01 - 1.99} = 12.0001$$

(c) For $[1.999, 2.001]$ we get: $(1.999, 7.988005999)$ and $(2.001, 8.012006001) \Rightarrow$ the average rate of

change is $\dfrac{8.012006001 - 7.988005999}{2.001 - 1.999} = 12.000001$

As the interval decreases in length the average rate of change is approaching 12.

73. (a) For $[1.9, 2.1]$ we get: $\frac{1}{4}(1.9)^4 - \frac{1}{3}(1.9)^3 = 0.971691\overline{6}$ or $(1.9, 0.971691\overline{6})$ and

$\frac{1}{4}(2.1)^4 - \frac{1}{3}(2.1)^3 = 1.775025$ or $(2.1, 1.775025) \Rightarrow$ the average rate of change is:

$$\frac{1.775025 - 0.971691\overline{6}}{2.1 - 1.9} = 4.01\overline{6}$$

(b)   For [1.99, 2.01] we get: $\frac{1}{4}(1.99)^4 - \frac{1}{3}(1.99)^3 = 1.293731\overline{6}$ or $(1.99, 1.293731\overline{6})$ and

$\frac{1}{4}(2.01)^4 - \frac{1}{3}(2.01)^3 = 1.373735003$ or $(2.01, 1.373735003) \Rightarrow$ the average rate of change is:

$\frac{1.373735003 - 1.293731\overline{6}}{201 - 1.99} = 4.00016$

(c)   For [1.999, 2.001] we get: $\frac{1}{4}(1.999)^4 - \frac{1}{3}(1.999)^3 = 1.329337332$ or $(1.999, 1.329337332)$ and

$\frac{1}{4}(2.001)^4 - \frac{1}{3}(2.001)^3 = 1.337337335$ or $(2.001, 1.337337335) \Rightarrow$ the average rate of change is:

$\frac{1.337337335 - 1.329337332}{2.001 - 1999} = 4.000001\overline{6}$

As the interval decreases in length the average rate of change is approaching 4.

75.   Use $\frac{f(x+h) - f(x)}{h}$, then $\frac{3(x+h)^3 - 3x^3}{h} = \frac{3x^3 + 9x^2 h + 9xh^2 + 3h^3 - 3x^3}{h} = 9x^2 + 9xh + 3h^2$

77.   Use $\frac{f(x+h) - f(x)}{h}$, then

$\frac{1 + (x+h) - (x+h)^3 - (1 + x - x^3)}{h} = \frac{1 + x + h - x^3 - 3x^2 h - 3xh^2 - h^3 - 1 - x + x^3}{h} = -3x^2 - 3xh - h^2 + 1$

79. $f(-2) \approx 5$ and $f(1) \approx 0$

81. $f(-1) \approx -1$, $f(1) \approx 1$, and $f(2) \approx -2$

83. $f(-3) = (-3)^3 - 4(-3)^2 = -27 - 36 = -63$, $f(1) = 3(1)^2 = 3$, and $f(4) = (4)^3 - 54 = 64 - 54 = 10$

85. $f(-2) = (-2)^2 + 2(-2) + 6 = 6$, $f(1) = 1 + 6 = 7$, and $f(2) = 2^3 + 1 = 9$

87.   (a)   The graph of f is shown in Figure 87.

      (b)   The function f is discontinuous at $x = 0$.

      (c)   First solve: $4 - x^2 = 0 \Rightarrow -x^2 = -4 \Rightarrow x^2 = 4 \Rightarrow x = -2, 2$; only $x = -2$ is in the domain for the
            equation. Now solve: $x^2 - 4 = 0 \Rightarrow x^2 = 4 \Rightarrow x = -2, 2$; only $x = 2$ is in the domain for the equation
            $\Rightarrow x = -2, 2$.

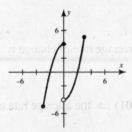

Figure 87

89.   (a)   The graph of f is shown in Figure 89.

      (b) f is continuous.

      (c)   First solve: $2x + 1 = 0 \Rightarrow 2x = -1 \Rightarrow x = -\frac{1}{2}$; this is in the domain of the equation.

            Now solve: $1 - x = 0 \Rightarrow 1 = x$; this is in the domain of the equation $\Rightarrow x = -\frac{1}{2}, 1$.

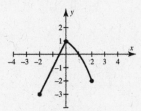

Figure 89

91. (a) The graph of f is shown in Figure 91.

    (b) f is not continuous.

    (c) First solve: $2 + x^2 = 0 \Rightarrow x^2 = -2$; there is no real solution for this equation.

    Now solve: $4 - x^2 = 0 \Rightarrow x^2 = 4 \Rightarrow x = \pm 2$; only $x = 2$ is in the domain of the equation $\Rightarrow x = 2$.

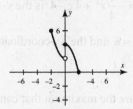

Figure 91

93. (a) The graph of f is shown in Figure 93.

    (b) f is continuous.

    (c) First solve: $2x = 0 \Rightarrow x = 0$; this is not in the domain for the equation.

    $-2 \neq 0 \Rightarrow$ no solutions $-1 \leq x \leq 0$. Finally solve: $x^2 - 2 = 0 \Rightarrow x^2 = 2 \Rightarrow x = \pm\sqrt{2}$; only $\sqrt{2}$ is in the domain for the equation $\Rightarrow x = \sqrt{2}$.

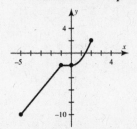

Figure 93

95. (a) The graph of f is shown in Figure 95.

    (b) f is continuous on its domain.

    (c) First solve: $x^3 + 3 = 0 \Rightarrow x^3 = -3 \Rightarrow x = \sqrt[3]{-3}$ or $-\sqrt[3]{3}$; this is in the domain for the equation.

    Now solve: $x + 3 = 0 \Rightarrow x = -3$; this is not in the domain for the equation.

    Finally solve: $4 + x - x^2 = 0 \Rightarrow$ using quadratic formula for $-x^2 + x + 4 = 0$ we get:

    $\dfrac{-1 \pm \sqrt{1 - 4(-1)(4)}}{2(-1)} = \dfrac{-1 \pm \sqrt{17}}{-2}$; only $\dfrac{1 + \sqrt{17}}{2}$ is in the domain of the equation $\Rightarrow x = -\sqrt[3]{3}, \dfrac{\sqrt{17}+1}{2}$.

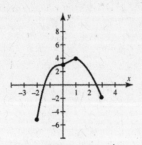

Figure 95

97. (a) To evaluate $F(3)$ we use the formula $F(x) = 2x$ since 3 is in the interval $0 \le x \le 4$.

   $F(3) = 2(3) = 6$; 300 boats are able to harvest 6 thousand tons of fish.

   (b) The maximum of $F(x) = 2x$ is 8 when $x = 4$. The maximum of $F(x) = -\dfrac{1}{4}x^2 + 4x - 4$ is the y-

   coordinate of the vertex. The x-coordinate of the vertex is $x = \dfrac{-4}{2\left(-\dfrac{1}{4}\right)} = 8$ and the y-coordinate is

   $F(8) = -\dfrac{1}{4}(8)^2 + 4(8) - 4 = -16 + 32 - 4 = 12$. 12 thousand tons of fish are the maximum that can be
   caught.

99. (a) $H(-2) = 0$, because $t < 0$; $H(0) = 1$, because $t \ge 0$; $H(3.5) = 1$, because $t \ge 0$

   (b) See Figure 99.

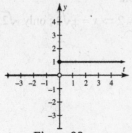

Figure 99

101. (a) There are two turning points that occur near (1, 13) and (7, 72).

   (b) The point (1, 13) means that in January the average temperature is 13°F. This is the minimum average
   monthly temperature in Minneapolis. After this the average temperature begins to increase. The point
   (7, 72) means that in July the average temperature is 72°F. This is the maximum average monthly
   temperature in Minneapolis. After this the average temperature begins to decrease.

103. (a) The height begins to decrease after 4 seconds. The object appears to have dropped after 4 seconds.

   (b) From 0 to 4 seconds, when the object is being lifted, the height could be modeled with a linear function.
   From 4 to 7 seconds, when the object begins to fall, the height can be modeled with a nonlinear
   function.

   (c) The altitude increases by 36 feet each second, during the first four seconds. Therefore, let $m = 36$.
   When $x = 4$, the height is 144 feet. Let $b = 144$. When $x = 7$, the height was 0,

   $a(7-4)^2 + 144 = 0 \Rightarrow a = -16$. Thus, $f(x) = \begin{cases} 36x, & \text{if } 0 \le x \le 4 \\ -16(x-4)^2 + 144, & \text{if } 4 < x \le 7 \end{cases}$

   (d) Using $f(x) = 36x$, $f(x) = 100 \Rightarrow 100 = 36x \Rightarrow x \approx 2.8$; using $f(x) = -16(x-4)^2 + 144$,

   $f(x) = 100 \Rightarrow 100 = -16(x-4)^2 + 144 \Rightarrow -44 = -16(x-4)^2 \Rightarrow x \approx 5.7$; the object is 100 feet high at
   about 2.8 seconds and at about 5.7 second

105. (a)   Using the cubic regression function on the calculator we find
$f(x) \approx -0.0006296x^3 + 0.06544x^2 - 0.368x + 2.8$.

(b)   See Figure 105.

(c)   Since x represents the number of years after 1960 we will let $x = 34$ and 60.

$f(34) \approx -0.0006296(34)^3 + 0.06544(34)^2 - 0.368(34) + 2.8 \approx 41.2$

$f(60) \approx -0.0006296(60)^3 + 0.06544(60)^2 - 0.368(60) + 2.8 \approx 80.3$

(d)   The computation for 1994 involved interpolation and the computation for 2020 involved extrapolation.

[−5, 60, 10] by 0, 80, 20]

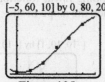

Figure 105

107. (a)   Using the regression function on the calculator, we find
$f(x) = 0.001383x^3 - 8.2235x^2 + 16295x - 10762916$.

(b)   See Figure 107.

[1955, 2005, 5] by [10, 25, 1]

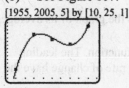

Figure 107

109. f is concave up on $(-\infty, \infty)$ and is not concave down.

111. f is concave up on $(-\infty, 0)$ and is concave down on $(0, \infty)$.

113. f is concave up on $(-\infty, -1)$; $(1, \infty)$ and is concave down on $(-1, 1)$.

Extended and Discovery Exercise for Section 4.2

1.   (a) $D = [0, 10]$

(b) $A(1) = 500\left(1 - \dfrac{1}{10}\right)^2 \Rightarrow A(1) = 500\left(\dfrac{9}{10}\right)^2 \Rightarrow A(1) = 500\left(\dfrac{81}{100}\right) \Rightarrow A(1) = 405$ After one minute of

draining the tank it contains 405 gallons of water.

(c) $500\left(1 - \dfrac{t}{10}\right)^2 = 500\left(1 - \dfrac{2t}{10} + \dfrac{t^2}{100}\right) = 5t^2 - 100t + 500 \Rightarrow$ the degree is: 2. The leading coefficient is: 5.

(d) $A(5) = 500\left(1 - \dfrac{5}{10}\right)^2 \Rightarrow A(5) = 500\left(\dfrac{1}{2}\right)^2 \Rightarrow A(5) = 500\left(\dfrac{1}{4}\right) \Rightarrow A(5) = \dfrac{500}{4} \Rightarrow A(5) = 125$ gallons

remaining. No more than half is drained. This is reasonable because the water will drain faster at first.

3.   (a)   See Figure 3.

(b)   The velocity of the bike rider is 0.25 ft/sec at 4 seconds.

$f(t) = \overline{t}$	$t_1 = 4$ $t_2 = 5$	$t_1 = 4$ $t_2 = 4.1$	$t_1 = 4$ $t_2 = 4.01$	$t_1 = 4$ $t_2 = 4.001$
average velocity (ft/sec)	0.236	0.248	0.2498	0.24998

Figure 3

5. (a)   See Figures 5a-d.
    (b)   The graph of each quadratic function has one turning point, whereas the graph of its average rate of change has no turning points.
    (c)   For any quadratic function, the graph of its average rate of change is a linear function. If the leading coefficient of the quadratic function is negative, the slope of the corresponding linear function is negative. If the leading coefficient of the quadratic function is positive, then the slope of the corresponding linear function is positive.

[−10, 10, 1] by [−10, 10, 1]   [−10, 10, 1] by [−10, 10, 1]   [−10, 10, 1] by [−10, 10, 1]   [−10, 10, 1] by [−10, 10, 1]

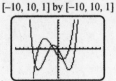 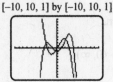 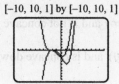 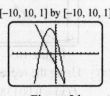

Figure 5a   Figure 5b   Figure 5c   Figure 5d

7. (a)   See Figures 7a-c.
    (b)   The graph of each quartic function has one or three turning points, whereas the graph of its average rate of change has two turning point.
    (c)   For any quartic function, the graph of its average rate of change is a cubic function. The leading coefficient of the quartic function and the leading coefficient of its average rate of change have the same sign.

[−10, 10, 1] by [−10, 10, 1]   [−10, 10, 1] by [−10, 10, 1]   [−10, 10, 1] by [−10, 10, 1]

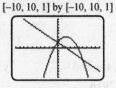 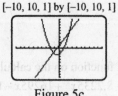

Figure 7a   Figure 7b   Figure 7c

Checking Basic Concepts for Section 4.1 and 4.2

1. (a)   Increasing: $(-2,1),(3,\infty)$
           Decreasing: $(-\infty,-2),(1,3)$
    (b)   Local maximum: approximately 3; local minimum; approximately −13 , −2 .
    (c)   Absolute minimum: approximately −13 ; no absolute maximum.
    (d)   From the graph they are approximately −3.1 , 0, 2.2, and 3.6; they are the same values.
3. (a)   Cubic graphs have a range from −∞ to ∞ ⇒ must have an x -intercept ⇒ not possible.
    (b)   See Figure 3b.
    (c)   See Figure 3c.
    (d)   Cubic graphs have at most two turning points ⇒ have at most 3 x -intercept ⇒ not possible.

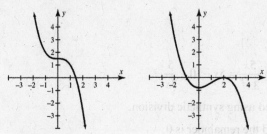

Figure 3b   Figure 3c

5. $f(x) \approx -1.01725x^4 + 10.319x^2 - 10$

4.3: Division of Polynomials

1. $\dfrac{5x^4 - 15}{10x} = \dfrac{5x^4}{10x} - \dfrac{15}{10x} = \dfrac{x^3}{2} - \dfrac{3}{2x}$

3. $\dfrac{3x^4 - 2x^2 - 1}{3x^3} = \dfrac{3x^4}{3x^3} - \dfrac{2x^2}{3x^3} - \dfrac{1}{3x^3} = x - \dfrac{2}{3x} - \dfrac{1}{3x^3}$

5. $\dfrac{x^3-4}{4x^3}=\dfrac{x^3}{4x^3}-\dfrac{4}{4x^3}=\dfrac{1}{4}-\dfrac{1}{x^3}$

7. $\dfrac{5x(3x^2-6x+1)}{3x^2}=\dfrac{15x^3-30x^2+5x}{3x^2}=\dfrac{15x^3}{3x^2}-\dfrac{30x^2}{3x^2}+\dfrac{5x}{3x^2}=5x-10+\dfrac{5}{3x}$

9. x^3-2x^2-5x+6 divided by $x-3$ can be performed using synthetic division.

$$
\begin{array}{r|rrrr}
3 & 1 & -2 & -5 & 6 \\
  &   & 3 & 3 & -6 \\
\hline
  & 1 & 1 & -2 & 0
\end{array}
$$
The quotient is x^2+x-2 and the remainder is 0.

11. $2x^4-7x^3-5x^2-19x+17$ divided by $x+1$ can be performed using synthetic division.

$$
\begin{array}{r|rrrrr}
-1 & 2 & -7 & -5 & -19 & 17 \\
   &   & -2 & 9 & -4 & 23 \\
\hline
   & 2 & -9 & 4 & -23 & 40
\end{array}
$$
The quotient is $2x^3-9x^2+4x-23$ and the remainder is 40.

13. $3x^3-7x+10$ divided by $x-1$ can be performed using synthetic division.

$$
\begin{array}{r|rrrr}
1 & 3 & 0 & -7 & 10 \\
  &   & 3 & 3 & -4 \\
\hline
  & 3 & 3 & -4 & 6
\end{array}
$$
The quotient is $3x^2+3x-4$ and the remainder is 6.

This can also be found using long division as shown.

$$
\require{enclose}
\begin{array}{r}
3x^2+3x-4 \\
x-1 \enclose{longdiv}{3x^3-0x^2-7x+10} \\
\underline{3x^3-3x^2} \\
3x^2-7x \\
\underline{3x^2-3x} \\
-4x+10 \\
\underline{-4x+4} \\
6
\end{array}
$$

15.   We can use synthetic division to divide x^4-3x^3-x+3 by $x-3$.

$$
\begin{array}{r|rrrrr}
3 & 1 & -3 & 0 & -1 & 3 \\
  &   & 3 & 0 & 0 & -3 \\
\hline
  & 1 & 0 & 0 & -1 & 0
\end{array}
$$

The quotient is x^3-1 and the remainder is 0.

17.   We can use long division to divide $4x^3-x^2-5x+6$ by $x-1$.

$$
\require{enclose}
\begin{array}{r}
4x^2+3x-2 \\
x-1 \enclose{longdiv}{4x^3-x^2-5x+6} \\
\underline{4x^3-4x^2} \\
3x^2-5x \\
\underline{3x^2-3x} \\
-2x+6 \\
\underline{-2x+2} \\
4
\end{array}
$$

The quotient is $4x^2 + 3x - 2 + \dfrac{4}{x-1}$.

19. We can use synthetic division to divide $x^3 + 1$ by $x + 1$.

$$
\begin{array}{r|rrrr}
-1 & 1 & 0 & 0 & 1 \\
   &   & -1 & 1 & -1 \\
\hline
   & 1 & -1 & 1 & 0
\end{array}
$$

The quotient is $x^2 - x + 1$ and the remainder is 0.

21. We can use long division to divide $6x^3 + 5x^2 - 8x + 4$ by $2x - 1$.

$$
\begin{array}{r}
3x^2 + 4x - 2 \\
2x-1{\overline{\smash{\big)}\,6x^3 + 5x^2 - 8x + 4}} \\
\underline{6x^3 - 3x^2} \\
8x^2 - 8x + 4 \\
\underline{8x^2 - 4x} \\
-4x + 4 \\
\underline{-4x + 2} \\
2
\end{array}
$$

The quotient is $3x^2 + 4x - 2 + \dfrac{2}{2x-1}$.

23. We can use long division to divide $3x^4 - 7x^3 + 6x - 16$ by $3x - 7$.

$$
\begin{array}{r}
x^3 + 2 \\
3x-7{\overline{\smash{\big)}\,3x^4 - 7x^3 + 0x^2 + 6x - 16}} \\
\underline{3x^4 - 7x^3} \\
6x - 16 \\
\underline{6x - 14} \\
-2
\end{array}
$$

The quotient is $x^3 + 2 + \dfrac{-2}{3x-7}$.

25. We can use long division to divide $5x^4 - 2x^2 + 6$ by $x^2 + 2$.

$$
\begin{array}{r}
5x^2 - 12 \\
x^2+2{\overline{\smash{\big)}\,5x^4 - 2x^2 + 6}} \\
\underline{5x^4 + 10x^2} \\
-12x^2 + 6 \\
\underline{-12x^2 - 24} \\
30
\end{array}
$$

The quotient is $5x^2 - 12 + \dfrac{30}{x^2+2}$.

27. We can use long division to divide $8x^3 + 10x^2 - 12x - 15$ by $2x^2 - 3$.

$$\begin{array}{r} 4x+5 \\ 2x^2-3{\overline{\smash{\big)}\,8x^3+10x^2-12x-15}} \\[-0.5ex] \underline{8x^3-12x} \\ 10x^2-15 \\ \underline{10x^2-15} \\ 0 \end{array}$$

The quotient is $4x+5$.

29. We can use long division to divide $2x^4-x^3+4x^2+8x+7$ by $2x^2+3x+2$.

$$\begin{array}{r} x^2-2x+4 \\ 2x^2+3x+2{\overline{\smash{\big)}\,2x^4-x^3+4x^2+8x+7}} \\[-0.5ex] \underline{2x^4+3x^3+2x^2} \\ -4x^3+2x^2+8x \\ \underline{-4x^3-6x^2-4x} \\ 8x^2+12x+7 \\ \underline{8x^2+12x+8} \\ -1 \end{array}$$

The quotient is $x^2-2x+4+\dfrac{-1}{2x^3+3x+2}$.

31. The divisor times the quotient will be equal to the dividend, $(x-2)(x^2-6x+3)=x^3-8x^2+15x-6$.

33. $$\begin{array}{r} x-1 \\ x-2{\overline{\smash{\big)}\,x^2-3x+1}} \\[-0.5ex] \underline{x^2-2x} \\ -x+1 \\ \underline{-x+2} \\ -1\quad (x-2)(x-1)-1 \end{array}$$

35. $$\begin{array}{r} x^2-1 \\ 2x+1{\overline{\smash{\big)}\,2x^3-x^2-2x}} \\[-0.5ex] \underline{2x^3+x^2} \\ -2x+0 \\ \underline{-2x-1} \\ 1\quad (2x+1)(x^2-1)+1 \end{array}$$

37. $$\begin{array}{r} x-1 \\ x^2+0x+1{\overline{\smash{\big)}\,x^3-x^2+x+1}} \\[-0.5ex] \underline{x^3+0x^2+x} \\ -x^2+1 \\ \underline{-x^2-1} \\ 2\quad (x^2+1)(x-1)+2 \end{array}$$

39.

$$\begin{array}{r|rrrr} 5 & 1 & 2 & -17 & -10 \\ & & -5 & 15 & 10 \\ \hline & 1 & -3 & -2 & 0 \end{array}$$

The quotient is $x^2 - 3x - 2$.

41.

$$\begin{array}{r|rrrr} 5 & 3 & -11 & -20 & 3 \\ & & 15 & 20 & 0 \\ \hline & 3 & 4 & 0 & 3 \end{array}$$

The quotient is $3x^2 + 4x + \dfrac{3}{x-5}$.

43.

$$\begin{array}{r|rrrrr} 2 & 1 & -3 & -4 & 12 & 0 \\ & & 2 & -2 & -12 & 0 \\ \hline & 1 & -1 & -6 & 0 & \end{array}$$

The quotient is $x^3 + x^2 - 6x$.

45.

$$\begin{array}{r|rrrrr} -1 & 1 & 1 & 0 & 0 & 2 & 2 \\ & & -1 & 0 & 0 & 0 & -2 \\ \hline & 1 & 0 & 0 & 0 & 2 & 0 \end{array}$$

The quotient is $x^4 + 2$.

47.

$$\begin{array}{r|rrrrrr} -\dfrac{1}{2} & 2 & -1 & -1 & 0 & 4 & 3 \\ & & -1 & 1 & 0 & 0 & -2 \\ \hline & 2 & -2 & 0 & 0 & 4 & 1 \end{array}$$

The quotient is $2x^4 - 2x^3 + 4 + \dfrac{1}{x+0.5}$.

49. Using the remainder theorem we find: $f(1), \Rightarrow 5(1)^2 - 3(1) + 1 = 5 - 3 + 1 = 3$

51. Using the remainder theorem we find:

$f(-2), \Rightarrow 4(-2)^3 - (-2)^2 + 4(-2) + 2 = -32 - 4 - 8 + 2 = -42$

53. If we divide the Area by the Width we will find the Length:

$$\begin{array}{r} 4x+3 \\ 3x+1 \overline{\smash{\big)}\ 12x^2 + 13x + 3} \\ \underline{12x^2 + \ 4x} \\ 9x + 3 \\ \underline{9x + 3} \\ 0 \end{array}$$

The length is $4x + 3$. When $x = 10$, the Length is $4(10) + 3 = 43$ feet.

4.4: Real Zeros of Polynomial Functions

1. The x-intercepts of f are -1, 1, and 2. Since $f(-1) = 0$, the factor theorem states that $(x+2)$ is a factor of $f(x)$. Similarly, $f(-1) = 0$ implies that $(x+1)$ is a factor, and $f(1) = 0$ implies that $(x-1)$ is a factor.

3. The x-intercepts of f are -2, -1, 1, and 2. Since $f(-2) = 0$, the factor theorem states that $(x+2)$ is a factor of $f(x)$. Similarly, $f(-1) = 0$ implies that $(x+1)$ is a factor, $f(1) = 0$ implies that $(x-1)$ is a factor and $f(2) = 0$ implies that $(x-2)$ is a factor.

5. $f(x) = 2x^2 - 25x + 77$ and zeros: $\dfrac{11}{2}$ and $7 \Rightarrow f(x) = 2\left(x - \dfrac{11}{2}\right)(x-7)$

7. $f(x) = x^3 - 2x^2 - 5x + 6$ and zeros: -2, 1, and $3 \Rightarrow f(x) = (x+2)(x-1)(x-3)$

9. $f(x) = -2x^3 + 3x^2 + 59x - 30$ and zeros: -5, $\dfrac{1}{2}$, and $6 \Rightarrow f(x) = -2(x+5)\left(x - \dfrac{1}{2}\right)(x-6)$

11. If $f(-3) = 0$ then the quadratic equation has a factor of $x - (-3)$ or $x+3$, likewise if $f(2) = 0$ then the quadratic equation has a factor $x-2$. If this quadratic equation has a leading coefficient 7, the complete factored form of $f(x)$ is $f(x) = 7(x+3)(x-2)$.

13. To factor $f(x)$ we need to determine the leading coefficient and zeros of f. The leading coefficient is -2 and the zeros are -1, 0, and 1. The complete factorization is $f(x) = -2x(x+1)(x-1)$.

15. From the graph the zeros are -4, 2, and 8. $f(x)$ is a cubic with a positive leading coefficient. Therefore, $f(x) = (x+4)(x-2)(x-8)$.

17. From the graph the zeros are -8, -4, -2, and 4. $f(x)$ is a quartic polynomial with a negative leading coefficient. Therefore, $f(x) = -1(x+8)(x+4)(x+2)(x-4)$.

19. Since the polynomial has zeros of -1, 2, and 3, it has factors $(x+1)(x-2)(x-3)$.

    If f passes through $(0, 3)$ then $a(0+1)(0-2)(0-3) = 3 \Rightarrow a(1)(-2)(-3) = 3 \Rightarrow 6a = 3 \Rightarrow a = \dfrac{1}{2}$. The complete factored form is: $f(x) = \dfrac{1}{2}(x+1)(x-2)(x-3)$.

21. Since f has zeros -1, 1, and 2, it has factors $(x+1)(x-1)(x-2)$. If $f(0) = 1$ then $a(1)(-1)(-2) = 1$ or $2a = 1 \Rightarrow a = \dfrac{1}{2}$. The complete factored form is: $f(x) = \dfrac{1}{2}(x+1)(x-1)(x-2)$.

23. Since f has zeros -2, -1, 1, and 2, it has factors $(x+2)(x+1)(x-1)(x-2)$. If $f(0) = -8$ then $a(2)(1)(-1)(-2) = -8 \Rightarrow 4a = -8 \Rightarrow a = -2$.
    The complete factored form is: $f(x) = -2(x+2)(x+1)(x-1)(x-2)$.

25. A graph of $Y_1 = 10X^2 + 17X - 6$ is shown in Figure 25. Its zeros are -2 and 0.3. Since the leading coefficient is 10, the complete factorization is $f(x) = 10(x+2)\left(x - \dfrac{3}{10}\right)$.

27. A graph of $Y_1 = -3X^3 - 3X^2 + 18X$ is shown in Figure 27. Its zeros are -3, 0, and 2. Since the leading coefficient is -3, the complete factorization is $f(x) = -3(x-0)(x-2)(x+3) = -3x(x-2)(x+3)$.

29. A graph of $Y_1 = X^4 + (5/2)X^3 - 3X^2 - (9/2)X$ is shown in Figure 29. Its zeros are -3, -1, 0 and $\dfrac{3}{2}$. Since the leading coefficient is 1, the complete factorization is

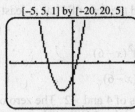
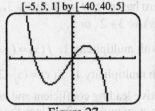
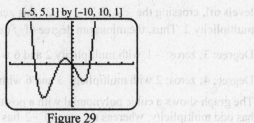

[-5, 5, 1] by [-20, 20, 5]   [-5, 5, 1] by [-40, 40, 5]   [-5, 5, 1] by [-10, 10, 1]

Figure 25   Figure 27   Figure 29

31. By the factor theorem, since 1 is a zero, $(x-1)$ is a factor. $x^3 - 9x^2 + 23x - 15$ divided by $x - 1$ can be
     performed using synthetic division.

     $\underline{1|}$ 1 −9 23 −15
     　　　　 1 −8   15
     　　─────────────────
     　 1 −8 15　　0

     The quotient is $x^2 - 8x + 15$ and the remainder is 0.

     Thus, $x^3 - 9x^2 + 23x - 15 = (x-1)(x^2 - 8x + 15) = (x-1)(x-3)(x-5)$.

     The complete factored form of $f(x) = x^3 - 9x^2 + 23x - 15$ is $f(x) = (x-1)(x-3)(x-5)$.

33. By the factor theorem, since -4 is a zero, $(x+4)$ is a factor. $-4x^3 - x^2 + 51x - 36$ divided by $x+4$ can be
     performed using synthetic division.

     $\underline{-4|}$ −4 −1   51 −36
     　　　　　 16 −60   36
     　　─────────────────────
     　 −4 15 −9　　0

     The quotient is $-4x^2 + 15x - 9$ and the remainder is 0.

     Thus, $-4x^3 - x^2 + 51x - 36 = (x+4)(-4x^2 + 15x - 9) = (x+4)(-4x+3)(x-3)$.

     The complete factored form of $f(x) = -4x^3 - x^2 + 51x - 36$ is $f(x) = -4(x+4)\left(x - \dfrac{3}{4}\right)(x-3)$.

35. By the factor theorem, since -2 is a zero, $(x+2)$ is a factor. $2x^4 - x^3 - 13x^2 - 6x$ divided by $x+2$ can be
     performed using synthetic division.

     $\underline{-2|}$ 2 −1 −13 −6 0
     　　　　　 −4   10   6 0
     　　──────────────────────
     　 2 −5 −3　　0 0

     The quotient is $2x^3 - 5x^2 - 3x$ and the remainder is 0. Thus,

     $2x^4 - x^3 - 13x^2 - 6x = (x+2)(2x^3 - 5x^2 - 3x) = (x+2)x(2x^2 - 5x - 3) =$

     $x(x+2)(2x+1)(x-3)$.

     The complete factored form of $f(x) = 2x^4 - x^3 - 13x^2 - 6x$ is $f(x) = 2x(x+2)\left(x + \dfrac{1}{2}\right)(x-3)$.

37. $f(2) = (2)^3 - 6(2)^2 + 11(2) - 6 = 0$; since $f(2) = 0$, by the factor theorem, $x - 2$ is a factor of
     $f(x) = x^3 - 6x^2 + 11x - 6$.

39. $f(3) = (3)^4 - 2(3)^3 - 13(3)^2 - 10(3) = -120$; since $f(-3) \neq 0$, by the factor theorem, $x - 3$ is not a factor of
     $x^4 - 2x^3 - 13x^2 - 10x$.

41. The zeros of $f(x)$ are 4 and -2. Since the graph does not cross the x-axis at $x = 4$, the zero of 4 has even
     multiplicity. The graph crosses the x-axis at $x = -2$. The zero of -2 has odd multiplicity. Since the graph

levels off, crossing the x-axis at $x = -2$, this zero has at least multiplicity 3, and the zero of 4 has at least multiplicity 2. Thus, the minimum degree of $f(x)$ is $3 + 2$, or 5.

43.   Degree: 3; zeros: -1 with multiplicity 2 and 6 with multiplicity 1. $f(x) = (x+1)^2(x-6)$

45.   Degree: 4; zeros: 2 with multiplicity 3 and 6 with multiplicity 1. $f(x) = (x-2)^3(x-6)$

47.   The graph shows a cubic polynomial with a positive leading coefficient and zeros of 4 and -2. The zero of 4 has odd multiplicity, whereas the zero of -2 has even multiplicity. Since the graph has a degree three, its complete factored form is $f(x) = (x-4)(x+2)^2$.

49.   The graph shows a quartic polynomial with a negative leading coefficient and zeros of -3 and 3. Both zeros have even multiplicity. Since the graph has a degree four polynomial, its complete factored form is $f(x) = -1(x+3)^2(x-3)^2$.

51.   The graph shows a fifth degree polynomial with a positive leading coefficient and zeros of -1 and 1. The -1 has an even multiplicity, whereas the zero of 1 has an odd multiplicity. Since the graph shows a fifth degree polynomial the factors are $(x+1)^2(x-1)^3$. To find the leading coefficient we use
$f(0) = -2 \Rightarrow a(1)^2(-1)^3 = -2 \Rightarrow -a = -2 \Rightarrow a = 2$. Its factored form is $f(x) = 2(x+1)^2(x-1)^3$.

53.   (a)   From the factors the x-intercepts are: $(-2,0)$ and $(1,0)$. To find the y-intercept set
$x = 0 \Rightarrow (0-1)(0+2) = (-1)(2) = -2 \Rightarrow (0,-2)$

      (b)   The zero -2 has multiplicity 1. The zero 1 has multiplicity 1.

      (c)   See Figure 53.

55.   (a)   From the factors the x-intercept is: $(2,0)$. To find the y-intercept set
$x = 0 \Rightarrow -(0-2)^2 = -(4) = -4 \Rightarrow (0,-4)$

      (b)   The zero 2 has multiplicity 2.

      (c)   See Figure 55.

57.   (a)   From the factors the x-intercept is $(1,0)$. To find the y-intercept set
$x = 0 \Rightarrow (0-1)^3 = (-1)^3 = -1 \Rightarrow (0,-1)$

      (b)   The zero 1 has multiplicity 3.

      (c)   See Figure 57.

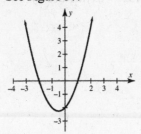   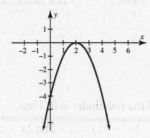   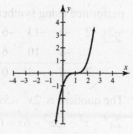

Figure 53   Figure 55   Figure 57

59.   (a)   From the factors the x-intercepts are: $(0,0)$ and $(2,0)$. To find the y-intercept set
$x = 0 \Rightarrow -(0)(0-2)^2 = (0)(4) = 0 \Rightarrow (0,0)$

      (b)   The zero 0 has multiplicity 1. The zero 2 has multiplicity 2.

      (c)   See Figure 59.

61.   (a)   From the factors the x-intercepts are: $(-2,0)$, $(0,0)$ and $(1,0)$. To find the y-intercept set
$x = 0 \Rightarrow -(0)(0-1)(0+2) = (0)(-1)(2) = 0 \Rightarrow (0,0)$

      (b)   The zero -2 has multiplicity 1. The zero 0 has multiplicity 1. The zero 1 has multiplicity 1.

      (c)   See Figure 61.

63. (a) From the factors the x-intercepts are: $(-2,0)$ and $(1,0)$. To find the y-intercept set

$$x = 0 \Rightarrow -(0-1)^2(0+2)^2 = -(1)(4) = -4 \Rightarrow (0,-4)$$

(b) The zero -2 has multiplicity 2. The zero 1 has multiplicity 2.

(c) See Figure 63.

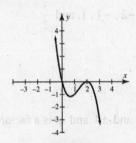

Figure 59

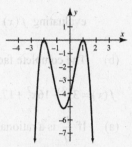

Figure 61

Figure 63

65. (a) From the factors the x-intercepts are: $(-2,0)$ and $(-1,0)$. To find the y-intercept set

$$x = 0 \Rightarrow 2(0+2)(0+1)^2 = 2(2)(1)^2 = 4 \Rightarrow (0,4)$$

(b) The zero -2 has multiplicity 1. The zero -1 has multiplicity 2.

(c) See Figure 65.

67. (a) From the factors the x-intercepts are: $(-2,0)$, $(0,0)$, and $(2,0)$. To find the y-intercept set

$$x = 0 \Rightarrow 0(0+2)(0-2) = 0(2)(-2) = 0$$

(b) The zero -2 has multiplicity 1. The zero 0 has multiplicity 2. The zero 2 has multiplicity 1.

(c) See Figure 67.

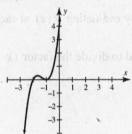

Figure 65

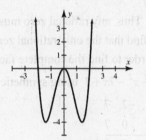

Figure 67

69. $f(x) = 2x^3 + 3x^2 - 8x + 3$

(a) If $\dfrac{p}{q}$ is a rational zero, then p is a factor of 3, which are ± 1 and ± 3 and q is a factor of 2, which are

± 1 or ± 2. Thus, any rational zero must be in the list $\pm\dfrac{1}{2}$, ± 1, $\pm\dfrac{3}{2}$, or ± 3. From Figure 69 we see that

there are three rational zeros of $\dfrac{1}{2}$, 1, and -3.

(b) The complete factored form is $f(x) = 2\left(x - \dfrac{1}{2}\right)(x-1)(x+3)$.

x	$\frac{1}{2}$	$-\frac{1}{2}$	1	-1	$\frac{3}{2}$	$-\frac{3}{2}$	3	-3
$f(x)$	0	$\frac{15}{2}$	0	12	$\frac{9}{2}$	15	60	0

Figure 69

71. $f(x) = 2x^4 + x^3 - 8x^2 - x + 6$

(a)   If $\dfrac{p}{q}$ is a rational zero, then p is a factor of 6, which are ± 1, ± 2, ± 3, and ± 6 and q is a factor of 2,

which are ± 1 and ± 2. Thus, any rational zero must be in the list $\pm\dfrac{1}{2}$, ± 1, $\pm\dfrac{3}{2}$, ± 2, ± 3, or ± 6. By

evaluating $f(x)$ at each of these values, we find that the zeros are -2, -1, 1, and $\dfrac{3}{2}$.

(b)   The complete factored form is $f(x) = 2(x+2)(x+1)(x-1)\left(x-\dfrac{3}{2}\right)$.

73. $f(x) = 3x^3 - 16x^2 + 17x - 4$

(a)   If $\dfrac{p}{q}$ is a rational zero, then p is a factor of 4, which are ± 1, ± 2, and ± 4 and q is a factor of 3,

which are ± 1 and ± 3. Thus, any rational zero must be in the list $\pm\dfrac{1}{3}$, ± 1, $\pm\dfrac{2}{3}$, ± 2, $\pm\dfrac{4}{3}$, or ± 4. By

evaluating $f(x)$ at each of these values, we find that the zeros are $\dfrac{1}{3}$, 1, and 4.

(b) The complete factored form is $f(x) = 3\left(x-\dfrac{1}{3}\right)(x-1)(x-4)$.

75. $f(x) = x^3 - x^2 - 7x + 7$

(a)   If $\dfrac{p}{q}$ is a rational zero, then p is a factor of 7, which are ± 1, and ± 7 and q is a factor of 1, which are

± 1. Thus, any rational zero must be in the list ± 1 or ± 7. By evaluating $f(x)$ at each of these values,
we find that the only rational zero is 1.

(b)   In order to find the complete factored form of $f(x)$ we need to divide the factor $(x-1)$ into

$x^3 - x^2 - 7x + 7$ using synthetic division.

$$\begin{array}{r|rrrr} 1 & 1 & -1 & -7 & 7 \\ & & 1 & 0 & 7 \\ \hline & 1 & 0 & -7 & 0 \end{array}$$

Thus $x^3 - x^2 - 7x + 7 = (x-1)(x^2 - 7)$.

The complete factored form is $f(x) = (x-1)(x-\sqrt{7})(x+\sqrt{7})$.

77. $P(x) = 2x^3 - 4x^2 + 2x + 7$, $P(x)$ has two sign changes, therefore are 2 or $2-2=0$ possible positive zeros.
$P(-x) = 2(-x)^3 - 4(-x)^2 + 2(-x) + 7 = -2(x)^3 - 4x^2 - 2x + 7$, $P(-x)$ has one sign change, therefore there is
one possible negative zero. From the graph of $P(x)$ in Figure 77, we see that the actual number of positive
and negative zeros are 0 and 1 respectively.

79. $P(x) = 5x^4 + 3x^2 + 2x - 9$, $P(x)$ has one sign change, therefore there is one possible positive zero.
$P(-x) = 5(-x)^4 + 3(-x)^2 + 2(-x) - 9 = 5x^4 + 3x - 2x - 9$, $P(-x)$ has one sign change, therefore there is one
possible negative zero. From the graph of $P(x)$ in Figure 79, we see that the actual number of positive and
negative zeros are 1 and 1 respectively.

81. $P(x) = x^5 + 3x^4 - x^3 + 2x + 3$, $P(x)$ has two sign changes, therefore there are 2 or $2-2=0$ possible positive
zeros. $P(-x) = (-x)^5 + 3(-x)^4 - (-x)^3 + 2(-x) + 3 = -x^5 + 3x + x^3 - 2x + 3$, $P(-x)$ has three sign changes,

therefore there are 3 or $3-2=1$ possible negative zeros. From the graph of $P(x)$ in Figure 81, we see that the actual number of positive and negative zeros are 0 and 1 respectively.

[-4, 4, 1] by [-25, 25, 5]   [-4, 4, 1] by [-25, 25, 5]   [-4, 4, 1] by [-50, 50, 10]

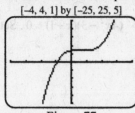   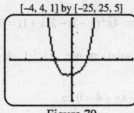   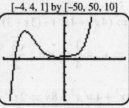

Figure 77   Figure 79   Figure 81

83. (a) $f(x) = x^3 + x^2 - 6x = 0 \Rightarrow x(x^2 + x - 6) = x(x+3)(x-2) = 0 \Rightarrow x = 0,\ -3,\ \text{or } 2.$

(b) Graph $Y_1 = X^3 + X^2 - 6X$ in $[-5, 5, 1]$ by $[-10, 10, 1]$. The x-intercepts are $(-3, 0)$, $(0, 0)$, and $(2, 0)$.

(c) Table $Y_1 = X^3 + X^2 - 6X$ starting at $x = -4$, incrementing by 1. The zeros or x-intercepts are $(-3, 0)$, $(0, 0)$, and $(2, 0)$.

85. (a) $f(x) = x^4 - 1 = 0 \Rightarrow x^4 = 1 \Rightarrow x = \pm 1.$

(b) Graph $Y_1 = X^4 - 1$ in $[-2, 2, 1]$ by $[-10, 10, 1]$. The x-intercepts are $(-1, 0)$ and $(1, 0)$.

(c) Table $Y_1 = X^4 - 1$ starting at $x = -2$, incrementing by 1. The zeros or x-intercepts are $(-1, 0)$ and $(1, 0)$.

87. (a) $f(x) = -x^3 + 4x = 0 \Rightarrow -x(x^2 - 4) = x(x+2)(x-2) = 0 \Rightarrow x = 0,\ \text{or } \pm 2.$

(b) Graph $Y_1 = -X^3 + 4X$ in $[-5, 5, 1]$ by $[-5, 5, 1]$. The x-intercepts are $(-2, 0)$, $(0, 0)$, and $(2, 0)$.

(c) Table $Y_1 = -X^3 + 4X$ starting at $x = -3$, incrementing by 1. The zeros or x-intercepts are $(-2, 0)$, $(0, 0)$, and $(2, 0)$.

89. $x^3 - 1 = 0 \Rightarrow x^3 = 1 \Rightarrow x = 1.$

91. $x^2 - x^3 = 0 \Rightarrow x^2(1-x) = 0 \Rightarrow x^2 = 0 \Rightarrow x = 0$ and $1 - x = 0 \Rightarrow x = 1$. The solutions are $x = 0, 1$.

93. $8x - x^4 = 0 \Rightarrow x(8 - x^3) = 0 \Rightarrow x = 0$ and $8 - x^3 = 0 \Rightarrow x^3 = 8 \Rightarrow x = 2$. The solutions are $x = 0, 2$.

95. $x^3 - 25x = 0 \Rightarrow x(x^2 - 25) = 0 \Rightarrow x(x-5)(x+5) = 0 \Rightarrow x = 0,\ \text{or } \pm 5$

97. $x^4 - x^2 = 2x^2 + 4 \Rightarrow x^4 - 3x^2 - 4 = 0 \Rightarrow (x^2 - 4)(x^2 + 1) = 0 \Rightarrow (x-2)(x+2)(x^2+1) = 0 \Rightarrow x = \pm 2$

99. $x^3 - 3x^2 - 18x = 0 \Rightarrow x(x^2 - 3x - 18) = 0 \Rightarrow x(x-6)(x+3) = 0 \Rightarrow x = 0, 6,\ \text{or } -3$

101. $2x^3 = 4x^2 - 2x \Rightarrow 2x^3 - 4x^2 + 2x = 0 \Rightarrow 2x(x^2 - 2x + 1) = 0 \Rightarrow 2x(x-1)(x-1) = 0 \Rightarrow x = 0$ or 1

103. $12x^3 = 17x^2 + 5x \Rightarrow 12x^3 - 17x^2 - 5x = 0 \Rightarrow x(12x^2 - 17x - 5) = 0 \Rightarrow x(4x+1)(3x-5) = 0 \Rightarrow x = 0,\ -\dfrac{1}{4},\ \text{or}$

$\dfrac{5}{3}$

105. $9x^4 + 4 = 13x^2 \Rightarrow 9x^4 - 13x^2 + 4 = 0 \Rightarrow (9x^2 - 4)(x^2 - 1) = 0 \Rightarrow x = \pm\dfrac{2}{3},\ x = \pm 1$

107. $4x^3 + 4x^2 - 3x - 3 = 0 \Rightarrow (4x^3 + 4x^2) - (3x + 3) = 0 \Rightarrow 4x^2(x+1) - 3(x+1) = 0 \Rightarrow (4x^2 - 3)(x+1) = 0$. Set

$4x^2 - 3 = 0 \Rightarrow 4x^2 = 3 \Rightarrow x^2 = \dfrac{3}{4} \Rightarrow x = \pm\dfrac{\sqrt{3}}{2}$. The solutions are; $x = -1,\ \pm\dfrac{\sqrt{3}}{2}$.

109. $2x^3 + 4 = x(x+8) \Rightarrow 2x^3 + 4 = x^2 + 8x \Rightarrow 2x^3 - x^2 - 8x + 4 = 0 \Rightarrow$

$(2x^3 - x^2) - (8x - 4) = 0 \Rightarrow x^2(2x-1) - 4(2x-1) = 0 \Rightarrow (x^2 - 4)(2x-1) = 0 \Rightarrow$

$(x+2)(x-2)(2x-1) = 0 \Rightarrow x = -2, 2, \dfrac{1}{2}$

111. $8x^4 - 30x^2 + 27 = 0 \Rightarrow (4x^2 - 9)(2x^2 - 3) = 0 \Rightarrow (2x+3)(2x-3)(2x^2 - 3) = 0$. Set

$2x^2 - 3 = 0 \Rightarrow 2x^2 = 3 \Rightarrow x^2 = \dfrac{3}{2} \Rightarrow x = \pm\sqrt{\dfrac{3}{2}} \Rightarrow x = \pm\dfrac{\sqrt{6}}{2}$. The solutions are; $x = \pm\dfrac{3}{2},\ \pm\dfrac{\sqrt{6}}{2}$

113. $x^6 - 19x^3 - 216 = 0 \Rightarrow (x^3 + 8)(x^3 - 27) = 0$. set $x^3 + 8 = 0 \Rightarrow x^3 = -8 \Rightarrow x = -2$. Also set

$x^3 - 27 = 0 \Rightarrow x^3 = 27 \Rightarrow x = 3$. The solutions are; $x = -2, 3$.

115. The graph of $f(x) = x^3 - 1.1x^2 - 5.9x + 0.7$ is shown in Figure 115. Its zeros are approximately -2.0095, 0.11639, and 2.9931. The solutions are $x \approx -2.01$, 0.12, or 2.99.

117. The graph of $f(x) = -0.7x^3 - 2x^2 + 4x + 2.5$ is shown in Figure 117. Its zeros are approximately -4.0503, -0.51594 and 1.7091. The solutions are $x \approx -4.05$, -0.52, or 1.71.

119. The graph of $f(x) = 2x^4 - 1.5x^3 - 24x^2 - 10x + 13$ is shown in Figure 119. Its zeros are approximately -2.6878, -1.0957, 0.55475, and 3.9787. The solutions are $x \approx -2.69$, -1.10, 0.55 or 3.98.

[–10, 10, 1] by [–10, 10, 11]   [–10, 10, 1] by [–10, 10, 1]   [–10, 10, 1] by [–120, 120, 20]

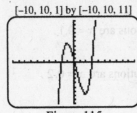
Figure 115

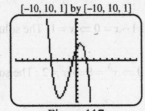
Figure 117

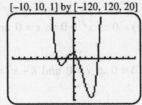
Figure 119

121. $3(x^2 + 4) + 2x(3x - 12) = 0 \Rightarrow 3x + 12 + 6x^2 - 24x = 0 \Rightarrow 9x^2 - 24x + 12 = 0 \Rightarrow$

$3(3x^2 - 8x + 4) = 0 \Rightarrow 3(3x - 2)(x - 2) = 0 \Rightarrow x = \dfrac{2}{3}, 2$

123. $3(x+1)^2(2x-1)^4 + 8(x+1)^3(2x-1)^3 = 0 \Rightarrow (x+1)^2(2x-1)^3[3(2x-1) + 8(x+1)] = 0 \Rightarrow$

$(x+1)^2(2x-1)^3[6x - 3 + 8x + 8] = 0 \Rightarrow (x+1)^2(2x-1)^3(14x + 5) = 0 \Rightarrow x = -1, -\dfrac{5}{14}, \dfrac{1}{2}$

125. $3kx^2 - 7x = 0 \Rightarrow x(3kx - 7) = 0 \Rightarrow x = 0, \dfrac{7}{3k}, k > 0$

127. $f(x) = x^2 - 5 \Rightarrow f(2) = 2^2 - 5 = -1$ and $f(3) = 3^2 - 5 = 4$; because $f(2) < 0$ and $f(3) > 0$, by the intermediate value property, there exists an x-value between 2 and 3 such that $f(x) = 0$.

129. $f(x) = 2x^3 - 1 \Rightarrow f(0) = 2(0)^3 - 1 = -1$ and $f(1) = 2(1)^3 - 1 = 1$; because $f(0) < 0$ and $f(1) > 0$, by the intermediate value property, there exists an x-value between 0 and 1 such that $f(x) = 0$.

131. $f(x) = x^5 - x^2 + 4 \Rightarrow f(1) = 1^5 - 1^2 + 4 = 4$ and $f(2) = 2^5 - 2^2 + 4 = 32$. Because $f(1) < 20$ and $f(2) > 20$, by the intermediate value property, there exists a number K such that $f(K) = 20$.

133. $T(x) = x^3 - 6x^2 + 8x$ when $0 \le x \le 4$. Graph $Y_1 = X^3 - 6X^2 + 8X$ in the window $[0, 4, 1]$ by $[-5, 5, 1]$. The zeros are $x = 0$, $x = 2$, and $x = 4$. Since $x = 0, 2,$ and 4 correspond to the hours after midnight, then the temperature was 0°F at 12 am, 2 am and 4 am.

135. The graph of $Y_1 = (\pi/3)X^3 - 10\pi X^2 + ((4000\pi)(0.6))/3$ and the smallest positive zero are shown in Figure 135. The ball with a 20-centimeter diameter will sink approximately 11.34 centimeters into the water.

[−20, 40, 5] by [−2500, 2500, 500]

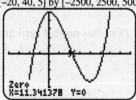

Figure 135

137. $f(x) = x^3 - 66x^2 + 1052x + 652$ and

$f(x) = 2500 \Rightarrow 2500 = x^3 - 66x^2 + 1052x + 652 \Rightarrow x^3 - 66x^2 + 1052x - 1848 = 0$; graph

$Y_1 = X^3 - 66X^2 + 1052X - 1848$ in the window $[0, 45, 5]$ by $[-5000, 5000, 1000]$. The zeros are at $x = 2$, $x = 22$, and $x = 42$. Since $x = 1$ corresponds to June 1, there were 2500 birds on approximately June 2, June 22, and July 12.

139. (a) Graph f in $[-10, 15, 1]$ by $[-70, 70, 10]$. It has three zeros of approximately -6.01, 2.15, and 11.7. The approximate complete factored form is $-0.184(x + 6.01)(x - 2.15)(x - 11.7)$.

   (b) The zeros represent the months when the average temperature is 0°F. The zero of -6.01 has no significance since it does not correspond to a month. The zeros of 2.15 and 11.7 mean that in approximately February and November the average temperature in Trout Lake is 0°F.

141. (a) The greater the distance downstream from the plant the less the concentration of copper. This agrees with intuition.

   (b) Using the cubic regression function on your calculator we find
   $f(x) \approx -0.000068x^3 + 0.0099x^2 - 0.653x + 23$.

   (c) See Figure 141c.

   (d) We must approximate the distance where the concentration of copper first drops to 10. Graph $Y_1 = C(x)$ and $Y_2 = 10$. The point of intersection is near $(32.1, 10)$ as shown in Figure 141d. Mussels would not be expected to live between the plant and approximately 32.1 miles downstream, that is when $0 \le x < 32.1$ (approximately).

[0, 70, 10] by [0, 22, 5]

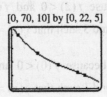

Figure 141c

[0, 70, 10] by [0, 22, 5]

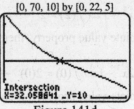

Intersection
X=32.058841 Y=10

Figure 141d

Extended and Discovery Exercises for Section 4.4

1. $P(x) = x^4 - x^3 + 3x^2 - 8x + 8; \ c = 2$

    $$\begin{array}{r|rrrrr} 2 & 1 & -1 & 3 & -8 & 8 \\ & & 2 & 2 & 10 & 4 \\ \hline & 1 & 1 & 5 & 2 & 12 \end{array}$$

    Since the bottom row of the synthetic division is all non-negative and $c > 0$, $P(x)$ has no real zero greater than 2.

3. $P(x) = x^4 + x^3 - x^2 + 3; \ c = -2$

    $$\begin{array}{r|rrrrr} -2 & 1 & 1 & -1 & 0 & 3 \\ & & -2 & 2 & -2 & 4 \\ \hline & 1 & -1 & 1 & -2 & 7 \end{array}$$

    Since the bottom row of the synthetic division alternates in sign and $c < 0$, $P(x)$ has no real zero less than -2.

5. $P(x) = 3x^4 + 2x^3 - 4x^2 + x - 1; \ c = 1$

    $$\begin{array}{r|rrrrr} 1 & 3 & 2 & -4 & 1 & -1 \\ & & 3 & 5 & 1 & 2 \\ \hline & 3 & 5 & 1 & 2 & 1 \end{array}$$

    Since the bottom row of the synthetic division are all non-negative and $c > 0$, $P(x)$ has no real zero greater than 1.

Checking Basic Concepts for Sections 4.3 and 4.4

1. $\dfrac{5x^4 - 10x^3 + 5x^2}{5x^2} = \dfrac{5x^4}{5x^2} - \dfrac{10x^3}{5x^2} + \dfrac{5x^2}{5x^2} = x^2 - 2x + 1$

3. Since the graph of the cubic polynomial has zeros -2 and 1, the multiplicities of the degree 3 equation is the zero -2 is even $\Rightarrow 2$ and the zero 1 is odd $\Rightarrow 1$. The factors are $(x+2)(x+2)(x-1)$. To find the leading coefficient use:

    $$f(0) = 2 \Rightarrow a(0+2)(0+2)(0-1) = 2 \Rightarrow a(2)(2)(-1) = 2 \Rightarrow -4a = 2 \Rightarrow a = -\frac{1}{2}.$$

    The complete factored form is: $f(x) = -\dfrac{1}{2}(x+2)^2(x-1)$.

5. Graph $Y_1 = X^4 - X^3 - 18X^2 + 16X + 32$ in $[-6, 6, 1]$ by $[-150, 150, 20]$. The x-intercepts are -4, -1, 2, and 4. See Figure 5. The factored form of f is $f(x) = (x+4)(x+1)(x-2)(x-4)$.

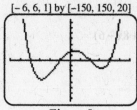

[$-6, 6, 1$] by [$-150, 150, 20$]

Figure 5

4.5: The Fundamental Theorem of Algebra

1. The graph of $f(x)$ does not intersect the x-axis. Therefore, f has no real zeros. Since f is degree 2, there must be two non-real, complex zeros.

3. The graph of $f(x)$ intersects the x-axis once. Therefore, f has one real zero. Since f is degree 3, there must be two non-real, complex zeros.

5. The graph of $f(x)$ intersects the x-axis twice. Therefore, f has two real zeros. Since f is degree 4, there must be two non-real, complex zeros.

7. The graph of $f(x)$ intersects the x-axis three times. Therefore, f has three real zeros. Since f is degree 5, there must be two non-real, complex zeros.

9. Degree: 2; leading coefficient: 1; zeros: $6i$ and $-6i$

    (a) $f(x) = (x - 6i)(x + 6i)$

    (b) $(x - 6i)(x + 6i) = x^2 - 36i^2 = x^2 + 36$. Thus, $f(x) = x^2 + 36$.

11. Degree: 3; leading coefficient: -1; zeros: -1, $2i$, and $-2i$

    (a) $f(x) = -1(x + 1)(x - 2i)(x + 2i)$

    (b) $-1(x + 1)(x - 2i)(x + 2i) = -1(x + 1)(x^2 - 4i^2) = -1(x + 1)(x^2 + 4) = $
         $-(x^3 + 4x + x^2 + 4) = -x^3 - x^2 - 4x - 4$. Thus, $f(x) = -x^3 - x^2 - 4x - 4$.

13. Degree: 4; leading coefficient: 10; zeros: 1, -1, $3i$, and $-3i$

    (a) $f(x) = 10(x - 1)(x + 1)(x - 3i)(x + 3i)$

    (b) $10(x - 1)(x + 1)(x - 3i)(x + 3i) = 10(x^2 - 1)(x^2 + 9) = 10(x^4 + 8x^2 - 9) = 10x^4 + 80x^2 - 90$. Thus,
         $f(x) = 10x^4 + 80x^2 - 90$.

15. Degree: 4; leading coefficient: $\dfrac{1}{2}$; zeros: $-i$ and $2i$

    (a) Since $f(x)$ has real coefficients, it must also have a third and fourth zero of i and $-2i$ the conjugate
         of $-i$ and $2i$. Therefore the complete factored form is: $f(x) = \dfrac{1}{2}(x + i)(x - i)(x + 2i)(x - 2i)$

    (b) $\dfrac{1}{2}(x + i)(x - i)(x + 2i)(x - 2i) = \dfrac{1}{2}(x^2 - i^2)(x^2 - 4i^2) = \dfrac{1}{2}(x^2 + 1)(x^2 + 4) = $
         $\dfrac{1}{2}(x^4 + 4x^2 + x^2 + 4) = \dfrac{1}{2}x^4 + \dfrac{5}{2}x^2 + 2$. Thus, $f(x) = \dfrac{1}{2}x^4 + \dfrac{5}{2}x^2 + 2$.

17. Degree: 3; leading coefficient: -2; zeros: $1 - i$ and 3

(a) Since $f(x)$ has real coefficients, it must also have a third zero of $1+i$ the conjugate of $1-i$. Therefore the complete factored form is: $f(x) = -2(x-(1+i))(x-(1-i))(x-3)$.

(b) $-2(x-(1+i))(x-(1-i))(x-3) = -2(x^2 - x + xi - x - xi + 1 - i^2)(x-3) =$
$-2(x^2 - 2x + 2)(x-3) = -2(x^3 - 2x^2 + 2x - 3x^2 + 6x - 6) = -2(x^3 - 5x^2 + 8x - 6) =$
$-2x^3 + 10x^2 - 16x + 12$. Thus, $f(x) = -2x^3 + 10x^2 - 16x + 12$.

19. First divide:

$$
\begin{array}{r}
3x^2 + 75 \\
x - \tfrac{5}{3} \overline{)3x^3 - 5x^2 + 75x - 125} \\
\underline{3x^3 - 5x^2} \quad\quad\quad\quad \\
75x - 125 \\
\underline{75x - 125} \\
0
\end{array}
$$

Now set $3x^2 + 75 = 0$ and solve. $3x^2 + 75 = 0 \Rightarrow 3x^2 = -75 \Rightarrow x^2 = -25 \Rightarrow x = \pm\sqrt{-25} \Rightarrow x = \pm 5i$. The solutions are $x = \dfrac{5}{3}, \pm 5i$.

21. If $-3i$ is a zero then $3i$ is also a zero $\Rightarrow (x+3i)(x-3i) = x^2 - 9i^2 = x^2 + 9$. Now divide this into the equation:

$$
\begin{array}{r}
2x^2 - x + 1 \\
x^2 + 9 \overline{)2x^4 - x^3 + 19x^2 - 9x + 9} \\
\underline{2x^4 \quad\quad + 18x^2} \quad\quad\quad\quad \\
-x^3 + x^2 - 9x + 9 \\
\underline{-x^3 \quad\quad -9x} \\
x^2 \quad\quad + 9 \\
\underline{x^2 \quad\quad + 9} \\
0
\end{array}
$$

Then use the quadratic formula to solve $2x^2 - x + 1$. $\dfrac{1 \pm \sqrt{1 - 4(2)(1)}}{2(2)} = \dfrac{1 \pm \sqrt{-7}}{4} = \dfrac{1}{4} \pm \dfrac{i\sqrt{7}}{4}$. Therefore the solutions are: $x = \pm 3i, \dfrac{1}{4} \pm \dfrac{i\sqrt{7}}{4}$.

23. $x^2 + 25 = 0 \Rightarrow x^2 = -25 \Rightarrow x = \pm 5i$. Thus, $f(x) = (x-5i)(x+5i)$.

25. $3x^3 + 3x = 0 \Rightarrow 3x(x^2 + 1) = 0 \Rightarrow x = 0$ or $\pm i$. Thus, $f(x) = 3(x-0)(x-i)(x+i)$.

27. $x^4 + 5x^2 + 4 = 0 \Rightarrow (x^2 + 1)(x^2 + 4) = 0 \Rightarrow x = \pm i$ or $\pm 2i$.
Thus, $f(x) = (x-i)(x+i)(x-2i)(x+2i)$.

29. The graph of $y = x^3 + 2x^2 + 16x + 32$ is shown in Figure 29. The x-intercept appears to be -2. We can use synthetic division to help factor f.

$$\begin{array}{r|rrrr} -2 & 1 & 2 & 16 & 32 \\ & & -2 & 0 & -32 \\ \hline & 1 & 0 & 16 & 0 \end{array}$$

$x^3 + 2x^2 + 16x + 32 = (x+2)(x^2+16) = (x+2)(x+4i)(x-4i)$.

Thus, $f(x) = (x+2)(x+4i)(x-4i)$.

[-10, 10, 1] by [-50, 50, 5]

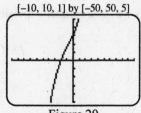

Figure 29

31. $x^3 + x = 0 \Rightarrow x(x^2+1) = 0 \Rightarrow x(x+i)(x-i) = 0 \Rightarrow x = 0, \pm i$

33. Factor or find the zero of 2 graphically.

$x^3 = 2x^2 - 7x + 14 \Rightarrow x^3 - 2x^2 + 7x - 14 = 0 \Rightarrow x^2(x-2) + 7(x-2) = 0 \Rightarrow$
$(x-2)(x^2+7) = 0 \Rightarrow (x-2)(x+i\sqrt{7})(x-i\sqrt{7}) = 0 \Rightarrow x = 2, \pm i\sqrt{7}$

35. $x^4 + 5x^2 = 0 \Rightarrow x^2(x^2+5) = x^2(x+i\sqrt{5})(x-i\sqrt{5}) = 0 \Rightarrow x = 0, \pm i\sqrt{5}$

37. $x^4 = x^3 - 4x^2 \Rightarrow x^4 - x^3 + 4x^2 = 0 \Rightarrow x^2(x^2 - x + 4) = 0$

Use the quadratic formula to find the zeros of $x^2 - x + 4$. The solutions are $x = 0, \dfrac{1}{2} \pm \dfrac{i\sqrt{15}}{2}$.

39. Find the zeros of 1 and 2 using the rational zero test or a graph. See Figures 39a & 39b.
$x^4 + x^3 = 16 - 8x - 6x^2 \Rightarrow x^4 + x^3 + 6x^2 + 8x - 16 = 0$

The graph of $x^4 + x^3 + 6x^2 + 8x - 16$ shows that -2 and 1 are zeros, so $(x+2)$ and $(x-1)$ are factors; using synthetic division twice gives the missing factor.

$$\begin{array}{r|rrrrr} -2 & 1 & 1 & 6 & 8 & -16 \\ & & -2 & 2 & -16 & 16 \\ \hline & 1 & -1 & 8 & -8 & 0 \end{array}$$

$$\begin{array}{r|rrrr} 1 & 1 & -1 & 8 & -8 \\ & & 1 & 0 & 8 \\ \hline & 1 & 0 & 8 & 0 \end{array}$$

$x^4 + x^3 + 6x^2 + 8x - 16 = (x+2)(x-1)(x^2+8) = (x+2)(x-1)(x+i\sqrt{8})(x-i\sqrt{8})$

The solutions are $x = -2, 1, \pm i\sqrt{8}$.

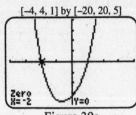
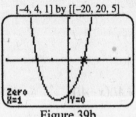

[-4, 4, 1] by [-20, 20, 5]   [-4, 4, 1] by [[-20, 20, 5]

Figure 39a   Figure 39b

41.   Find the zero of -2 using the rational zero test or a graph of $y = 3x^3 + 4x^2 - x + 6$. Use synthetic division to factor $3x^3 + 4x^2 - x + 6$.

$$
\begin{array}{r|rrrr}
-2 & 3 & 4 & -1 & 6 \\
   &   & -6 & 4 & -6 \\
\hline
   & 3 & -2 & 3 & 0
\end{array}
$$

$3x^3 + 4x^2 - x + 6 = (x + 2)(3x^2 - 2x + 3)$

Use the quadratic formula to find the zeros of $3x^2 - 2x + 3$. The solutions are $x = -2, \dfrac{1}{3} \pm \dfrac{i\sqrt{8}}{3}$.

43.

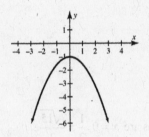

45.   This graph is not possible.

47.

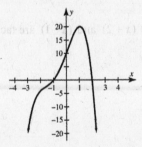

49.   This graph is not possible.

4.6: Rational Functions and Models

1.   Yes, since the numerator and denominator are both polynomials. Since $4x - 5 \neq 0$, $D = \left\{ x \mid x \neq \dfrac{5}{4} \right\}$.

3. Yes, since f can be written as $f(x) = \dfrac{x^2 - x - 2}{1}$, and $g(x) = 1$ is a polynomial. $D =$ all real numbers.

5. No, since the numerator is not a polynomial. $D = \{x \mid x \neq -1\}$

7. Yes, since the numerator and denominator are both polynomials. Since $x^2 + 1 \neq 0$, $D =$ all real numbers.

9. No, since the numerator is not a polynomial. Since $x^2 + x \neq 0 \Rightarrow x(x+1) \neq 0$, $D = \{x \mid x \neq -1, x \neq 0\}$.

11. Yes, since f can be written as $f(x) = \dfrac{4(x+1) - 3}{x+1} \Rightarrow f(x) = \dfrac{4x + 1}{x+1}$.

    Since $x + 1 \neq 0$, $D = \{x \mid x \neq -1\}$.

13. There is a horizontal asymptote of $y = 4$ and a vertical asymptote of $x = 2$; $D = \{x \mid x \neq 2\}$.

15. There is a horizontal asymptote of $y = -4$ and a vertical asymptote of $x = \pm 2$; $D = \{x \mid x \neq \pm 2\}$.

17. There is a horizontal asymptote of $y = 0$ and there is no vertical asymptote; $D =$ all real numbers.

19. There is a horizontal asymptote of $y = 1$ and two vertical asymptotes at $x = \pm 2$. Since $x \neq \pm 2$;
    $D = \{x \mid x \neq \pm 2\}$.

21. Since the output of Y_1 gets closer to 3 as the input x gets larger, it is reasonable to conjecture that the equation for the horizontal asymptote is $y = 3$.

23. Horizontal Asymptotes: The degree of the numerator is equal to the degree of the denominator and the ratio of the leading coefficients is $\dfrac{1}{1} = 1$. Therefore, $y = 1$ is a horizontal asymptote.

    Vertical Asymptotes: Find the zeros of the denominator, $x + 1 = 0 \Rightarrow x = -1$. Therefore $x = -1$ is a vertical asymptote. (Note that -1 is not a zero of the numerator.)

25. Horizontal Asymptotes: The degree of the numerator is equal to the degree of the denominator and the ratio of the leading coefficients is $\dfrac{4}{2} = 2$. Therefore, $y = 2$ is a horizontal asymptote.

    Vertical Asymptotes: Find the zeros of the denominator, $2x - 6 = 0 \Rightarrow x = 3$. Therefore $x = 3$ is a vertical asymptote. (Note that 3 is not a zero of the numerator.)

27. Horizontal Asymptotes: The degree of the numerator is less than the degree of the denominator, therefore, the x-axis, $y = 0$, is a horizontal asymptote.

    Vertical Asymptotes: Find the zeros of the denominator, $x^2 + 1 = 0 \Rightarrow x^2 = -1$. Therefore there are no vertical asymptotes.

29. Horizontal Asymptotes: The degree of the numerator is equal to the degree of the denominator and the ratio of the leading coefficients is $\dfrac{-1}{-1} = 1$. Therefore, $y = 1$ is a horizontal asymptote.

Vertical Asymptotes: Find the zeros of the denominator, $9 - x^2 = 0 \Rightarrow x^2 = 9 \Rightarrow x = \pm 3$. Therefore $x = \pm 3$ are vertical asymptotes. (Note that ± 3 are not zeros of the numerator.)

31. Horizontal Asymptotes: The degree of the numerator is equal to the degree of the denominator and the ratio of the leading coefficients is $\dfrac{4}{2} = 2$. Therefore, $y = 2$ is a horizontal asymptote.

Vertical Asymptotes: Find the zeros of the denominator, $2x - 6 = 0 \Rightarrow 2x = 6 \Rightarrow x = 3$. Therefore $x = 3$ is a vertical asymptote. (Note that 3 is not a zero of the numerator.)

33. Horizontal Asymptotes: The degree of the numerator is less than the degree of the denominator, therefore, the x-axis, $y = 0$, is a horizontal asymptote.

Vertical Asymptotes: Find the zeros of the denominator, $x^2 - 5 = 0 \Rightarrow x^2 = 5 \Rightarrow x = \pm\sqrt{5}$. Therefore $x = \pm\sqrt{5}$ are vertical asymptotes. (Note that $\pm\sqrt{5}$ are not zeros of the numerator.)

35. Horizontal Asymptotes: The degree of the numerator is greater than the degree of the denominator, therefore there is no horizontal asymptote.

Vertical Asymptotes: Find the zeros of the denominator,

$x^2 + 3x - 10 = 0 \Rightarrow (x + 5)(x - 2) = 0 \Rightarrow x = -5, 2$. Therefore $x = -5$ and $x = 2$ are the vertical asymptotes. (Note that -5 and 2 are not a zeros of the numerator.)

37. Horizontal Asymptotes: The degree of the numerator is equal to the degree of the denominator and the ratio of the leading coefficients is $\dfrac{1}{2}$. Therefore, $y = \dfrac{1}{2}$ is a horizontal asymptote.

Vertical Asymptotes: Find the zeros of the denominator, $(2x - 5)(x + 1) = 0 \Rightarrow x = \dfrac{5}{2}$ and $x = -1$.

Therefore $x = \dfrac{5}{2}$ is a vertical asymptote. (Note that $\dfrac{5}{2}$ is not a zero of the numerator and $x = -1$ is a zero of the numerator).

39. Horizontal Asymptotes: The degree of the numerator is equal to the degree of the denominator and the ratio of the leading coefficients is $\dfrac{3}{1} = 3$. Therefore, $y = 3$ is a horizontal asymptote.

Vertical Asymptotes: Find the zeros of the denominator, $(x + 2)(x - 1) = 0 \Rightarrow x = -2, 1$. Here, only $x = 1$ is a vertical asymptote. (Note that -2 is a zero of the numerator.)

41. Horizontal Asymptotes: The degree of the numerator is greater than the degree of the denominator, therefore there is no horizontal asymptote.

Vertical Asymptotes: Find the zeros of the denominator, $x + 3 = 0 \Rightarrow x = -3$. But $x = -3$ is not a vertical asymptote since -3 is a zero of the numerator and $f(x) = x - 3$ for $x \neq -3$. There are no vertical asymptotes.

43. $f(x) = \dfrac{a}{x - 1}$ has a vertical asymptote of $x = 1$. It has a horizontal asymptote of $y = 0$, since the degree of the numerator is less than the degree of the denominator. The best choice is graph b.

45. $f(x) = \dfrac{x-a}{x+2}$ has a vertical asymptote of $x = -2$ and a horizontal asymptote of $y = 1$, since the degree of the

numerator equals the degree of the denominator and the ratio of the leading coefficients is $\dfrac{1}{1}$.

The best choice is graph d.

47.   One example of a symbolic representation of a rational function with a vertical asymptote of $x = -3$ and a

horizontal asymptote of $y = 1$ is $f(x) = \dfrac{x+1}{x+3}$. *Answers may vary.*

49.   One example of a symbolic representation of a rational function with vertical asymptotes of $x = \pm 3$ and a

horizontal asymptote of $y = 0$ is $f(x) = \dfrac{1}{x^2 - 9}$. *Answers may vary.*

51.   See Figure 51. Since the degree of the numerator is less than the degree of the denominator, the horizontal
asymptote is $y = 0$. To find the vertical asymptote we find the zero of the denominator, $x^2 = 0 \Rightarrow x = 0$.

53.   See Figure 53. Since the degree of the numerator is less than the degree of the denominator, the horizontal
asymptote is $y = 0$. To find the vertical asymptote we find the zero of the denominator, $2x = 0 \Rightarrow x = 0$.

55. $g(x) = \dfrac{1}{x-3}$ is $y = \dfrac{1}{x}$ transformed 3 units right $\Rightarrow$ it has a vertical asymptote of $x = 3$. It still has a

horizontal asymptote of $y = 0$. See Figure 55. Then $g(x) = f(x-3)$.

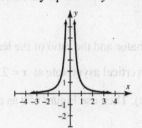

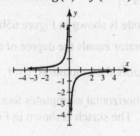

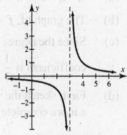

Figure 51   Figure 53   Figure 55

57. $g(x) = \dfrac{1}{x} + 2$ is $y = \dfrac{1}{x}$ transformed 2 units up $\Rightarrow$ it has a horizontal asymptote of $y = 2$. It still has a vertical

asymptote of $x = 0$. See Figure 57. Then $g(x) = f(x) + 2$.

59. $g(x) = \dfrac{1}{x+1} - 2$ is $y = \dfrac{1}{x}$ transformed 1 unit left and 2 units down. It now has a vertical asymptote of $x = -1$,

and a horizontal asymptote of $y = -2$. See Figure 59. Then $g(x) = f(x+1) - 2$.

61. $g(x) = -\dfrac{2}{(x-1)^2}$ is $y = \dfrac{1}{x^2}$ transformed by reflecting it across the x-axis, then shifting 1 unit right, and

vertically stretching by a factor of 2. It now has a vertical asymptote of $x = 1$, and it still has a horizontal
asymptote of $y = 0$. See Figure 61. Then $g(x) = -2h(x-1)$.

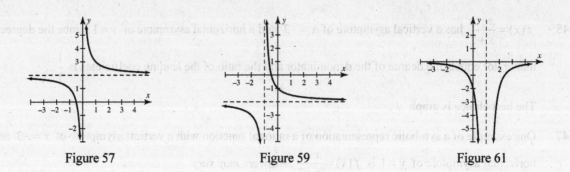

Figure 57   Figure 59   Figure 61

63. $g(x) = \dfrac{1}{(x+1)^2} - 2$ is $y = \dfrac{1}{x^2}$ transformed 1 unit left and 2 units down. It now has a vertical asymptote of

$x = -1$, and a horizontal asymptote of $y = -2$. See Figure 63. Then $g(x) = h(x+1) - 2$.

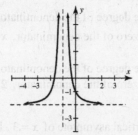

Figure 63

65.   (a) $x - 2 = 0 \Rightarrow x = 2;\ D = \{x \mid x \neq 2\}$

       (b)   The graph of f using dot mode is shown in Figure 65b.

       (c)   Since the degree of the numerator equals the degree of the denominator and the ratio of the leading
              coefficients is $\dfrac{1}{1} = 1$, the horizontal asymptote is $y = 1$. There is a vertical asymptote at $x = 2$.

       (d)   First, sketch the vertical and horizontal asymptotes found in part (c). Then use Figure 65b as a guide to
              a more complete graph of f. The sketch is shown in Figure 65d.

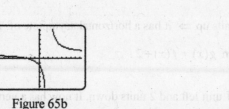

Figure 65b   Figure 65d

67.   (a) $x^2 - 4 = 0 \Rightarrow x^2 = 4 \Rightarrow x = \pm 2;\ D = \{x \mid x \neq \pm 2\}$

       (b)   The graph of f using dot mode is shown in Figure 67b.

       (c)   Since the degree of the numerator is less than the degree of the denominator there is a horizontal
              asymptote at $y = 0$. There are vertical asymptotes at $x = \pm 2$.

       (d)   First, sketch the vertical and horizontal asymptotes found in part (c). Then use Figure 67b as a guide to
              a more complete graph of f. The sketch is shown in Figure 67b.

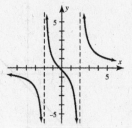

Figure 67d

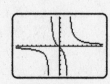

Figure 67b

69. (a) $1 - 0.25x^2 = 0 \Rightarrow 0.25x^2 = 1 \Rightarrow x^2 = 4 \Rightarrow x = \pm 2; D = \{x \mid x \neq \pm 2\}$

     (b)   The graph of f using dot mode is shown in Figure 69b.

     (c)   Since the degree of the numerator is less than the degree of the denominator there is a horizontal
           asymptote at $y = 0$. There are vertical asymptotes at $x = \pm 2$.

     (d)   First, sketch the vertical and horizontal asymptotes found in part (c). Then use Figure 69b as a guide to
           a more complete graph of f. The sketch is shown in Figure 69d.

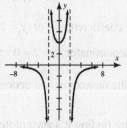

Figure 69d

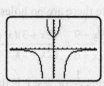

Figure 69b

71. (a) $x - 2 = 0 \Rightarrow x = 2; D = \{x \mid x \neq 2\}$

     (b)   The graph of f using dot mode is shown in Figure 71b.

     (c)   Since the degree of the numerator is greater than the degree of the denominator there is no horizontal
           asymptote. There are no vertical asymptotes. (The numerator equals 2 when $x = 2$).

     (d)   First, sketch the vertical and horizontal asymptotes found in part (c). Then use Figure 71b as a guide to
           a more complete graph of f. The sketch is shown in Figure 71d.

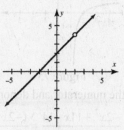

Figure 71d

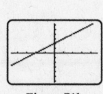

Figure 71b

73. First divide:

$$\begin{array}{r} x - 1 \\ x-1\overline{\smash{\big)}\,x^2 - 2x + 1} \\ \underline{x^2 - x} \\ -x + 1 \\ \underline{-x + 1} \\ 0 \end{array}$$

Now graph $x - 1$; since the denominator $x - 1 \neq 0$ there is a hole at $x = 1$ in the graph. See Figure 73.

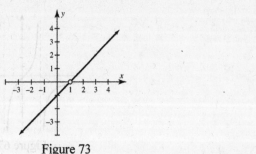

Figure 73

75. First we find the asymptotes. Since the degree of the numerator and denominator are the same, the ratio of the

     leading coefficients $\frac{1}{1}$ or $y=1$ is the horizontal asymptote. To find the vertical asymptote we find the zero of

     the denominator. $x+1=0 \Rightarrow x=-1$. Since $x=-1$ is a vertical asymptote it has no holes.
     See Figure 75.

77. First divide: $g(x)=\frac{(2x+1)(x-2)}{(x-2)(x-2)}=\frac{2x+1}{x-2}$. Now graph $\frac{2x+1}{x-2}$ by finding it's asymptotes. The ratio of the

     leading coefficients $\frac{2}{1}$ or $y=2$ is the horizontal asymptote. To find the vertical asymptote we find the zero

     of the denominator. $x-2=0 \Rightarrow x=2$. Since $x=2$ is an asymptote there are no holes. See Figure 77.

79. Factor the numerator and denominator, then divide: $f(x)=\frac{2x^2+9x+9}{2x^2+7x+6}=\frac{(2x+3)(x+3)}{(2x+3)(x+2)}=\frac{x+3}{x+2}$. Now graph

     $\frac{x+3}{x+2}$ by finding it's asymptotes. The ratio of the leading coefficients $\frac{1}{1}$ or $y=1$ is the horizontal asymptote.

     To find the vertical asymptote we find the zero of the denominator. $x+2=0 \Rightarrow x=-2$. Since $x=-2$ is an

     asymptote there is only a hole is at $2x+3=0$ or $x=\frac{-3}{2}$. See Figure 79.

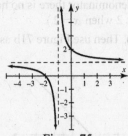   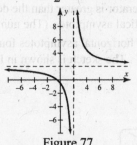   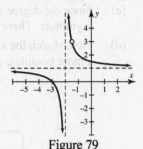

Figure 75   Figure 77   Figure 79

81. Factor the numerator and denominator, then divide:

     $$f(x)=\frac{-2x^2+11x-14}{x^2-5x+6}=\frac{(-2x+7)(x-2)}{(x-3)(x-2)}=\frac{-2x+7}{x-3}.$$

     Now graph $\frac{-2x+7}{x-3}$ by finding it's asymptotes. The ratio of the leading coefficients $\frac{-2}{1}$ or $y=-2$ is the

     horizontal asymptote. To find the vertical asymptote we find the zero of the denominator. $x-3=0 \Rightarrow x=3$.
     Since $x=3$ is an asymptote there is only a hole is at $x=2$. See Figure 81.

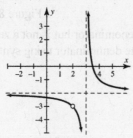

Figure 81

83.   There is a vertical asymptote at $x = -1$ since -1 is a zero of the denominator but is not a zero of the numerator. To find any slant asymptote we will divide the numerator by the denominator using synthetic division:

$$
\begin{array}{r|rrr}
-1 & 1 & 0 & 1 \\
   &   & -1 & 1 \\
\hline
   & 1 & -1 & 2
\end{array}
$$

Therefore, $f(x) = x - 1 + \dfrac{2}{x+1}$ and as $|x|$ becomes large, $f(x)$ approaches $y = x - 1$. There is a slant

asymptote at $y = x - 1$. The graph of f using dot mode is shown in Figure 83a. To sketch this graph, first sketch the vertical and slant asymptotes found above. Then use Figure 83a as a guide to a more complete graph of f. The sketch is shown in Figure 83b.

[−8, 8, 1] by [−8, 8, 1]

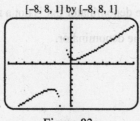

Figure 83a

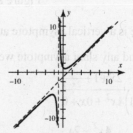

Figure 83b

85.   There is a vertical asymptote at $x = -2$ since -2 is a zero of the denominator but is not a zero of the numerator. To find any slant asymptote we will divide the numerator by the denominator using synthetic division:

$$
\begin{array}{r|rrr}
-2 & 0.5 & -2 & 2 \\
   &   & -1 & 6 \\
\hline
   & 0.5 & -3 & 8
\end{array}
$$

Therefore, $f(x) = 0.5x - 3 + \dfrac{8}{x+2}$ and as $|x|$ becomes large, $f(x)$ approaches $y = 0.5x - 3$. There is a slant

asymptote at $y = 0.5x - 3$. The graph of f using dot mode is shown in Figure 85a. To sketch this graph, first sketch the vertical and slant asymptotes found above. Then use Figure 85a as a guide to a more complete graph of f. The sketch is shown in Figure 85b.

[−14, 14, 2] by [−14, 14, 2]

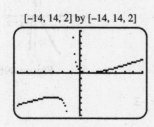

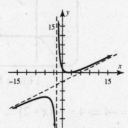

Figure 85a   Figure 85b

87.   There is a vertical asymptote at $x = 1$ since 1 is a zero of the denominator but is not a zero of the numerator. To find any slant asymptote we will divide the numerator by the denominator using synthetic division:

$$\begin{array}{r|rrr} 1 & 1 & 2 & 1 \\ & & 1 & 3 \\ \hline & 1 & 3 & 4 \end{array}$$

Therefore, $f(x) = x + 3 + \dfrac{4}{x-1}$ and as $|x|$ becomes large, $f(x)$ approaches $y = x + 3$. There is a slant asymptote at $y = x + 3$. The graph of f using dot mode is shown in Figure 87a. To sketch this graph, first sketch the vertical and slant asymptotes found above. Then use Figure 87a as a guide to a more complete graph of f. The sketch is shown in Figure 87b.

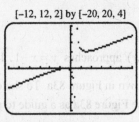

[–12, 12, 2] by [–20, 20, 4]

Figure 87a

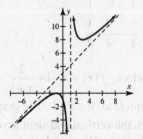

Figure 87b

89.   There is a vertical asymptote at $x = \dfrac{1}{2}$ since $\dfrac{1}{2}$ is a zero of the denominator but is not a zero of the numerator. To find any slant asymptote we will divide the numerator by the denominator.

$$\require{enclose}\begin{array}{r} 2x + 1 + \frac{1}{2x-1} \\ 2x - 1 \enclose{longdiv}{4x^2 + 0x + 0} \\ \underline{4x^2 - 2x} \\ 2x + 0 \\ \underline{2x - 1} \\ 1 \end{array}$$

Therefore, $f(x) = 2x + 1 + \dfrac{1}{2x-1}$ and as $|x|$ becomes large, $f(x)$ approaches $y = 2x + 1$. There is a slant asymptote at $y = 2x + 1$. The graph of f using dot mode is shown in Figure 89a. To sketch this graph, first sketch the vertical and slant asymptotes found above. Then use Figure 89a as a guide to a more complete graph of f. The sketch is shown in Figure 89b.

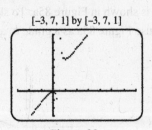

[–3, 7, 1] by [–3, 7, 1]

Figure 89a

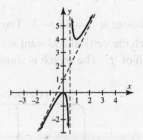

Figure 89b

91. $\dfrac{4}{x+2} = -4 \Rightarrow 4 = -4(x+2) \Rightarrow 4 = -4x - 8 \Rightarrow 12 = -4x \Rightarrow x = -3$

93. $\dfrac{x+1}{x} = 2 \Rightarrow x+1 = 2x \Rightarrow x = 1$

95. $\dfrac{1-x}{3x-1} = -\dfrac{3}{5} \Rightarrow -3(3x-1) = 5(1-x) \Rightarrow -9x+3 = 5-5x \Rightarrow -2 = 4x \Rightarrow x = -\dfrac{1}{2}$

97. $f(x) = \dfrac{2x-4}{x-1}$ is in lowest terms.

   1)   To find the vertical asymptote we find the zero of the denominator, $x-1 = 0 \Rightarrow x = 1$.

   2)   Since the degree of the numerator is equal to the degree of the denominator the equation of the

         horizontal asymptote is $y = \dfrac{2}{1} = 2$.

   3) $f(0) = \dfrac{2(0)-4}{0-1} = 4$; $(0, 4)$

   4) $\dfrac{2x-4}{x-1} = 0 \Rightarrow 2x-4 = 0 \Rightarrow 2x = 4 \Rightarrow x = 2$; $(2, 0)$

   5) $\dfrac{2x-4}{x-1} = 2 \Rightarrow 2x-4 = 2(x-1) \Rightarrow 2x-4 = 2x-2 \Rightarrow 0 = 2 \Rightarrow f$ does not cross the horizontal asymptote.

   6)   See Figure 97.

   7)   See Figure 97.

99. $f(x) = \dfrac{x+1}{x-4}$ is in lowest terms.

   1)   To find the vertical asymptote we find the zero of the denominator, $x-4 = 0 \Rightarrow x = 4$.

   2)   Since the degree of the numerator is equal to the degree of the denominator the equation of the
         horizontal asymptote is $y = 1$.

   3) $f(0) = \dfrac{(0)+1}{(0)-4} = \dfrac{1}{-4}$; $\left(0, \dfrac{-1}{4}\right)$

   4) $\dfrac{x+1}{x-4} = 0 \Rightarrow x+1 = 0 \Rightarrow x = -1$; $(-1, 0)$.

   5) $\dfrac{x+1}{x-4} = 1 \Rightarrow x+1 = x-4 \Rightarrow 1 = -4 \Rightarrow f$ does not cross the horizontal asymptote.

   6)   See Figure 99.

   7)   See Figure 99.

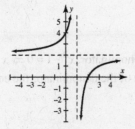

Figure 97

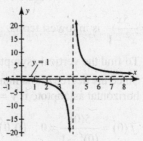

Figure 99

101. $f(x) = \dfrac{x+2}{x-3}$ is in lowest terms.

   1)   To find the vertical asymptote we find the zero of the denominator, $x-3 = 0 \Rightarrow x = 3$.

   2)   Since the degree of the numerator is equal to the degree of the denominator the equation of the
         horizontal asymptote is $y = 1$.

3) $f(0) = \dfrac{(0)+2}{(0)-3} = \dfrac{2}{-3}$; $\left(0, -\dfrac{2}{3}\right)$

4) $\dfrac{x+2}{x-3} = 0 \Rightarrow x+2 = 0 \Rightarrow x = -2$; $(-2, 0)$.

5) $\dfrac{x+2}{x-3} = 1 \Rightarrow x+2 = x-3 \Rightarrow 2 = -3 \Rightarrow f$ does not cross the horizontal asymptote.

6)   See Figure 101.

7)   See Figure 101.

103. $f(x) = \dfrac{4-2x}{8-x}$ is in lowest terms.

1)   To find the vertical asymptote we find the zero of the denominator, $8-x = 0 \Rightarrow x = 8$.

2)   Since the degree of the numerator is equal to the degree of the denominator the equation of the horizontal asymptote is $y = 2$.

3) $f(0) = \dfrac{4-2(0)}{8-(0)} = \dfrac{4}{8} = \dfrac{1}{2}$; $\left(0, \dfrac{1}{2}\right)$

4) $\dfrac{4-2x}{8-x} = 0 \Rightarrow 4-2x = 0 \Rightarrow x = 2$; $(2, 0)$.

5) $\dfrac{4-2x}{8-x} = 2 \Rightarrow 4-2x = 2(8-x) \Rightarrow 4-2x = 16-2x \Rightarrow 4 = 16 \Rightarrow f$ does not cross the horizontal asymptote.

6)   See Figure 103.

7)   See Figure 103.

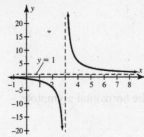

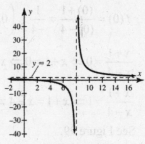

Figure 101   Figure 103

105. $f(x) = \dfrac{5x}{x^2 - 1}$ is in lowest terms.

1)   To find the vertical asymptote we find the zero of the denominator, $x^2 - 1 = 0 \Rightarrow x = \pm 1$.

2)   Since the degree of the numerator is less than the degree of the denominator the equation of the horizontal asymptote is $y = 0$.

3) $f(0) = \dfrac{5(0)}{(0)^2 - 1} = 0$; $(0, 0)$

4) $\dfrac{5x}{x^2 - 1} = 0 \Rightarrow 5x = 0 \Rightarrow x = 0$; $(0, 0)$.

5) $\dfrac{5x}{x^2 - 1} = 0 \Rightarrow 5x = 0 \Rightarrow x = 0 \Rightarrow f$ crosses the horizontal asymptote at $(0, 0)$.

6)   See Figure 105.

7)   See Figure 105.

107. $f(x) = \dfrac{(x+6)(x-2)}{(x+3)(x-4)}$ is in lowest terms.

1)   To find the vertical asymptote we find the zero of the denominator, $(x+3)(x-4) = 0 \Rightarrow x = -3, 4$.

2)   Since the degree of the numerator is equal to the degree of the denominator the equation of the horizontal asymptote is $y = 1$.

3) $f(0) = \dfrac{(0+6)(0-2)}{(0+3)(0-4)} = 1$; $(0,1)$

4) $\dfrac{(x+6)(x-2)}{(x+3)(x-4)} = 0 \Rightarrow (x+6)(x-2) = 0 \Rightarrow x = -6, 2$; $(-6,0), (2,0)$.

5) $\dfrac{(x+6)(x-2)}{(x+3)(x-4)} = 1 \Rightarrow (x+6)(x-2) = (x+3)(x-4) \Rightarrow x^2 + 4x - 12 = x^2 - x - 12 \Rightarrow 4x = -x \Rightarrow 5x = 0 \Rightarrow$

      $x = 0 \Rightarrow f$ crosses the horizontal asymptote at $(0,1)$.

6)   See Figure 107.

7)   See Figure 107.

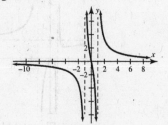

Figure 105

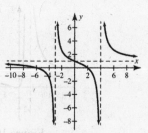

Figure 107

109. $f(x) = \dfrac{x^2 - 2x}{x^2 + 6x + 9}$ is in lowest terms.

1)   To find the vertical asymptote we find the zero of the denominator,
      $x^2 + 6x + 9 = 0 \Rightarrow (x+3)^2 = 0 \Rightarrow x + 3 = 0 \Rightarrow x = -3$.

2)   Since the degree of the numerator is equal to the degree of the denominator the equation of the

      horizontal asymptote is $y = \dfrac{1}{1} = 1$.

3) $f(0) = \dfrac{0^2 - 2(0)}{0^2 + 6(0) + 9} = \dfrac{0}{9} = 0$; $(0, 0)$

4) $\dfrac{x^2 - 2x}{x^2 + 6x + 9} = 0 \Rightarrow x^2 - 2x = 0 \Rightarrow x(x-2) = 0 \Rightarrow x = 0, x = 2$; $(0, 0), (2, 0)$

5) $\dfrac{x^2 - 2x}{x^2 + 6x + 9} = 1 \Rightarrow x^2 - 2x = x^2 + 6x + 9 \Rightarrow -8x = 9 \Rightarrow x = -\dfrac{9}{8} \Rightarrow f$ crosses the horizontal asymptote at

      $\left(-\dfrac{9}{8}, 1\right)$

6)   See Figure 109.

7)   See Figure 109.

111. $f(x) = \dfrac{x^2 + 2x + 1}{x^2 - x - 6}$ is in lowest terms.

1)   To find the vertical asymptotes we find the zeros of the
      denominator, $x^2 - x - 6 = 0 \Rightarrow (x-3)(x+2) = 0 \Rightarrow x = 3, -2$.

2)   Since the degree of the numerator is equal to the degree of the denominator the equation of the
      horizontal asymptote is $y = \dfrac{1}{1} = 1$.

3) $f(0) = \dfrac{0^2 + 2(0) + 1}{0^2 - 0 - 6} = -\dfrac{1}{6} = 0; \left(0, -\dfrac{1}{6}\right)$

4) $\dfrac{x^2 + 2x + 1}{x^2 - x - 6} = 0 \Rightarrow x^2 + 2x + 1 = 0 \Rightarrow (x+1)^2 = 0 \Rightarrow x = -1; (-1, 0)$

5) $\dfrac{x^2 + 2x + 1}{x^2 - x - 6} = 1 \Rightarrow x^2 + 2x + 1 = x^2 - x - 6 \Rightarrow 3x = -7 \Rightarrow x = -\dfrac{7}{3} \Rightarrow f$ crosses the horizontal asymptote at
      $-\dfrac{7}{3}$.

6)   See Figure 111.
7)   See Figure 111.

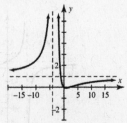

Figure 109

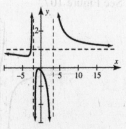

Figure 111

113. See Figure 113. $D: (-\infty, \infty); R: (0, 0.5]$; horizontal asymptote at $y = 0$; no vertical asymptote

115. See Figure 115. $D: (-\infty, \infty); R: (0, 1]$; horizontal asymptote at $y = 0$ (crosses the horizontal asymptote at
      $(0,0)$; no vertical asymptote

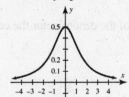

Figure 113

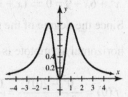

Figure 115

117. (a) $f(98) = \dfrac{100}{101 - 98} = \dfrac{100}{3} \approx 33\%$, the least active 98% post about 33% of the postings.

      (b) $f(100) = \dfrac{100}{101 - 100} = \dfrac{100}{1} = 100\%$, from part (a) we found $f(98) \approx 33\%$, so
             $f(100) - f(98) = 100 - 33 = 67\%$. The most active 2% post about 67% of the postings.

119. (a) $P(1) = \dfrac{5}{0.03(1) + 0.97} = \dfrac{5}{1} = 5\%$, at 1 second there is a 5% chance that the visitor is abandoning the
             website.

      (b) $P(60) = \dfrac{5}{0.03(60) + 0.97} = \dfrac{5}{2.77} \approx 1.8\%$, at 60 seconds there is a 1.8% chance the visitor is abandoning
             the website.

121. (a) $T(4) = -\dfrac{1}{4-8} \Rightarrow T(4) = \dfrac{1}{4} \Rightarrow T(4) = 0.25$; when vehicles leave the ramp at an average rate of 4

vehicles per minute, the wait is 0.25 minutes or 15 seconds.

$T(7.5) = \dfrac{1}{7.5-8} \Rightarrow T(7.5) = \dfrac{1}{0.5} \Rightarrow T(7.5) = 2.0$; when vehicles leave the ramp at an average rate of 7.5

vehicles per minute, the wait is 2 minutes.

(b)   The wait increases dramatically.

123. (a) $N(20) = \dfrac{20^2}{1600-40(20)} = \dfrac{400}{800} = \dfrac{1}{2}$; $N(39) = \dfrac{39^2}{1600-40(39)} = \dfrac{1521}{40} = 38.025$

(b)   The wait increases dramatically.

(c)   We find the vertical asymptote by finding the zero of the denominator,
$1600 - 40x = 0 \Rightarrow 1600 = 40x \Rightarrow x = 40$.

125. (a)   Graph $Y_1 = (10X+1)/(X+1)$ and $Y_2 = 10$ in $[0, 14, 1]$ by $[0, 14, 1]$. Since the degree of the numerator

and denominator are equal, there is a horizontal asymptote at $y = \dfrac{10}{1} = 10$. See Figure 125.

(b)   The initial population would be $f(0) = 1$ million insects.

(c)   After many months the population starts to level off at 10 million.

(d)   The horizontal asymptote $y = 10$ represents the limiting population after a very long time.

127. (a) $f(x) = \dfrac{2540}{x} \Rightarrow f(400) = \dfrac{2540}{400} = 6.35$ inches . A curve designed for 60 mph with a radius of 400 feet

should have the outer rail elevated 6.35 inches.

(b)   Figure 127 shows the graph of $Y_1 = 2540/X$. This means that a sharper curve must be banked more. As
the radius increases, the curve is not as sharp and the elevation of the outer rail decreases.

(c)   Since the degree of the numerator is less than the degree of the denominator, the graph of $f(x) = \dfrac{2540}{x}$

has a horizontal asymptote of $y = 0$. As the radius of the curve increases without bound $(x \to \infty)$, the
tracks become straight and no elevation or banking $(y \to 0)$ of the outer rail is necessary.

(d) $f(x) = 12.7 \Rightarrow 12.7 = \dfrac{2540}{x} \Rightarrow x = \dfrac{2540}{12.7} = 200$; a radius of 200 feet requires an elevation of 12.7

inches.

[0, 14, 1] by [0, 14, 1]

Figure 125

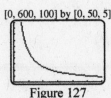

[0, 600, 100] by [0, 50, 5]

Figure 127

Extended and Discovery Exercise for Section 4.6

1. $f(x) = \dfrac{1}{x}$

$\dfrac{f(3)-f(1)}{3-1} = \dfrac{\frac{1}{3}-1}{2} = -\dfrac{1}{3}$, $\dfrac{f(x+h)-f(x)}{h} = \dfrac{\frac{1}{x+h}-\frac{1}{x}}{h} = \dfrac{\frac{x-x-h}{(x+h)(x)}}{h} = \dfrac{\frac{-h}{x(x+h)}}{h} = -\dfrac{1}{x(x+h)}$

3. $f(x) = \dfrac{3}{2x}$

$$\frac{f(3) - f(1)}{3 - 1} = \frac{\frac{3}{6} - \frac{3}{2}}{2} = -\frac{1}{2}, \quad \frac{f(x+h) - f(x)}{h} = \frac{\frac{3}{2(x+h)} - \frac{3}{2x}}{h} = \frac{\frac{3x - 3x - 3h}{2x(x+h)}}{h} =$$

$$\frac{x}{x} \cdot \frac{3}{2(x+h)} - \frac{3}{2x} \cdot \frac{x+h}{x+h} = \frac{3x}{2x(x+h)} - \frac{3(x+h)}{2x(x+h)} = -\frac{3}{2x(x+h)}$$

Checking Basic Concepts for Sections 4.5 and 4.6

1.   Since the polynomial is quadratic, it can have at most two zeros, which are given as $\pm 4i$. Thus, there no real zeros; since the leading coefficient is 3, $f(x) = 3(x + 4i)(x - 4i) = 3(x^2 + 16) = 3x^2 + 48$.

3.   Use a graph of $y = x^3 - x^2 + 4x - 4$ or use the rational zero test to find the zero of 1. See Figure 3. The graph indicates that 1 is a zero, so $(x - 1)$ is a factor. Use synthetic division to find the other factor.

```
1|  1  -1   4  -4
        1   0   4
    1   0   4   0
```

$$f(x) = (x - 1)(x^2 + 4) = (x - 1)(x + 2i)(x - 2i)$$

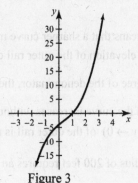

Figure 3

5.   (a)   The denominator $x-1 \neq 0 \Rightarrow D\{x \mid x \neq 1\}$, or $(-\infty, 1) \cup (1, \infty)$.

(b) $f(x) = \dfrac{1}{x-1} + 2 \Rightarrow f(x) = \dfrac{1}{x-1} + \dfrac{2(x-1)}{x-1} \Rightarrow f(x) = \dfrac{2x-1}{x-1}$. To find the vertical asymptote we find the

zero of the denominator. $x-1 = 0 \Rightarrow x = 1$ is the vertical asymptote. Since the degree of the numerator

and denominator is the same, we use the ratio of the leading coefficients $\dfrac{2}{1} \Rightarrow y = 2$ to find the

horizontal asymptote.

(c)   See Figure 5.

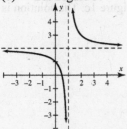

Figure 5

7.   (a) $f(x) = \dfrac{3x-1}{2x-2}$

Since $2x - 2 = 0 \Rightarrow x = 1$, the vertical asymptote is the line $x = 1$. Since the degree of the numerator is

equal to the degree of the denominator the equation of the horizontal asymptote is $y = \dfrac{3}{2}$. See Figure

7a.

(b) $f(x) = \dfrac{1}{(x+1)^2}$

Since $(x+1)^2 = 0 \Rightarrow x = -1$, the vertical asymptote is the line $x = -1$. Since the degree of the
denominator is greater than the degree of the numerator the equation of the horizontal asymptote is
$y = 0$. See Figure 7b.

(c) $f(x) = \dfrac{x+2}{x^2-4} = \dfrac{(x+2)}{(x+2)(x-2)} \Rightarrow f(x)$ has a common factor of $(x+2)$ shows that $f(x)$ has a hole at

$x = -2$. Since $x - 2 = 0 \Rightarrow x = 2$, the equation of the vertical asymptote is the line $x = 2$. Since the
degree of the numerator is less than the degree of the denominator the equation of the horizontal
asymptote is $y = 0$. See Figure 7c.

(d) $f(x) = \dfrac{(x^2+1)}{x^2-1}$

Since $x^2 - 1 = 0 \Rightarrow x = \pm 1$, the equations of the vertical asymptotes are the lines $x = \pm 1$. Since the
degree of the numerator is equal to the degree of the denominator the equation of the horizontal
asymptote is $y = 1$. See Figure 7d.

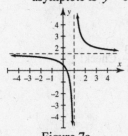

Figure 7a

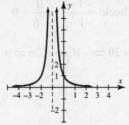

Figure 7b

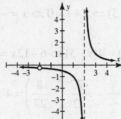

Figure 7c

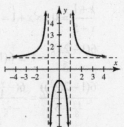

Figure 7d

4.7: More Equations and Inequalities

1. (a) $\dfrac{2x}{x+2} = 6 \Rightarrow 2x = 6(x+2) \Rightarrow 2x = 6x+12 \Rightarrow 4x = -12 \Rightarrow x = -3$

    (b) Graph $Y_1 = (2X)/(X+2)$ and $Y_2 = 6$ in $[-10, 10, 1]$ by $[-10, 10, 1]$ using dot mode. See Figure 1b.
        The intersection point is $(-3, 6)$. The solution is $x = -3$.

    (c) Table $Y_1 = (2X)/(X+2)$ starting at $x = -5$ and incrementing by 1. See Figure 1c. The solution is
        $x = -3$.

$[-10, 10, 1]$ by $[-10, 10, 1]$

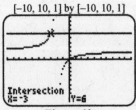

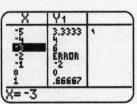

  Figure 1b   Figure 1c

3. (a) $2 - \dfrac{5}{x} + \dfrac{2}{x^2} = 0 \Rightarrow \dfrac{2x^2}{x^2} - \dfrac{5x}{x^2} + \dfrac{2}{x^2} = 0 \Rightarrow \dfrac{2x^2 - 5x + 2}{x^2} = 0 \Rightarrow 2x^2 - 5x + 2 = 0 \Rightarrow$

        $(2x-1)(x-2) = 0 \Rightarrow x = \dfrac{1}{2}, 2$

    (b) Graph $Y_1 = 2 - 5/X + 2/X^2$ in $[-10, 10, 1]$ by $[-10, 10, 1]$ using dot mode. The x-intercepts are the
        points is $(0.5, 0)$ and $(2, 0)$. The solutions are $x = \dfrac{1}{2}, 2$.

    (c) Table $Y_1 = 2 - 5/X + 2/X^2$ starting at $x = -0.5$ and incrementing by 0.5. The solutions are
        $x = \dfrac{1}{2}, 2$.

5. (a) $\dfrac{1}{x+1} + \dfrac{1}{x-1} = \dfrac{1}{x^2-1} \Rightarrow \dfrac{x-1}{x^2-1} + \dfrac{x+1}{x^2-1} = \dfrac{1}{x^2-1} \Rightarrow \dfrac{2x}{x^2-1} = \dfrac{1}{x^2-1} \Rightarrow 2x = 1 \Rightarrow x = \dfrac{1}{2}$

    (b) Graph $Y_1 = 1/(X+1) + 1/(X-1)$ and $Y_2 = 1/(X^2-1)$ in $[-2, 2, 1]$ by $[-5, 5, 1]$ using dot mode. The
        intersection point is $\left(\dfrac{1}{2}, -\dfrac{4}{3}\right)$. The solution is $x = \dfrac{1}{2}$.

    (c) Table $Y_1 = 1/(X+1) + 1/(X-1) - 1/(X^2-1)$ starting at $x = -1$ and incrementing by 0.5. Find the x-
        value where $Y_1 = 0$. The solution is $x = \dfrac{1}{2}$

7. $\dfrac{x+1}{x-5} = 0 \Rightarrow x+1 = 0(x-5) \Rightarrow x+1 = 0 \Rightarrow x = -1$; Check: $\dfrac{-1+1}{-1-5} = \dfrac{0}{-6} = 0$

9. $\dfrac{6(1-2x)}{x-5} = 4 \Rightarrow 6(1-2x) = 4(x-5) \Rightarrow 6-12x = 4x-20 \Rightarrow -16x = -26 \Rightarrow x = \dfrac{13}{8}$ Check:

    $\dfrac{6(1-2(\frac{13}{8}))}{\frac{13}{8}-5} = \dfrac{6(-\frac{9}{4})}{-\frac{27}{8}} = \dfrac{-\frac{27}{2}}{-\frac{27}{8}} = -\dfrac{27}{2} \cdot \left(-\dfrac{8}{27}\right) = 4$

11. $\dfrac{1}{x+2} + \dfrac{1}{x} = 1 \Rightarrow x + (x+2) = x(x+2) \Rightarrow 2x+2 = x^2 + 2x \Rightarrow x^2 = 2 \Rightarrow x = \pm\sqrt{2}$

Check: $\dfrac{1}{\sqrt{2}+2}+\dfrac{1}{\sqrt{2}}=\dfrac{\sqrt{2}}{\sqrt{2}(\sqrt{2}+2)}+\dfrac{\sqrt{2}+2}{\sqrt{2}(\sqrt{2}+2)}=\dfrac{2\sqrt{2}+2}{2+2\sqrt{2}}=1$

Check: $\dfrac{1}{-\sqrt{2}+2}+\dfrac{1}{\sqrt{2}}=\dfrac{-\sqrt{2}}{-\sqrt{2}(-\sqrt{2}+2)}+\dfrac{-\sqrt{2}+2}{-\sqrt{2}(-\sqrt{2}+2)}=\dfrac{2-2\sqrt{2}}{2-2\sqrt{2}}=1$

13. $\dfrac{1}{x}-\dfrac{2}{x^2}=5\Rightarrow x-2=5x^2\Rightarrow 5x^2-x+2=0$; This quadratic equation has no solutions since the discriminant

is negative: $(-1)^2-4(5)(2)=1-40=-39<0$. No real solution.

15. $\dfrac{x^3-4x}{x^2+1}=0\Rightarrow x^3-4x=0(x^2+1)\Rightarrow x^3-4x=0\Rightarrow x(x+2)(x-2)=0\Rightarrow x=0,-2,2$

Check: $\dfrac{(0)^3-4(0)}{(0)^2+1}=\dfrac{0}{1}=0$; Check: $\dfrac{(-2)^3-4(-2)}{(-2)^2+1}=\dfrac{0}{5}=0$; Check: $\dfrac{(2)^3-4(2)}{(2)^2+1}=\dfrac{0}{5}=0$

17. $\dfrac{35}{x^2}=\dfrac{4}{x}+15\Rightarrow 35=4x+15x^2\Rightarrow 15x^2+4x-35=0\Rightarrow (5x-7)(3x+5)=0\Rightarrow x=\dfrac{-5}{3},\dfrac{7}{5}$ Check:

$\dfrac{35}{(\frac{-5}{3})^2}=\dfrac{4}{\frac{-5}{3}}+15\Rightarrow\dfrac{63}{5}=\dfrac{-12}{5}+\dfrac{75}{5}\Rightarrow\dfrac{63}{5}=\dfrac{63}{5}$;

Check: $\dfrac{35}{(\frac{7}{5})^2}=\dfrac{4}{\frac{7}{5}}+15\Rightarrow\dfrac{125}{7}=\dfrac{20}{7}+\dfrac{105}{7}\Rightarrow\dfrac{125}{7}=\dfrac{125}{7}$

19. $\dfrac{x+5}{x+2}=\dfrac{x-4}{x-10}\Rightarrow (x+5)(x-10)=(x-4)(x+2)\Rightarrow x^2-5x-50=x^2-2x-8\Rightarrow 3x+42=0\Rightarrow x=-14$;

Check: $\dfrac{(-14)+5}{(-14)+2}=\dfrac{(-14)-4}{(-14)-10}\Rightarrow\dfrac{-9}{-12}=\dfrac{-18}{-24}\Rightarrow\dfrac{3}{4}=\dfrac{3}{4}$

21. $\dfrac{1}{x-2}-\dfrac{2}{x-3}=\dfrac{-1}{x^2-5x+6}\Rightarrow\dfrac{1}{x-2}-\dfrac{2}{x-3}=\dfrac{-1}{(x-2)(x-3)}\Rightarrow x-3-2(x-2)=-1\Rightarrow -x+1=-1\Rightarrow x=2$; No

solution. Check: There is no solution since $x=2$ is not defined in the original equation.

23. $\dfrac{2x}{x^2-1}=\dfrac{2}{x+1}-\dfrac{1}{x-1}\Rightarrow\dfrac{2x}{(x+1)(x-1)}=\dfrac{2}{x+1}-\dfrac{1}{x-1}\Rightarrow 2x=2(x-1)-1(x+1)\Rightarrow 2x=x-3\Rightarrow x=-3$

Check: $\dfrac{2(-3)}{(-3)^2-1}=\dfrac{2}{(-3)+1}-\dfrac{1}{(-3)-1}=\dfrac{-3}{4}=-\dfrac{3}{4}$

25. $\dfrac{4}{x^2-3x}=\dfrac{1}{x^2-9}\Rightarrow\dfrac{4}{x(x-3)}=\dfrac{1}{(x-3)(x+3)}\Rightarrow 4(x+3)=1(x)\Rightarrow 4x+12=x\Rightarrow x=-4$

Check: $\dfrac{4}{(-4)^2-3(-4)}=\dfrac{1}{(-4)^2-9}\Rightarrow\dfrac{4}{28}=\dfrac{1}{7}\Rightarrow\dfrac{1}{7}=\dfrac{1}{7}$.

27. $\dfrac{2}{x-1}+1=\dfrac{4}{x^2-1}\Rightarrow\dfrac{2}{x-1}+1=\dfrac{4}{(x+1)(x-1)}\Rightarrow 2(x+1)+(x+1)(x-1)=4\Rightarrow$

$2x+2+x^2-1=4\Rightarrow x^2+2x-3=0\Rightarrow (x+3)(x-1)=0\Rightarrow x=-3,1$. Since $x=1$ is not defined in the original

equation $x=-3$. Check: $\dfrac{2}{-3-1}+1=\dfrac{4}{(-3)^2-1}\Rightarrow -\dfrac{1}{2}+1=\dfrac{4}{8}\Rightarrow\dfrac{1}{2}=\dfrac{1}{2}$

29. $\dfrac{1}{x+2} = \dfrac{4}{4-x^2} - 1 \Rightarrow \dfrac{1}{x+2} = \dfrac{4}{(2+x)(2-x)} - 1 \Rightarrow 2-x = 4 - (2+x)(2-x) \Rightarrow$

$2-x = 4-(4-x^2) \Rightarrow 2-x = 4-4+x^2 \Rightarrow x^2 + x - 2 = 0 \Rightarrow (x+2)(x-1) = 0 \Rightarrow x = -2, 1$. Since $x = -2$ is

not defined in the original equation $x = 1$.

Check: $\dfrac{1}{1+2} = \dfrac{4}{4-(1)^2} - 1 \Rightarrow \dfrac{1}{3} = \dfrac{4}{3} - 1 \Rightarrow \dfrac{1}{3} = \dfrac{1}{3}$

31. $\dfrac{1}{x-1} + \dfrac{1}{x+1} = \dfrac{2}{x^2-1} \Rightarrow \dfrac{1}{x-1} + \dfrac{1}{x+1} = \dfrac{2}{(x+1)(x-1)} \Rightarrow x+1+x-1 = 2 \Rightarrow 2x = 2 \Rightarrow x = 1$. Since $x = 1$ is not

defined in the original equation, there is no solution to the equation.

33. (a)   The boundary numbers, the x-values for which $f(x) = 0$, are $-4, -2$, and 2.

   (b) $f(x) > 0$ on the interval for which the graph of f is above the x-axis. That is $(-4, -2) \cup (2, \infty)$. In
         set builder notation the intervals are $\{x \mid -4 < x < -2 \text{ or } x > 2\}$.

   (c) $f(x) < 0$ on the interval for which the graph of f is below the x-axis. That is $(-\infty, -4) \cup (-2, 2)$. In
         set builder notation the intervals are $\{x \mid x < -4 \text{ or } -2 < x < 2\}$.

35. (a)   The boundary numbers, the x-values for which $f(x) = 0$, are $-4, -2, 0$ and 2.

   (b) $f(x) > 0$ on the interval for which the graph of f is above the x-axis. That is $(-4, -2) \cup (0, 2)$. In set
         builder notation the intervals are $\{x \mid -4 < x < -2 \text{ or } 0 < x < 2\}$.

   (c) $f(x) < 0$ on the interval for which the graph of f is below the x-axis. That is
         $(-\infty, -4) \cup (-2, 0) \cup (2, \infty)$. In set builder notation the intervals are $\{x \mid x < -4 \text{ or } -2 < x < 0 \text{ or } x > 2\}$.

37. (a)   The boundary numbers, the x-values for which $f(x) = 0$, are $-2, 1$, and 2.

   (b) $f(x) > 0$ on the interval for which the graph of f is above the x-axis. That is $(-\infty, -2) \cup (-2, 1)$. In
         set builder notation the intervals are $\{x \mid x < -2 \text{ or } -2 < x < 1\}$.

   (c) $f(x) < 0$ on the interval for which the graph of f is below the x-axis. That is $(1, 2) \cup (2, \infty)$. In set
         builder notation the intervals are $\{x \mid 1 < x < 2 \text{ or } x > 2\}$.

39. (a) $f(x)$ is undefined at $x = 0$.

   (b) $f(x) > 0$ on the interval for which the graph of f is above the x-axis. That is $(-\infty, 0) \cup (0, \infty)$. In set
         builder notation the intervals are $\{x \mid x < 0 \text{ or } x > 0\}$.

   (c) $f(x) < 0$ on the interval for which the graph of f is below the x-axis. There are no solutions.

41. (a) $f(x)$ is undefined at $x = 1$, $f(x) = 0$ at $x = 0$.

   (b) $f(x) > 0$ on the interval for which the graph of f is above the x-axis. That is $(-\infty, 0) \cup (1, \infty)$. In set
         builder notation the intervals are $\{x \mid x < 0 \text{ or } x > 1\}$.

   (c) $f(x) < 0$ on the interval for which the graph of f is below the x-axis. That is $(0, 1)$. In set builder
         notation the interval is $\{x \mid 0 < x < 1\}$.

43. (a) $f(x)$ is undefined at $x = \pm 2$, $f(x) = 0$ at $x = 0$.

   (b) $f(x) > 0$ on the interval for which the graph of f is above the x-axis. That is $(-\infty, -2) \cup (2, \infty)$. In
         set builder notation the intervals are $\{x \mid x < -2 \text{ or } x > 2\}$.

   (c) $f(x) < 0$ on the interval for which the graph of f is below the x-axis. That is $(-2, 0) \cup (0, 2)$. In set
         builder notation the intervals are $\{x \mid -2 < x < 0 \text{ or } 0 < x < 2\}$.

45. (a)   First find the boundary numbers by solving $x^3 - x = 0$.

         $x^3 - x = 0 \Rightarrow x(x^2 - 1) = 0 \Rightarrow x(x+1)(x-1) = 0 \Rightarrow x = 0, -1, \text{ or } 1$

The boundary values divide the number line into four intervals. Choose a test value from each of these intervals and evaluate $x^3 - x$ for these values. See Figure 45a. The solution is $(-1, 0) \cup (1, \infty)$. In set builder notation the intervals are $\{x \mid -1 < x < 0 \text{ or } x > 1\}$.

(b) Graph $Y_1 = X^3 - X$ as shown in Figure 45b. The x-intercepts are $-1, 0$, and 1. The graph is above the x-axis on the interval $(-1, 0) \cup (1, \infty)$.

Interval	Test Value x	$x^3 - x$	Positive or Negative?
$(-\infty, -1)$	-2	-6	Negative
$(-1, 0)$	-0.5	0.375	Positive
$(0, 1)$	0.5	-0.375	Negative
$(1, \infty)$	2	6	Positive

Figure 45a

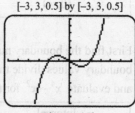

[-3, 3, 0.5] by [-3, 3, 0.5]

Figure 45b

47. (a) Write the inequality as $x^3 + x^2 - 2x \geq 0$. Then find the boundary numbers by solving $x^3 + x^2 - 2x = 0$.
$$x^3 + x^2 - 2x = 0 \Rightarrow x(x^2 + x - 2) = 0 \Rightarrow x(x+2)(x-1) = 0 \Rightarrow x = 0, -2, \text{ or } 1$$
The boundary values divide the number line into four intervals. Choose a test value from each of these intervals and evaluate $x^3 + x^2 - 2x$ for these values. See Figure 47a. The solution is $[-2, 0] \cup [1, \infty)$. In set builder notation the intervals are $\{x \mid -2 \leq x \leq 0 \text{ or } x \geq 1\}$.

(b) Graph $Y_1 = X^3 + X^2 - 2X$ as shown in Figure 47b. The x-intercepts are $-1, 0$, and 1. The graph intersects or is above the x-axis on the interval $[-2, 0] \cup [1, \infty)$.

Interval	Test Value x	$x^3 + x^2 - 2x$	Positive or Negative?
$(-\infty, -2)$	-3	-12	Negative
$(-2, 0)$	-1	2	Positive
$(0, 1)$	0.5	-0.625	Negative
$(1, \infty)$	2	8	Positive

Figure 47a

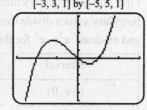

[-3, 3, 1] by [-5, 5, 1]

Figure 47b

49. (a) First find the boundary numbers by solving $x^4 - 13x^2 + 36 = 0$.
$$x^4 - 13x^2 + 36 = 0 \Rightarrow (x^2 - 4)(x^2 - 9) = 0 \Rightarrow (x+2)(x-2)(x+3)(x-3) \Rightarrow$$
$x = -3, -2, 2$ or 3 The boundary values divide the number line into five intervals. Choose a test value from each of these intervals and evaluate $x^4 - 13x^2 + 36$ for these values. See Figure 49a. The solution is $(-3, -2) \cup (2, 3)$. In set builder notation the intervals are $\{x \mid -3 < x < -2 \text{ or } 2 < x < 3\}$.

(b) Graph $Y_1 = X^4 - 13X^2 + 36$ as shown in Figure 49b. The x-intercepts are $-3, -2, 2$ and 3. The graph is below the x-axis on the interval $(-3, -2) \cup (2, 3)$.

Interval	Test Value x	$x^4 - 13x^2 + 36$	Positive or Negative?
$(-\infty, -3)$	-4	84	Positive
$(-3, -2)$	-2.5	-6.1875	Negative
$(-2, 2)$	0	36	Positive
$(2, 3)$	2.5	-6.1875	Negative
$(3, \infty)$	4	84	Positive

Figure 49a

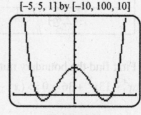

[-5, 5, 1] by [-10, 100, 10]

Figure 49b

51. First find the boundary numbers by solving $x^3 = 0$. $x^3 = 0 \Rightarrow x = 0$. The boundary value divides the number line into two intervals. Choose a test value from each of these intervals and evaluate x^3 for these values. See Figure 51. The solution is $(0, \infty)$.

Interval	Test Value x	x^3	Positive or Negative?
$(-\infty, 0)$	-2	-8	Negative
$(0, \infty)$	2	8	Positive

Figure 51

53. First find the boundary numbers by solving $x^4 - x^2 = 0$. $x^4 - x^2 = 0 \Rightarrow x^2(x^2 - 1) = 0 \Rightarrow x = 0, \pm 1$. The boundary values divide the number line into four intervals. Choose a test value from each of these intervals and evaluate $x^4 - x^2$ for these values. See Figure 53. The solution is $(-1, 0) \cup (0, 1)$.

Interval	Test Value x	$x^4 - x^2$	Positive or Negative?
$(-\infty, -1)$	-2	12	Positive
$(-1, 0)$	$-\frac{1}{2}$	-0.188	Negative
$(0, 1)$	$\frac{1}{2}$	-0.188	Negative
$(1, \infty)$	2	12	Positive

Figure 53

55. First find the boundary numbers by solving $x^3 - x^2 = 0$. $x^3 - x^2 = 0 \Rightarrow x^2(x - 1) = 0 \Rightarrow x = 0, 1$. The boundary values divide the number line into three intervals. Choose a test value from each of these intervals and evaluate $x^3 - x^2$ for these values. See Figure 55. The solution is $(1, \infty)$.

Interval	Test Value x	$x^3 - x^2$	Positive or Negative?
$(-\infty, 0)$	-1	-2	Negative
$(0, 1)$	0.5	-0.125	Negative
$(1, \infty)$	2	4	Positive

Figure 55

57. First find the boundary numbers by solving $(x-1)(x-2)^2 = 0$. $(x-1)(x-2)^2 = 0 \Rightarrow x = 1, 2$. The boundary values divide the number line into three intervals. Choose a test value from each of these intervals and evaluate $(x-1)(x-2)^2$ for these values. See Figure 57. The solution is $(-\infty, 1)$.

Interval	Test Value x	$(x-1)(x-2)^2$	Positive or Negative?
$(-\infty, 1)$	0	-4	Negative
$(1, 2)$	1.5	0.125	Positive
$(2, \infty)$	3	2	Positive

Figure 57

59. First find the boundary numbers by solving $x^4 - 13x^2 + 36 = 0$.
$x^4 - 13x^2 + 36 = 0 \Rightarrow (x^2 - 4)(x^2 - 9) = 0 \Rightarrow x = \pm 2, \pm 3$. The boundary values divide the number line into five

intervals. Choose a test value from each of these intervals and evaluate $x^4 - 13x^2 + 36$ for these values. See Figure 59. Because the boundary numbers are included, the solution is $[-3, -2] \cup [2, 3]$.

Interval	Test Value x	$x^4 - 13x^2 + 36$	Positive or Negative?
$(-\infty, -3)$	-5	336	Positive
$(-3, -2)$	-2.5	-6.188	Negative
$(-2, 2)$	0	36	Positive
$(2, 3)$	2.5	-6.188	Negative
$(3, \infty)$	4	84	Positive

Figure 59

61. First find the boundary numbers by solving $(x-1)(x-2)(x+2) = 0$. $(x-1)(x-2)(x+2) = 0 \Rightarrow x = -2, 1$ or 2

The boundary values divide the number line into four intervals. Choose a test value from each of these intervals and evaluate $(x-1)(x-2)(x+2)$ for these values. See Figure 61. The solution is $[-2, 1] \cup [2, \infty)$.

Interval	Test value x	$(x-1)(x-2)(x+2)$	Positive or Negative?
$(-\infty, -2)$	-3	-20	Negative
$(-2, 1)$	0	4	Positive
$(1, 2)$	1.5	-0.875	Negative
$(2, \infty)$	3	10	Positive

Figure 61

63. Write the inequality as $2x^4 + 2x^3 - 12x^2 \leq 0$. Find the boundary numbers by solving $2x^4 + 2x^3 - 12x^2 = 0$.
$2x^4 + 2x^3 - 12x^2 = 0 \Rightarrow 2x^2(x^2 + x - 6) = 0 \Rightarrow 2x^2(x+3)(x-2) = 0 \Rightarrow x = -3, 0$ or 2

The boundary values divide the number line into four intervals. Choose a test value from each of these intervals and evaluate $2x^4 + 2x^3 - 12x^2$ for these values. See Figure 63. The solution is $[-3, 2]$.

Interval	Test value x	$2x^4 + 2x^3 - 12x^2$	Positive or Negative?
$(-\infty, -3)$	-4	192	Positive
$(-3, 0)$	-1	-12	Negative
$(0, 2)$	1	-8	Negative
$(2, \infty)$	3	108	Positive

Figure 63

65. Given: $P(x) = (x-4)(x-1)(x+2)$

(a) $P(x) = 0 \Rightarrow (x-4)(x-1)(x+2) = 0 \Rightarrow x = -2, 1, 4$.

(b) $P(x) < 0$. Since the boundary values are at $x = -2, 1, 4$, there are four disjoint intervals to analyze.
        See Figure 65. The solution is $(-\infty, -2) \cup (1, 4)$.

(c) $P(x) > 0$. Since the boundary values are at $x = -2, 1, 4$, there are four disjoint intervals to analyze.
        See Figure 65. The solution is $(-2, 1) \cup (4, \infty)$.

Interval	Test value x	$P(x)$	Positive or Negative?
$(-\infty, -2)$	-4	-80	Negative
$(-2, 1)$	0	8	Positive
$(1, 4)$	2	-8	Negative
$(4, \infty)$	10	648	Positive

Figure 65

67. Write the inequality as $x^3 - 7x^2 + 14x - 8 \le 0$.

Then graph $Y_1 = X^{\wedge}3 - 7X^{\wedge}2 + 14X - 8$ as shown in Figure 67. The x-intercepts are 1, 2, and 4.

The graph intersects or is below the x-axis on the interval $(-\infty, 1] \cup [2, 4]$. In set builder notation the intervals are $\{x \mid x \le 1 \text{ or } 2 \le x \le 4\}$.

[-2, 5, 1] by [-10, 10, 1]

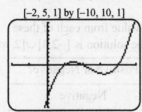

Figure 67

69. (a) Find the zeros of the numerator and the denominator.
Numerator: No zero; Denominator: $x = 0$
This boundary number divides the number line into two intervals. Choose a test value from each of these intervals and evaluate $\dfrac{1}{x}$ for these values. See Figure 69a. The solution is $(-\infty, 0)$. In set builder notation the interval is $\{x \mid x < 0\}$.

(b) Graph $Y_1 = 1/X$ as shown in Figure 69b. The graph has a vertical asymptote at $x = 0$ and no x-intercepts. The graph is below the x-axis on the interval $(-\infty, 0)$.

[-5, 5, 1] by [-10, 100, 10]

Interval	Test Value x	$1/x$	Positive or Negative?
$(-\infty, 0)$	-1	-1	Negative
$(0, \infty)$	1	1	Positive

Figure 69a

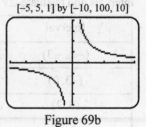

Figure 69b

71. (a) Find the zeros of the numerator and the denominator.
Numerator: No zero; Denominator: $x + 3 = 0 \Rightarrow x = -3$
This boundary number divides the number line into two intervals. Choose a test value from each of these intervals and evaluate $\dfrac{4}{x+3}$ for these values. See Figure 71a. The solution is $(-3, \infty)$. In set builder notation the interval is $\{x \mid x > -3\}$.

(b) Graph $Y_1 = 4/(X + 3)$ using dot mode as shown in Figure 71b. The graph has a vertical asymptote at $x = -3$ and no x-intercepts. The graph is above the x-axis on the interval $(-3, \infty)$. Note that the value $x = -3$ cannot be included in the solution set since $\dfrac{4}{x+3}$ is undefined for $x = -3$.

Interval	Test Value x	$4/(x + 3)$	Positive or Negative?
$(-\infty, -3)$	-4	-4	Negative
$(-3, \infty)$	1	1	Positive

Figure 71a

Figure 71b

73. (a) Find the zeros of the numerator and the denominator.

Numerator: No zero; Denominator: $x^2 - 4 = 0 \Rightarrow x = \pm 2$

These boundary numbers divide the number line into three intervals. Choose a test value from each of these intervals and evaluate $\dfrac{5}{x^2 - 4}$ for these values. See Figure 73a. The solution is $(-2, 2)$. In set builder notation the interval is $\{x \mid -2 < x < 2\}$.

(b) Graph $Y_1 = 5/(X^2 - 4)$ using dot mode as shown in Figure 73b. The graph has vertical asymptotes at $x = \pm 2$ and no x-intercepts. The graph is below the x-axis on the interval $(-2, 2)$.

Interval	Test Value x	$5/(x^2 - 4)$	Positive or Negative?
$(-\infty, -2)$	-3	1	Positive
$(-2, 2)$	0	-1.25	Negative
$(2, \infty)$	3	1	Positive

Figure 73a

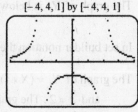

Figure 73b

75. First split the function and set $p(x)$ and $q(x)$ both equal to zero. So:

$\dfrac{1}{x^2 + 1} = 0 \Rightarrow 1 = 0; x^2 + 1 = 0 \Rightarrow x^2 = -1 \Rightarrow$ there are no boundary numbers. See Figure 75. The solution is $(-\infty, \infty)$.

Interval	Test value x	$\dfrac{1}{x^2 + 1}$	Positive or Negative?
$(-\infty, \infty)$	0	1	Positive

Figure 75

77. First, split the function and set $p(x)$ and $q(x)$ both equal to zero. So:

$\dfrac{1}{(x-1)^2} = 0 \Rightarrow 1 = 0; (x-1)^2 = 0 \Rightarrow x = 1$. See Figure 77. Since both intervals are greater than zero, there are no real solutions.

Interval	Test value x	$\dfrac{1}{(x-1)^2}$	Positive or Negative?
$(-\infty, 1)$	0	1	Positive
$(1, \infty)$	10	$\dfrac{1}{81}$	Positive

Figure 77

79. First, split the function and set $p(x)$ and $q(x)$ both equal to zero. So: $-\dfrac{1}{x} = 0 \Rightarrow -1 = 0; x = 0$. See Figure 79.

The solution is $(-\infty, 0)$.

Interval	Test value x	$-\dfrac{1}{x}$	Positive or Negative?
$(-\infty, 0)$	-1	1	Positive
$(0, \infty)$	1	-1	Negative

Figure 79

81. The graph of $Y_1 = (X+1)^{\wedge}2/(X-2)$ has a vertical asymptote at $x = 2$ and an x-intercept at $x = -1$. The graph of Y_1 intersects or is below the x-axis on the interval $(-\infty, 2)$. Note that 2 cannot be included. In set builder notation the interval is $\{x \mid x > 2\}$.

83. The graph of $Y_1 = (3-2X)/(1+X)$ has a vertical asymptote at $x = -1$ and an x-intercept at $x = \dfrac{3}{2}$.

The graph of Y_1 is below the x-axis on the interval $(-\infty, -1) \cup \left(\dfrac{3}{2}, \infty\right)$.

In set builder notation the intervals are $\left\{x \mid x < -1 \text{ or } x > \dfrac{3}{2}\right\}$.

85. The graph of $Y_1 = (X+1)(X-2)/(X+3)$ has a vertical asymptote of at $x = -3$ and x-intercepts at $x = -1$ and $x = 2$. The graph of Y_1 is below the x-axis on the interval $(-\infty, -3) \cup (-1, 2)$.

In set builder notation the intervals are $\{x \mid x < -3 \text{ or } -1 < x < 2\}$.

87. The graph of $Y_1 = 2X - 5/((X+1)(X-1))$ has vertical asymptotes at $x = -1$ and $x = 1$ and an x-intercept at $x = \dfrac{5}{2}$. The graph of Y_1 is on or above the x-axis on the interval $(-1, 1) \cup \left[\dfrac{5}{2}, \infty\right)$.

In set builder notation the intervals are $\left\{x \mid -1 < x < 1 \text{ or } x \geq \dfrac{5}{2}\right\}$.

89. First, rewrite the inequality:

$\dfrac{1}{x-3} \leq \dfrac{5}{x-3} \Rightarrow \dfrac{1}{x-3} - \dfrac{5}{x-3} \leq 0 \Rightarrow \dfrac{-4}{x-3} \leq 0$

The graph of $Y_1 = -4/(X-3)$ has a vertical asymptote at $x = 3$ and no x-intercept.

The graph of Y_1 intersects or is below the x-axis on the interval $(3, \infty)$. Note that 3 can not be included.

In set builder notation the interval is $\{x \mid x > 3\}$.

91. First, rewrite the inequality:

$2 - \dfrac{5}{x} + \dfrac{2}{x^2} \geq 0 \Rightarrow \dfrac{2x^2 - 5x + 2}{x^2} \geq 0 \Rightarrow \dfrac{(2x-1)(x-2)}{x^2} \geq 0$

The graph of $Y_1 = (2X-1)(X-2)/X^{\wedge}2$ has a vertical asymptote at $x = 0$ and x-intercepts $x = \dfrac{1}{2}$ and

$x = 2$. The graph of Y_1 intersects or is above the x-axis on the interval $(-\infty, 0) \cup \left(0, \dfrac{1}{2}\right] \cup [2, \infty)$. In

set builder notation the intervals are $\left\{x \mid x < 2 \text{ or } 0 < x \leq \dfrac{1}{2} \text{ or } x \geq 2\right\}$.

93. First, rewrite the inequality:

$\dfrac{1}{x} \le \dfrac{2}{x+2} \Rightarrow \dfrac{1}{x} - \dfrac{2}{x+2} \le 0 \Rightarrow \dfrac{x+2-2x}{x(x+2)} \le 0 \Rightarrow \dfrac{2-x}{x(x+2)} \le 0$. The graph of $Y_1 = (2-X)/((X(+2))$ has vertical

asymptotes at $x = 0$ and $x = -2$, and x-intercept of $x = 2$. The graph of Y_1 is below the x-axis on

the interval $(-2, 0) \cup [2, \infty)$. In set builder notation the intervals are $\{x \mid -2 < x < 0 \text{ or } x \ge 2\}$.

95. (a) $\dfrac{x-3}{x+5} = 0 \Rightarrow x - 3 = 0 \Rightarrow x = 3; x = -5$ is a vertical asymptote/boundary number.

The solution is $[3]$.

(b) $\dfrac{x-3}{x+5} < 0$. Since the boundary values are at $x = -5, 3$, there are three disjoint intervals to analyze.

See Figure 95. The solution is $(-5, 3]$.

(c) $\dfrac{x-3}{x+5} > 0$. Since the boundary values are at $x = -5, 3$, there are three disjoint intervals to analyze.

See Figure 95. The solution is $(-\infty, -5) \cup [3, \infty)$.

Interval	Test value x	$\dfrac{x-3}{x+5}$	Positive or Negative?
$(-\infty, -5)$	-6	9	Positive
$(-5, 3)$	0	-0.6	Negative
$(3, \infty)$	6	$.2727$	Positive

Figure 95

97. (a) $\dfrac{x-1}{x+2} = 1 \Rightarrow \dfrac{x-1}{x+2} - 1 = 0 \Rightarrow \dfrac{x-1}{x+2} - \dfrac{(x+2)}{(x+2)} = 0 \Rightarrow \dfrac{-3}{x+2} = 0 \Rightarrow -3 = 0$; so there are no real

solutions. $x + 2 = 0 \Rightarrow x = -2$ is a vertical asymptote/boundary number.

(b) $\dfrac{x-1}{x+2} > 1$. Since the boundary value is at $x = -2$, there are two disjoint intervals to analyze.

See Figure 97. The solution is $(-\infty, -2)$.

(c) $\dfrac{x-1}{x+2} < 1$. Since the boundary value is at $x = -2$, there are two disjoint intervals to analyze.

See Figure 97. The solution is $(-2, \infty)$.

Interval	Test value x	$\dfrac{x-1}{x+2} - 1$	Positive or Negative?
$(-\infty, -2)$	-3	3	Positive
$(-2, \infty)$	0	-1.5	Negative

Figure 97

99. (a) $\dfrac{(x-2)(2) - (2x+1)(1)}{(x-2)^2} = 0 \Rightarrow \dfrac{-5}{(x-2)^2} = 0 \Rightarrow -5 = 0$. Therefore there are no real solutions.

(b) $\dfrac{(x-2)(2) - (2x+1)(1)}{(x-2)^2} < 0 \Rightarrow \dfrac{-5}{(x-2)^2} < 0 \Rightarrow x = 2$ is a boundary value. Using test values, the

solution is $(-\infty, 2) \cup (2, \infty)$.

101. (a) $\dfrac{(x^2+1)(2x)-(x^2-1)(2x)}{(x^2+1)^2}=0 \Rightarrow \dfrac{4x}{(x^2+1)^2}=0 \Rightarrow 4x=0 \Rightarrow x=0$. The solution is $[0]$.

    (b) $\dfrac{(x^2+1)(2x)-(x^2-1)(2x)}{(x^2+1)^2}\ge 0 \Rightarrow \dfrac{4x}{(x^2+1)^2}\ge 0 \Rightarrow x=0$ is a boundary number. Using test values,

    the solution is $[0,\infty)$.

103. (a) $\dfrac{(x-1)(2x)-(x^2)(1)}{(x-1)^2}=0 \Rightarrow \dfrac{x(x-2)}{(x-1)^2}=0 \Rightarrow x(x-2)=0 \Rightarrow x=0,2$. The solution is $\{0,2\}$.

    (b) $\dfrac{(x-1)(2x)-(x^2)(1)}{(x-1)^2}>0 \Rightarrow \dfrac{x(x-2)}{(x-1)^2}>0 \Rightarrow x=0,1,2$ are boundary numbers. Using test values, the

    solution is $(-\infty,0)\cup(2,\infty)$.

105. (a) Since 15 seconds equals 0.25 minute, graph $Y_1=(X-5)/(X^2-10X)$ and $Y_2=0.25$ as shown in
    Figure 105. The intersection point is near (12.4, 0.25). The gate should admit 12.4 cars per minute on
    average to keep the wait less than 15 seconds. Note: The reason the answer is greater than 10 cars per
    minute is because cars are arriving randomly. For some minutes, more than 10 vehicles might arrive at
    the gate.

    (b) $\dfrac{12.4}{5}=2.48$ or 3 parking attendants must be on duty to keep the average wait less than 15 seconds.

[10, 15, 1] by [0, 1, 0.1]

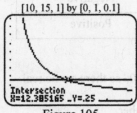

Figure 105

107. Let $y=\text{height}$, $x=\text{width}$, and $2x=\text{length of the box}$. To relate the variables, we use the volume formula

    for a box and the fact that $V=196$ cubic inches. $V=(2x)xy=2x^2y \Rightarrow y=\dfrac{V}{2x^2}=\dfrac{196}{2x^2}=\dfrac{98}{x^2}$.

    The surface area of the box is the sum of the areas of the 6 rectangular sides. We are given that $A=280\,\text{in}^2$.

    $A=2(x\cdot y)+2(2x\cdot y)+2(x\cdot 2x) \Rightarrow A=6xy+4x^2=280$; and since $y=\dfrac{98}{x^2}$ we get the equation

    $6x\left(\dfrac{98}{x^2}\right)+4x^2=280 \Rightarrow \dfrac{588}{x}+4x^2=280$.

    Figures 107a and 107b show the intersection points when graphing $Y_1=588/X+4X^2$ and $Y_2=280$
    There are two possible solutions (in inches):

    $\text{width}=7$, $\text{length}=14$, $\text{height}=\dfrac{98}{7^2}=2$ or $\text{width}\approx 2.266$, $\text{length}\approx 4.532$, $\text{height}\approx \dfrac{98}{2.266^2}\approx 19.086$.

[0, 10, 2] by [100, 400, 20]   [0, 10, 2] by [100, 400, 20]

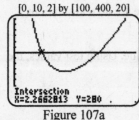   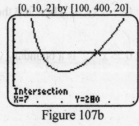

Figure 107a   Figure 107b

109. (a) $V = L \cdot W \cdot H \Rightarrow V = x \cdot x \cdot h \Rightarrow 108 = x^2 h \Rightarrow \dfrac{108}{x^2} = h$. The surface area

$A(x) = x^2 + 4xh \Rightarrow A(x) = x^2 + 4x\left(\dfrac{108}{x^2}\right) \Rightarrow A(x) = \left(x^2 + \dfrac{432}{x}\right)$. This finds surface area in square

inches. To find square feet we must divide by 144 (the number of square inches in a square foot) $\Rightarrow$

$A(x) = \left(x^2 + \dfrac{432}{x}\right) \div 144 \Rightarrow A(x) = \dfrac{x^2}{144} + \dfrac{3}{x}$.

(b) $C = 0.10\left(\dfrac{x^2}{144} + \dfrac{3}{x}\right)$

(c)   Graph the equation: $C = 0.10\left(\dfrac{x^2}{144} + \dfrac{3}{x}\right)$. The minimum cost is when $x \approx 6 \Rightarrow h = \dfrac{108}{(6)^2} \Rightarrow h = 3$. The

dimensions are $6 \times 6 \times 3$ inches.

111. (a) $D(0.05) = \dfrac{2500}{30(0.3 + 0.05)} \approx 238$

The braking distance far a car traveling at 50 mph on a 5% uphill grade is about 238 feet.

(b)   Table $Y_1 = 2500 / (30(0.3 + X))$ starting at $x = 0$ and incrementing by 0.05 as shown in Figure 111.

The table indicates that as the grade x increases, the braking distance decreases. This agrees with
driving experience.

(c) $220 = \dfrac{2.500}{30(0.3 + x)} \Rightarrow 0.3 + x = \dfrac{2500}{30(220)} \Rightarrow x = \dfrac{2500}{30(220)} - 0.3 \approx 0.079$, or 7.9%

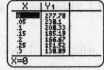

Figure 111

113. (a)   Graph $Y_1 = X^{\wedge}2 / (1600 - 40X)$ and $Y_2 = 8$ (not shown). The graphs intersect approximately at $(36, 8)$.
The graph of Y_1 is equal to or is below the graph of Y_2 when $x \le 36$ (approximately).

(b)   The average line length is less than or equal to 8 cars when the average arrival rate is 36 cars per hour
or less.

115. (a)   The graph of $D(x) = \dfrac{120}{x}$ increases as x decreases, which means the braking distance increases as the
coefficient of friction becomes smaller.

(b) $\dfrac{120}{x} \ge 400 \Rightarrow \dfrac{120}{400} \ge x \Rightarrow 0.3 \ge x$; the braking distance is 400 feet or more when $0 < x \le 0.3$.

117. The volume of a cube is $V = x^3$, where x is the length of a side.

$212.8 \le V \le 213.2 \Rightarrow 212.8 \le x^3 \le 213.2 \Rightarrow \sqrt[3]{212.8} \le x \le \sqrt[3]{213.2}$ inches . This is approximately
$5.97022 \le x \le 5.97396$ inches.

119. If $y = \dfrac{k}{x}$ and $y = 2$ when $x = 3$, then $2 = \dfrac{k}{3} \Rightarrow k = 6$.

121. If $y = kx^3$ and $y = 64$ when $x = 2$, then $64 = k(8) \Rightarrow 8k = 64 \Rightarrow k = 8$.

123. If $T = kx^{3/2}$ and $T = 20$ when $x = 4$, then $20 = k(4)^{3/2} \Rightarrow 20 = 8k \Rightarrow k = 2.5$.

The variation equation becomes $T = 2.5x^{3/2}$. When $x = 16$, $T = 2.5(16)^{3/2} = 2.5(64) = 160$.

125. If $y = \dfrac{k}{x}$ and $y = 5$ when $x = 6$, then $5 = \dfrac{k}{6} \Rightarrow k = 30$.

The variation equation becomes $y = \dfrac{30}{x}$. When $x = 15$, $y = \dfrac{30}{15} = 2$.

127. If y is inversely proportional to x, then $y = \dfrac{k}{x}$ must hold. So if the value of x is doubled, the right side of

this equation becomes $\dfrac{k}{2x} = \dfrac{1}{2} \cdot \dfrac{k}{x} = \dfrac{1}{2}y$. Therefore y becomes half its original value.

129. If y is directly proportional to x^3, then $y = kx^3$ must hold. So if the value of x is tripled, the right side of

this equation becomes $k(3x)^3 = 27 \cdot kx^3 = 27y$. Therefore y becomes 27 times its original value.

131. Since $y = kx^n$, we know that $k = \dfrac{y}{x^n}$ where k is a constant. Using trial and error, let $n = 1$ while calculating

various values for k using x- and y-values from the table. For example, when $x = 2$ and $y = 2$, the value

of k is 1. But for $x = 4$ and $y = 8$, the value of k is 2. Since the value of k did not remain constant we

know that the value of n is not 1. Repeat this process for $n = 2$. For each x and y pair in the table, the value

of k is 0.5. Therefore $k = 0.5$ and $n = 2$.

133. Since $y = \dfrac{k}{x^n}$, we know that $k = yx^n$ where k is a constant. Using trial and error, let $n = 1$ while calculating

various values for k using x- and y-values from the table. For example, when $x = 2$ and $y = 1.5$, the value

of k is 3. And for $x = 3$ and $y = 1$, the value of k is also 3. Since the value of k remained constant we

know that the value of n is 1. For each x and y pair in the table, the value of k is 3. Therefore $k = 3$ and

$n = 1$.

135. If y is directly proportional to $x^{1.25}$, then $y = kx^{1.25}$ must hold. Given that $y = 1.9$ when $x = 1.1$, we may

calculate the value of $k = \dfrac{y}{x^{1.25}} = \dfrac{1.9}{1.1^{1.25}} \approx 1.69$. The variation equation can be written as $y \approx 1.69x^{1.25}$. When

a fiddler crab has claws weighing 0.75 grams, its body weight will be $y \approx 1.69(0.75)^{1.25} \approx 1.18$ grams.

137. Let I = the intensity of the light, and let d = the distance from a star to the earth. If I is inversely

proportional to d^2, then $I = \dfrac{k}{d^2}$ must hold. Suppose a ground-based telescope can see a star with intensity

I_g. Then the Hubble Telescope can see stars of intensity $\dfrac{1}{50} \cdot I_g = \dfrac{1}{50} \cdot \dfrac{k}{d^2} = \dfrac{k}{(d\sqrt{50})^2}$. That is, the Hubble

Telescope can see $\sqrt{50} \approx 7$ times as far as ground-based telescopes.

139. Since the resistance varies inversely as the square of the diameter of the wire, increasing the diameter of the wire by a factor of 1.5 will decrease the resistance by a factor of $\dfrac{1}{1.5^2} = \dfrac{4}{9}$. Hence, a 25 foot wire with a diameter of 3 millimeters (1.5 times 2 millimeters) will have a resistance of $\dfrac{4}{9}(0.5 \text{ ohm}) = \dfrac{2}{9} \text{ ohm}$.

141. In this exercise $F = \dfrac{K\sqrt{T}}{L}$. If both T and L are doubled, $\dfrac{K\sqrt{2T}}{2L} = \dfrac{\sqrt{2}}{2} \cdot \dfrac{K\sqrt{T}}{L} = \dfrac{\sqrt{2}}{2}F$. Therefore F decreases by a factor of $\dfrac{\sqrt{2}}{2}$.

4.8: Radical Equations and Power Functions

1. $8^{2/3} = (8^{1/3})^2 = 2^2 = 4$

3. $16^{-3/4} = (16^{1/4})^{-3} = (2)^{-3} = \dfrac{1}{2^3} = \dfrac{1}{8}$

5. $-81^{0.5} = -81^{1/2} = -\sqrt{81} = -9$

7. $(9^{3/4})^2 = 9^{3/2} = (\sqrt{9})^3 = 3^3 = 27$

9. $\dfrac{8^{5/6}}{8^{1/2}} = 8^{5/6 - 1/2} = 8^{1/3} = 2$

11. $27^{5/6} \cdot 27^{-1/6} = 27^{5/6 + (-1/6)} = 27^{2/3} = \sqrt[3]{27^2} = 3^2 = 9$

13. $(-27)^{-5/3} = \dfrac{1}{(-27)^{5/3}} = \dfrac{1}{\sqrt[3]{(-27)^5}} = \dfrac{1}{(-3)^5} = -\dfrac{1}{243}$

15. $(0.5^{-2})^2 = \left(\dfrac{1}{2}\right)^{-4} = 2^4 = 16$

17. $\left(\dfrac{2}{3}\right)^{-2} = \left(\dfrac{3}{2}\right)^2 = \dfrac{9}{4}$

19. $\sqrt{2x} = (2x)^{1/2}$

21. $\sqrt[3]{z^5} = z^{5/3}$

23. $(\sqrt[4]{y})^{-3} = (y^{1/4})^{-3} = y^{-3/4} = \dfrac{1}{y^{3/4}}$

25. $\sqrt{x} \cdot \sqrt[3]{x} = x^{1/2} \cdot x^{1/3} = x^{1/2 + 1/3} = x^{5/6}$

27. $\sqrt{y \cdot \sqrt{y}} = (y \cdot y^{1/2})^{1/2} = (y^{3/2})^{1/2} = y^{3/4}$

29. $a^{-3/4} b^{1/2} = \dfrac{b^{1/2}}{a^{3/4}} = \dfrac{\sqrt{b}}{\sqrt[4]{a^3}}$

31. $(a^{1/2} + b^{1/2})^{1/2} = \sqrt{\sqrt{a} + \sqrt{b}}$

33. $f(32) = \sqrt[3]{2(32)} = \sqrt[3]{64} = 4$

35. $f(4) = 2\sqrt{(4)^5} = 2\sqrt{1024} = 2(32) = 64$

37. $f(125) = \sqrt[4]{5(125)} = \sqrt[4]{625} = 5$

39. $f(-1) = \sqrt[5]{32(-1)} = \sqrt[5]{-32} = -2$

41.   Reflect the graph of $y = \sqrt[3]{x}$ through the y-axis to get the graph of $f(x) = \sqrt[3]{-x}$. See Figure 41.

43.   Shift the graph of $y = \sqrt[3]{x}$ right 1 unit and up 1 unit to get the graph of $f(x) = \sqrt[3]{x-1}+1$. See Figure 43.

45.   Shift the graph of $y = \sqrt[4]{x}$ left 2 units and down 1 unit to get the graph of $f(x) = \sqrt[4]{x+2}-1$. See Figure 45.

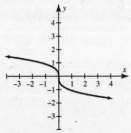
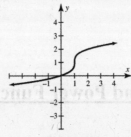
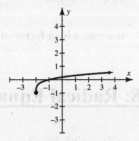

Figure 41   Figure 43   Figure 45

47.   Shift the graph of $y = \sqrt[4]{x}$ right 1 unit and vertically stretch by a multiple of 2 units to get the graph of $f(x) = 2\sqrt[4]{x-1}$. See Figure 47.

49.   Shift the graph of $y = \sqrt{x}$ left 3 units and up 2 units to get the graph of $f(x) = \sqrt{x+3}+2$. See Figure 49.

51.   Vertically stretch the graph of $y = \sqrt{x}$ by a multiple of 2 units to get the graph of $f(x) = 2\sqrt{x}$. See Figure 51.

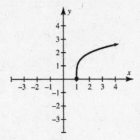
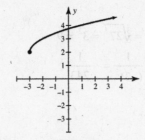
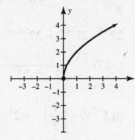

Figure 47   Figure 49   Figure 51

53. $\sqrt{x}-4=0 \Rightarrow \sqrt{x}=4 \Rightarrow (\sqrt{x})^2 = (4)^2 \Rightarrow x = 16$. Check: $\sqrt{16}-4=0 \Rightarrow 4-4=0$.

55. $2\sqrt{x}+3=9 \Rightarrow 2\sqrt{x}=6 \Rightarrow \sqrt{x}=3 \Rightarrow (\sqrt{x})^2 = (3)^2 \Rightarrow x = 9$. Check: $2\sqrt{9}+3=9 \Rightarrow 6+3=9$.

57. $\sqrt{2-x}=4 \Rightarrow (\sqrt{2-x})^2 = (4)^2 \Rightarrow 2-x=16 \Rightarrow -14=x$. Check: $\sqrt{2-x}=4 \Rightarrow \sqrt{2-(-14)}=4 \Rightarrow \sqrt{16}=4$.

59. $\sqrt{x+2}=x-4 \Rightarrow x+2 = x^2-8x+16 \Rightarrow x^2-9x+14=0 \Rightarrow (x-2)(x-7)=0 \Rightarrow x=2$ or 7.

      Check: $\sqrt{2+2}=2 \neq 2-4$ (not a solution); $\sqrt{7+2}=3=7-4$. The only solution is $x=7$.

61. $\sqrt{3x+7}=3x+5 \Rightarrow 3x+7=9x^2+30x+25 \Rightarrow 9x^2+27x+18=0 \Rightarrow 9(x^2+3x+2)=0 \Rightarrow$
      $9(x+2)(x+1)=0 \Rightarrow x=-2$ or -1

      Check: $\sqrt{3(-2)+7} \neq 3(-2)+5$ (not a solution); $\sqrt{3(-1)+7}=2=3(-1)+5$. The only solution is $x=-1$.

63. $\sqrt{5x-6}=x \Rightarrow 5x-6=x^2 \Rightarrow x^2-5x+6=0 \Rightarrow (x-2)(x-3)=0 \Rightarrow x=2$ or 3

      Check: $\sqrt{5(2)-6}=2$; $\sqrt{5(3)-6}=3$

65. $\sqrt{x+5}+1=x \Rightarrow \sqrt{x+5}=x-1 \Rightarrow x+5=(x-1)^2 \Rightarrow x+5=x^2-2x+1 \Rightarrow$
      $x^2-3x-4=0 \Rightarrow (x-4)(x+1)=0 \Rightarrow x=-1$ or $x=4$

Check: $\sqrt{-1+5}+1=-1 \Rightarrow \sqrt{4}+1=-1 \Rightarrow 3 \neq -1$ (not a solution).

Check: $\sqrt{4+5}+1=4 \Rightarrow \sqrt{9}+1=4 \Rightarrow 4=4$. The only solution is $x=4$.

67. $\sqrt{x+1}+3=\sqrt{3x+4} \Rightarrow (\sqrt{x+1}+3)^2 = 3x+4 \Rightarrow x+1+6\sqrt{x+1}+9 = 3x+4 \Rightarrow$

$6\sqrt{x+1} = 2x-6 \Rightarrow 3\sqrt{x+1} = x-3 \Rightarrow (3\sqrt{x+1})^2 = (x-3)^2 \Rightarrow$

$9(x+1) = x^2-6x+9 \Rightarrow x^2-15x = 0 \Rightarrow x(x-15) = 0 \Rightarrow x=0$ or 15

Check: $\sqrt{(0)+1}+3 \neq \sqrt{3(0)+4}$ (not a solution; $\sqrt{(15)+1}+3 = 7 = \sqrt{3(15)+4}$. The only solution is $x=15$.

69. $\sqrt{2x}-\sqrt{x+1}=1 \Rightarrow \sqrt{2x} = 1+\sqrt{x+1} \Rightarrow 2x = (1+\sqrt{x+1})^2 \Rightarrow$

$2x = 1+2\sqrt{x+1}+x+1 \Rightarrow x-2 = 2\sqrt{x+1} \Rightarrow x^2-4x+4 = 4(x+1) \Rightarrow$

$x^2-4x+4 = 4x+4 \Rightarrow x^2-8x = 0 \Rightarrow x(x-8) = 0 \Rightarrow x=0$ or 8.

Check: $\sqrt{2(0)}-\sqrt{0+1} = 1 \Rightarrow \sqrt{0}-\sqrt{1} = 1 \Rightarrow -1=1$ (not a solution);

$\sqrt{2(8)}-\sqrt{8+1} = 1 \Rightarrow \sqrt{16}-\sqrt{9} = 1 \Rightarrow 4-3 = 1 \Rightarrow 1=1$ The only solution is 8.

71. $\sqrt[3]{z+1}=-3 \Rightarrow z+1=(-3)^3 \Rightarrow z+1 = -27 \Rightarrow z = -28$

Check: $\sqrt[3]{-28-1} = \sqrt[3]{-27} = -3$

73. $\sqrt[3]{x+1} = \sqrt[3]{2x-1} \Rightarrow x+1 = 2x-1 \Rightarrow x=2$

Check: $\sqrt[3]{2+1} = \sqrt[3]{3} = \sqrt[3]{2(2)-1}$

75. $\sqrt[4]{x-2}+4 = 20 \Rightarrow \sqrt[4]{x-2} = 16 \Rightarrow x-2 = (16)^4 \Rightarrow x-2 = 65,536 \Rightarrow x = 65,538$

Check: $\sqrt[4]{65,538-2}+4 = 20 \Rightarrow \sqrt[4]{65,536}+4 = 20 \Rightarrow 20 = 20$

77. $f(x) = x^{3/2}-x^{1/2} \Rightarrow f(50) = 50^{3/2}-50^{1/2} \approx 346.48$

79. The graph of $f(x) = x^a$ is increasing, since $a>0$. However, since $a<b$ its graph increases more slowly than the graph of $f(x) = x^b$. The best choice is graph b.

81. Shift the graph of $f(x) = x^{1/2}$ up 1 unit to get the graph of $f(x) = x^{1/2}+1$. See Figure 81.

83. Shift the graph of $f(x) = x^{2/3}$ down 1 unit to get the graph of $f(x) = x^{2/3}-1$. See Figure 83.

85. Shift the graph of $f(x) = x^{2/3}$ to the left 1 unit and down 2 units to get the graph of $f(x) = (x+1)^{2/3}-2$. See Figure 85.

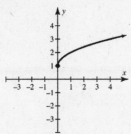

Figure 81

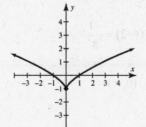

Figure 83

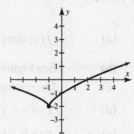

Figure 85

87. $f(x) = -x^3$

    (a) $f(0) = 0; f(-2) = 8$

    (b)   See Figure 87.

    (c)   Decreasing: $(-\infty, \infty)$

    (d) $x \to -\infty, f(x) \to \infty$

    (e) $x \to \infty, f(x) \to -\infty$

(f) $f(x) = -x^3; f(-x) = -(-x)^3 = x^3 \Rightarrow f(x) \neq f(-x)$. This means the function is not even. Since $f(-x) = -f(x)$, the function is odd.

89. $f(x) = x^4$

    (a) $f(0) = 0; f(-2) = 16$

    (b)   See Figure 89.

    (c)   Decreasing: $(-\infty, 0)$; Increasing: $(0, \infty)$

    (d) $x \to -\infty, f(x) \to \infty$

    (e) $x \to \infty, f(x) \to \infty$

    (f) $f(x) = x^4; f(-x) = (-x)^4 = x^4 \Rightarrow f(x) = f(-x)$. This means the function is even.

91. $f(x) = -x^{-4} = \dfrac{-1}{x^4}$

    (a) $f(0)$ does not exist; $f(-2) = \dfrac{-1}{16}$

    (b)   See Figure 91.

    (c)   Decreasing: $(-\infty, 0)$; Increasing: $(0, \infty)$

    (d) $x \to -\infty, f(x) \to 0$

    (e) $x \to \infty, f(x) \to 0$

    (f) $f(x) = -x^{-4}; f(-x) = -(-x)^{-4} = -x^{-4} \Rightarrow f(x) = f(-x)$. This means the function is even.

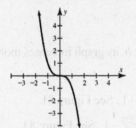 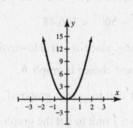 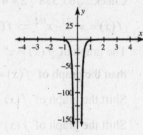

Figure 87   Figure 89   Figure 91

93. $f(x) = x^{-3} = \dfrac{1}{x^3}$

    (a) $f(0)$ does not exist; $f(-2) = \dfrac{-1}{8}$

    (b)   See Figure 93.

    (c)   Decreasing: $(-\infty, 0) \cup (0, \infty)$

    (d) $x \to -\infty, f(x) \to 0$

    (e) $x \to \infty, f(x) \to 0$

    (f) $f(x) = x^{-3}; f(-x) = (-x)^{-3} = -x^{-3} \Rightarrow f(x) \neq f(-x)$. This means the function is not even. Since $f(-x) = -f(x)$, the function is odd.

95. $f(x) = \dfrac{1}{x}$

    (a) $f(0)$ does not exist; $f(-2) = \dfrac{-1}{2}$

    (b)   See Figure 95.

(c)   Decreasing: $(-\infty, 0) \cup (0, \infty)$

(d) $x \to -\infty, f(x) \to 0$

(e) $x \to \infty, f(x) \to 0$

(f) $f(x) = \dfrac{1}{x}; f(-x) = (\dfrac{1}{(-x)}) = \dfrac{-1}{x} \Rightarrow f(x) \ne f(-x)$. This means the function is not even. Since

$f(-x) = -f(x)$, the function is odd.

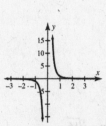
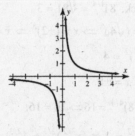

Figure 93   Figure 95

97.   Given: $5x^3$. If we replace x with $3x$, then $5(3x)^3 \Rightarrow 5x^3(27)$, so it increases by a factor of 27.

If we replace x with $\dfrac{1}{2}x$, then $5\left(\dfrac{1}{2}x\right)^3 \Rightarrow 5x^3\left(\dfrac{1}{8}\right)$, so it decreases by a factor of $\dfrac{1}{8}$.

99.   Given: $\dfrac{3}{x^2}$. If we replace x with $3x$, then $\dfrac{3}{(3x)^2} \Rightarrow \dfrac{3}{x^2} \cdot \dfrac{1}{9}$, so it decreases by a factor of $\dfrac{1}{9}$.

If we replace x with $\dfrac{1}{2}x$, then $\dfrac{3}{\left(\dfrac{1}{2}x\right)^2} \Rightarrow \dfrac{3}{x^2} \cdot 4$, so it increases by a factor of 4.

101. (a)   For the given graph, n is even and positive.

(b)   For the given graph, a is negative.

(c)   For the given graph, $f(x)$ is negative on $(-\infty, 0) \cup (0, \infty)$.

(d)   For the given graph, $f(x)$ is increasing on $(-\infty, 0)$ and decreasing on $(0, \infty)$.

(e)   The graph has symmetry at the y-axis, so the function is even.

(f)   There are no asymptotes.

103. (a)   For the given graph, n is odd and negative.

(b)   For the given graph, a is positive.

(c)   For the given graph, $f(x)$ is negative on $(-\infty, 0)$ and positive on $(0, \infty)$.

(d)   For the given graph, $f(x)$ is decreasing on $(-\infty, 0) \cup (0, \infty)$.

(e)   The graph has symmetry at the origin, so the function is odd.

(f)   There is a vertical asymptote at $x = 0$ and a horizontal asymptote at $y = 0$.

105. (a)   For the given graph, n is even and negative.

(b)   For the given graph, a is negative.

(c)   For the given graph, $f(x)$ is negative on $(-\infty, 0) \cup (0, \infty)$.

(d)   For the given graph, $f(x)$ is decreasing on $(-\infty, 0)$ and increasing on $(0, \infty)$.

(e)   The graph has symmetry at the y-axis, so the function is even.

(f)   There is a vertical asymptote at $x = 0$ and a horizontal asymptote at $y = 0$.

107. (a)   For the given graph, n is odd and positive.

(b)	For the given graph, a is positive.

(c)	For the given graph, $f(x)$ is negative on $(-\infty,0)$ and positive on $(0,\infty)$.

(d)	For the given graph, $f(x)$ is increasing on $(-\infty,\infty)$.

(e)	The graph has symmetry at the origin, so the function is odd.

(f)	There are no asymptotes.

109.	$x^3 = 8 \Rightarrow x = \sqrt[3]{8} \Rightarrow x = 2$; Check: $2^3 = 8$

111.	$x^{1/4} = 3 \Rightarrow x = 3^4 \Rightarrow x = 81$; Check: $81^{1/4} = \sqrt[4]{81} = 3$

113.	$x^{2/5} = 4 \Rightarrow (x^{2/5})^{5/2} = 4^{5/2} \Rightarrow x = (\sqrt{4})^5 \Rightarrow x = (\pm 2)^5 \Rightarrow x = \pm 32$

	Check: $(\pm 32)^{2/5} = (\sqrt[5]{\pm 32})^2 = (\pm 2)^2 = 4$

115.	$x^{4/3} = 16 \Rightarrow (x^{4/3})^{3/4} = (16)^{3/4} \Rightarrow x = \pm 8$

	Check: $(8)^{4/3} = 16 \Rightarrow 16 = 16$; $(-8)^{4/3} = 16 \Rightarrow 16 = 16$.

117.	$4x^{3/2} + 5 = 21 \Rightarrow 4x^{3/2} = 16 \Rightarrow x^{3/2} = 4 \Rightarrow (x^{3/2})^{2/3} = 4^{2/3} \Rightarrow x = \sqrt[3]{4^2} \Rightarrow x = \sqrt[3]{16}$

	Check: $4(\sqrt[3]{16})^{3/2} + 5 = 4(16^{1/3})^{3/2} + 5 = 4(16^{1/2}) + 5 = 4(\sqrt{16}) + 5 = 4(4) + 5 = 16 + 5 = 21$

119.	$n^{-2} + 3n^{-1} + 2 = 0 \Rightarrow (n^{-1})^2 + 3(n^{-1}) + 2 = 0$, let $u = n^{-1}$, then $u^2 + 3u + 2 = 0 \Rightarrow (u+2)(u+1) = 0 \Rightarrow u = -2$ or $u = -1$. Because $u = n^{-1}$, it follows that $n = u^{-1}$.

	Thus $n = (-2)^{-1} \Rightarrow n = \dfrac{1}{(-2)} \Rightarrow n = -\dfrac{1}{2}$ or $n = (-1)^{-1} \Rightarrow n = \dfrac{1}{(-1)} = -1$. Therefore, $n = -1, -\dfrac{1}{2}$.

121.	$5n^{-2} + 13n^{-1} = 28 \Rightarrow 5(n^{-1})^2 + 13(n^{-1}) - 28 = 0$, let $u = n^{-1}$, then $5u^2 + 13u - 28 = 0 \Rightarrow$

	$(5u - 7)(u + 4) = 0 \Rightarrow u = \dfrac{7}{5}$ or $u = -4$. Because $u = n^{-1}$, it follows that $n = u^{-1}$.

	Thus $n = \left(\dfrac{7}{5}\right)^{-1} \Rightarrow n = \dfrac{5}{7}$ or $n = (-4)^{-1} \Rightarrow n = -\dfrac{1}{4}$. Therefore, $n = -\dfrac{1}{4}, \dfrac{5}{7}$.

123.	$x^{2/3} - x^{1/3} - 6 = 0 \Rightarrow (x^{1/3})^2 - x^{1/3} - 6 = 0$, let $u = x^{1/3}$, then $u^2 - u - 6 = 0 \Rightarrow (u-3)(u+2) = 0 \Rightarrow u = -2$ or $u = 3$. Because $u = x^{1/3}$, it follows that $x = u^3$.

	Thus $x = (-2)^3 \Rightarrow x = -8$ or $x = 3^3 \Rightarrow x = 27$. Therefore, $x = -8, 27$.

125.	$6x^{2/3} - 11x^{1/3} + 4 = 0 \Rightarrow 6(x^{1/3})^2 - 11(x^{1/3}) + 4 = 0$, let $u = x^{1/3}$, then $6u^2 - 11u + 4 = 0$. Using the quadratic

	formula to solve we get: $u = \dfrac{11 \pm \sqrt{121 - 4(6)(4)}}{2(6)} = \dfrac{11 \pm \sqrt{25}}{12} = \dfrac{11 \pm 5}{12} \Rightarrow u = \dfrac{16}{12} \Rightarrow u = \dfrac{4}{3}$ or $u = \dfrac{6}{12} \Rightarrow u = \dfrac{1}{2}$.

	Because $u = x^{1/3}$, it follows that $x = u^3$.

	Thus $x = \left(\dfrac{4}{3}\right)^3 \Rightarrow x = \dfrac{64}{27}$ or $x = \left(\dfrac{1}{2}\right)^3 \Rightarrow x = \dfrac{1}{8}$. Therefore, $x = \dfrac{1}{8}, \dfrac{64}{27}$.

127.	$x^{3/4} - x^{1/2} - x^{1/4} + 1 = 0 \Rightarrow (x^{1/4})^3 - (x^{1/4})^2 - (x^{1/4}) + 1 = 0$, let $u = x^{1/4}$, then

	$u^3 - u^2 - u + 1 = 0 \Rightarrow (u^3 - u^2) - (u - 1) = 0 \Rightarrow u^2(u-1) - 1(u-1) = 0 \Rightarrow$

	$(u^2 - 1)(u - 1) = 0 \Rightarrow (u+1)(u-1)(u-1) = 0 \Rightarrow u = -1$ or $u = 1$.

	Because $u = x^{1/4}$ it follows that $x = u^4$.

	Thus $x = (-1)^4 \Rightarrow x = 1$ or $x = (1)^4 \Rightarrow x = 1$. Therefore, $x = 1$.

129. $x^{-2/3} - 2x^{-1/3} - 3 = 0 \Rightarrow (x^{-1/3})^2 - 2x^{-1/3} - 3 = 0$, let $u = x^{-1/3}$, then $u^2 - 2u - 3 = 0 \Rightarrow$

$(u-3)(u+1) = 0 \Rightarrow u = 3$ or -1. Because $u = x^{-1/3}$ it follows that $x = u^{-3}$.

Thus $x = 3^{-3} = \dfrac{1}{27}$, $x = -1^{-3} = -1$.

131. Average rate of change $= \dfrac{s\left(\frac{9}{2}\right) - s\left(\frac{1}{2}\right)}{\frac{9}{2} - \frac{1}{2}} = \dfrac{\sqrt{96\left(\frac{9}{2}\right)} - \sqrt{96\left(\frac{1}{2}\right)}}{4} = \dfrac{\sqrt{432} - \sqrt{48}}{4} \approx 3.5$. The average speed over the

time interval is about 3.5 mph.

133. *Answers may vary.* All of these functions have an x-intercept at $(0,0)$. Their graphs are similar in shape. They are all increasing over their domain and are odd functions with symmetry at the origin.

135. $S(w) = 3 \Rightarrow 1.27w^{2/3} = 3 \Rightarrow w^{2/3} = \dfrac{3}{1.27} \Rightarrow w = \left(\dfrac{3}{1.27}\right)^{3/2} \Rightarrow w \approx 3.63$

The bird weighs about 3.63 pounds

137. $f(15) = 15^{1.5} \approx 58.1$; The planet would take about 58.1 years to orbit the sun.

139. (a) Following the hint, $f(1) = a(1)^b = 1960 \Rightarrow a = 1960$

(b) Plot the four data points (0.5, 4500), (1, 1960), (2, 850), and (3, 525) together with $f(x) = 1960x^b$ for different values of b. Through trial and error, a value of $b \approx -1.2$ can be found. Figure 139 shows a graph of $Y_1 = 1960X^{\wedge}(-1.2)$ along with the data.

(c) $f(x) = 1960x^{-1.2} \Rightarrow f(4) = 1960(4)^{-1.2} \approx 371$. This means that if the zinc ion concentration in the water reaches 371 milligrams per liter, the rainbow trout will live only 4 minutes on average.

141. (a) $f(2) = 0.445(2)^{5/4} \approx 1.06$ grams

(b) We must solve the equation $0.445x^{5/4} = 0.5$ for x. Graph $Y_1 = 0.445X^{\wedge}1.25$ and $Y_2 = 0.5$ Their graphs intersect near (1.1, 0.5). See Figure 141. When the claw weighs 0.5 grams, the crab weighs about 1.1 grams.

(c) $0.445x^{1.25} = 0.5 \Rightarrow 0.445x^{5/4} = 0.5 \Rightarrow x^{5/4} = \dfrac{0.5}{0.445} \Rightarrow x = \left(\dfrac{0.5}{0.445}\right)^{4/5} \Rightarrow x \approx 1.1$ grams

143. Use the power regression feature of your graphing calculator to find the values of a and b in the equation $f(x) = ax^b$. For this exercise, $a \approx 3.20$ and $b \approx 0.20$.

The graph of $Y_1 = 3.20X^{\wedge}0.20$ is shown together with the data in Figure 143.

[0, 5, 1] by [0, 5000, 1000]   [0, 2, 0.2] by [0, 1, 0.2]   [1, 9, 1] by [0, 6, 1]

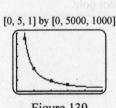
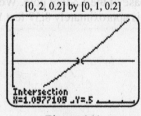
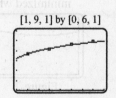

Figure 139   Figure 141   Figure 143

145. (a) Using the power regression function on the graphing calculator we see that $f(x) \approx 0.005192x^{1.7902}$. *Answers may vary.*

(b) The year 2012 is 32 years after 1980. $f(32) \approx 0.005192(32)^{1.7902} \approx 2.6$ million. It appears that the value is about a half million to high. The estimate involved extrapolation.

(c) See Figure 145. We can see from the graph of $f(x)$ that the number of employees reached 1 .million in about 1999.

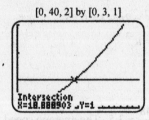

[0, 40, 2] by [0, 3, 1]

Intersection
X=18.888903 Y=1

Figure 145

147. Use the power regression feature of your graphing calculator to find the values of a and b in the equation $f(x) = ax^b$. For this exercise, $a \approx 874.54$ and $b \approx -0.49789$.

Extended and Discovery Exercises for Section 4.8

1. $\dfrac{f(x+h)-f(x)}{h} = \dfrac{\sqrt{x+h}-\sqrt{x}}{h} = \dfrac{\sqrt{x+h}-\sqrt{x}}{h} \cdot \dfrac{\sqrt{x+h}+\sqrt{x}}{\sqrt{x+h}+\sqrt{x}} = \dfrac{x+h-x}{h(\sqrt{x+h}+\sqrt{x})} = \dfrac{1}{\sqrt{x+h}+\sqrt{x}}$

3. $\dfrac{x^{-2/3}+x^{1/3}}{x} = \dfrac{x^{-2/3}(1+x)}{x} = \dfrac{1+x}{x^{2/3}(x)} = \dfrac{1+x}{x^{5/3}}$

5. $\dfrac{\frac{2}{3}(x+1)x^{-1/3}-x^{2/3}}{(x+1)^2} = \dfrac{x^{-1/3}[\frac{2}{3}x+\frac{2}{3}-x]}{(x+1)^2} = \dfrac{-\frac{1}{3}x+\frac{2}{3}}{x^{1/3}(x+1)^2} = \dfrac{\frac{-x+2}{3}}{x^{1/3}(x+1)^2} = \dfrac{2-x}{3x^{1/3}(x+1)^2}$

7. (a) Since the distance from C to D is 20 feet, the distance from P to C is $20-x$

    (b) The value must be between 0 and 20. That is, $0 < x < 20$.

    (c) $(AP)^2 = (DP)^2 + (AD)^2 \Rightarrow (AP)^2 = x^2 + 12^2 \Rightarrow AP = \sqrt{x^2 + 12^2}$;

    $(BP)^2 = (CP)^2 + (BC)^2 \Rightarrow (BP)^2 = (20-x)^2 + 16^2 \Rightarrow BP = \sqrt{(20-x)^2 + 16^2}$

    (d) $f(x) = \sqrt{x^2 + 12^2} + \sqrt{(20-x)^2 + 16^2}, 0 < x < 20$.

    (e) The graph of $y_1 = \sqrt{x^2 + 12^2} + \sqrt{(20-x)^2 + 16^2}$ is shown in Figure 73. Here $f(4) \approx 35.28$. When the stake is 4 feet from the 12-foot pole, approximately 35.28 feet of wire will be required.

    (f) Using the calculator, $f(x)$ is a minimum (about 34.41 feet) when $x \approx 8.57$ feet.

    (g) This problem examined how the total amount of wire used can be expressed in terms of the distance from the stake at P to the base of the 12-foot pole. We find that the amount of wire used can be minimized when the stake is approximately 8.57 feet from the 12-foot pole.

Figure 7

Checking Basic Concepts for Sections 4.7 and 4.8

1. (a) $\dfrac{3x-1}{1-x}=1 \Rightarrow 3x-1=1-x \Rightarrow 4x=2 \Rightarrow x=\dfrac{1}{2}$; Check: $\dfrac{3(\frac{1}{2})-1}{1-(\frac{1}{2})}=1 \Rightarrow \dfrac{\frac{1}{2}}{\frac{1}{2}}=1 \Rightarrow 1=1$

   (b) $3+\dfrac{8}{x}=\dfrac{35}{x^2} \Rightarrow 3x^2+8x=35 \Rightarrow 3x^2+8x-35=0 \Rightarrow (3x-7)(x+5)=0 \Rightarrow x=-5, \dfrac{7}{3}$

   Check: $3+\dfrac{8}{(-5)}=\dfrac{35}{(-5)^2} \Rightarrow 3-\dfrac{8}{5}=\dfrac{35}{25} \Rightarrow \dfrac{75}{25}-\dfrac{40}{25}=\dfrac{35}{25} \Rightarrow \dfrac{35}{25}=\dfrac{35}{25}$

   Check: $3+\dfrac{8}{(\frac{7}{3})}=\dfrac{35}{(\frac{7}{3})^2} \Rightarrow 3+\dfrac{24}{7}=\dfrac{45}{7} \Rightarrow \dfrac{21}{7}+\dfrac{24}{7}=\dfrac{45}{7} \Rightarrow \dfrac{45}{7}=\dfrac{45}{7}$

   (c) $\dfrac{1}{x-1}-\dfrac{1}{3(x+2)}=\dfrac{1}{(x+2)(x-1)} \Rightarrow 3(x+2)-(x-1)=3 \Rightarrow 3x+6-x+1=3 \Rightarrow 2x=-4 \Rightarrow x=-2$;

   Check: $\dfrac{1}{(-2)-1}-\dfrac{1}{3(-2+2)}=\dfrac{1}{(-2+2)(-2-1)} \Rightarrow \dfrac{1}{-3}-\dfrac{1}{0}=\dfrac{1}{0}$ Since -2 is not defined in the original

   equation there is no solution.

3. The graph of $Y_1=(X^2-1)/(X+2)$ has a vertical asymptote at $x=-2$ and x-intercepts at $x=\pm 1$.

   The graph of Y_1 intersects or is above the x-axis on the interval $(-2,-1] \cup [1,\infty)$.

   In set builder notation the interval is $\{x|-2<x\le -1 \text{ or } x\ge 1\}$.

5. (a) $-4^{3/2}=-(\sqrt{4})^3=-(2)^3=-8$

   (b) $(8^{-2})^{1/3}=8^{-2/3}=\dfrac{1}{8^{2/3}}=\dfrac{1}{(\sqrt[3]{8})^2}=\dfrac{1}{2^2}=\dfrac{1}{4}$

   (c) $\sqrt[3]{27^2}=(\sqrt[3]{27})^2=3^2=9$

7. $\sqrt{5x-4}=x-2 \Rightarrow 5x-4=x^2-4x+4 \Rightarrow x^2-9x+8=0 \Rightarrow (x-1)(x-8)=0 \Rightarrow x=1 \text{ or } 8$

   Check: $\sqrt{5(1)-4}\ne 1-2$ (not a solution); $\sqrt{5(8)-4}=6=8-2$. The only solution is $x=8$.

9. Use the power regression feature of your graphing calculator to find the values of a and b in the equation $f(x)=ax^b$. For this exercise, $a=2$ and $b=0.5$. That is, $f(x)=2x^{1/2}$ or $f(x)=2\sqrt{x}$.

Chapter 4 Review Exercises

1. First put in standard order:
   $f(x)=-7x^3-2x^2+x+4$, now the degree is: 3, and the leading coefficient is: -7.

3. (a) There is a local minimum of -2. There is a local maximum of 4.
   (b) There is neither an absolute minimum nor an absolute maximum.

5. $f(x)=2x^6-5x^4-x^2 \Rightarrow f$ is even since each power of x is even.

7. $f(x)=7x^5+3x^3-x \Rightarrow f$ is odd since each power of x is odd.

9. f is odd since $f(-x)=-f(x)$ for all x-values in the table.

11. See Figure 11. *Answers may vary.*

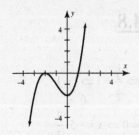

Figure 11

13.   (a)   2 turning points; 3 x-intercepts at: $\approx (-2,0),(0,0),(1,0)$.

      (b)   This is an odd degree function going down to the right $\Rightarrow$ negative.

      (c)   Since it has 2 turning points, it is a minimum of 3rd degree.

15.   A negative cubic function $\Rightarrow$ up on the left end and down on the right end; $f(x) \to \infty$ as $x \to -\infty$;
      $f(x) \to -\infty$ as $x \to \infty$.

17. $f(-2) = (-2)^3 + 1 \Rightarrow f(-2) = -7 \Rightarrow (-2, -7)$ and $f(-1) = (-1)^3 + 1 \Rightarrow f(-1) = 0 \Rightarrow (-1, 0)$.

      To find the average rate of change we use: $\dfrac{y_2 - y_1}{x_2 - x_1} \Rightarrow \dfrac{0 - (-7)}{(-1) - (-2)} \Rightarrow \dfrac{7}{1} \Rightarrow 7$

19.   (a)   The graph of f has of two different pieces: $y = 2x$ when $0 \le x < 2$ and $y = 8 - x^2$ when $2 \le x \le 4$.
              The graph is shown in Figure 19. The graph of f is continuous.

      (b)   To evaluate $f(1)$ we must use the first piece of the function: $f(1) = 2(1) = 2$

              To evaluate $f(3)$ we must use the second piece of the function: $f(3) = 8 - (3)^2 = 8 - 9 = -1$

      (c)   We must solve the equation $f(x) = 2$ for each piece of the piecewise-defined function f.

              $2x = 2 \Rightarrow x = 1$ and $8 - x^2 = 2 \Rightarrow x^2 = 6 \Rightarrow x = \pm\sqrt{6}$

              Since $x = -\sqrt{6}$ is not in the domain of f, the solutions are $x = 1$ or $\sqrt{6}$.

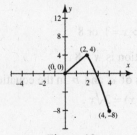

Figure 19

21. $\dfrac{14x^3 - 21x^2 - 7x}{7x} = 2x^2 - 3x - 1$

23.
$$
\begin{array}{r}
2x^2 - 3x + 1 \\
2x+3 \overline{)\ 4^3 \qquad\quad -7x + 4} \\
\underline{4x^3 + 6x^2} \\
-6x^2 - 7x + 4 \\
\underline{-6x^2 - 9x} \\
2x + 4 \\
\underline{2x + 3} \\
1
\end{array}
$$

Therefore the quotient is: $2x^2 - 3x + 1 + \dfrac{1}{2x+3}$

25. Since the leading coefficient of $f(x)$ is $\dfrac{1}{2}$ and the degree of the polynomial is 3, all of the zeros are given,

thus: $f(x) = \dfrac{1}{2}(x-1)(x-2)(x-3)$.

27. The graph has x-intercepts or zeros of $x = -2, 1, 3 \Rightarrow$ it has factors $(x+2)(x-1)(x-3)$. Since $f(0) = 3$ we

can solve for the leading coefficient a by setting $a(0+2)(0-1)(0-3) = 3$ and solving

$\Rightarrow a(2)(-1)(-3) = 3 \Rightarrow 6a = 3 \Rightarrow a = \dfrac{1}{2}$.

The complete factored form is: $f(x) = \dfrac{1}{2}(x+2)(x-1)(x-3)$.

29. By the rational zeros test, any rational zero of $2x^3 + x^2 - 13x + 6$ must be one of the following:

$\pm 6, \pm 3, \pm 2, \pm 1, \pm \dfrac{3}{2}, \pm \dfrac{1}{2}$. By evaluating f at each of these values we find that the three rational zeros are

$-3, \dfrac{1}{2}$, and 2.

31. $9x = 3x^3 \Rightarrow 3x^3 - 9x = 0 \Rightarrow 3x(x^2 - 3) = 0 \Rightarrow x = 0, \pm\sqrt{3}$

33. The x-intercepts on the graph of $Y_1 = X^3 - 3X + 1$ are approximately -1.88, 0.35 and 1.53. See Figure 33.

[-10, 10, 1] by [-10, 10, 1]

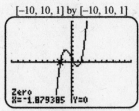

Figure 33

35. $x^3 + x = 0 \Rightarrow x(x^2 + 1) = 0 \Rightarrow x = 0$ or $x = \pm i$

37. $f(x) = 4(x-1)(x-3i)(x+3i)$; the expanded form is $f(x) = 4x^3 - 4x^2 + 36x - 36$.

39. If i is a zero then $-i$ is also a zero $\Rightarrow (x+i)(x-i)$ or $x^2 + 1$ is a factor. By polynomial long division:

$$
\begin{array}{r}
x^2 + x + 1 \\
x^2 + 1 \overline{\smash{\big)}\, x^4 + x^3 + 2x^2 + x + 1} \\
\underline{x^4 \quad\;\; + x^2} \\
x^3 + x^2 + x + 1 \\
\underline{x^3 \quad\quad\; + x} \\
x^2 \quad\;\; + 1 \\
\underline{x^2 \quad\;\; + 1} \\
0
\end{array}
$$

Solve for $x^2 + x + 1 = 0$ using the quadratic formula: $x = \dfrac{-1 \pm \sqrt{1 - 4(1)(1)}}{2(1)} = \dfrac{-1 \pm \sqrt{-3}}{2} = -\dfrac{1}{2} \pm \dfrac{i\sqrt{3}}{2}$. Therefore

the zeros are $\pm i, -\dfrac{1}{2} \pm \dfrac{i\sqrt{3}}{2}$ and the complete factored form is:

$$f(x) = (x+i)(x-i)\left(x - \left(-\dfrac{1}{2} + \dfrac{i\sqrt{3}}{2}\right)\right)\left(x - \left(-\dfrac{1}{2} - \dfrac{i\sqrt{3}}{2}\right)\right)$$

41.   Horizontal asymptote: since the degree of the numerator and denominator is the same the horizontal

asymptote is the leading coefficient ratio $y = \dfrac{2}{3}$. Vertical asymptotes: they can be found by finding the zeros

of the denominator that are not zeros of the numerator

$\Rightarrow$ solve: $3x^2 + 8x - 3 \Rightarrow (3x-1)(x+3) = 0 \Rightarrow x = -3, \dfrac{1}{3}$. However, $x = -3$ is not a vertical asymptote because

-3 is a zero of the numerator.

The horizontal asymptote is $y = \dfrac{2}{3}$ and the vertical asymptote is $x = \dfrac{1}{3}$.

43. $g(x)$ is the function $f(x) = \dfrac{1}{x}$ shifted 1 left and 2 down. See Figure 43.

45. $g(x) = \dfrac{x^2 - 1}{x^2 + 2x + 1} \Rightarrow g(x) = \dfrac{(x+1)(x-1)}{(x+1)(x+1)} \Rightarrow g(x) = \dfrac{(x-1)}{(x+1)} \Rightarrow g(x) = \dfrac{x-1}{x+1}$. This is similar to $f(x) = \dfrac{1}{x}$ with a

horizontal asymptote of $y = 1$ and a vertical asymptote of $x = -1$. See Figure 45.

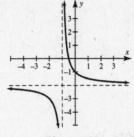

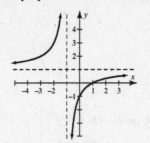

Figure 43   Figure 45

47. $g(x) = 2x^{-1} \Rightarrow \dfrac{2}{x}$ is like the function $f(x) = \dfrac{1}{x}$ with a horizontal asymptote of $y = 0$ and a vertical

asymptote of $x = 0$. See Figure 47.

49.   One example of a function that has a vertical asymptote at $x = -2$ and a horizontal asymptote at $y = 2$ is

shown in Figure 49.

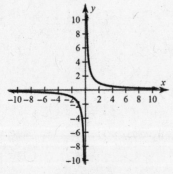

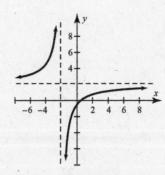

Figure 47   Figure 49

51. $\dfrac{5x+1}{x+3}=3 \Rightarrow 5x+1=3(x+3) \Rightarrow 5x+1=3x+9 \Rightarrow 2x=8 \Rightarrow x=4$

This answer is supported graphically by graphing $Y_1=(5X+1)/(X+3)$ and $Y_2=3$ The intersection point is $(4,3)$. Making a table of Y_1 starting at 0 and incrementing by 1 will support this result numerically.

53. $\dfrac{1}{x+2}+\dfrac{1}{x-2}=\dfrac{4}{x^2-4} \Rightarrow x-2+x+2=4 \Rightarrow 2x-4 \Rightarrow x=2$. Since 2 is defined in the original equation, there is no solution.

55. (a) $(-\infty,-4)\cup(-2,3)$; In set builder notation the interval is $\{x|x<-4 \text{ or } -2<x<3\}$.

    (b) $(-4,-2)\cup(3,\infty)$; In set builder notation the interval is $\{x|-4<x<-2 \text{ or } x>3\}$.

57. Find the boundary numbers by solving
$x^3+x^2-6x=0$. $x^3+x^2-6x=0 \Rightarrow x(x^2+x-6)=0 \Rightarrow x(x+3)(x-2)=0 \Rightarrow x=-3, 0 \text{ or } 2$.

The boundary values divide the number line into four intervals. Choose a test value from each of these intervals and evaluate x^3+x^2-6x for these values. See Figure 57. The solution is $(-3,0)\cup(2,\infty)$; In set builder notation the interval is $\{x|-3<x<0 \text{ or } x>2\}$.

Interval	Test Value x	x^3+x^2-6x	Positive or Negative?
$(-\infty,-3)$	-4	-24	Negative
$(-3,0)$	-1	6	Positive
$(0,2)$	1	-4	Negative
$(2,\infty)$	3	18	Positive

Figure 57

59. Graph $Y_1=(2X-1)/(X+2)$ using dot mode as shown in Figure 59. The graph has a vertical asymptote at $x=-2$ and an x-intercept at $x=\dfrac{1}{2}$. The graph is above the x-axis on the interval $(-\infty,-2)\cup\left(\dfrac{1}{2},\infty\right)$; In set builder notation the interval is $\left\{x|x<-2 \text{ or } x>\dfrac{1}{2}\right\}$

[−8, 4, 2] by [−5, 5, 1]

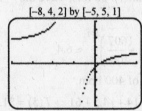

Figure 59

61. $(36^{3/4})^2=36^{3/2}=(\sqrt{36})^3=6^3=216$

63. $(2^{-3/2}\cdot2^{1/2})^{-3}=(2^{-3/2+1/2})^{-3}=(2^{-1})^{-3}=2^3=8$

65. $\sqrt[3]{x^4}=x^{4/3}$

67. $\sqrt[3]{y\cdot\sqrt{y}}=(y\cdot y^{1/2})^{1/3}=(y^{3/2})^{1/3}=y^{1/2}$

69. $D=\{x|x\ge0\}; f(3)=3^{5/2}\approx15.59$

71. $x^5=1024 \Rightarrow x=\sqrt[5]{1024}=4$ Check: $4^5=1024$

73. $\sqrt{x-2}=x-4 \Rightarrow x-2=(x-4)^2 \Rightarrow x-2=x^2-8x+16 \Rightarrow x^2-9x+18=0 \Rightarrow (x-6)(x-3)=0 \Rightarrow x=3 \text{ or } 6$

Check: $\sqrt{3-2}\ne3-4$ (not a solution); $\sqrt{6-2}=2=6-4$. The only solution is $x=6$.

75. $2x^{1/4} + 3 = 6 \Rightarrow 2x^{1/4} = 3 \Rightarrow x^{1/4} = \frac{3}{2} \Rightarrow x = \left(\frac{3}{2}\right)^4 \Rightarrow x = \frac{81}{16}$

Check: $2\left(\frac{81}{16}\right)^{1/4} + 3 = 2\left(\frac{3}{2}\right) + 3 = 3 + 3 = 6$

77. $\sqrt[3]{2x-3} + 1 = 4 \Rightarrow \sqrt[3]{2x-3} = 3 \Rightarrow 2x-3 = 3^3 \Rightarrow 2x-3 = 27 \Rightarrow 2x = 30 \Rightarrow x = 15$

Check: $\sqrt[3]{2(15)-3} + 1 = \sqrt[3]{27} + 1 = 3 + 1 = 4$

79. $2n^{-2} - 5n^{-1} = 3 \Rightarrow 2(n^{-1})^2 - 5(n^{-1}) = 3$, let $u = n^{-1}$, then $2u^2 - 5u - 3 = 0 \Rightarrow (2u+1)(u-3) = 0 \Rightarrow u = -\frac{1}{2}, 3$.

If $u = n^{-1}$, then it follows that $n = u^{-1}$, thus $n = \left(-\frac{1}{2}\right)^{-1} \Rightarrow n = -2$ or $n = (3)^{-1} \Rightarrow n = \frac{1}{3}$

Check: $2(-2)^{-2} - 5(-2)^{-1} = 3 \Rightarrow 2\left(\frac{1}{4}\right) - 5\left(-\frac{1}{2}\right) = 3 \Rightarrow \frac{1}{2} + \frac{5}{2} = 3 \Rightarrow 3 = 3$

Check: $2\left(\frac{1}{3}\right)^{-2} - 5\left(\frac{1}{3}\right)^{-1} = 3 \Rightarrow 2(9) - 5(3) = 3 \Rightarrow 18 - 15 = 3 \Rightarrow 3 = 3$

The solution are: $n = -2, \frac{1}{3}$.

81. $k^{2/3} - 4k^{1/3} - 5 = 0 \Rightarrow (k^{1/3})^2 - 4(k^{1/3}) - 5 = 0$, let $u = k^{1/3}$, then

$u^2 - 4u - 5 = 0 \Rightarrow (u-5)(u+1) = 0 \Rightarrow u = -1, 5$. If $u = k^{1/3}$, then it follows that $k = u^3$, thus

$k = (-1)^3 \Rightarrow k = -1$ or $k = (5)^3 \Rightarrow k = 125$.

Check: $(-1)^{2/3} - 4(-1)^{1/3} - 5 = 0 \Rightarrow 1 - (-4) - 5 = 0 \Rightarrow 0 = 0$

Check: $(125)^{2/3} - 4(125)^{1/3} - 5 = 0 \Rightarrow 25 - 4(5) - 5 = 0 \Rightarrow 0 = 0$. The solutions are: $k = -1, 125$.

83. (a) Dog: $f(24) = 1607(24)^{-0.75} \approx 148$ bpm; Person: $f(66) = 1607(66)^{-0.75} \approx 69$ bpm.

(b) We must approximate the length of an animal with a heart rate of 400 bpm. To do this, solve the equation $f(x) = 400$ or $1607x^{-0.75} = 400$ bpm.

$1607x^{-0.75} = 400 \Rightarrow x^{-0.75} = \frac{400}{1607} \Rightarrow x^{0.75} = \frac{1607}{400} \Rightarrow x^{3/4} = \frac{1607}{400} \Rightarrow x = \left(\frac{1607}{400}\right)^{4/3} \approx 6.4$

A six-inch animal such as a large bird or smaller rodent might have a heart rate of 400 bpm.

85. (a) May is $T(5) \Rightarrow T(5) = -0.064(5)^3 + 0.56(5)^2 + 2.9(5) + 61 \Rightarrow T(5) = -8 + 14 + 14.5 + 61 \Rightarrow T(5) = 81.5$. The temperature in May is 81.5°F.

(b) Graph the function. See Figure 85. From the graph the ocean reaches a maximum temperature of about 87.3°F in July.

[0, 13, 1] by [0, 100, 10]

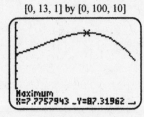

Figure 85

87. Since the time is directly proportional to the square root of the height, the equation $t = k\sqrt{h}$ must hold. If it takes 1 second to drop 16 feet then $1 = k\sqrt{16} \Rightarrow k = 0.25$. Therefore, from a height of 256 feet the object will take $t = 0.25\sqrt{256} = 4$ seconds to strike the ground.

Chapters 1-4 Cumulative Review Exercises

1. (a) $D = \{-3, -1, 0, 1\}$; $R = \{4, -2, 5\}$
    (b) No, all input must have unique outputs and -1 has an output of -2 and 5.

3. $D = \{x | -2 \le x \le 2\}$; $R = \{x | 0 \le x \le 3\}$

5. $c(x) = 0.25x + 200$; $c(2000) = 0.25(2000) + 200 \Rightarrow c(2000) = 700$. The monthly cost of driving and maintaining a car driven 2000 miles is $700.

7. $f(-3) = (-3)^3 - (-3) \Rightarrow f(-3) = -24$ or $(-3, -24)$; $f(-2) = (-2)^3 - (-2) \Rightarrow f(-2) = -6$ or $(-2, -6)$. The average rate of change $= \dfrac{-6 - (-24)}{-2 - (-3)} = \dfrac{18}{1} = 18$.

9. Slope $= \dfrac{(-4) - 5}{3 - (-2)} = \dfrac{-9}{5}$. Using point slope form $y = m(x - x_1) + y$ and point $(-2, 5)$ we get:

$$y = \frac{-9}{5}(x + 2) + 5 \Rightarrow y = \frac{-9}{5}x - \frac{18}{5} + 5 \Rightarrow y = \frac{-9}{5}x + \frac{7}{5}$$

11. All lines parallel to the x-axis have 0 slope and will have graphs of $y =$ (the y-coordinate) $\Rightarrow y = -5$.

13. Each radio costs $15 to manufacture. The fixed cost for this company to produce radios is $2000.

15. $-3(2 - 3x) - (-x - 1) = 1 \Rightarrow -6 + 9x + x + 1 = 1 \Rightarrow 10x - 5 = 1 \Rightarrow 10x = 6 \Rightarrow x = \dfrac{3}{5}$

17. $|3x - 4| + 1 = 5 \Rightarrow |3x - 4| = 4 \Rightarrow 3x - 4 = 4 \Rightarrow 3x = 8 \Rightarrow x = \dfrac{8}{3}$ or $3x - 4 = -4 \Rightarrow 3x = 0 \Rightarrow x = 0$.

Therefore $x = 0, \dfrac{8}{3}$.

19. $7x^2 + 9x = 10 \Rightarrow 7x^2 + 9x - 10 = 0 \Rightarrow (7x - 5)(x + 2) = 0 \Rightarrow x = -2, \dfrac{5}{7}$

21. $3x^{2/3} + 5x^{1/3} - 2 = 0 \Rightarrow 3(x^{1/3})^2 + 5(x^{1/3}) - 2 = 0$. Let $u = x^{1/3}$, then

$3u^2 + 5u - 2 = 0 \Rightarrow (3u - 1)(u + 2) = 0 \Rightarrow u = -2, \dfrac{1}{3}$. If $u = x^{1/3}$ then it follows $x = u^3$, thus

$x = (-2)^3 \Rightarrow x = -8$ or $x = \left(\dfrac{1}{3}\right)^3 \Rightarrow x = \dfrac{1}{27}$. Therefore $x = -8, \dfrac{1}{27}$.

23. $\dfrac{2x - 3}{5 - x} = \dfrac{4x - 3}{1 - 2x} \Rightarrow (2x - 3)(1 - 2x) = (4x - 3)(5 - x) \Rightarrow$

$2x - 4x^2 - 3 + 6x = 20x - 4x^2 - 15 + 3x \Rightarrow 15x = 12 \Rightarrow x = \dfrac{4}{5}$

25. $\dfrac{1}{2}x - (4 - x) + 1 = \dfrac{3}{2}x - 5 \Rightarrow \dfrac{1}{2}x - 4 + x + 1 = \dfrac{3}{2}x - 5 \Rightarrow \dfrac{3}{2}x - 3 = \dfrac{3}{2}x - 5 \Rightarrow -3 = -5$, since this is false the equation has no solutions and is a contradiction.

27. $-\dfrac{1}{3}x-(1+x)>\dfrac{2}{3}x \Rightarrow -\dfrac{1}{3}x-1-x>\dfrac{2}{3}x \Rightarrow -\dfrac{4}{3}x-1>\dfrac{2}{3}x \Rightarrow -2x>1 \Rightarrow x<-\dfrac{1}{2} \Rightarrow \left(-\infty,\,-\dfrac{1}{2}\right)$; In set builder

notation the interval is $\left\{x\,\middle|\,x<-\dfrac{1}{2}\right\}$.

29. The solutions to $|5x-7|\geq 3$ satisfy $x\leq s_1$ or $x\geq s_2$ where s_1 and s_2 are the solutions to $|5x-7|=3$.

$|5x-7|=3$ is equivalent to $5x-7=-3 \Rightarrow x=\dfrac{4}{5}$ and $5x-7=3 \Rightarrow x=2$.

The interval is $\left[-\infty,\,\dfrac{4}{5}\right)\cup[2,\,\infty)$; In set builder notation the interval is $\left\{x\,\middle|\,x\leq\dfrac{4}{5}\text{ or }x\geq 2\right\}$.

31. For $x^3-9x\leq 0$, first set $x^3-9x=0 \Rightarrow x(x^2-9)=0 \Rightarrow x(x+3)(x-3)=0 \Rightarrow x=-3,\,0,\,3$. These boundary numbers separate the number line into these intervals: $(-\infty,\,-3)$, $(-3,\,0)$, $(0,\,3)$ and $(3,\,\infty)$. Checking an x value in each interval we get:

for $x=-5$, $(-5)^3-9(-5)\leq 0 \Rightarrow -125+45\leq 0 \Rightarrow -80\leq 0$, which is true $\Rightarrow$ a solution;

for $x=-1$, $(-1)^3-9(-1)\leq 0 \Rightarrow -1+9\leq 0 \Rightarrow 8\leq 0$, which is false $\Rightarrow$ not a solution; for $x=1$,

   $(1)^3-9(1)\leq 0 \Rightarrow 1-9\leq 0 \Rightarrow -8\leq 0$, which is true $\Rightarrow$ a solution.

for $x=5$, $(5)^3-9(5)\leq 0 \Rightarrow 125-45\leq 0 \Rightarrow 80\leq 0$, which is false $\Rightarrow$ not a solution.

Therefore $(-\infty,\,-3]\cup[0,\,3]$. In set builder notation the intervals are $\{x\,|\,x\leq -3\text{ or }0\leq x\leq 3\}$.

33. (a) $x=-3,\,-1,\,1,\,2$

   (b) $(-\infty,\,-3)\cup(-1,\,1)\cup(2,\,\infty)$; In set builder notation the intervals are $\{x\,|\,x<-3\text{ or }-1<x<1\text{ or }x>2\}$.

   (c) $[-3,\,-1]\cup[1,\,2]$; In set builder notation the intervals are $\{x\,|-3\leq x\leq -1\text{ or }1\leq x\leq 2\}$.

35. (a) Shift $f(x)$ 2 units left and 1 unit down. See Figure 35a.

   (b) Multiply the y-coordinate of each point by -2 and plot. The new graph will open up. See Figure 35b.

   (c) Shift $f(x)$ up one unit and reflect across the y-axis. See Figure 35c.

   (d) Multiply the x-coordinate of each point by 2 and plot. See Figure 35d

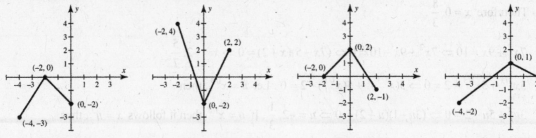

Figure 35a   Figure 35b   Figure 35c   Figure 35d

37. (a) It is increasing: $(-\infty,\,-2)$ and $(1,\,\infty)$; it is decreasing: $(-2,\,1)$. In set builder notation the intervals are $\{x\,|\,x<-2\text{ or }x>1\}$ and $\{x\,|-2<x<1\}$.

   (b) The zeros are approximately $x=-3.3,\,0$ and 1.8.

   (c) The turning points are: $(-2,\,3)$ and $(1,\,-1)$.

   (d) It has local extrema, maximum: 3, and minimum: -1.

39. Answers may vary. See Figure 39.

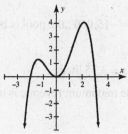

Figure 39

41. (a)
$$4a^2 \overline{\smash{\big)}\, 4a^3 - 8a^2 + 12}$$
with quotient $a - 2 + \dfrac{12}{4a^2}$

$$\underline{4a^3}$$
$$-8a^2 + 12$$
$$\underline{-8a^2 +}$$
$$12$$

Therefore the quotient is: $a - 2 + \dfrac{3}{a^2}$

(b)
$$x - 1 \overline{\smash{\big)}\, 2x^3 \qquad -4x + 1}$$
with quotient $2x^2 + 2x - 2 + \dfrac{-1}{x-1}$

$$\underline{2x^3 - 2x^2}$$
$$2x^2 - 4x + 1$$
$$\underline{2x^2 - 2x}$$
$$-2x + 1$$
$$\underline{-2x + 2}$$
$$-1$$

Therefore the quotient is: $2x^2 + 2x - 2 + \dfrac{-1}{x-1}$

43. $f(x) = 4(x+3)(x-1)^2(x-4)^3$

45. $\dfrac{3+4i}{1-i} \cdot \dfrac{1+i}{1+i} = \dfrac{3+3i+4i+4i^2}{1-(i)^2} = \dfrac{3+7i+4(-1)}{1-(-1)} = \dfrac{-1+7i}{2}$ or $\dfrac{-1}{2} + \dfrac{7i}{2}$

47. For the domain the denominator cannot equal zero $\Rightarrow$ set, $x^2 - 3x - 4 = 0 \Rightarrow (x+1)(x-4) = 0$. Then $D = \{x \,|\, x \neq -1, x \neq 4\}$. Since the degree of the numerator is less then the degree of the denominator the horizontal asymptote is $y = 0$. The vertical asymptotes are values when the denominator equals $0 \Rightarrow$ the vertical asymptotes are: $x = -1$, $x = 4$.

49. For m_1 use $(0, 0)$ and $(2, 40{,}000) \Rightarrow \dfrac{40{,}000 - 0}{2-0} \Rightarrow \dfrac{40{,}000}{2} \Rightarrow m_1 = 20{,}000$; the pool is being filled at a rate of 20,000 gallons per hour.

For m_2 use $(2, 40{,}000)$ and $(3, 50{,}000) \Rightarrow \dfrac{50{,}000 - 40{,}000}{3-2} \Rightarrow \dfrac{10{,}000}{1} \Rightarrow m_2 = 10{,}000$; now the pool is being filled at a rate of 10,000 gallons per hour.

For m_3 use $(3, 50{,}000)$ and $(4, 50{,}000) \Rightarrow \dfrac{50{,}000 - 50{,}000}{4-3} \Rightarrow \dfrac{0}{1} \Rightarrow m_3 = 0$; the amount of water in the pool is remaining constant.

For m_4 use $(4, 50,000)$ and $(6, 20,000) \Rightarrow \dfrac{20,000 - 50,000}{6 - 4} \Rightarrow \dfrac{-30,000}{2} \Rightarrow m_4 = -15,000$; the pool is being drained at a rate of 15,000 gallons per hour.

51. $0.35(2) + 0.12x = 0.20(x + 2) \Rightarrow 0.70 + 0.12x = 0.20x + 0.40 \Rightarrow 0.30 = 0.08x \Rightarrow x = 3.75$ liters

53. Graph $R(x) = x(800 - x)$ with x by the 100's from 0 to 800. See Figure 53. The maximum revenue is made when 400 toy figures are sold.

55. (a) $C(t) = (805 - 5t)$

     (b) $C(t) = 17,000 \Rightarrow 17,000 = t(805 - 5t) \Rightarrow 17,000 = 805t - 5t^2 \Rightarrow 5t^2 - 805t + 17,000 = 0 \Rightarrow$
           $5(t^2 - 161t + 3400) = 0 \Rightarrow 5(t - 25)(t - 136) = 0 \Rightarrow t = 25, 136$. The cost is \$17,000 when either 25 or 136 tickets are purchased.

     (c)   Graph $C(t) = t(805 - 5t)$. See Figure 55. The maximum cost of \$32,400 is reached at 80 to 81 tickets are sold.

57. Since volume of a cylinder is: $V = \pi r^2 h$. Then $10\pi = \pi r^2 h \Rightarrow h = \dfrac{10}{r^2}$.

     The surface area of the can is: $S(r) = 2\pi r^2 + 2\pi rh \Rightarrow S(r) = 2\pi r^2 + 2\pi r \cdot \dfrac{10}{r^2} = 2\pi r^2 + \dfrac{20\pi}{r}$. Graph this

     equation. See Figure 57. The minimum surface area occurs when $r \approx 1.7$ inches $\Rightarrow$ then $h = \dfrac{10}{(1.7)^2}$ or

     $h \approx 3.5$ inches.

[0, 1000, 100] by [0, 200,000, 20,000]   [0, 300, 100] by [0, 50,000, 10,000]   [0, 5, 1] by [0, 100, 10]

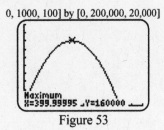

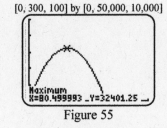

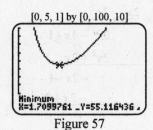

Figure 53   Figure 55   Figure 57

Chapter 5: Exponential and Logarithmic Functions

5.1: Combining Functions

1. $(f+g)(3) = f(3)+g(3) = 2+5 = 7$

3. $(fg)(x) = f(x)g(x) = (x^2)(4x) = 4x^3$

5. $(g \circ f)(x) = g(f(x))$. Therefore, $(g \circ f)(x)$ calculates the cost of x square yards of carpet.

7. (a) $f(3) = 2(3)-3 = 3$ and $g(3) = 1-3^2 = -8$. Thus, $(f+g)(3) = f(3)+g(3) = 3+(-8) = -5$.

   (b) $f(-1) = 2(-1)-3 = -5$ and $g(-1) = 1-(-1)^2 = 0$.

   Thus, $(f-g)(-1) = f(-1)-g(-1) = -5-0 = -5$.

   (c) $f(0) = 2(0)-3 = -3$ and $g(0) = 1-0^2 = 1$. Thus, $(fg)(0) = f(0) \cdot g(0) = (-3)(1) = -3$.

   (d) $f(2) = 2(2)-3 = 1$ and $g(2) = 1-2^2 = -3$. Thus, $(f/g)(2) = \dfrac{f(2)}{g(2)} = \dfrac{1}{-3} = -\dfrac{1}{3}$.

9. (a) $f(2) = 2(2)+1 = 5$ and $g(2) = \dfrac{1}{2}$. Thus, $(f+g)(2) = f(2)+g(2) = 5+\dfrac{1}{2} = \dfrac{11}{2}$.

   (b) $f\left(\dfrac{1}{2}\right) = 2\left(\dfrac{1}{2}\right)+1 = 2$ and $g\left(\dfrac{1}{2}\right) = \dfrac{1}{\frac{1}{2}} = 2$. Thus, $(f-g)\left(\dfrac{1}{2}\right) = f\left(\dfrac{1}{2}\right)-g\left(\dfrac{1}{2}\right) = 2-2 = 0$.

   (c) $f(4) = 2(4)+1 = 9$ and $g(4) = \dfrac{1}{4}$. Thus, $(fg)(4) = f(4) \cdot g(4) = (9)\left(\dfrac{1}{4}\right) = \dfrac{9}{4}$.

   (d) $f(0) = 2(0)+1 = 1$ and $g(0) = \dfrac{1}{0} \Rightarrow$ undefined. Thus, $(f/g)(0) \Rightarrow$ undefined.

11. (a) $(f+g)(x) = f(x)+g(x) = 2x+x^2$; Domain: all real numbers.

    (b) $(f-g)(x) = f(x)-g(x) = 2x-x^2$; Domain: all real numbers.

    (c) $(fg)(x) = f(x) \cdot g(x) = (2x)(x^2) = 2x^3$; Domain: all real numbers.

    (d) $(f/g)(x) = \dfrac{f(x)}{g(x)} = \dfrac{2x}{x^2} = \dfrac{2}{x}$; Domain: $\{x \mid x \neq 0\}$.

13. (a) $(f+g)(x) = f(x)+g(x) = (3x)+(1-x) = 2x+1$; Domain: all real numbers.

    (b) $(f-g)(x) = f(x)-g(x) = (3x)-(1-x) = 4x-1$; Domain: all real numbers.

    (c) $(fg)(x) = f(x) \cdot g(x) = (3x)(1-x) = 3x-3x^2$; Domain: all real numbers.

    (d) $(f/g)(x) = \dfrac{f(x)}{g(x)} = \dfrac{3x}{1-x}$; Domain: $\{x \mid x \neq 1\}$.

15. (a) $(f+g)(x) = f(x)+g(x) = \left(\dfrac{1}{2}x+2\right)+\left(4-\dfrac{1}{2}x\right) = 6$; Domain: all real numbers.

    (b) $(f-g)(x) = f(x)-g(x) = \left(\dfrac{1}{2}x+2\right)-\left(4-\dfrac{1}{2}x\right) = x-2$; Domain: all real numbers.

(c) $(fg)(x) = f(x) \cdot g(x) = \left(\dfrac{1}{2}x + 2\right)\left(4 - \dfrac{1}{2}x\right) = -\dfrac{1}{4}x^2 + x + 8$; Domain: all real numbers.

(d) $(f/g)(x) = \dfrac{f(x)}{g(x)} = \dfrac{\left(\dfrac{1}{2}x + 2\right)}{\left(4 - \dfrac{1}{2}x\right)} = \dfrac{x+4}{8-x}$; Domain: $\{x \mid x \neq 8\}$.

17. (a) $(f+g)(x) = f(x) + g(x) = (x^2 - 1) + (x^2 + 1) = 2x^2$; Domain: all real numbers.

     (b) $(f-g)(x) = f(x) - g(x) = (x^2 - 1) - (x^2 + 1) = -2$; Domain: all real numbers.

     (c) $(fg)(x) = f(x) \cdot g(x) = (x^2 - 1)(x^2 + 1) = x^4 - 1$; Domain: all real numbers.

     (d) $(f/g)(x) = \dfrac{f(x)}{g(x)} = \dfrac{x^2 - 1}{x^2 + 1}$; Domain: all real numbers.

19. (a) $(f+g)(x) = f(x) + g(x) = (x - \sqrt{x-1}) + (x + \sqrt{x-1}) = 2x$; Domain: $\{x \mid x \geq 1\}$.

     (b) $(f-g)(x) = f(x) - g(x) = (x - \sqrt{x-1}) - (x + \sqrt{x-1}) = -2\sqrt{x-1}$; Domain: $\{x \mid x \geq 1\}$.

     (c) $(fg)(x) = f(x) \cdot g(x) = (x - \sqrt{x-1})(x + \sqrt{x-1}) = x^2 - (x-1) = x^2 - x + 1$; Domain: $\{x \mid x \geq 1\}$.

     (d) $(f/g)(x) = \dfrac{f(x)}{g(x)} = \dfrac{x - \sqrt{x-1}}{x + \sqrt{x-1}}$; Domain: $\{x \mid x \geq 1\}$.

21. (a) $(f+g)(x) = f(x) + g(x) = (\sqrt{x} - 1) + (\sqrt{x} + 1) = 2\sqrt{x}$; Domain: $\{x \mid x \geq 0\}$.

     (b) $(f-g)(x) = f(x) - g(x) = (\sqrt{x} - 1) - (\sqrt{x} + 1) = -2$; Domain: $\{x \mid x \geq 0\}$.

     (c) $(fg)(x) = f(x) \cdot g(x) = (\sqrt{x} - 1)(\sqrt{x} + 1) = x - 1$; Domain: $\{x \mid x \geq 0\}$.

     (d) $(f/g)(x) = \dfrac{f(x)}{g(x)} = \dfrac{\sqrt{x} - 1}{\sqrt{x} + 1}$; Domain: $\{x \mid x \geq 0\}$.

23. (a) $(f+g)(x) = f(x) + g(x) = \dfrac{1}{x+1} + \dfrac{3}{x+1} = \dfrac{4}{x+1}$; Domain: $\{x \mid x \neq -1\}$.

     (b) $(f-g)(x) = f(x) - g(x) = \dfrac{1}{x+1} - \dfrac{3}{x+1} = -\dfrac{2}{x+1}$; Domain: $\{x \mid x \neq -1\}$.

     (c) $(fg)(x) = f(x) \cdot g(x) = \left(\dfrac{1}{x+1}\right)\left(\dfrac{3}{x+1}\right) = \dfrac{3}{(x+1)^2}$; Domain: $\{x \mid x \neq -1\}$.

     (d) $(f/g)(x) = \dfrac{f(x)}{g(x)} = \dfrac{\frac{1}{x+1}}{\frac{3}{x+1}} = \dfrac{1}{x+1} \cdot \dfrac{x+1}{3} = \dfrac{1}{3}$; Domain: $\{x \mid x \neq -1\}$.

25. (a) $(f+g)(x) = f(x) + g(x) = \dfrac{1}{2x-4} + \dfrac{x}{2x-4} = \dfrac{1+x}{2x-4}$; Domain: $\{x \mid x \neq 2\}$.

     (b) $(f-g)(x) = f(x) - g(x) = \dfrac{1}{2x-4} - \dfrac{x}{2x-4} = \dfrac{1-x}{2x-4}$; Domain: $\{x \mid x \neq 2\}$.

     (c) $(fg)(x) = f(x) \cdot g(x) = \left(\dfrac{1}{2x-4}\right)\left(\dfrac{x}{2x-4}\right) = \dfrac{x}{(2x-4)^2}$; Domain: $\{x \mid x \neq 2\}$.

     (d) $(f/g)(x) = \dfrac{f(x)}{g(x)} = \dfrac{\frac{1}{2x-4}}{\frac{x}{2x-4}} = \dfrac{1}{2x-4} \cdot \dfrac{2x-4}{x} = \dfrac{1}{x}$; Domain: $\{x \mid x \neq 0 \text{ and } x \neq 2\}$.

27. (a) $(f+g)(x) = f(x) + g(x) = (x^2 - 1) + (|x+1|) = x^2 - 1 + |x+1|$; Domain: all real numbers.

(b) $(f-g)(x) = f(x) - g(x) = (x^2 - 1) - (|x+1|) = x^2 - 1 - |x+1|$; Domain: all real numbers.

(c) $(fg)(x) = f(x) \cdot g(x) = (x^2 - 1)(|x+1|) = (x^2 - 1)(|x+1|)$; Domain: all real numbers.

(d) $(f/g)(x) = \dfrac{f(x)}{g(x)} = \dfrac{x^2 - 1}{|x+1|}$; Domain: $\{x \mid x \neq -1\}$.

29.   (a) $(f+g)(x) = f(x) + g(x) = \dfrac{(x-1)(x-2)}{x+1} + \dfrac{(x+1)(x-1)}{x-2} =$

$\dfrac{(x-1)(x-2)(x-2)}{(x+1)(x-2)} + \dfrac{(x-1)(x+1)(x+1)}{(x+1)(x-2)} =$

$\dfrac{(x-1)(x^2 - 4x + 4)}{(x+1)(x-2)} + \dfrac{(x-1)(x^2 + 2x + 1)}{(x+1)(x-2)} = \dfrac{(x-1)(2x^2 - 2x + 5)}{(x+1)(x-2)}$;

Domain: $\{x \mid x \neq -1, x \neq 2\}$

(b) $(f-g)(x) = f(x) - g(x) = \dfrac{(x-1)(x-2)}{x+1} - \dfrac{(x+1)(x-1)}{x-2} =$

$\dfrac{(x-1)(x-2)(x-2)}{(x+1)(x-2)} - \dfrac{(x-1)(x+1)(x+1)}{(x+1)(x-2)} =$

$\dfrac{(x-1)(x^2 - 4x + 4)}{(x+1)(x-2)} - \dfrac{(x-1)(x^2 + 2x + 1)}{(x+1)(x-2)} = \dfrac{(x-1)(-6x + 3)}{(x+1)(x-2)} = \dfrac{-3(x-1)(2x-1)}{(x+1)(x-2)}$;

Domain: $\{x \mid x \neq -1, x \neq 2\}$

(c) $(fg)(x) = f(x) \cdot g(x) = \dfrac{(x-1)(x-2)}{x+1} \cdot \dfrac{(x+1)(x-1)}{x-2} =$

$\dfrac{(x-2)(x-1)(x+1)(x-1)}{(x+1)(x-2)} = (x-1)(x-1) = (x-1)^2$; Domain: $\{x \mid x \neq -1, x \neq 2\}$

(d) $(f/g)(x) = \dfrac{f(x)}{g(x)} = \dfrac{(x-2)(x-1)}{x+1} \div \dfrac{(x+1)(x-1)}{x-2} = \dfrac{(x-2)(x-1)}{x+1} \cdot \dfrac{x-2}{(x+1)(x-1)} = \dfrac{(x-2)^2}{(x+1)^2}$; Domain:

$\{x \mid x \neq 1, x \neq -1 \text{ and } x \neq 2\}$

31.   (a) $(f+g)(x) = f(x) + g(x) = \dfrac{2}{(x+1)(x-1)} + \dfrac{(x+1)}{(x-1)(x-1)} =$

$\dfrac{2(x-1)}{(x+1)(x-1)^2} + \dfrac{(x+1)^2}{(x+1)(x-1)^2} = \dfrac{2x-2}{(x+1)(x-1)^2} + \dfrac{x^2 + 2x + 1}{(x+1)(x-1)^2} = \dfrac{x^2 + 4x - 1}{(x+1)(x-1)^2}$; Domain:

$\{x \mid x \neq -1, x \neq 1\}$

(b) $(f-g)(x) = f(x) - g(x) = \dfrac{2}{(x+1)(x-1)} - \dfrac{(x+1)}{(x-1)(x-1)} =$

$\dfrac{2(x-1)}{(x+1)(x-1)^2} - \dfrac{(x+1)^2}{(x+1)(x-1)} = \dfrac{2x-2}{(x+1)(x-1)^2} - \dfrac{x^2 + 2x + 1}{(x+1)(x-1)^2} = \dfrac{-x^2 - 3}{(x+1)(x-1)^2}$; Domain:

$\{x \mid x \neq -1, x \neq 1\}$

(c) $(fg)(x) = f(x) \cdot g(x) = \dfrac{2}{(x+1)(x-1)} \cdot \dfrac{(x+1)}{(x-1)(x-1)} = \dfrac{2}{(x-1)^3}$; Domain: $\{x \mid x \neq -1, x \neq 1\}$

(d) $(f/g)(x) = f(x) \div g(x) = \dfrac{2}{(x+1)(x-1)} \div \dfrac{(x+1)}{(x-1)(x-1)} = \dfrac{2}{(x+1)(x-1)} \cdot \dfrac{(x-1)(x-1)}{(x+1)} = \dfrac{2(x-1)}{(x+1)^2}$; Domain:

        $\{x | x \neq -1, x \neq 1\}$

33. (a) $(f+g)(x) = f(x) + g(x) = (x^{5/2} - x^{3/2}) + (x^{1/2}) = x^{5/2} - x^{3/2} + x^{1/2} = x^{1/2}(x^2 - x + 1)$; Domain:

           $\{x | x \geq 0\}$

     (b) $(f-g)(x) = f(x) - g(x) = (x^{5/2} - x^{3/2}) - (x^{1/2}) = x^{5/2} - x^{3/2} - x^{1/2} = x^{1/2}(x^2 - x - 1)$; Domain:

           $\{x | x \geq 0\}$

     (c) $(fg)(x) = f(x) \cdot g(x) = (x^{5/2} - x^{3/2})(x^{1/2}) = x^3 - x^2 = x^2(x-1)$; Domain: $\{x | x \geq 0\}$

     (d) $(f/g)(x) = f(x) \div g(x) = \left(\dfrac{x^{5/2} - x^{3/2}}{x^{1/2}}\right) = x^2 - x = x(x-1)$; Domain: $\{x | x > 0\}$

35. (a)   From the graph $f(-2) = 2$ and $g(-2) = -3$. Thus, $(f+g)(-2) = f(-2) + g(-2) = 2 + (-3) = -1$.

     (b)   From the graph $f(-1) = 1.5$ and $g(-1) = -2$. Thus, $(f-g)(-1) = f(-1) - g(-1) = 1.5 - (-2) = 3.5$.

     (c)   From the graph $f(2) = 0$ and $g(2) = 1$. Thus, $(fg)(2) = f(2) \cdot g(2) = (0)(1) = 0$.

     (d)   From the graph $f(0) = 1$ and $g(0) = -1$. Thus, $(f/g)(0) = \dfrac{f(0)}{g(0)} = \dfrac{1}{-1} = -1$.

37. (a)   From the graph $f(2) = 4$ and $g(2) = -2$. Thus, $(f+g)(2) = f(2) + g(2) = 4 + (-2) = 2$.

     (b)   From the graph $f(1) = 1$ and $g(1) = -3$. Thus, $(f-g)(1) = f(1) - g(1) = 1 - (-3) = 4$.

     (c)   From the graph $f(0) = 0$ and $g(0) = -4$. Thus, $(fg)(0) = f(0) \cdot g(0) = (0)(-4) = 0$.

     (d)   From the graph $f(1) = 1$ and $g(1) = -3$. Thus, $(f/g)(1) = \dfrac{f(1)}{g(1)} = \dfrac{1}{-3} = -\dfrac{1}{3}$.

39. (a)   From the graph $f(0) = 0$ and $g(0) = 2$. Thus, $(f+g)(0) = f(0) + g(0) = 0 + 2 = 2$.

     (b)   From the graph $f(-1) = -2$ and $g(-1) = 1$. Thus, $(f-g)(-1) = f(-1) - g(-1) = -2 - 1 = -3$.

     (c)   From the graph $f(1) = 2$ and $g(1) = 1$. Thus, $(fg)(1) = f(1) \cdot g(1) = (2)(1) = 2$.

     (d)   From the graph $f(2) = 4$ and $g(2) = -2$. Thus, $(f/g)(2) = \dfrac{f(2)}{g(2)} = \dfrac{4}{-2} = -2$.

41. (a) $(f+g)(-1) = f(-1) + g(-1) = -3 + -2 = -5$.

     (b) $(g-f)(0) = g(0) - f(0) = 3 - 5 = -2$.

     (c) $(gf)(2) = g(2) \cdot f(2) = 0 \cdot 1 = 0$.

     (d) $(f/g)(2) = \dfrac{f(2)}{g(2)} = \dfrac{1}{0} \Rightarrow$ undefined .

43. (a) $(f+g)(0) = f(0) + g(0) = 4 + 2 = 6$.

     (b) $(g-f)(5) = g(5) - f(5) = 1 - 3 = -2$.

     (c) $(gf)(-5) = g(-5) \cdot f(-5) = (-1) \cdot (-3) = 3$.

     (d) $(f/g)(5) = \dfrac{f(5)}{g(5)} = \dfrac{3}{1} = 3$.

45. (a) $(f+g)(2) = f(2) + g(2) = 7 + (-2) = 5$

     (b) $(f-g)(4) = f(4) - g(4) = 10 - 5 = 5$

     (c) $(fg)(-2) = f(-2) \cdot g(-2) = (0)(6) = 0$

(d) $(f/g)(0) = \dfrac{f(0)}{g(0)} = \dfrac{5}{0} \Rightarrow$ undefined

47.   For example: $(f-g)(2) = f(2) - g(2) = 7 - (-2) = 9$. See Figure 47. A dash (—) indicates that the value of
       the function is undefined.

x	−2	0	2	4
$(f+g)(x)$	6	5	5	15
$(f-g)(x)$	−6	5	9	5
$(fg)(x)$	0	0	−14	50
$(f/g)(x)$	0	—	−3.5	2

Figure 47

49.   (a) $g(x) = 2x+1 \Rightarrow g(-3) = 2(-3)+1 \Rightarrow g(-3) = -5$

       (b) $g(x) = 2x+1 \Rightarrow g(b) = 2(b)+1 \Rightarrow g(b) = 2b+1$

       (c) $g(x) = 2x+1 \Rightarrow g(x^3) = 2(x^3)+1 \Rightarrow g(x^3) = 2x^3+1$

       (d) $g(x) = 2x+1 \Rightarrow g(2x-3) = 2(2x-3)+1 \Rightarrow g(2x-3) = 4x-6+1 \Rightarrow g(2x-3) = 4x-5$

51.   (a) $g(x) = 2(x+3)^2-4 \Rightarrow g(-3) = 2((-3)+3)^2-4 \Rightarrow g(-3) = 2(0)-4 \Rightarrow g(-3) = -4$

       (b) $g(x) = 2(x+3)^2-4 \Rightarrow g(b) = 2((b)+3)^2-4 \Rightarrow g(b) = 2(b+3)^2-4$

       (c) $g(x) = 2(x+3)^2-4 \Rightarrow g(x^3) = 2((x^3)+3)^2-4 \Rightarrow g(x^3) = 2(x^3+3)^2-4$

       (d) $g(x) = 2(x+3)^2-4 \Rightarrow g(2x-3) = 2((2x-3)+3)^2-4 \Rightarrow g(2x-3) = 2(2x)^2-4 \Rightarrow$

              $g(2x-3) = 2(4x^2)-4 \Rightarrow g(2x-3) = 8x^2-4$

53.   (a) $g(x) = \dfrac{1}{2}x^2+3x-1 \Rightarrow g(-3) = \dfrac{1}{2}(-3)^2+3(-3)-1 \Rightarrow g(-3) = \dfrac{9}{2}-9-1 \Rightarrow$

              $g(-3) = \dfrac{9}{2}-\dfrac{20}{2} \Rightarrow g(-3) = -\dfrac{11}{2}$

       (b) $g(x) = \dfrac{1}{2}x^2+3x-1 \Rightarrow g(b) = \dfrac{1}{2}(b)^2+3(b)-1 \Rightarrow g(b) = \dfrac{1}{2}b^2+3b-1$

       (c) $g(x) = \dfrac{1}{2}x^2+3x-1 \Rightarrow g(x^3) = \dfrac{1}{2}(x^3)^2+3(x^3)-1 \Rightarrow g(x^3) = \dfrac{1}{2}x^6+3x^3-1$

       (d) $g(x) = \dfrac{1}{2}x^2+3x-1 \Rightarrow g(2x-3) = \dfrac{1}{2}(2x-3)^2+3(2x-3)-1 \Rightarrow$

              $g(2x-3) = \dfrac{4x^2-12x+9}{2}+6x-9-1 \Rightarrow g(2x-3) = 2x^2-6x+\dfrac{9}{2}+6x-\dfrac{20}{2} \Rightarrow g(2x-3) = 2x^2-\dfrac{11}{2}$

55.   (a) $g(x) = \sqrt{x+4} \Rightarrow g(-3) = \sqrt{(-3)+4} \Rightarrow g(-3) = \sqrt{1} \Rightarrow g(-3) = 1$

       (b) $g(x) = \sqrt{x+4} \Rightarrow g(b) = \sqrt{(b)+4} \Rightarrow g(b) = \sqrt{b+4}$

       (c) $g(x) = \sqrt{x+4} \Rightarrow g(x^3) = \sqrt{(x^3)+4} \Rightarrow g(x^3) = \sqrt{x^3+4}$

       (d) $g(x) = \sqrt{x+4} \Rightarrow g(2x-3) = \sqrt{(2x-3)+4} \Rightarrow g(2x-3) = \sqrt{2x+1}$

57.   (a) $g(x) = |3x-1|+4 \Rightarrow g(-3) = |3(-3)-1|+4 \Rightarrow g(-3) = |-10|+4 \Rightarrow g(-3) = 14$

       (b) $g(x) = |3x-1|+4 \Rightarrow g(b) = |3(b)-1|+4 \Rightarrow g(b) = |3b-1|+4$

       (c) $g(x) = |3x-1|+4 \Rightarrow g(x^3) = |3(x^3)-1|+4 \Rightarrow g(x^3) = |3x^3-1|+4$

       (d) $g(x) = |3x-1|+4 \Rightarrow g(2x-3) = |3(2x-3)-1|+4 \Rightarrow g(2x-3) = |6x-10|+4$

59. (a) $g(x) = \dfrac{4x}{x+3} \Rightarrow g(-3) = \dfrac{4(-3)}{(-3)+3} \Rightarrow g(-3) = \dfrac{-12}{0} \Rightarrow g(-3) = $ undefined

   (b) $g(x) = \dfrac{4x}{x+3} \Rightarrow g(b) = \dfrac{4(b)}{b+3} \Rightarrow g(b) = \dfrac{4b}{b+3}$

   (c) $g(x) = \dfrac{4x}{x+3} \Rightarrow g(x^3) = \dfrac{4(x^3)}{x^3+3} \Rightarrow g(x^3) = \dfrac{4x^3}{x^3+3}$

   (d) $g(x) = \dfrac{4x}{x+3} \Rightarrow g(2x-3) = \dfrac{4(2x-3)}{(2x-3)+3} \Rightarrow g(2x-3) = \dfrac{8x-12}{2x} \Rightarrow$

   $g(2x-3) = \dfrac{4x-6}{x} \Rightarrow g(2x-3) = \dfrac{2(2x-3)}{x}$

61. (a) $(f \circ g)(2) = f(g(2)) = f(2^2) = f(4) = \sqrt{4+5} = \sqrt{9} = 3$

   (b) $(g \circ f)(-1) = g(f(-1)) = g(\sqrt{-1+5}) = g(2) = 2^2 = 4$

63. (a) $(f \circ g)(-4) = f(g(-4)) = f(|-4|) = f(4) = 5(4) - 2 = 18$

   (b) $(g \circ f)(5) = g(f(5)) = g(5(5) - 2) = g(23) = |23| = 23$

65. (a) $(f \circ g)(x) = f(g(x)) = f(x+1) = 2(x+1) = 2x+2$; D : all real numbers

   (b) $(g \circ f)(x) = g(f(x)) = g(2x) = (2x)+1 = 2x+1$; D : all real numbers

   (c) $(f \circ f)(x) = f(f(x)) = f(2x) = 2(2x) = 4x$; D : all real numbers

67. (a) $(f \circ g)(x) = f(g(x)) = f(-4x-1) = 2(-4x-1)+1 = -8x-1$; D : all real numbers

   (b) $(g \circ f)(x) = g(f(x)) = g(2x+1) = -4(2x+1)-1 = -8x-5$; D : all real numbers

   (c) $(f \circ f)(x) = f(f(x)) = f(2x+1) = 2(2x+1)+1 = 4x+3$; D : all real numbers

69. (a) $(f \circ g)(x) = f(g(x)) = f(x^2+3x-1) = (x^2+3x-1)^3$; D : all real numbers

   (b) $(g \circ f)(x) = g(f(x)) = g(x^3) = (x^3)^2 + 3(x^3) - 1 = x^6 + 3x^3 - 1$; D : all real numbers

   (c) $(f \circ f)(x) = f(f(x)) = f(x^3) = (x^3)^3 = x^9$; D : all real numbers

71. (a) $(f \circ g)(x) = f(g(x)) = f(x^4 + x^2 - 3x - 4) = x^4 + x^2 - 3x - 2$; D : all real numbers

   (b) $(g \circ f)(x) = g(f(x)) = g(x+2) = (x+2)^4 + (x+2)^2 - 3(x+2) - 4$; D : all real numbers

   (c) $(f \circ f)(x) = f(f(x)) = f(x+2) = (x+2)+2 = x+4$; D : all real numbers

73. (a) $(f \circ g)(x) = f(g(x)) = f(x^3) = 2 - 3x^3$; D : all real numbers

   (b) $(g \circ f)(x) = g(f(x)) = g(2-3x) = (2-3x)^3$; D : all real numbers

   (c) $(f \circ f)(x) = f(f(x)) = f(2-3x) = 2 - 3(2-3x) = 9x - 4$; D : all real numbers

75. (a) $(f \circ g)(x) = f(g(x)) = f(5x) = \dfrac{1}{5x+1}$; $D = \left\{x \mid x \neq -\dfrac{1}{5}\right\}$

   (b) $(g \circ f)(x) = g(f(x)) = g\left(\dfrac{1}{x+1}\right) = 5\left(\dfrac{1}{x+1}\right) = \dfrac{5}{x+1}$; $D = \{x \mid x \neq -1\}$

   (c) $(f \circ f)(x) = f(f(x)) = f\left(\dfrac{1}{x+1}\right) = \dfrac{1}{\dfrac{1}{x+1}+1} = \dfrac{1}{\dfrac{1+x+1}{x+1}} = \dfrac{x+1}{x+2}$; $D = \{x \mid x \neq -1 \text{ and } x \neq -2\}$

77. (a) $(f \circ g)(x) = f(g(x)) = f(\sqrt{4-x^2}) = \sqrt{4-x^2} + 4$; $D = \{x \mid -2 \leq x \leq 2\}$

   (b) $(g \circ f)(x) = g(f(x)) = g(x+4) = \sqrt{4-(x+4)^2}$; $D = \{x \mid -6 \leq x \leq -2\}$

(c) $(f \circ f)(x) = f(f(x)) = f(x+4) = (x+4)+4 = x+8$; D : all real numbers

79. (a) $(f \circ g)(x) = f(g(x)) = f(3x) = \sqrt{3x-1}$; $D = \left\{x \mid x \ge \dfrac{1}{3}\right\}$

   (b) $(g \circ f)(x) = g(f(x)) = g(\sqrt{x-1}) = 3\sqrt{x-1}$; $D = \{x \mid x \ge 1\}$

   (c) $(f \circ f)(x) = f(f(x)) = f(\sqrt{x-1}) = \sqrt{\sqrt{x-1}-1}$; $D = \{x \mid x \ge 2\}$

81. (a) $(f \circ g)(x) = f(g(x)) = f\left(\dfrac{1-x}{5}\right) = 1 - 5\left(\dfrac{1-x}{5}\right) = 1 - (1-x) = x$; D : all real numbers

   (b) $(g \circ f)(x) = g(f(x)) = g(1-5x) = \dfrac{1-(1-5x)}{5} = \dfrac{5x}{5} = x$; D : all real numbers

   (c) $(f \circ f)(x) = f(f(x)) = f(1-5x) = 1 - 5(1-5x) = 1 - 5 + 25x = 25x - 4$; D : all real numbers

83. (a) $(f \circ g)(x) = f(g(x)) = f\left(\dfrac{1}{kx}\right) = \dfrac{1}{k\left(\frac{1}{kx}\right)} = \dfrac{1}{\frac{1}{x}} = x$; D: $\{x \mid x \ne 0\}$

   (b) $(g \circ f)(x) = g(f(x)) = g\left(\dfrac{1}{kx}\right) = \dfrac{1}{k\left(\frac{1}{kx}\right)} = \dfrac{1}{\frac{1}{x}} = x$; D: $\{x \mid x \ne 0\}$

   (c) $(f \circ f)(x) = f(f(x)) = f\left(\dfrac{1}{kx}\right) = \dfrac{1}{k\left(\frac{1}{kx}\right)} = \dfrac{1}{\frac{1}{x}} = x$; D: $\{x \mid x \ne 0\}$

85. (a) $(f \circ g)(4) = f(g(4)) = f(0) = -4$
   (b) $(g \circ f)(3) = g(f(3)) = g(2) = 2$
   (c) $(f \circ f)(2) = f(f(2)) = f(0) = -4$

87. (a) $(f \circ g)(1) = f(g(1)) = f(2) = -3$
   (b) $(g \circ f)(-2) = g(f(-2)) = g(-3) = -2$
   (c) $(g \circ g)(-2) = g(g(-2)) = g(-1) = 0$

89. (a) $(g \circ f)(1) = g(f(1)) = g(4) = 5$
   (b) $(f \circ g)(4) = f(g(4)) = f(5) \Rightarrow$ undefined
   (c) $(f \circ f)(3) = f(f(3)) = f(1) = 4$

91. (a) $(g \circ f)(0) = g(f(0)) = g(5) = 10$
   (b) $(f \circ g)(5) = f(g(5)) = f(10) = 10$
   (c) $(g \circ g)(10) = g(g(10)) = g(5) = 10$

93. $g(3) = 4$; $f(4) = 2$

95. $h(x) = \sqrt{x-2} \Rightarrow g(x) = \sqrt{x}$ and $f(x) = x-2$. *Answers may vary.*

97. $h(x) = 4(2x+1)^3 \Rightarrow g(x) = 4x^3$ and $f(x) = 2x+1$. *Answers may vary.*

99. $h(x) = (x^3-1)^2 \Rightarrow g(x) = x^2$ and $f(x) = x^3-1$. *Answers may vary.*

101. $h(x) = -4|x+2|-3 \Rightarrow g(x) = -4|x|-3$ and $f(x) = x+2$. *Answers may vary.*

103. $h(x) = \dfrac{1}{(x-1)^2} \Rightarrow g(x) = \dfrac{1}{x^2}$ and $f(x) = x-1$. *Answers may vary.*

105. $h(x) = x^{3/4} - x^{1/4} \Rightarrow g(x) = x^3 - x$ and $f(x) = x^{1/4}$. *Answers may vary.*

107. Since each album sells for $15 each, the revenue function would be $R(x) = 15x$. Therefore,
$P(x) = R(x) - C(x) = 15x - (5x + 12,000) = 10x - 12,000$; $P(3000) = 10(3000) - 12,000 = \$18,000$

109. (a) $I(x) = 36x$

   (b) $C(x) = 2.54x$

   (c) $F(x) = (C \circ I)(x)$

   (d) $F(x) = 36(2.54x) = 91.44x$

111. (a) $d(x) = f(x) + g(x) = \dfrac{11}{6}x + \dfrac{1}{9}x^2$

   (b) $d(60) = \dfrac{11}{6}(60) + \dfrac{1}{9}(60)^2 = 510$; this car requires 510 feet to stop at 60 mph.

113. (a) $(g \circ f)(1) = g(f(1)) = g(1.5) = 5.25$;
       A 1% decrease in the ozone layer could result in a 5.25% increase in skin cancer.

   (b) $(f \circ g)(21) = f(g(21))$, but $g(21)$ is not given by the table. So $(f \circ g)(21)$ is not possible using these tables.

115. (a) $(g \circ f)(1975) = g(f(1975)) = g(3) = 4.5\%$

   (b) $(g \circ f)(x)$ computes the percent increase in peak demand during year x.

117. (a) $(g \circ f)(2) = g(f(2)) = g(77) \approx 25°C$. After 2 hours the temperature is about 25°C. *Answers may vary.*

   (b) $(g \circ f)(x)$ computes the Celsius temperature after x hours.

119. (a)   For $A(s) = \dfrac{\sqrt{3}}{4}s^2$, $A(4s) = \dfrac{\sqrt{3}}{4}(4s)^2 = 16\dfrac{\sqrt{3}}{4}s^2 = 16A(s)$; if the length of a side is quadrupled, the area increases by a factor of 16.

   (b)   For $A(s) = \dfrac{\sqrt{3}}{4}s^2$, $A(s+2) = \dfrac{\sqrt{3}}{4}(s+2)^2 = \dfrac{\sqrt{3}}{4}(s^2 + 4s + 4) = \dfrac{\sqrt{3}}{4}s^2 + \sqrt{3}s + \sqrt{3} = A(s) + \sqrt{3}(s+1)$; if the length of a side increases by 2, the area increases by $\sqrt{3}(s+1)$.

121. The radius of the circular wave at the end of t seconds would be $6t$ inches. $C = 2\pi r \Rightarrow C(t) = 2\pi(6t) = 12\pi t$

123. (a)   To find the total emissions we must add the developed and developing countries' emissions for each year. The resulting table is shown in Figure 123.

   (b) $h(x) = f(x) + g(x)$

x	1990	2000	2010	2020	2030
$h(x)$	32	35.5	39	42.5	46

Figure 123

125. $S = \pi r \sqrt{r^2 + h^2}$ and $h = 2r \Rightarrow S = \pi r \sqrt{r^2 + (2r)^2} \Rightarrow S = \pi r \sqrt{5r^2} \Rightarrow S = \pi r^2 \sqrt{5}$

127. If $h = 3r$, then the volume of the cylinder $V = \pi r^2 h$ can be rewritten as $V = \pi\left(\dfrac{h}{3}\right)^2 h \Rightarrow V = \dfrac{\pi h^3}{9}$.

129. $L = 3W \Rightarrow W = \dfrac{L}{3}$; $H = \dfrac{W}{2} = \dfrac{L}{6}$ $V = LWH = L\left(\dfrac{L}{3}\right)\left(\dfrac{L}{6}\right) = \dfrac{L^3}{18}$

131. Let $f(x) = ax + b$ and $g(x) = cx + d$. Then $f(x) + g(x) = (ax + b) + (cx + d) = (a+c)x + (b+d)$, which is linear.

133. (a) $f(x) = k$ and $g(x) = ax + b$. Thus, $(f \circ g)(x) = f(g(x)) = f(ax + b) = k$; $f \circ g$ is a constant function.

   (b) $f(x) = k$ and $g(x) = ax + b$. Thus, $(g \circ f)(x) = g(f(x)) = g(k) = ak + b$; $g \circ f$ is a constant function.

5.2: Inverse Functions and Their Representations

1.   Closing a window.

3.   Closing a book, standing up, and walking out of the classroom.

5.   Subtract 2 from x; $x+2$ and $x-2$

7.   Divide x by 3 and add 2; $3(x-2)$ and $\dfrac{x}{3}+2$

9.   Subtract 1 from x and cube the result; $\sqrt[3]{x}+1$ and $(x-1)^3$

11.   Take the reciprocal of x; $\dfrac{1}{x}$ and $\dfrac{1}{x}$

13.   Since a horizontal line will intersect the graph at most once, f is one-to-one.

15.   Since a horizontal line will intersect the graph two times, f is not one-to-one.

17.   Since a horizontal line will intersect the graph two times, f is not one-to-one.

19.   Since a horizontal line will intersect the graph at most once, f is one-to-one.

21.   Since $f(2)=3$ and $f(3)=3$, different inputs result in the same output. Therefore f is not one-to-one. Does not have an inverse.

23.   Since different inputs always result in different outputs, f is one-to-one. Does have an inverse.

25.   Since the graph of f is a line sloping upward from left to right, a horizontal line can intersect it at most once. Therefore, f is one-to-one.

27.   Since the graph of f is a parabola, a horizontal line can intersect it more than once. Therefore, f is not one-to-one.

29.   Since the graph of f is a parabola, a horizontal line can intersect it more than once. Therefore, f is not one-to-one.

31.   Since the graph of f is a V-shape, a horizontal line can intersect it more than once. Therefore, f is not one-to-one.

33.   Since the graph of f is both increasing and decreasing, a horizontal line can intersect it more than once. Therefore, f is not one-to-one.

35.   Since the graph of f is both increasing and decreasing, a horizontal line can intersect it more than once. Therefore, f is not one-to-one.

37.   Since the graph of f is a line sloping upward from left to right, a horizontal line can intersect it at most once. Therefore f is one-to-one.

39.   Since the person goes up and down, there would be several times when the person attained a particular height above the ground. A one-to-one function would not model this situation.

41.   Since the population of the United States increased each of these years, a horizontal line would intersect a graph of this population at most once. A one-to-one function would model this situation.

43.   The inverse operation of adding by 5 is subtracting by 5. Therefore, $f^{-1}(x)=x-5$.

45.   The inverse operation of multiplying by 6 is dividing by 6. Therefore, $f^{-1}(x)=\dfrac{x}{6}$.

47.   The inverse operations of multiplying by -7 and adding 2 are subtracting 2 and dividing by -7, Therefore,
$f^{-1}(x)=-\dfrac{x-2}{7}=\dfrac{2-x}{7}$.

49. The inverse operation of taking the cube root of x is cubing x. Therefore, $f^{-1}(x) = x^3$.

51. The inverse operations of multiplying by -2 and adding 10 are subtracting 10 and dividing by -2. Therefore,
$$f^{-1}(x) = \frac{x-10}{-2} = -\frac{1}{2}x + 5.$$

53. The inverse operations of multiplying by 3 and subtracting 1 are adding 1 and dividing by 3. Therefore,
$$f^{-1}(x) = \frac{x+1}{3}.$$

55. The inverse operations of cubing a number, multiplying by 2 and subtracting 5 are adding 5, dividing by 2 and taking the cube root of the result. Therefore, $f^{-1}(x) = \sqrt[3]{\dfrac{x+5}{2}}$.

57. The inverse operations of squaring a number and subtracting 1 are adding 1 and taking the square root. Therefore, $f^{-1}(x) = \sqrt{x+1}$.

59. The inverse operations of multiplying by 2 and taking the reciprocal are taking the reciprocal and dividing by 2. Therefore, $f^{-1}(x) = \dfrac{1}{x} \div 2 = \dfrac{1}{2x}$.

61. The inverse operations of multiplying by -5, adding 4, multiplying by $\dfrac{1}{2}$, and adding 1 are subtracting 1, dividing by $\dfrac{1}{2}$ (or multiplying by 2), subtracting 4, and dividing by
$$-5 \Rightarrow f^{-1}(x) = \frac{2(x-1)-4}{-5} = \frac{2x-6}{-5} = -\frac{2(x-3)}{5}.$$

63. $y = \dfrac{x}{x+2} \Rightarrow y(x+2) = x \Rightarrow xy + 2y = x \Rightarrow 2y = x - xy \Rightarrow 2y = x(1-y) \Rightarrow x = \dfrac{2y}{1-y} \Rightarrow$
$$f^{-1}(x) = \frac{2x}{1-x} \Rightarrow f^{-1}(x) = -\frac{2x}{x-1}$$

65. $y = \dfrac{2x+1}{x-1} \Rightarrow y(x-1) = 2x+1 \Rightarrow xy - y = 2x+1 \Rightarrow xy - 2x = y+1 \Rightarrow$
$$x(y-2) = y+1 \Rightarrow x = \frac{y+1}{y-2} \Rightarrow f^{-1}(x) = \frac{x+1}{x-2}$$

67. $y = \dfrac{1}{x} - 3 \Rightarrow y + 3 = \dfrac{1}{x} \Rightarrow x = \dfrac{1}{y+3} \Rightarrow f^{-1}(x) = \dfrac{1}{x+3}$

69. $y = \dfrac{1}{x^3 - 1} \Rightarrow \dfrac{1}{y} = x^3 - 1 \Rightarrow \dfrac{1}{y} + 1 = x^3 \Rightarrow \dfrac{1}{y} + \dfrac{y}{y} = x^3 \Rightarrow \dfrac{y+1}{y} = x^3 \Rightarrow f^{-1}(x) = \sqrt[3]{\dfrac{x+1}{x}}$

71. $y = 4 - x^2$, $x \geq 0 \Rightarrow x^2 = 4 - y \Rightarrow x = \sqrt{4-y}$; therefore $f^{-1}(x) = \sqrt{4-x}$

73. $y = (x-2)^2 + 4$, $x \geq 2 \Rightarrow y - 4 = (x-2)^2 \Rightarrow \sqrt{y-4} = x - 2 \Rightarrow x = \sqrt{y-4} + 2$; therefore $f^{-1}(x) = \sqrt{x-4} + 2$.

75. $y = x^{2/3} + 1$, $x \geq 0 \Rightarrow x^{2/3} = y - 1 \Rightarrow x = (y-1)^{3/2}$; $\Rightarrow f^{-1}(x) = (x-1)^{3/2}$

77. $y = \sqrt{9 - 2x^2}$, $-\dfrac{3}{12} \leq x \leq \dfrac{3}{\sqrt{2}} \Rightarrow y^2 = 9 - 2x^2 \Rightarrow y^2 - 9 = -2x^2 \Rightarrow \dfrac{9 - y^2}{2} = x^2 \Rightarrow$
$$\sqrt{\frac{9-y^2}{2}} = x; \Rightarrow f^{-1}(x) = \sqrt{\frac{9-x^2}{2}}$$

79. $y = -4x \Rightarrow -\dfrac{y}{4} = x \Rightarrow f^{-1}(x) = -\dfrac{x}{4}$; D and R are all real numbers.

81. $y = 6 - x \Rightarrow 6 - y = x \Rightarrow f^{-1}(x) = 6 - x$; D and R are all real numbers.

83. $y = 5x + 7 \Rightarrow y - 7 = 5x \Rightarrow \dfrac{y-7}{5} = x \Rightarrow f^{-1}(x) = \dfrac{x-7}{5}$; D and R are all real numbers.

85. $y = 5x - 15 \Rightarrow y + 15 = 5x \Rightarrow \dfrac{y+15}{5} = x \Rightarrow f^{-1}(x) = \dfrac{x+15}{5}$; D and R are all real numbers.

87. $y = \sqrt[3]{x-5} \Rightarrow y^3 = x - 5 \Rightarrow y^3 + 5 = x \Rightarrow f^{-1}(x) = x^3 + 5$; D and R are all real numbers

89. $y = \dfrac{x-5}{4} \Rightarrow 4y = x - 5 \Rightarrow 4y + 5 = x \Rightarrow f^{-1}(x) = 4x + 5$; D and R are all real numbers.

91. $y = \sqrt{x-5}$, $x \geq 5 \Rightarrow y^2 = x - 5 \Rightarrow y^2 + 5 = x \Rightarrow f^{-1}(x) = x^2 + 5$; $D = \{x | x \geq 0\}$ and $R = \{y | y \geq 5\}$

93. $y = \dfrac{1}{x+3} \Rightarrow \dfrac{1}{y} = x + 3 \Rightarrow \dfrac{1}{y} - 3 = x \Rightarrow f^{-1}(x) = \dfrac{1}{x} - 3$; $D = \{x | x \neq 0\}$ and $D = \{y | y \neq -3\}$

95. $y = 2x^3 \Rightarrow \dfrac{y}{2} = x^3 \Rightarrow \sqrt[3]{\dfrac{y}{2}} = x \Rightarrow f^{-1}(x) = \sqrt[3]{\dfrac{x}{2}}$; D and R are all real numbers.

97. $y = x^2$, $x \geq 0 \Rightarrow \sqrt{y} = x \Rightarrow f^{-1}(x) = \sqrt{x}$; $D = \{x | x \geq 0\}$ and $D = \{y | y \geq 0\}$

99. The domain and range of f are $D = \{1, 2, 3\}$ and $R = \{5, 7, 9\}$. Interchange these to get the domain and range of f^{-1}; $D = \{5, 7, 9\}$ and $R = \{1, 2, 3\}$. See Figure 99.

x	5	7	9
$f^{-1}(x)$	1	2	3

Figure 99

101. The domain and range of f are $D = \{0, 2, 4\}$ and $R = \{0, 4, 16\}$. Interchange these to get the domain and range of f^{-1}; $D = \{0, 4, 16\}$ and $R = \{0, 2, 4\}$. See Figure 101.

x	0	4	16
$f^{-1}(x)$	0	2	4

Figure 101

103. Since f multiplies x by 4, f^{-1} divides x by 4. See Figure 103.

x	0	2	4	6
$f^{-1}(x)$	0	$\frac{1}{2}$	1	$\frac{3}{2}$

Figure 103

105. $f(1) = 3 \Rightarrow f^{-1}(3) = 1$

107. $g(3) = 4 \Rightarrow g^{-1}(4) = 3$

109. $(f \circ g^{-1})(1) = f(g^{-1}(1)) = f(2) = 5$. Note that $g(2) = 1 \Rightarrow g^{-1}(1) = 2$.

111. $(g \circ f^{-1})(5) = g(f^{-1}(5)) = g(2) = 1$. Note that $f(2) = 5 \Rightarrow f^{-1}(5) = 2$.

113. $f(4) = 0 \Rightarrow f^{-1}(0) = 4$

115. $g(0) = 4 \Rightarrow g^{-1}(4) = 0$

117. $(f \circ g^{-1})(2) = f(g^{-1}(2)) = f(6) = 6$. Note that $g(6) = 2 \Rightarrow g^{-1}(2) = 6$.

119. $(f^{-1} \circ f^{-1})(8) = f^{-1}(f^{-1}(8)) = f^{-1}(0) = 4$. Note that $f(0) = 8 \Rightarrow f^{-1}(8) = 0$ and $f(4) = 0 \Rightarrow f^{-1}(0) = 4$.

121. (a) $f(1) \approx \$110$

  (b) $f^{-1}(110) \approx 1$ year

(c) $f^{-1}(160) \approx 5$ years

The function f^{-1} computes the number of years it takes for this savings account to accumulate x dollars.

123. (a) $f(-1) = 2$

(b) $f^{-1}(-2) = 3$

(c) $f^{-1}(0) = 1$

(d) $(f^{-1} \circ f)(3) = f^{-1}(f(3)) = f^{-1}(-2) = 3$

125. (a) $f(4) = 4$

(b) $f^{-1}(0) = 0$

(c) $f^{-1}(6) = 9$

(d) $(f^{-1} \circ f)(4) = f^{-1}(f(4)) = f^{-1}(4) = 4$

127. The graph of f passes through the points $(-2, -4)$ and $(1, 2)$. Thus, the graph of f^{-1} passes through the points $(-4, -2)$ and $(2, 1)$. Plot these points and sketch the reflection of f in the line $y = x$ to obtain the graph of f^{-1}. Notice that since the graph of f is a line, its reflection will also be a line. See Figure 127.

129. The graph of f passes through the points $\left(-2, \frac{1}{4}\right)$, $(0, 1)$, $(1, 2)$ and $(2, 4)$. Thus, the graph of f^{-1} passes through the points $\left(\frac{1}{4}, -2\right)$, $(1, 0)$, $(2, 1)$ and $(4, 2)$. Plot these points and sketch the reflection of f in the line $y = x$ to obtain the graph of f^{-1}. See Figure 129.

131. The graph of f passes through the points $(-3, -2)$, $(-1, 1)$ and $(2, 2)$. Thus, the graph of f^{-1} passes through the points $(-2, -3)$, $(1, -1)$ and $(2, 2)$. Plot these points and sketch the reflection of f in the line $y = x$ to obtain the graph of f^{-1}. See Figure 131.

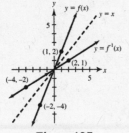

Figure 127

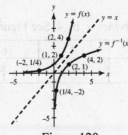

Figure 129

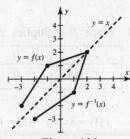

Figure 131

133. The graphs of $y = 2x - 1$, $y = \dfrac{x + 1}{2}$, and $y = x$ are shown in Figure 133.

135. The graphs of $y = x^3 - 1$, $y = \sqrt[3]{x + 1}$, and $y = x$ are shown in Figure 135.

137. The graphs of $y = (x + 1)^2$, where $x \ge -1$, $y = \sqrt{x} - 1$, and $y = x$ are shown in Figure 137.

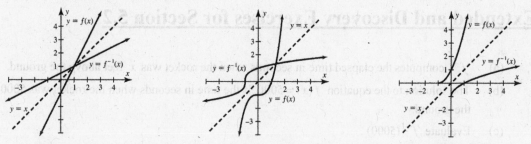

Figure 133   Figure 135   Figure 137

139. The graphs of $Y_1 = 3X - 1$, $Y_2 = (X+1)/3$ and $Y_3 = X$ are shown in Figure 139.

141. The graphs of $Y_1 = (X^3)/3 - 1$, $Y_2 = \sqrt[3]{(3X+3)}$ and $Y_3 = X$ are shown in Figure 141.

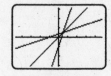 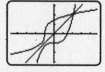

Figure 139   Figure 141

143. (a) Since each volume value is the result of exactly one radius value, V represents a one-to-one function.

   (b) The inverse of V computes the radius r of a sphere with volume V.

   (c) $V = \dfrac{4}{3}\pi r^3 \Rightarrow r^3 = \dfrac{3V}{4\pi} \Rightarrow r = \sqrt[3]{\dfrac{3V}{4\pi}}$

   (d) No; if V and r were interchanged, then r would represent the volume and V would represent the radius.

145. (a) $W = \dfrac{25}{7}(70) - \dfrac{800}{7} = \dfrac{950}{7} \approx 135.7$ pounds

   (b) Yes, since no weight value corresponds to more than one height value.

   (c) $W = \dfrac{25}{7}h - \dfrac{800}{7} \Rightarrow \dfrac{25}{7}h = W + \dfrac{800}{7} \Rightarrow h = \dfrac{7}{25}\left(W + \dfrac{800}{7}\right) = \dfrac{7}{25}W + 32 \Rightarrow W^{-1} = \dfrac{7}{25}W + 32$

   (d) $W^{-1}(150) = \dfrac{7}{25}(150) + 32 \Rightarrow W^{-1}(150) = 74$; the maximum recommended height for a person weighing 150 pounds is 74 inches.

   (e) The inverse computes the maximum recommended height of a person with a given weight.

147. (a) $(F \circ Y)(2) = F(Y(2)) = F(3520) = 10{,}560$. $(F \circ Y)(2)$ computes the number of feet in 2 miles.

   (b) $F^{-1}(26{,}400) = 8800$. F^{-1} converts feet to yards. There are 8800 yards in 26,400 feet.

   (c) $(Y^{-1} \circ F^{-1})(21{,}120) = Y^{-1}(F^{-1}(21{,}120)) = Y^{-1}(7040) = 4$. $(Y^{-1} \circ F^{-1})(21{,}120)$ computes the number of miles in 21,120 feet.

149. (a) $(Q \circ C)(96) = Q(C(96)) = Q(6) = 1.5.$ $(Q \circ C)(96)$ computes the number of quarts in 96 tablespoons.

   (b) $Q^{-1}(2) = 8$. Q^{-1} converts quarts into cups. There are 8 cups in 2 quarts.

   (c) $(C^{-1} \circ Q^{-1})(1.5) = C^{-1}(Q^{-1}(1.5)) = C^{-1}(6) = 96$. $(C^{-1} \circ Q^{-1})(1.5)$ computes the number of tablespoons in 1.5 quarts.

Extended and Discovery Exercises for Section 5.2

1.　(a)　f^{-1} computes the elapsed time in seconds when the rocket was x feet above the ground.

　　(b)　The solution to the equation $f(x) = 5000$ is the time in seconds when the rocket was 5000 feet above the ground.

　　(c)　Evaluate $f^{-1}(5000)$.

Checking Basic Concepts for Sections 5.1 and 5.2

1.　(a)　$(f+g)(1) = f(1) + g(1) = -1 + 2 = 1$

　　(b)　$(f-g)(-1) = f(-1) - g(-1) = 1 - (-2) = 3$

　　(c)　$(fg)(0) = f(0) \cdot g(0) = (-2)(-1) = 2$

　　(d)　$(f/g)(2) = \dfrac{f(2)}{g(2)} = \dfrac{2}{0} \Rightarrow$ undefined

　　(e)　$(f \circ g)(2) = f(g(2)) = f(0) = -2$

　　(f)　$(g \circ f)(-2) = g(f(-2)) = g(0) = -1$

3.　(a)　$(f+g)(x) = f(x) + g(x) = (x^2 + 3x - 2) + (3x - 1) = x^2 + 6x - 3$

　　(b)　$(f/g)(x) = \dfrac{f(x)}{g(x)} = \dfrac{x^2 + 3x - 2}{3x - 1},\ x \neq \dfrac{1}{3}$

　　(c)　$(f \circ g)(x) = f(g(x)) = f(3x-1) = (3x-1)^2 + 3(3x-1) - 2 = 9x^2 + 3x - 4$

5.　(a)　All horizontal lines intersect once $\Rightarrow$ yes; yes; $y = x + 1 \Rightarrow y - 1 = x$, therefore $f^{-1}(x) = x - 1$.

　　(b)　A parabola $\Rightarrow$ no; no.

7.　(a)　Since $f(0) = -2$, then $f^{-1}(-2) = 0$.

　　(b)　$(f^{-1} \circ g)(1) \Rightarrow f^{-1}(g(1)) \Rightarrow f^{-1}(2) = 2$

5.3: Exponential Functions and Models

1.　$2^{-3} = \dfrac{1}{2^3} = \dfrac{1}{8}$

3.　$3(4)^{1/2} = 3\sqrt{4} = 3(2) = 6$

5.　$-2(27)^{2/3} = -2(\sqrt[3]{27})^2 = -2(3)^2 = -2(9) = -18$

7.　$4^{1/6}4^{1/3} = 4^{1/6 + 1/3} = 4^{1/2} = \sqrt{4} = 2$

9.　$3^0 = 1$

11.　$(5^{101})^{1/101} = 5^1 = 5$

13. For each unit increase in x, the y-values decrease by 1.2, so the data is linear. Since $y = 2$ when $x = 0$, the function $f(x) = -1.2x + 2$ can model the data.

15. For each unit increase in x, the y-values are multiplied by $\dfrac{1}{2}$, so the data is exponential. Since the initial value is $C = 8$ and $a = \dfrac{1}{2}$, the function $f(x) = 8\left(\dfrac{1}{2}\right)^x$ can model the data.

17. For each 2-unit increase in x, the y-values are multiplied by 4. That is, for each unit increase in x, the y-values are multiplied by 2, so the data is exponential. Since the initial value is $C = 5$ and $a = 2$, the function $f(x) = 5(2^x)$ can model the data.

19. For each unit increase in x, the y-values increase by a fixed amount, so the data is linear. The slope is $m = \dfrac{16 - (-20)}{6 - (-6)} = \dfrac{36}{12} = 3$. We have $y = 3x + b \Rightarrow 4 = 3(2) + b \Rightarrow 4 = 6 + b \Rightarrow b = -2$, it follows that the data can be modeled by $f(x) = 3x - 2$.

21. Since the data do not change a fixed amount for each unit increase in x, the data are not linear. Because we are not given $f(0)$, we cannot immediately determine C. Instead, find a by evaluating the following ratios.

$\dfrac{f(5)}{f(2)} = \dfrac{\frac{1}{3}}{9} = \dfrac{1}{27}$ It follows that $a^3 = \dfrac{1}{27} \Rightarrow a = \dfrac{1}{3}$. Next find C by using the fact that

$f(2) = 9 = C\left(\dfrac{1}{3}\right)^2 \Rightarrow \dfrac{1}{9}C = 9 \Rightarrow C = 81$. Thus $f(x) = 81\left(\dfrac{1}{3}\right)^x$.

23. For $x > 4$, $f(x) > g(x)$; for example $f(10) = 1024$ whereas $g(10) = 100$. That is, $f(x) = 2^x$ becomes larger.

25. $f(x) = 2x + 1$, $f(0) = 1$, $f(10) = 21$. $g(x) = 2^{-x}$, $g(0) = 2$, $g(10) = 9.76 \times 10^{-4}$.

     Therefore, $f(x)$ becomes larger for $0 \le x \le 10$.

27. In the first option, each amount increases by 2¢ for each week, so the growth is linear. Since there are 52 weeks in a year, the total amount paid at the end of a year would be only about $1¢ + 51(2¢) = 1¢ + 102¢ = \$1.03$. It would not be a good idea to accept this offer. In the second option, the amount is multiplied by 2 for each week, so the growth is exponential. The total amount paid at the end of the year would be $1¢(2¢)^{51}$. This is about \$22,517,998,136,900. It would be an extremely good idea to accept this offer.

29. $f(0) = 5 \Rightarrow C = 5$ and $a = 1.5$

31. $f(0) = 10$ and $f(1) = 20 \Rightarrow C = 10$ and $a = \dfrac{20}{10} = 2$

33. $f(1) = 9$ and $f(2) = 27 \Rightarrow a = \dfrac{27}{9} = 3$; $C = f(0) = 9 \div 3 = 3$

35. $f(-2) = \frac{9}{2}$ and $f(2) = \frac{1}{18} \Rightarrow Ca^{-2} = \frac{9}{2} \Rightarrow \frac{C}{a^2} = \frac{9}{2} \Rightarrow C = \frac{9a^2}{2}$; $Ca^2 = \frac{1}{18} \Rightarrow C = \frac{1}{18a^2}$ thus

$\frac{1}{18a^2} = \frac{9a^2}{2} \Rightarrow 162a^4 = 2 \Rightarrow a^4 = \frac{1}{81} \Rightarrow a = \frac{1}{3}$; then $C = \frac{1}{18(\frac{1}{3})^2} \Rightarrow C = \frac{1}{18(\frac{1}{9})} \Rightarrow C = \frac{1}{2}$.

37. Linear: the slope is $m = \frac{4-8}{0-1} = \frac{-4}{-1} = 4$. Since the y-intercept is given as $(0, 4)$ it follows that the data can

be modeled by $f(x) = 4x + 4$.

Exponential: Using the point $(0, 4)$ and the exponential model $g(x) = Ca^x$ we have $4 = Ca^0 \Rightarrow C = 4$. Now

using the point $(1, 8)$ we have $8 = 4a^1 \Rightarrow a = 2$. The function is $g(x) = 4(2)^x$.

39. Linear: The slope is $m = \frac{12-1.5}{-2-1} = \frac{10.5}{-3} = -\frac{7}{2}$. We have $y = -\frac{7}{2}x + b \Rightarrow 12 = -\frac{7}{2}(-2) + b \Rightarrow$,

$12 = 7 + b \Rightarrow b = 5$ it follows that the data can be modeled by $f(x) = -\frac{7}{2}x + 5$.

Exponential: Find a by evaluating the following ratios. $\frac{g(1)}{g(-2)} = \frac{1.5}{12} = \frac{1}{8}$ and $\frac{g(1)}{g(-2)} = \frac{Ca^1}{Ca^{-2}} = a^3$. it follows

that $a^3 = \frac{1}{8} \Rightarrow a = \frac{1}{2}$. Thus $g(x) = C\left(\frac{1}{2}\right)^1 \Rightarrow \frac{1}{2}C = 1.5 \Rightarrow C = 3$. Thus, $g(x) = 3\left(\frac{1}{2}\right)^x$.

41. $C = 5000$, $a = 2$; x represents time in hours

43. $C = 200,000$, $a = 0.95$, x represents the number of years after 2008

45. $f(9.5) = 30(0.9)^{9.5} \approx 11$; the tire's pressure is about 11 pounds per square inch after 9.5 minutes.

47. $f(x) = 4e^{-1.2x} \Rightarrow f(-2.4) = 4e^{-1.2(-2.4)} = 4e^{2.88} \approx 71.2571$

49. $f(x) = \frac{e^x - e^{-x}}{2} \Rightarrow f(-0.7) = \frac{e^{-0.7} - e^{0.7}}{2} \approx -0.7586$

51. See Figure 51.

53. See Figure 53.

55. See Figure 55.

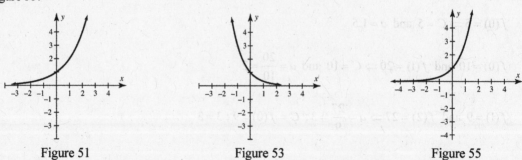

Figure 51   Figure 53   Figure 55

57. See Figure 57.

59. See Figure 59.

61. See Figure 61.

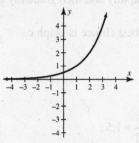

Figure 57

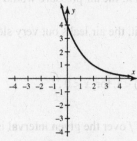

Figure 59

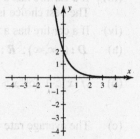

Figure 61

63. See Figure 63.
65. See Figure 65.

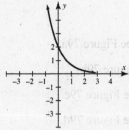

Figure 63

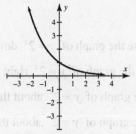

Figure 65

67. Since $y = 1$ when $x = 0$, $C = 1$ and so $y = a^x$.

Since $y = 4$ when $x = -2$, $4 = a^{-2} \Rightarrow \dfrac{1}{a^2} = 4 \Rightarrow a^2 = \dfrac{1}{4} \Rightarrow a = \dfrac{1}{2}$. That is $C = 1$ and $a = \dfrac{1}{2}$.

69. Since $y = \dfrac{1}{2}$ when $x = 0$, $C = \dfrac{1}{2}$ and so $y = \dfrac{1}{2} a^x$.

Since $y = 8$ when $x = 2$, $8 = \dfrac{1}{2} a^2 \Rightarrow a^2 = 16 \Rightarrow a = 4$. That is $C = \dfrac{1}{2}$ and $a = 4$.

71. (a) $D : (-\infty, \infty)$; $R : (0, \infty)$

    (b) Decreasing, as x increases $\left(\dfrac{1}{8}\right)^x$ decreases.

    (c) $y = 0$ as x increases, $7\left(\dfrac{1}{8}\right)^x$ gets closer and closer to 0.

    (d) y-intercept: 7; no x-intercept.

    (e) All horizontal lines intersect once $\Rightarrow$ yes; yes.

73. (i)   The graph of $y = e^x$ increases faster than the graph of $y = 1.5^x$. The best choice is graph b.

    (ii) The graph of $y = 3^{-x}$ decreases faster than the graph of $y = 0.99^x$. The best choice is graph d.

    (iii) The graph of $y = 1.5^x$ increases slower than the graph of $y = e^x$. The best choice is graph a.

    (iv) The graph of $y = 0.99^x$ is almost a horizontal line since $y = 1^x$ is horizontal. The best choice is graph c.

75. (i)   The amount of money would increase faster at a 10% rate with continuous compounding than at a 5% rate with annual compounding. The best choice is graph b.

    (ii)   The amount of money would increase slower at a 5% rate with annual compounding than at a 10% rate with continuous compounding. The best choice is graph d.

    (iii)   If a car tire has a large hole in it, the air pressure would decrease rapidly and then gradually decrease. The best choice is graph a.

    (iv)   If a car tire has a tiny hole in it, the air leaks out very slowly. The best choice is graph c.

77.   (a) $D:(-\infty, \infty)$; $R:(0, \infty)$

    (b) $f(x) > 1$ for $x > 0$, and $f(x) < 1$ for $x < 0$.

    (c)   The average rate of change of f over the given interval is $\dfrac{4-1}{2-0} = \dfrac{3}{2} = 1.5$.

    (d) $f(x)$ is increasing on $(-\infty, \infty)$, and does not decrease.

    (e)   no x-intercepts; y-intercept: $(0,1)$

    (f)   It doubles.

79.   (a)   To graph $y = 2^x - 2$, translate the graph of $y = 2^x$ down 2 units. See Figure 79a.

    (b)   To graph $y = 2^{x-1}$, translate the graph of $y = 2^x$ right 1 unit. See Figure 79b.

    (c)   To graph $y = 2^{-x}$, reflect the graph of $y = 2^x$ about the y-axis. See Figure 79c.

    (d)   To graph $y = -2^x$, reflect the graph of $y = 2^x$ about the x-axis. See Figure 79d.

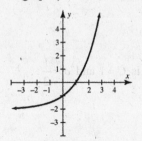

Figure 79a

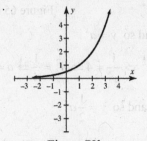

Figure 79b

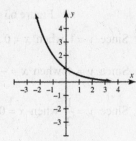

Figure 79c

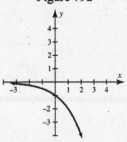

Figure 79d

81.   See Figure 81.

83.   See Figure 83.

85.   See Figure 85.

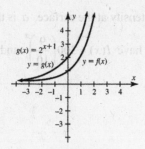

Figure 81

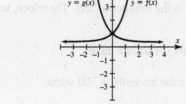

Figure 83

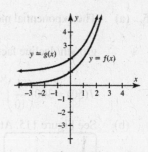

Figure 85

87.   See Figure 87.
89.   See Figure 89.
91.   See Figure 91.

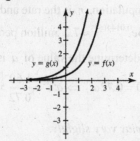

Figure 87

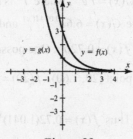

Figure 89

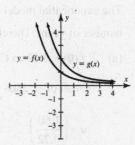

Figure 91

93. $A_n = A_0(1+r)^n \Rightarrow A_5 = 600(1+0.07)^5 \approx \$841.53$

95. $A_n = A_0\left(1+\dfrac{r}{m}\right)^{mn} \Rightarrow A_{20} = 950\left(1+\dfrac{0.03}{365}\right)^{365(20)} \approx \$1730.97$

97. $A_n = A_0 e^{rn} \Rightarrow A_8 = 2000e^{0.10(8)} \approx \$4451.08$

99. $A_n = A_0\left(1+\dfrac{r}{m}\right)^{mn} \Rightarrow A_{2.5} = 1600\left(1+\dfrac{0.013}{12}\right)^{12(2.5)} \approx \$1652.83$

101. $A = A_0\left(1+\dfrac{r}{m}\right)^{mn} \Rightarrow A_{20} = 500\left(1+\dfrac{0.008}{2}\right)^{2(20)} \approx \$586.57$

103. $A = A_0 e^{rt} \Rightarrow A_8 = 1500e^{0.0075(8)} \approx \$1592.75$

105. $A_{10} = 8000(1+0.06)^{10} \approx \$14,326.78$

107.   We will use the graph of the graph of $y = e^{0.06x}$ and $y = (1.063)^x$ to represent the continuous and annual growth respectively. From the graph we see that the $y = (1.063)^x$ grows at a faster rate making the annual investment better.

109.   (a)   The exponential model is given as $B(x) = Ca^x$, where C is the initial concentration, a is the growth factor, and x is the number of weeks. therefore, we have $B(x) = 2.5(3)^x$.

       (b) $B(1.5) = 2.5(3)^{1.5} \approx 13$ million per millimeter.

111.   A decrease of 25% per week means the density of flies is reduced by 75% each week

       $D(t) = \left(2\times10^6\right)\left(\dfrac{3}{4}\right)^{3.2} \approx 0.8$ million per acre.

113.   A decrease of 2% per week per share $\Rightarrow$ share value $= 40(0.98)^{6.5} \approx \$35.08$

115. (a) The exponential model is given as $I(x) = Ca^x$, where C is the intensity at the surface, a is the growth/decline factor, and x is the number of feet. Therefore, we have $I(x) = 300\left(\dfrac{9}{10}\right)^x$ and

$I(50) = 300\left(\dfrac{9}{10}\right)^{50} \approx 1.5$ watts per square meter.

(b) See Figure 115. At the surface the intensity is 300 watts.

Figure 115

117. The exponential model is given as $G(x) = Pe^{rt}$, where P is the initial population, r is the rate and t is the number of years. Therefore, we have $G(x) = 6.6e^{0.0144(x)}$ and $G(x) = 6.6e^{0.0144(6)} \approx 7.2$ million people.

119. (a) $f(0) = 0.72 \Rightarrow C \approx 0.72$, so $f(x) = 0.72a^x$. One possible way to determine the value of a is to let the graph of f pass through the point $(20, 1.60)$. $f(20) = 1.60 \Rightarrow 1.60 = 0.72a^{20} \Rightarrow a^{20} = \dfrac{1.60}{0.72} \Rightarrow$

$a = \left(\dfrac{1.60}{0.72}\right)^{1/20} \Rightarrow a \approx 1.041$; Thus $f(x) = 0.72(1.041)^x$. *Answers may vary slightly.*

(b) Let $x = 13$ correspond to 2013, so; $f(13) = 0.72(1.041)^{13} \approx 1.21$ the CFC-12 concentration in 2013 is about 1.21 ppb. *Answers may vary slightly.*

121. (a) The number of *E. coli* was modeled by $N(x) = N_0\, e^{0.014x}$, where x is in minutes and $N_0 = 500,000$. Since 3 hours is 180 minutes, $N(180) = 500,000e^{0.014(180)} \approx 6,214,000$ bacteria per milliliter.

(b) Solve the equation $500,000\, e^{0.014x} = 10,000,000$. Graph $Y_1 = 500000e^{\wedge}(0.014X)$ and $Y_2 = 10E6$ as shown in Figure 121. The point of intersection is near $(214, 10,000,000)$. Thus, there will be 10 million *E. coli* after about 214 minutes, or about 3.6 hours.

[0, 300, 100] by [0, 20,000,000, 5,000,000]

Figure 121

123. (a) $p(x) = 1 - e^{-5x/6} \Rightarrow p(3) = 1 - e^{-5(3)/6} \Rightarrow p(3) = 1 - e^{-15/6} \approx 0.92$, or 92%

(b) Find x when $p(x) = 0.5$. That is, solve $0.5 = 1 - e^{-5x/6}$ for x.

Graph $Y_1 = 0.5$ and $Y_2 = 1 - e^{\wedge}(-5X/6)$. as shown in Figure 123. There is a 50-50 chance of at least one car entering the intersection during an interval of about 0.83 minutes.

[0, 5, 1] by [0, 1, 0.1]

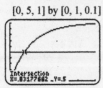

Figure 123

125. Let $T(t) = Ca^t$, where t is the number of hours. Initially, $T(0) = 100$, so $C = 100$ and $T(t) = 100a^t$. Next we will find the value of a. Because the half-life is 2.8 hours, 50% of its hits remain after 2.8 hours, so

$T(2.8) = 50$. $100a^{2.8} = 50 \Rightarrow a^{2.8} = \dfrac{1}{2} \Rightarrow a = \left(\dfrac{1}{2}\right)^{1/2.8}$ Thus we have $T(t) = 100\left(\left(\dfrac{1}{2}\right)^{1/2.8}\right)^t = 100\left(\dfrac{1}{2}\right)^{t/2.8}$ and

$T(5.5) = 100\left(\dfrac{1}{2}\right)^{5.5/2.8} \approx 25.6\%$.

127. (a) The exponential model is given as $P(x) = Ca^x$, where C is the initial percentage of putts made, a is the growth/decline factor, and x is the number of feet. Therefore, we have $P(x) = 95(0.9)^x$.

   (b) $P(20) = 95(0.9)^{20} \approx 11.5\%$

   (c) Since the percentage decreases by a factor of 0.9 we use the graph of $y = 0.9^x$ to see that with every increase in distance of 6.6 feet the percentage of putts made will decrease by half. See Figure 127.

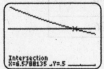

Figure 127

129. To find the age of the fossil solve $0.10 = 1\left(\dfrac{1}{2}\right)^{x/5700}$. Graph $Y_1 = 0.10$ and $Y_2 = 0.5^{\wedge}(X/5700)$. The graphs intersect near $(18934.99, 0.1)$, so the fossil is about 18,935 years old.

131. $P = 1\left(\dfrac{1}{2}\right)^{x/1600} \Rightarrow P = 1\left(\dfrac{1}{2}\right)^{3000/1600} \Rightarrow P \approx 0.273$, or 27.3%

133. (a) Since the initial amount is 2.5 ppm, $C = 2.5$. Since 30% of the chlorine dissipates each day, 70% remains. So, $a = 0.7$.

   (b) $f(2) = 2.5(0.7)^2 = 1.225$ parts per million.

135. (a) $H(30) = 0.157(1.033)^{30} \approx 0.42$. This means that approximately 0.42 horsepower are required for each ton that the locomotive is pulling at 30 mph.

   (b) To pull a 5000-ton train at 30 mph, approximately a $0.42(5000) = 2100$ horsepower engine is needed.

   (c) $\dfrac{2100}{1350} \approx 1.56$. This value must be rounded up. Two locomotives having 1350 horsepower would move a 5000-ton train at 30 mph.

Extended and Discovery Exercises for Section 5.3

1. $P = A(1 + r/n)^{-nt} \Rightarrow P = \dfrac{A}{(1 + r/n)^{nt}} \Rightarrow A = P(1 + r/n)^{nt}$

3. $P = 15,000(1 + 0.05/12)^{-12(3)} = 12,914.64367$; The present value should be $12,914.64.

5. (a) $x = 0 \Rightarrow e^0 = 1 \Rightarrow e^{0+0.001} = 1.0010005 \Rightarrow \dfrac{1 + 1.0010005}{2} \approx 1.0005$

(b) $e^0 = 1$

(c)   They are very similar.

7.   (a) $x = -0.5 \Rightarrow e^{-0.5} = 0.6065 \Rightarrow e^{-0.5+0.001} = 0.6071 \Rightarrow \dfrac{0.6065 + 0.6071}{2} \approx 0.6068$

       (b) $e^{-0.5} \approx 0.6065$

       (c)   They are very similar.

9.   The average rate of change near x and the value of the function at x are approximately equal.

5.4: Logarithmic Functions and Models

1.   See Figure 1.

x	10^0	10^4	10^{-8}	$10^{1.26}$
$\log x$	0	4	-8	1.26

Figure 1

3. (a) $\log(-3)$ is undefined

    (b) $\log \dfrac{1}{100} = \log \dfrac{1}{10^2} = \log 1^{-2} = -2$

    (c) $\log \sqrt{0.1} = \log 10^{-1/2} = -\dfrac{1}{2}$

    (d) $\log 5^0 = 0$

5. (a) $\log 10 = \log 10^1 = 1$

    (b) $\log 10{,}000 = \log 10^4 = 4$

    (c) $20 \log 0.1 = 20 \log 10^{-1} = 20(-1) = -20$

    (d) $\log 10 + \log 0.001 = \log 10^1 + \log 10^{-3} = 1 + (-3) = -2$

7. (a) $2 \log 0.1 + 4 = 2 \log 10^{-1} + 4 = 2(-1) + 4 = -2 + 4 = 2$

    (b) $\log 10^{1/2} = \dfrac{1}{2}$

    (c) $3 \log 100 - \log 1000 = 3 \log 10^2 - \log 10^3 = 3(2) - 3 = 6 - 3 = 3$

    (d) $\log(-10)$ is undefined

9. (a)   Since $10^1 \le 79 \le 10^2$, $\log 10^1 \le \log 79 \le \log 10^2 \Rightarrow 1 \le \log 79 \le 2 \Rightarrow n = 1$; $\log 79 \approx 1.898$

    (b)   Since $10^2 \le 500 \le 10^3$, $\log 10^2 \le \log 500 \le \log 10^3 \Rightarrow 2 \le \log 500 \le 3 \Rightarrow n = 2$; $\log 500 \approx 2.699$

    (c)   Since $10^0 \le 5 \le 10^1$, $\log 10^0 \le \log 5 \le \log 10^1 \Rightarrow 0 \le \log 5 \le 1 \Rightarrow n = 0$; $\log 5 \approx 0.6990$

    (d)   Since $10^{-1} \le 0.5 \le 10^0$, $\log 10^{-1} \le \log 0.5 \le \log 10^0 \Rightarrow -1 \le \log 0.5 \le 0 \Rightarrow n = -1$; $\log 0.5 \approx -0.3010$

11. (a) $\dfrac{3}{2}$ since $\sqrt{1000} = 1000^{1/2} = (10^3)^{1/2} = 10^{3/2}$

    (b) $\dfrac{1}{3}$ since $\log \sqrt[3]{10} = \log 10^{1/3}$

    (c) $\log \sqrt[5]{0.1} = \log (10^{-1})^{1/5} = \log (10)^{-1/5} = -\dfrac{1}{5}$

    (d) $\log \sqrt{0.01} = \log (10^{-2})^{1/2} = \log 10^{-1} = -1$

13. The input to a logarithmic function must be positive. Thus, any element of the domain of f must satisfy $x + 3 > 0$, or equivalently, $x > -3$. Thus $D: (-3, \infty)$, or $\{x \mid x > -3\}$.

15. The input to a logarithmic function must be positive. Thus, any element of the domain of f must satisfy $x^2 - 1 > 0 \Rightarrow x^2 > 1 \Rightarrow x < -1$ or $x > 1$ Thus $D: (-\infty, -1) \cup (1, \infty)$, or $\{x \mid x < -1, \text{ or } x > 1\}$.

17. The input to a logarithmic function must be positive. Thus, any element of the domain of f must satisfy $4^x > 0 \Rightarrow x$ can be any real number. Thus $D: (-\infty, \infty)$, or $\{x \mid -\infty < x < \infty\}$.

19. The input to a logarithmic function must be positive. Thus, any element of the domain of f must satisfy $\sqrt{3 - x} - 1 > 0 \Rightarrow \sqrt{3 - x} > 1 \Rightarrow x < 2$. Thus $D: (-\infty, 2)$, or $\{x \mid x < 2\}$.

21.

x	2^0	2^4	2^{-8}	$2^{1.26}$
$\log_2 x$	0	4	-8	1.26

23. $\log_8 8^{-5.7} = -5.7$

25. $7^{\log_7 2x} = 2x$ for $x > 0$

27. $\log_{1/3}\left(\dfrac{1}{3}\right)^{64} = 64$

29. $\ln e^{-4} = -4$

31. $\log_5 5^{\pi} = \pi$

33. $3^{\log_3(x-1)} = x - 1$, for $x > 1$

35. $\log_2 64 = \log_2 2^6 = 6$

37. $\log_4 2 = \log_4 \sqrt{4} = \log_4 4^{1/2} = \dfrac{1}{2}$

39. $\ln e^{-3} = -3$

41. $\log_8 64 = \log_8 8^2 = 2$

43. $\log_{1/2}\left(\dfrac{1}{4}\right) = \log_{1/2}\left(\dfrac{1}{2}\right)^2 = 2$

45. $\log_{1/6} 36 = \log_{1/6}\left(\dfrac{1}{6}\right)^{-2} = -2$

47. $\log_a \dfrac{1}{a} = \log_a a^{-1} = -1$

49. $\log_5 5^0 = 0$

51. $\log_2 \dfrac{1}{16} = \log_2 2^{-4} = -4$

53.   (a) $n < \log_2 9 < n+1 \Rightarrow n < \log_2 2^3 < \log_2 9 < \log_2 2^4 < n+1$; $n = 3$

   (b) $n < \log_4 11 < n+1 \Rightarrow n < \log_4 4^1 < \log_4 11 < \log_4 4^2 < n+1$; $n = 1$

   (c) $n < \log_3 35 < n+1 \Rightarrow n < \log_3 3^3 < \log_3 35 < \log_3 3^4 < n+1$; $n = 3$

   (d) $n < \log_5 130 < n+1 \Rightarrow n < \log_5 5^3 < \log_5 130 < \log_5 5^4 < n+1$; $n = 3$

55.   See Figure 55.

x	6	7	21
$f(x)$	0	2	8

Figure 55

57.   (a) $7^{4x} = 4 \Rightarrow \log_7 4 = 4x$

   (b) $e^x = 7 \Rightarrow \ln(7) = x$

   (c) $c^x = b \Rightarrow \log_c(b) = x$

59.   (a) $\log_8(x) = 3 \Rightarrow x = 8^3$

(b) $\log_9(2+x) = 5 \Rightarrow 2 + x = 9^5$

(c) $\log_k b = c \Rightarrow b = k^c$

61. (a) $10^x = 0.01 \Rightarrow \log 10^{-2} = x \Rightarrow x = -2$

(b) $10^x = 7 \Rightarrow \log 7 = x \Rightarrow x \approx 0.85$

(c) $10^x = -4 \Rightarrow \log(-4) = x \Rightarrow$ no solution

63. (a) $4^x = \dfrac{1}{16} \Rightarrow 4^x = 4^{-2} \Rightarrow x = -2$

(b) $e^x = 2 \Rightarrow \ln(2) = x \Rightarrow x \approx 0.69$

(c) $5^x = 125 \Rightarrow 5^x = 5^3 \Rightarrow x = 3$

65. (a) $9^x = 1 \Rightarrow 9^x = 9^0 \Rightarrow x = 0$

(b) $10^x = \sqrt{10} \Rightarrow 10^x = 10^{1/2} \Rightarrow x = \dfrac{1}{2}$

(c) $4^x = \sqrt[3]{4} \Rightarrow 4^x = 4^{1/3} \Rightarrow x = \dfrac{1}{3}$

67. $e^{-x} = 3 \Rightarrow \ln e^{-x} = \ln 3 \Rightarrow -x = \ln 3 \Rightarrow x = -\ln 3 \approx -1.10$

69. $10^x - 5 = 95 \Rightarrow 10^x = 100 \Rightarrow 10^x = 10^2 \Rightarrow x = 2$

71. $10^{3x} = 100 \Rightarrow 10^{3x} = 10^2 \Rightarrow 3x = 2 \Rightarrow x = \dfrac{2}{3} \approx 0.67$

73. $5(10^{4x}) = 65 \Rightarrow 10^{4x} = 13 \Rightarrow \log(13) = 4x \Rightarrow x = \dfrac{\log(13)}{4} \approx 0.28$

75. $4(3^x) - 3 = 13 \Rightarrow 4(3^x) = 16 \Rightarrow 3^x = 4 \Rightarrow \log_3(4) = \dfrac{\log(4)}{\log(3)} \approx 1.26$

77. $e^x + 1 = 24 \Rightarrow e^x = 23 \Rightarrow \ln e^x = \ln 23 \Rightarrow x = \ln 23 \approx 3.14$

79. $2^x + 1 = 15 \Rightarrow 2^x = 14 \Rightarrow \log_2 2^x = \log_2 14 \Rightarrow x = \log_2 14$ or $x \approx 3.81$

81. $5e^x + 2 = 20 \Rightarrow 5e^x = 18 \Rightarrow e^x = \dfrac{18}{5} \Rightarrow \ln e^x = \ln \dfrac{18}{5} \Rightarrow x = \ln\left(\dfrac{18}{5}\right) \approx 1.28$

83. $8 - 3(2)^{0.5x} = -40 \Rightarrow -3(2)^{0.5x} = -48 \Rightarrow 2^{0.5x} = 16 \Rightarrow 2^{0.5x} = 2^4 \Rightarrow 0.5x = 4 \Rightarrow x = \dfrac{4}{0.5} \Rightarrow x = 8$

85. (a) $\log x = 2 \Rightarrow 10^2 = x \Rightarrow x = 100$

(b) $\log x = -3 \Rightarrow 10^{-3} = x \Rightarrow x = \dfrac{1}{1000}$

(c) $\log x = 1.2 \Rightarrow 10^{1.2} = x \Rightarrow x \approx 15.8489$

87. (a) $\log_2 x = 6 \Rightarrow x = 2^6 = x \Rightarrow 64$

(b) $\log_3 x = -2 \Rightarrow x = 3^{-2} = x \Rightarrow \dfrac{1}{9}$

(c) $\ln x = 2 \Rightarrow x = e^2 \Rightarrow x \approx 7.3891$

89. $\log_2 x = 1.2 \Rightarrow 2^{\log_2 x} = 2^{1.2} \Rightarrow x = 2^{1.2} \approx 2.2974$

91. $5\log_7(2x) = 10 \Rightarrow \log_7(2x) = 2 \Rightarrow 2x = 7^2 \Rightarrow 2x = 49 \Rightarrow x = \dfrac{49}{2}$

93. $2 \log x = 6 \Rightarrow \log x = 3 \Rightarrow 10^{\log x} = 10^3 \Rightarrow x = 10^3 = 1000$

95. $2 \log 5x = 4 \Rightarrow \log 5x = 2 \Rightarrow 10^{\log 5x} = 10^2 \Rightarrow 5x = 100 \Rightarrow x = 20$

97. $4 \ln x = 3 \Rightarrow \ln x = \dfrac{3}{4} \Rightarrow e^{\ln x} = e^{3/4} \Rightarrow x = e^{3/4} \approx 2.1170$

99. $5 \ln x - 1 = 6 \Rightarrow 5 \ln x = 7 \Rightarrow \ln x = \dfrac{7}{5} \Rightarrow e^{\ln x} = e^{7/5} \Rightarrow x = e^{7/5} \approx 4.0552$

101. $4 \log_2 x = 16 \Rightarrow \log_2 x = 4 \Rightarrow 2^{\log_2 x} = 2^4 \Rightarrow x = 2^4 = 16$

103. $5 \ln (2x) + 6 = 12 \Rightarrow 5 \ln (2x) = 6 \Rightarrow \ln (2x) = \dfrac{6}{5} \Rightarrow e^{\ln 2x} = e^{6/5} \Rightarrow 2x = e^{6/5} \Rightarrow x = \dfrac{e^{6/5}}{2} \approx 1.6601$

105. $9 - 3 \log_4 2x = 3 \Rightarrow -3 \log_4 2x = -6 \Rightarrow \log_4 2x = 2 \Rightarrow 4^{\log_4 2x} = 4^2 \Rightarrow 2x = 16 \Rightarrow x = 8$

107. $f(x) = a + b \log x$ and $f(1) = 5 \Rightarrow 5 = a + b \log 1 \Rightarrow 5 = a + b(0) \Rightarrow a = 5$

      $f(x) = 5 + b \log x$ and $f(10) = 7 \Rightarrow 7 = 5 + b \log 10 \Rightarrow 7 = 5 + b(1) \Rightarrow b = 2$

      The function is $f(x) = 5 + 2 \log x$.

109.   Since $f(x) = e^x$, $f^{-1}(x) = \ln x$. See Figure 109.

111.   See Figure 111.

113.   Decreasing. See Figure 113.

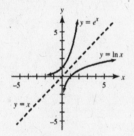
Figure 109

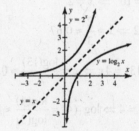
Figure 111

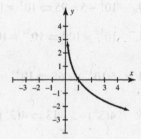
Figure 113

115.   (a)   See Figure 115.

       (b) f is increasing on $(0, \infty)$; f^{-1} is increasing on $(-\infty, \infty)$.

117.   The graph of $g(x) = \log(x-2)$ is similar to the graph of $f(x) = \log x$, except it is shifted to the right 2 units. The graph has a vertical asymptote of $x = 2$. See Figure 117.

119.   The graph of $g(x) = 3 \log x$ is similar to the graph of $f(x) = \log x$, except it is stretched by a factor of 3. The graph has a vertical asymptote of $x = 0$. See Figure 119.

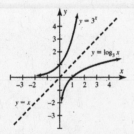
Figure 115

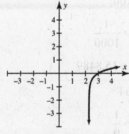
Figure 117

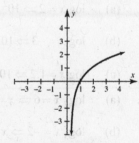
Figure 119

121.   The graph of $g(x) = \log_2 (-x)$ is similar to the graph of $f(x) = \log_2 x$, except it is reflected through the y-axis. The graph has a vertical asymptote of $x = 0$. See Figure 121.

123.   The graph of $g(x) = 2 + \ln(x-1)$ is similar to the graph of $f(x) = \ln x$, except it is shifted to the right one unit and up two units. The graph has a vertical asymptote of $x = 1$. See Figure 123.

125. The graph is shown in Figure 125. $D = \{x \mid x > -1\}$

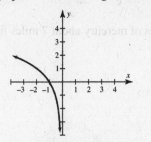

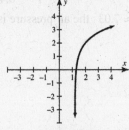

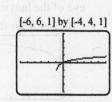

[-6, 6, 1] by [-4, 4, 1]

Figure 121   Figure 123   Figure 125

127. The graph is shown in Figure 127. $D = \{x \mid x < 0\}$

[-6, 6, 1] by [-4, 4, 1]

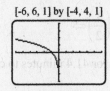

Figure 127

129. At $x = 0$ days, $C(0) = 4(2^0) = 4$ million bacteria/mL. Taking the logarithm of the concentration equation,

$$\ln C(x) = \ln(4(2^x)) = \ln 4 + x \ln 2 \Rightarrow \ln(20) = \ln 4 + x \ln 2 \Rightarrow x = \frac{\ln(20) - \ln 4}{\ln 2} \approx 2.3 \text{ days.}$$

131. $D(10^{-4}) = 10 \log(10^{16} \cdot 10^{-4}) = 10 \log(10^{12}) = 10 \cdot 12 = 120$ db

133. (a) $L(100) = 3 \log(100) = 3 \cdot 2 = 6$, a 100-thousand-pound airplane needs 6000 feet of runway.

   (b) Since the weight is measured in 1000-pound units, start by evaluating $L(10)$ and $L(100)$.

      $L(10) = 3 \log 10 = 3$ and $L(100) = 3 \log 100 = 6$. Thus, a 10,000-pound plane requires approximately 3000 feet of runway, whereas as 100,000-pound plane requires approximately 6000 feet. The distance does not increase by a factor of 10.

   (c) If the weight increases tenfold, the runway length increases by 3000 feet.

135. (a) $\text{pH} = -\log(x) \Rightarrow \text{pH} = -\log(10^{-4.7}) \Rightarrow \text{pH} = -(-4.7) = 4.7$

   (b) $-\log(x) = 8.2 \Rightarrow \log(x) = -8.2 \Rightarrow 10^{\log(x)} = 10^{-8.2} \Rightarrow x = 10^{-8.2}$. Since $\dfrac{10^{-4.7}}{10^{-8.2}} = 10^{3.5}$ the hydrogen

      concentration in the rainwater is $10^{3.5} \approx 3162$ times greater than it is in seawater.

137. (a) Since $I_0 = 1$, $R(x) = 6.0 \Rightarrow \log x = 6.0 \Rightarrow 10^{\log x} = 10^{6.0} \Rightarrow x = 10^6 = 1,000,000$

      Similarly, $R(x) = 8.0 \Rightarrow \log x = 8.0 \Rightarrow 10^{\log x} = 10^{8.0} \Rightarrow x = 10^8 = 100,000,000$

   (b) $\dfrac{10^8}{10^6} = 10^{8-6} = 10^2 = 100$. The Indonesian earthquake was 100 times more intense (powerful) than the

      Yugoslavian earthquake.

139. (a) $f(x) = 0.48 \ln(x+1) + 27 \Rightarrow f(0) = 0.48 \ln(0+1) + 27 = 0.48 \ln 1 + 27 = 0.48(0) + 27 = 27$ and

      $f(100) = 0.48 \ln(100+1) + 27 \approx 29.2$ inches. At the center, or eye, of the hurricane the pressure is 27 inches of mercury, while 100 miles from the eye the air pressure has risen to 29.2 inches of mercury.

   (b) Graph $Y_1 = 0.48 \ln(X+1) + 27$ as shown in Figure 139. At first, the air pressure rises rapidly as one moves away from the eye. Then, the air pressure starts to level off and does not increase significantly for distances greater than 200 miles.

(c) $f(x) = 28 \Rightarrow 28 = 0.48 \ln(x+1) + 27 \Rightarrow \ln(x+1) = \dfrac{1}{0.48} \Rightarrow e^{\ln(x+1)} = e^{1/0.48} \Rightarrow$

$x+1 = e^{1/0.48} \Rightarrow x = e^{1/0.48} - 1 \approx 7.03$; the air pressure is 28 inches of mercury about 7 miles from the eye of the hurricane.

[0, 250, 50] by [25, 30, 1]

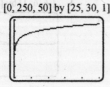

Figure 139

141. (a) $T(x) = 20 + 80e^{-x} \Rightarrow T(1) = 20 + 80e^{-1} \approx 49.4$; the temperature is about 49.4°C after 1 hour.

(b) $T(x) = 60 \Rightarrow 60 = 20 + 80e^{-x} \Rightarrow 40 = 80e^{-x} \Rightarrow e^{-x} = \dfrac{1}{2} \Rightarrow \ln e^{-x} = \ln\left(\dfrac{1}{2}\right) \Rightarrow$

$-x = \ln\left(\dfrac{1}{2}\right) \Rightarrow x = -\ln\left(\dfrac{1}{2}\right) \approx 0.693$; the water took about 0.69 hours, or 41.4 minutes to cool to 60°C.

143. (a) $f(x) = e^{-x/3} \Rightarrow f(5) = e^{-5/3} \approx 0.189$

The probability that no car enters the intersection during a 5-minute period is about 0.189 or 18.9%.

(b) $f(x) = 0.30 \Rightarrow 0.30 = e^{-x/3} \Rightarrow \ln e^{-x/3} = \ln 0.3 \Rightarrow -\dfrac{x}{3} = \ln 0.3 \Rightarrow x = -3 \ln 0.3 \approx 3.6$

The probability that no car enters the intersection will be 30% during a period of about 3.6 minutes.

145. $f(x) = a + b \log x$ and $f(1) = 7 \Rightarrow 7 = a + b \log 1 \Rightarrow 7 = a + b(0) \Rightarrow a = 7$ $f(x) = 7 + b \log x$ and $f(10) = 11 \Rightarrow 11 = 7 + b \log 10 \Rightarrow 11 = 7 + b(1) \Rightarrow b = 4$ The function that models the given data is $f(x) = 7 + 4 \log x$.

$f(x) = 16 \Rightarrow 16 = 7 + 4 \log x \Rightarrow 9 = 4 \log x \Rightarrow \log x = \dfrac{9}{4} \Rightarrow 10^{\log x} = 10^{9/4} \Rightarrow x = 10^{9/4} \approx 178$. The island would be about 178 square kilometers.

147. (a) $f(x) = Ca^x$ and $f(0) = 3 \Rightarrow 3 = Ca^0 \Rightarrow 3 = C(1) \Rightarrow C = 3$

$f(x) = 3a^x$ and $f(1) = 6 \Rightarrow 6 = 3a^1 \Rightarrow 2 = a^1 \Rightarrow a = 2$; thus $f(x) = 3(2^x)$ models the data.

(b) $f(x) = 16 \Rightarrow 16 = 3(2^x) \Rightarrow 2^x = \dfrac{16}{3}$. Graph $Y_1 = 2 \char94 X$ and $Y_2 = 16/3$ in [0,5,1] by [4, 20, 5].

The graphs (not shown) intersect near (2.41, 5.33), so there were 16 million bacteria after about 2.4 days.

Extended and Discovery Exercises for Section 5.4

1.   (a)   If $x = 1$ then $\ln(1) = 0$ and if $x = 1.001$ then $\ln(1.001) = 0.0009995 \Rightarrow (1, 0)$ and $(1.001, 0.0009995)$, the average rate of change is $\dfrac{0.0009995 - 0}{1.001 - 1} \approx 1.00$.

(b)   If $x = 2$ then $\ln(2) = 0.693147$ and if $x = 2.001$ then $\ln(2.001) = 0.693647 \Rightarrow (2, 0.693147)$ and $(2.001, 0.693647)$, the average rate of change is $\dfrac{0.693647 - 0.693147}{2.001 - 2} \approx 0.50$.

(c)   If $x = 3$ then $\ln(3) = 1.098612$ and if $x = 3.001$ then $\ln(3.001) = 1.098946 \Rightarrow (3, 1.098612)$ and

$(3.001, 1.098946)$, the average rate of change is $\dfrac{1.098946 - 1.098612}{3.001 - 2} \approx 0.33$.

(d)   If $x = 4$ then $\ln(4) = 1.386294$ and if $x = 4.001$ then $\ln(4.001) = 1.386544 \Rightarrow (4, 1.386294)$ and

$(4.001, 1.386544)$, the average rate of change is $\dfrac{1.386544 - 1.386294}{4.001 - 4} \approx 0.25$.

3.   (a) $T(C(x)) = 6.5\ln\left(\dfrac{364(1.005)^x}{280}\right) = 6.5\ln\left(1.3 \cdot 1.005^x\right)$; $T(100) = 6.5\ln\left(\dfrac{364(1.005)^{100}}{280}\right) \approx 4.95$

This model predicts an average global temperature increase of about 5°F in the year 2100.

(b)   Graph $Y_1 = 364(1.005)^{\wedge}X$ in $[0, 200, 50]$ by $[0, 1000, 100]$ and $Y_2 = 6.5\ln(364(1.005)^{\wedge}X/280)$ in $[0, 200, 50]$ by $[0, 10, 1]$. The graph of Y_1 is exponential while the graph of Y_2 appears linear over this time interval. See Figures 3a and 3b.

(c) C is an exponential function and T is approximately linear over the same time period. While the carbon dioxide levels in the atmosphere increase exponentially, the average global temperature rises at nearly a constant rate each year.

$[0, 200, 50]$ by $[0, 1000, 100]$ $[0, 200, 50]$ by $[0, 10, 1]$

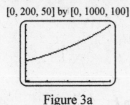   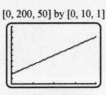

Figure 3a   Figure 3b

Checking Basic Concepts for Sections 5.3 and 5.4

1. $A = 1200\left(1 + \dfrac{0.095}{12}\right)^{12(4)} \approx 1752.12$; After four years the account balance will be \$1752.12. If compounding continuously, the account balance will be $A = 1200e^{0.095(4)} \approx \$1754.74$.

3. $\log_2 15$ represents the power of 2 resulting in 15. That is, $2^{\log_2 15} = 15$ by definition. There is no integer k such that $2^k = 15$, so $\log_2 15$ does not equal an integer. Since $2^4 = 16$, the value of $\log_2 15$ is slightly less than 4.

5.   (a) $e^x = 5 \Rightarrow \ln e^x = \ln 5 \Rightarrow x = \ln 5 \approx 1.609$

(b) $10^x = 25 \Rightarrow \log 10^x = \log 25 \Rightarrow x = \log 25 \approx 1.398$

(c) $\log x = 1.5 \Rightarrow 10^{\log x} = 10^{1.5} \Rightarrow x = 10^{1.5} \approx 31.623$

7.   (a) $P(2) = 37.3e^{0.01(2)} \approx 38.1$ million , in 2012 California's population was about 38.1 million.

(b)   The y-intercept is found by letting $x = 0$, $P(x) = 37.3e^{0.01(0)} = 37.3$, in 2010 California's population was about 37.3 million.

(c)   The function $P(x) = 37.3e^{0.01(x)}$ is in the form $A = Pe^{rt}$ where r is the rate of increase in the population. The annual increase is calculated as $e^r - 1 \Rightarrow e^{.001} - 1 \approx 0.01$.

(d) $40 = 37.3e^{0.01(x)} \Rightarrow \dfrac{40}{37.3} = e^{0.01x} \Rightarrow \ln\left(\dfrac{40}{37.3}\right) = 0.01x \Rightarrow x = \dfrac{\ln\left(\dfrac{40}{37.3}\right)}{0.01} \approx 7$, since x represents the

number of years after 2010 the result is 2017.

5.5: Properties of Logarithms

1. $\log 4 + \log 7 \approx 1.447$; $\log 28 \approx 1.447$; $\log 4 + \log 7 = \log 28 = \log(4 \cdot 7)$; Property 2

3. $\ln 72 - \ln 8 \approx 2.197$; $\ln 9 \approx 2.197$; $\ln 72 - \ln 8 = \ln 9 = \ln\left(\dfrac{72}{8}\right)$; Property 3

5. $10 \log 2 \approx 3.010$; $\log 1024 \approx 3.010$; $10 \log 2 = \log 1024 = \log 2^{10}$; Property 4

7. $\log_2 ab = \log_2 a + \log_2 b$

9. $\ln 7a^4 = \ln 7 + 4 \ln a$

11. $\log \dfrac{6}{z} = \log 6 - \log z$

13. $\log \dfrac{x^2}{3} = \log x^2 - \log 3 = 2 \log x - \log 3$

15. $\ln \dfrac{2x^7}{3k} = \ln 2x^7 - \ln 3k = \ln 2 + \ln x^7 - (\ln 3 + \ln k) = \ln 2 + 7 \ln x - \ln 3 - \ln k$

17. $\log_2 4k^2 x^3 = \log_2 4k^2 + \log_2 x^3 = \log_2 4 + \log_2 k^2 + \log_2 x^3 = 2 + 2 \log_2 k + 3 \log_2 x$

19. $\log_5 \dfrac{25x^3}{y^4} = \log_5 25x^3 - \log_5 y^4 = 2 + 3 \log_5 x - 4 \log_5 y$

21. $\ln \dfrac{x^4}{y^2 \sqrt{z^3}} = \ln(x^4) - (\ln y^2 + \ln(z^{3/2})) = 4 \ln(x) - 2 \ln(y) - \dfrac{3}{2} \ln(z)$

23. $\log_4 0.25(x+2)^3 = \log_4 0.25 + \log_4 (x+2)^3 = -1 + 3 \log_4 (x+2)$

25. $\log_5 \dfrac{x^3}{(x-4)^4} = \log_5 x^3 - \log_5 (x-4)^4 = 3 \log_5 x - 4 \log_5 (x-4)$

27. $\log_2 \dfrac{\sqrt{x}}{z^2} = \log_2 \sqrt{x} - \log_2 z^2 = \log_2 x^{1/2} - \log_2 z^2 = \dfrac{1}{2} \log_2 x - 2 \log_2 z$

29. $\ln \sqrt[3]{\dfrac{2x+6}{(x+1)^5}} = \ln\left(\dfrac{2x+6}{(x+1)^5}\right)^{1/3} = \dfrac{1}{3}[\ln(2x+6) - 5 \ln(x+1)] = \dfrac{1}{3} \ln(2x+6) - \dfrac{5}{3} \ln(x+1)$

31. $\log_2 \dfrac{\sqrt[3]{x^2 - 1}}{\sqrt{1+x^2}} = \log_2 \sqrt[3]{x^2 - 1} - \log_2 \sqrt{1+x^2} = \log_2 (x^2-1)^{1/3} - \log_2 (1+x^2)^{1/2} =$

        $\dfrac{1}{3} \log_2 (x^2 - 1) - \dfrac{1}{2} \log_2 (1+x^2)$

33. $\log 2 + \log 3 = \log(2 \cdot 3) = \log 6$

35. $\ln \sqrt{5} - \ln 25 = \ln\left(\dfrac{5^{1/2}}{5^2}\right) = \ln 5^{-3/2} = -\dfrac{3}{2} \ln 5$

37. $\log 20 + \log \dfrac{1}{10} = \log (20)\left(\dfrac{1}{10}\right) = \log \dfrac{20}{10} = \log 2$

39. $\log 4 + \log 3 - \log 2 = \log \dfrac{4(3)}{2} = \log \dfrac{12}{2} = \log 6$

41. $\log_7 5 + \log_7 k^2 = \log_7 (5)(k^2) = \log_7 5k^2$

43. $\ln x^6 - \ln x^3 = \ln \dfrac{x^6}{x^3} = \ln x^{6-3} = \ln x^3$

45. $\log \sqrt{x} + \log x^2 - \log x = \log \left(\dfrac{x^{1/2} \cdot x^2}{x}\right) = \log x^{3/2} = \dfrac{3}{2}\log x$

47. $3\ln(x) - \dfrac{3}{2}\ln(y) + 4\ln(z) = \ln(x^3) - \ln(\sqrt{y^3}) + \ln(z^2) = \ln\left(\dfrac{x^3 z^4}{\sqrt{y^3}}\right)$

49. $\ln \dfrac{1}{e^2} + \ln 2e = \ln\left(\dfrac{1}{e^2} \cdot 2e\right) = \ln \dfrac{2}{e}$

51. $2\ln x - 4\ln y + \dfrac{1}{2}\ln z = \ln x^2 - \ln y^4 + \ln z^{1/2} = \ln \dfrac{x^2}{y^4} + \ln \sqrt{z} = \ln \dfrac{x^2\sqrt{z}}{y^4}$

53. $\log 4 - \log x + 7\log \sqrt{x} = \log \dfrac{4}{x} + \log(\sqrt{x})^7 = \log\left(\dfrac{4x^{7/2}}{x}\right) = \log 4x^{5/2} = \log 4\sqrt{x^5}$

55. $2\log(x^2-1) + 4\log(x-2) - \dfrac{1}{2}\log y = \log(x^2-1)^2 + \log(x-2)^4 - \log y^{1/2} =$

    $\log \dfrac{(x^2-1)^2(x-2)^4}{\sqrt{y}}$

57. See Figure 57.
59. See Figure 59.

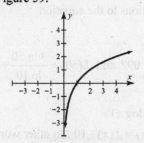

Figure 57   Figure 59

61. $f(x) = \log_8 \sqrt{2^{3x}} = \dfrac{x}{2}\log_8 2^3 = \dfrac{x}{2}\log_8 8 = \dfrac{1}{2}x;\ \boldsymbol{D}:(-\infty, \infty)$

63. $f(x) = \ln(1-2^x);\ \boldsymbol{D}:(-\infty, 0)$

65. $f(x) = \ln(x-1) + \ln(x+1) = \ln(x-1)(x+1) = \ln(x^2-1);\ \boldsymbol{D}:(1, \infty)$

67. $(f \circ g)(x) = f(g(x)) = f(\sqrt{x}) = \log \sqrt{x} = \dfrac{1}{2}\log x \Rightarrow \boldsymbol{D}:(0, \infty)$

    $(g \circ f)(x) = g(f(x)) = g(\log x) = \sqrt{\log x} \Rightarrow D:(0, \infty)$

69. $(f \circ g)(x) = f(g(x)) = f(1-e^x) = \ln(1-e^x) \Rightarrow \boldsymbol{D}:(-\infty, 0)$

    $(g \circ f)(x) = g(f(x)) = g(\ln x) = 1 - e^{\ln x} = 1 - x \Rightarrow D:(0, \infty)$

71. $\ln\left|x+\sqrt{x^2+3}\right|+\ln\left|x-\sqrt{x^2+3}\right|=\ln\left(\left|x+\sqrt{x^2+3}\right|\right)\left(\left|x-\sqrt{x^2+3}\right|\right)=\ln\left|x^2-\left(x^2+3\right)\right|=\ln\left|-3\right|=\ln 3$

73. $\dfrac{1}{3}\ln\left(\dfrac{x^2+1}{5}\right)-\dfrac{1}{3}\ln\left(\dfrac{x^2+4}{5}\right)=\dfrac{1}{3}\ln\left(\dfrac{\frac{x^2+1}{5}}{\frac{x^2+4}{5}}\right)\ln\left(\dfrac{(x^2+1)}{(x^2+4)}\right)^{\frac{1}{3}}=\ln\sqrt[3]{\dfrac{x^2+1}{x^2+4}}$

75. $\log_2 25=\dfrac{\log 25}{\log 2}\approx 4.644$

77. $\log_5 130=\dfrac{\log 130}{\log 5}\approx 3.024$

79. $5\log_4 25=5\dfrac{\log 25}{\log 4}\approx 11.610$

81. $-2\log_2 0.65=-2\dfrac{\log 0.65}{\log 2}\approx 1.243$

83. $\log_2 5+\log_2 7=\dfrac{\log 5}{\log 2}+\dfrac{\log 7}{\log 2}\approx 5.129$

85. $\sqrt{\log_4 46}=\sqrt{\dfrac{\log 46}{\log 4}}\approx 1.662$

87. $\dfrac{\log_2 12}{\log_2 3}=\dfrac{\frac{\log 12}{\log 2}}{\frac{\log 3}{\log 2}}=\dfrac{\log 12}{\log 2}\cdot\dfrac{\log 2}{\log 3}=\dfrac{\log 12}{\log 3}\approx 2.262$

89. Graph $Y_1=\log(X^3+X^2+1)/\log 2$ and $Y_2=7$ in $[0, 10, 2]$ by $[0, 10, 2]$. The graphs intersect near the point (4.714, 7), so the solution to $\log_2(x^3+x^2+1)=7$ is $x\approx 4.714$.

91. Graph $Y_1=\log(X^2+1)/\log 2$ and $Y_2=5-\log(X^4+1)/\log 3$ in $[-5, 5, 2]$ by $[-5, 10, 2]$. The graphs intersect near the points $(-2.035, 2.362)$ and $(2.035, 2.362)$, so the solutions to the equation $\log_2(x^2+1)=5-\log_3(x^4+1)$ are $x\approx\pm 2.035$.

93. Change base 10 to base e: $L(x)=3\log x=\dfrac{3\cdot\ln x}{\ln 10}$; $L(50)=3\log 50\approx 5.097$ and $L(50)=\dfrac{3\ln 50}{\ln 10}\approx 5.097$.
     Yes, the answers agree.

95. $f(x)=160+10\log x\Rightarrow f(10x)=160+10\log(10x)=160+10(\log 10+\log x)=$
     $160+10(1+\log x)=160+10+10\log x=160+10\log x+10$, thus $f(10x)=f(x)+10$. In other words, the decibel level increases by 10 decibels.

97. $\ln I-\ln I_0=-kx\Rightarrow\ln I=\ln I_0-kx\Rightarrow e^{\ln I}=e^{\ln I_0-kx}\Rightarrow e^{\ln I}=e^{\ln I_0}\cdot e^{-kx}\Rightarrow I=I_0 e^{-kx}$

99. (a) $P=37.3e^{0.01x}\Rightarrow\dfrac{P}{37.3}=e^{0.01x}\Rightarrow\ln\left(\dfrac{P}{37.3}\right)=0.01x\Rightarrow x=100\ln\left(\dfrac{P}{37.3}\right)$

     (b) $x=100\ln\dfrac{40}{37.3}\Rightarrow x\approx 7$, The population is expected to reach 40 million during 2017.

101. $r=p-k\ln t\Rightarrow t=e^{\frac{p-r}{k}}$

103. $A = P\left(1+\dfrac{r}{n}\right)^{nt} \Rightarrow t = \dfrac{\log\dfrac{A}{P}}{n\log\left(1+\dfrac{r}{n}\right)}$

105. $A = Pe^{rt} \Rightarrow \dfrac{A}{P} = e^{rt} \Rightarrow \ln\dfrac{A}{P} = \ln e^{rt} \Rightarrow \ln\dfrac{A}{P} = rt \Rightarrow \dfrac{\ln\frac{A}{P}}{r} = t \Rightarrow t = \dfrac{\ln\frac{A}{P}}{r}$

107. $\log 1 + 2\log 2 + 3\log 3 + 4\log 4 + 5\log 5 = \log 1 + \log 2^2 + \log 3^3 + \log 4^4 + \log 5^5 =$

$\log(1 \cdot 2^2 \cdot 3^3 \cdot 4^4 \cdot 5^5) = \log 86,400,000$

Extended Discovery for Section 5.5

1. $\left(\dfrac{1}{\pi}\ln(640320^3 + 744)\right)^2 \approx 163.00000000000000000000000000000232$; which is not an integer.

5.6: Exponential and Logarithmic Equations

1. (a) The graphs appear to intersect near the point (2, 7.5). Thus, the solution is $x \approx 2$.

   (b) $f(x) = g(x) \Rightarrow e^x = 7.5 \Rightarrow \ln e^x = \ln 7.5 \Rightarrow x = \ln 7.5$ about 2.015

3. (a) The graphs appear to intersect near the point (2, 2.5). Thus, the solution is $x \approx 2$.

   (b) $f(x) = g(x) \Rightarrow 10^{0.2x} = 2.5 \Rightarrow \log 10^{0.2x} = \log 2.5 \Rightarrow 0.2x = \log 2.5 \Rightarrow x = \dfrac{\log 2.5}{0.2}$ about 1.990

5. $4e^x = 5 \Rightarrow e^x = \dfrac{5}{4} \Rightarrow \ln e^x = \ln\dfrac{5}{4} \Rightarrow x = \ln\dfrac{5}{4} \approx 0.2231$

7. $2(10^x) = 200 \Rightarrow 10^x = 100 \Rightarrow x = \log 100 = 2$

9. $2^{x+3} = 128 \Rightarrow x + 3 = \log_2 128 = \log_2 2^7 = 7 \Rightarrow x = 7 - 3 = 4$

11. $2(10^x) + 5 = 45 \Rightarrow 2(10^x) = 40 \Rightarrow 10^x = 20 \Rightarrow \log 10^x = \log 20 \Rightarrow x = \log 20 \approx 1.301$

13. $2.5e^{-1.2x} = 1 \Rightarrow e^{-1.2x} = \dfrac{1}{2.5} \Rightarrow \ln e^{-1.2x} = \ln\dfrac{1}{2.5} \Rightarrow -1.2x = \ln\dfrac{1}{2.5} \Rightarrow x = \dfrac{\ln\frac{1}{2.5}}{-12} \approx 0.7636$

15. $1.2(0.9^x) = 0.6 \Rightarrow 0.9^x = 0.5 \Rightarrow \ln 0.9^x = \ln 0.5 \Rightarrow x\ln 0.9 = \ln 0.5 \Rightarrow x = \dfrac{\ln 0.5}{\ln 0.9} \approx 6.579$

17. $4(1.1^{x-1}) = 16 \Rightarrow 1.1^{x-1} = 4 \Rightarrow \ln 1.1^{x-1} = \ln 4 \Rightarrow (x-1)\ln 1.1 = \ln 4 \Rightarrow x - 1 = \dfrac{\ln 4}{\ln 1.1} \Rightarrow x = \dfrac{\ln 4}{\ln 1.1} + 1 \approx 15.55$

19. $5(1.2)^{3x-2} + 94 = 100 \Rightarrow 5(1.2)^{3x-2} = 6 \Rightarrow (1.2)^{3x-2} = 1.2 \Rightarrow \ln 1.2^{3x-2} = \ln 1.2 \Rightarrow$

$(3x-2)\ln 1.2 = \ln 1.2 \Rightarrow 3x - 2 = \dfrac{\ln 1.2}{\ln 1.2} \Rightarrow 3x - 2 = 1 \Rightarrow 3x = 3 \Rightarrow x = 1$

21. $5^{3x} = 5^{1-2x} \Rightarrow \log_5 5^{3x} = \log_5 5^{1-2x} \Rightarrow 3x = 1 - 2x \Rightarrow 5x = 1 \Rightarrow x = \dfrac{1}{5}$

23. $10^{x^2} = 10^{3x-2} \Rightarrow \log 10^{x^2} = \log 10^{3x-2} \Rightarrow x^2 = 3x - 2 \Rightarrow x^2 - 3x + 2 = 0 \Rightarrow (x-1)(x-2) = 0 \Rightarrow x = 1 \text{ or } x = 2$

25. No solution since no power of $\frac{1}{5}$ will result in a negative value.

27. $\left(\frac{2}{5}\right)^{x-2} = \frac{1}{3} \Rightarrow \log\left(\frac{2}{5}\right)^{x-2} = \log\left(\frac{1}{3}\right) \Rightarrow (x-2)\log\left(\frac{2}{5}\right) = \log\left(\frac{1}{3}\right) \Rightarrow x-2 = \frac{\log\left(\frac{1}{3}\right)}{\log\left(\frac{2}{5}\right)} \Rightarrow$

$x = 2 + \dfrac{\log\left(\frac{1}{3}\right)}{\log\left(\frac{2}{5}\right)} \approx 3.199$

29. $4^{x-1} = 3^{2x} \Rightarrow \log 4^{x-1} = \log 3^{2x} \Rightarrow (x-1)\log 4 = 2x\log 3 \Rightarrow x\log 4 - \log 4 = 2x\log 3 \Rightarrow$

$x\log 4 - 2x\log 3 = \log 4 \Rightarrow x(\log 4 - 2\log 3) = \log 4 \Rightarrow x = \dfrac{\log 4}{\log 4 - 2\log 3} \approx -1.710$

31. $e^{x-3} = 2^{3x} \Rightarrow \ln e^{x-3} = \ln 2^{3x} \Rightarrow x-3 = 3x\ln 2 \Rightarrow -3 = -x + 3x\ln 2 \Rightarrow$

$3 = x - 3x\ln 2 \Rightarrow 3 = x(1 - 3\ln 2) \Rightarrow x = \dfrac{3}{1 - 3\ln 2} \Rightarrow x = \dfrac{3}{1 - \ln 8} \Rightarrow x \approx -2.779$

33. $3(1.4)^x - 4 = 60 \Rightarrow 3(1.4)^x = 64 \Rightarrow 1.4^x = \dfrac{64}{3} \Rightarrow \log 1.4^x = \log\dfrac{64}{3} \Rightarrow x\log 1.4 = \log\dfrac{64}{3} \Rightarrow$

$x = \dfrac{\log\frac{64}{3}}{\log 1.4} \approx 9.095$

35. $5(1.015)^{x-1980} = 8 \Rightarrow 1.015^{x-1980} = \dfrac{8}{5} \Rightarrow \log 1.015^{x-1980} = \log\dfrac{8}{5} \Rightarrow (x-1980)\log 1.015 = \log\dfrac{8}{5} \Rightarrow$

$x\log 1.015 - 1980\log 1.015 = \log\dfrac{8}{5} \Rightarrow x\log 1.015 = \log\dfrac{8}{5} + 1980\log 1.015 \Rightarrow$

$x = \dfrac{\log\frac{8}{5} + 1980\log 1.015}{\log 1.015} = \dfrac{\log\frac{8}{5}}{\log 1.015} + 1980 \approx 2012$

37. $5^{2x} = 5^{x-3} \Rightarrow 2x = x-3 \Rightarrow x = -3$

39. $e^{-x} = e^{x^2} \Rightarrow -x = x^2 \Rightarrow 0 = x^2 + x \Rightarrow 0 = x(x+1) \Rightarrow x = 0$ or $x+1 = 0 \Rightarrow x = 0$ or $x = -1$

41. $2^{3x} = 8^{-x+2} \Rightarrow 2^{3x} = (2^3)^{-x+2} \Rightarrow 2^{3x} = 2^{-3x+6} \Rightarrow 3x = -3x+6 \Rightarrow 6x = 6 \Rightarrow x = 1$

43. $25^{2x} = 125^{2-x} \Rightarrow (5^2)^{2x} = (5^3)^{2-x} \Rightarrow 5^{4x} = 5^{6-3x} \Rightarrow 4x = 6-3x \Rightarrow 7x = 6 \Rightarrow x = \dfrac{6}{7}$

45. $32^{3x} = 16^{5x+3} \Rightarrow (2^5)^{3x} = (2^4)^{5x+3} \Rightarrow 2^{15x} = 2^{20x+12} \Rightarrow 15x = 20x+12 \Rightarrow -5x = 12 \Rightarrow x = -\dfrac{12}{5}$

47. $3\log x = 2 \Rightarrow \log x = \dfrac{2}{3} \Rightarrow 10^{\log x} = 10^{2/3} \Rightarrow x = 10^{2/3} \approx 4.642$

49. $\ln 2x = 5 \Rightarrow e^{\ln 2x} = e^5 \Rightarrow 2x = e^5 \Rightarrow x = \dfrac{e^5}{2} \approx 74.207$

51. $\log 2x^2 = 2 \Rightarrow 10^{\log 2x^2} = 10^2 \Rightarrow 2x^2 = 100 \Rightarrow x^2 = 50 \Rightarrow x = \pm\sqrt{50} \approx \pm 7.071$

53. $\log_2(3x-2) = 4 \Rightarrow 2^{\log_2(3x-2)} = 2^4 \Rightarrow 3x-2 = 16 \Rightarrow 3x = 18 \Rightarrow x = 6$

55. $\log_5(8-3x) = 3 \Rightarrow 5^{\log_5(8-3x)} = 5^3 \Rightarrow 8-3x = 125 \Rightarrow -3x = 117 \Rightarrow x = -39$

57. $160 + 10\log x = 50 \Rightarrow 10\log x = -110 \Rightarrow \log x = -11 \Rightarrow 10^{\log x} = 10^{-11} \Rightarrow x = 10^{-11}$

59. $\ln x + \ln x^2 = 3 \Rightarrow \ln x + 2\ln x = 3 \Rightarrow 3\ln x = 3 \Rightarrow \ln x = 1 \Rightarrow e^{\ln x} = e^1 \Rightarrow x = e \approx 2.718$

61. $2\log_2 x = 4.2 \Rightarrow \log_2 x = 2.1 \Rightarrow 2^{\log_2 x} = 2^{2.1} \Rightarrow x = 2^{2.1} \approx 4.287$

63. $\log x + \log 2x = 2 \Rightarrow \log(x \cdot 2x) = 2 \Rightarrow \log 2x^2 = 2 \Rightarrow 10^{\log 2x^2} = 10^2 \Rightarrow 2x^2 = 100 \Rightarrow x^2 = 50 \Rightarrow x = \pm\sqrt{50}$.

When $x = -\sqrt{50} < 0$, $\log x$ is undefined. Therefore, the only solution is $x = \sqrt{50} \approx 7.071$.

65. $\log(2 - 3x) = 3 \Rightarrow 10^3 = 2 - 3x \Rightarrow 1000 = 2 - 3x \Rightarrow 998 = -3x \Rightarrow x = -\dfrac{998}{3}$

67. $\ln(x) + \ln(3x - 1) = \ln(10) \Rightarrow \ln(x(3x - 1)) = \ln(10) \Rightarrow \ln(3x^2 - x) = \ln(10) \Rightarrow$

$3x^2 - x = 10 \Rightarrow 3x^2 - x - 10 = 0 \Rightarrow (3x + 5)(x - 2) = 0 \Rightarrow 3x + 5 = 0$ or $x - 2 = 0 \Rightarrow x = -\dfrac{5}{3}$ or $x = 2$. When

$x = -\dfrac{5}{3}$, both $\ln(3x - 1)$ and $\ln(x)$ are undefined, therefore the only solution is $x = 2$.

69. $2\ln x = \ln(2x + 1) \Rightarrow \ln x^2 = \ln(2x + 1) \Rightarrow e^{\ln x^2} = e^{\ln(2x+1)} \Rightarrow x^2 = 2x + 1 \Rightarrow$

$x^2 - 2x - 1 = 0 \Rightarrow x = \dfrac{2 \pm \sqrt{2^2 - 4(1)(-1)}}{2(1)} = \dfrac{2 \pm \sqrt{8}}{2} = \dfrac{2 \pm 2\sqrt{2}}{2} = 1 \pm \sqrt{2} \Rightarrow$

$x = 1 + \sqrt{2}$ or $x = 1 - \sqrt{2}$.

When $x = 1 - \sqrt{2} \approx -0.414$, $\ln x$ is undefined. The only solution is $x = 1 + \sqrt{2}$.

71. $\log(x + 1) + \log(x - 1) = \log 3 \Rightarrow \log[(x + 1)(x - 1)] = \log 3 \Rightarrow \log(x^2 - 1) = \log 3 \Rightarrow$

$10^{\log(x^2 - 1)} = 10^{\log 3} \Rightarrow x^2 - 1 = 3 \Rightarrow x^2 = 4 \Rightarrow x = -2$ or $x = 2$

When $x = -2$ both $\log(x - 1)$ and $\log(x + 1)$ are undefined. The only solution is $x = 2$.

73. $\log_2 2x = 4 - \log_2(x + 2) \Rightarrow \log_2 2x + \log_2(x + 2) = 4 \Rightarrow \log_2[2x(x + 2)] = 4 \Rightarrow$

$\log_2(2x^2 + 4x) = 4 \Rightarrow 2^{\log_2(2x^2 + 4x)} = 4 \Rightarrow 2x^2 + 4x - 16 = 0 \Rightarrow 2(x^2 + 2x - 8) = 0 \Rightarrow$

$2(x + 4)(x - 2) = 0 \Rightarrow x = -4$, $x = 2$ since we cannot take the log of a negative number, $x = 2$.

75. $\log_5(x + 1) + \log_5(x - 1) = \log_5 15 \Rightarrow \log_5(x + 1)(x - 1) = \log_5 15 \Rightarrow$

$\log_5(x^2 - 1) = \log_5 15 \Rightarrow 5^{\log_5(x^2 - 1)} = 5^{\log_5 15} \Rightarrow x^2 - 1 = 15 \Rightarrow x^2 - 16 = 0 \Rightarrow$

$(x + 4)(x - 4) = 0$, since we cannot take the log of a negative number, $x = 4$.

77. $e^{2x} - 6e^x + 8 = 0 \Rightarrow (e^x - 2)(e^x - 4) = 0 \Rightarrow e^x = 2$, $e^x = 4 \Rightarrow x = \ln 2, \ln 4$

79. $2e^{2x} + e^x - 6 = 0 \Rightarrow (2e^x - 3)(e^x + 2) = 0 \Rightarrow e^x = \dfrac{3}{2} \Rightarrow x = \ln\dfrac{3}{2}$

81. $(\log_2 x)^2 + \log_2 x - 2 = 0 \Rightarrow (\log_2 x + 2)(\log_2 x - 1) = 0 \Rightarrow \log_2 x = -2, \log_2 x = 1 \Rightarrow x = 2^{-2} = \dfrac{1}{4}; x = 2; x = \dfrac{1}{4}, 2$

83. $(\ln x)^2 + 16 - 10\ln x \Rightarrow (\ln x - 2)(\ln x - 8) = 0 \Rightarrow \ln x = 2, \ln x = 8 \Rightarrow x = e^2, x = e^8$

85. Graph $Y_1 = 2X + e^{\wedge}(X)$ and $Y_2 = 2$ as shown in Figure 85.

The graphs intersect near the point (0.31, 2), so the solution to $2x + e^x = 2$ is $x \approx 0.31$.

87. Graph $Y_1 = X^{\wedge}2 + X \ln(X)$ and $Y_2 = 2$ as shown in Figure 87.

The graphs intersect near the point (1.71, 2), so the solution to $x^2 - x \ln x = 2$ is $x \approx 1.71$.

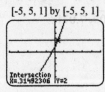

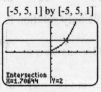

Figure 85   Figure 87

89. $-2e^x + 1 \geq -1 \Rightarrow -2e^x + 2 \geq 0 \Rightarrow e^x - 1 \leq 0 \Rightarrow e^x \leq 1 \Rightarrow x \leq 0;\ (-\infty, 0]$

91. $1 - 2^{2x} > -63 \Rightarrow 64 > 2^{2x} \Rightarrow 2^6 > 2^{2x} \Rightarrow 6 > 2x \Rightarrow 3 > x;\ (-\infty, 3)$

93. $\log x \geq 3 \Rightarrow x \geq 10^3;\ [1000, \infty)$

95. $5 - \ln 2x > 6 \Rightarrow -1 > \ln 2x \Rightarrow e^{-1} > 2x \Rightarrow x < \dfrac{1}{2e};\ \left(0, \dfrac{1}{2e}\right)$

97. $7.5 = 7(1.01)^{x-2011} \Rightarrow \dfrac{7.5}{7} = 1.01^{x-2011} \Rightarrow \ln\left(\dfrac{7.5}{7}\right) = \ln(1.01)^{x-2011} \Rightarrow$

$\ln\left(\dfrac{7.5}{7}\right) = (x - 2011)\ln(1.01) \Rightarrow \dfrac{\ln\left(\dfrac{7.5}{7}\right)}{\ln(1.01)} = x - 2011 \Rightarrow x = 2011 + \dfrac{\ln\left(\dfrac{7.5}{7}\right)}{\ln(1.01)} \Rightarrow x \approx 2018$

99. $10 = 100\left(\dfrac{1}{2}\right)^{\frac{t}{3}} \Rightarrow 0.1 = (0.5)^{\frac{t}{3}} \Rightarrow \ln(0.1) = \ln(0.5)^{\frac{t}{3}} \Rightarrow \dfrac{\ln(0.1)}{\ln(0.5)} = \dfrac{t}{3} \Rightarrow t \approx 10 \text{ hours}$

101. $250 = 1000e^{-0.12x} \Rightarrow 0.25 = e^{-0.12x} \Rightarrow \ln(0.25) = -0.12x \Rightarrow x = \dfrac{\ln(0.25)}{-0.12} \approx 11.55 \text{ ft}.$

103. (a)   Let $T(x) = Ca^x$, where x is the number of transistors. Initially, $T(0) = 2300$, so $C = 2300$ and

$T(x) = 2300a^x$. Next we will find the value of a. Because the number of transistors doubles every 2 years there will be 4600 transistors 2 years after 1971, so

$T(2) = 4600.\ 2300a^2 = 4600 \Rightarrow a^2 = 2 \Rightarrow a = (2)^{1/2}$. Thus, $T(x) = 2300\left((2)^{1/2}\right)^x = 2300(2)^{x/2}$

(b) $T(40) = 2300(2)^{40/2} = 2300(2)^{20} = 2,411,724,800$. In 2011 there were about 2.4 billion transistors on an integrated circuit.

(c) $10,000,000 = 2300(2)^{x/2} \Rightarrow \dfrac{10,000,000}{2300} = 2^{x/2} \Rightarrow \ln\left(\dfrac{10,000,000}{2300}\right) = \ln(2^{x/2}) \Rightarrow$

$\ln\left(\dfrac{10,000,000}{2300}\right) = \dfrac{x}{2}\ln(2) \Rightarrow \dfrac{\ln\left(\dfrac{10,000,000}{2300}\right)}{\ln(2)} = \dfrac{x}{2} \Rightarrow 2\left(\dfrac{\ln\left(\dfrac{10,000,000}{2300}\right)}{\ln(2)}\right) = x \Rightarrow x \approx 24$. Since x

represents the number of years after 1971 the year is about 1995.

105. (a) $H(8) = 0.3 + 0.28\ln(8) \approx 0.88$. When x is 8 the Human Development Index is 0.88.

(b) $0.68 = 0.3 + 0.28\ln(x) \Rightarrow 0.38 = 0.28\ln(x) \Rightarrow \dfrac{0.38}{0.28} = \ln x \Rightarrow e^{0.38/0.28} = x \Rightarrow x \approx 3.9$

107. We must solve the equation $f(x) = 95$.

$230(0.881)^x = 95 \Rightarrow 0.881^x = \dfrac{95}{230} \Rightarrow \log 0.881^x = \log\dfrac{95}{230} \Rightarrow x \log 0.881 = \log\dfrac{95}{230}$

$$x = \frac{\log \frac{95}{230}}{\log 0.881} \approx 7$$

Since $x = 0$ corresponds to 1974, $x \approx 7$ represents $1974 + 7 = 1981$, rounded to the nearest year.

109. (a) $P(x) = Ca^{x-2000} \Rightarrow P(2000) = Ca^{2000-2000} \Rightarrow 1 - Ca^0 \Rightarrow C = 1$, so $P(x) = a^{x-2000}$

     $P(2025) = a^{2025-2000} \Rightarrow 1.4 = a^{25} \Rightarrow a = 1.4^{1/25} \approx 1.01355$

     (b) $P(x) = (1.01355)^{x-2000} \Rightarrow P(2010) = (1.01355)^{10} \approx 1.144$; in 2010 the population if India will be
     about1.14 billion.

     (c) $P(x) = 1.5 \Rightarrow 1.01355^{x-2000} = 1.5 \Rightarrow \ln 1.01355^{x-2000} = \ln 1.5 \Rightarrow (x-2000)\ln 1.01355 = \ln 1.5 \Rightarrow$

     $x \ln 1.01355 - 2000 \ln 1.01355 = \ln 1.5 \Rightarrow x \ln 1.01355 = \ln 1.5 + 2000 \ln 1.01355 \Rightarrow$

     $x = \frac{\ln 1.5 + 2000 \ln 1.01355}{\ln 1.01355} = \frac{\ln 1.5}{\ln 1.01355} + 2000 \approx 2030.13$;

     India's population might reach 1.5 billion in 2030.

111. (a) $T_0 = 32$, $D = 212 - 32 \Rightarrow D = 180$. Solving for a use

     $$70 = 32 + 180a^{1/2} \Rightarrow 38 = 180a^{1/2} \Rightarrow \frac{38}{180} = a^{1/2} \Rightarrow a = 0.045$$

     (b) $T(t) = 32 + 180(0.045)^t \Rightarrow T\left(\frac{1}{6}\right) = 32 + 180(0.045)^{1/6} \Rightarrow T\left(\frac{1}{6}\right) \approx 139°F$

     (c) $40 = 32 + 180(0.045)^t \Rightarrow 8 = 180(0.045)^t \Rightarrow \frac{8}{180} = 0.045^t \Rightarrow \log \frac{8}{180} = \log 0.045^t \Rightarrow$

     $$\log \frac{8}{180} = t \log 0.045 \Rightarrow t = \frac{\log \frac{8}{180}}{\log 0.045} \Rightarrow t \approx 1$$

     Graph $Y_1 = 32 + 180(0.045)^t$ and $Y_2 = 40$. The lines intersect at $t \approx 1$. See Figure 111.

     [0, 5, 1] by [10, 50, 1]

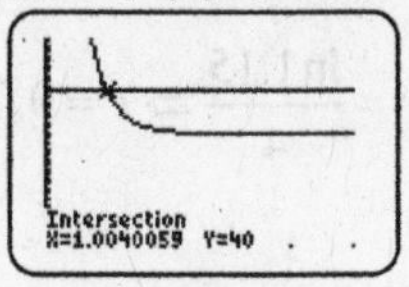

     Figure 111

113. (a) $f(1.5) = 20 - 15(0.365)^{1.5} \approx 16.7°C$

     (b) $15 = 20 - 15(0.365)^t \Rightarrow 15(0.365)^t = 5 \Rightarrow 0.365^t = \frac{1}{3} \Rightarrow \ln 0.365^t = \ln \frac{1}{3} \Rightarrow t \ln 0.365 = \ln \frac{1}{3} \Rightarrow$

     $t = \frac{\ln \frac{1}{3}}{\ln 0.365} \approx 1.09$; the soda can warmed to 15°C after about 1.09 minutes or 1 hour 5.4 minutes.

115. $280 \ln (x+1) + 1925 = 2300 \Rightarrow 280 \ln (x+1) = 375 \Rightarrow \ln (x+1) = \frac{375}{280} \Rightarrow e^{\ln (x+1)} = e^{375/280} \Rightarrow$

     $x + 1 = e^{375/280} \Rightarrow x = e^{375/280} - 1 \approx 2.8$; a person who consumes 2300 calories daily owns about 2.8 acres.

117. $\frac{2 - \log (100-x)}{0.42} = 2 \Rightarrow 2 - \log (100-x) = 0.84 \Rightarrow \log (100-x) = 1.16 \Rightarrow$

     $10^{\log(100-x)} = 10^{1.16} \Rightarrow 100 - x = 10^{1.16} \Rightarrow x = 100 - 10^{1.16} \approx 85.546$; after 2 years approximately 85.5% of the
     robins have died.

119. (a)   A concentration increase of 15% implies $B(t) = 1.15$ when $B_0 = 1$.

$$1.15 = e^{6k} \Rightarrow \ln(1.15) = 6k \Rightarrow k = \frac{\ln(1.15)}{6} \approx 0.0233 \, .$$

(b) $B(t) = 1.2e^{0.0233(8.2)} \approx 1.45$ billion per liter.

(c) $B(t) = 1e^{0.0233(1)} \approx 1.0236 \Rightarrow$ concentration increases by about 2.36% per hour.

121. $A = P\left(1 + \dfrac{r}{n}\right)^{nt} \Rightarrow 2000 = 1000\left(1 + \dfrac{0.085}{4}\right)^{0.4t} \Rightarrow 2 = (1.02125)^{4t} \Rightarrow \ln(2) = \ln(1.02125)^{4t} \Rightarrow$

$$\frac{\ln(2)}{\ln(1.02125)} = 4t \Rightarrow t = \frac{\frac{\ln(2)}{\ln(1.02125)}}{4} \approx 8.25 \text{ yrs} \, .$$

123.   (a) $750 = 500e^{0.03t} \Rightarrow \dfrac{750}{500} = e^{0.03t} \Rightarrow \ln(1.5) = 0.03t \Rightarrow \dfrac{\ln 1.5}{0.03} = t \Rightarrow t \approx 13.5$

(b)   Five hundred dollars invested at 3% compounded continuously results in $750 after 13.5 years.

125. $P = 100\left(\dfrac{1}{2}\right)^{t/5700} \Rightarrow 35 = 100\left(\dfrac{1}{2}\right)^{t/5700} \Rightarrow \left(\dfrac{1}{2}\right)^{t/5700} = 0.35 \Rightarrow \ln\left(\dfrac{1}{2}\right)^{t/5700} = \ln 0.35 \Rightarrow$

$$\frac{t}{5700} \ln \frac{1}{2} = \ln 0.35 \Rightarrow \frac{t}{5700} = \frac{\ln 0.35}{\ln \frac{1}{2}} \Rightarrow t = 5700 \left(\frac{\ln 0.35}{\ln \frac{1}{2}}\right) \approx 8633 \, . \text{ The fossil is about 8633 years old.}$$

127. $0.5 = 1 - e^{-0.5x} \Rightarrow -0.5 = -e^{-0.5x} \Rightarrow 0.5 = e^{-0.5x} \Rightarrow \ln(0.5) = -0.5x \Rightarrow x = \dfrac{\ln(0.5)}{-0.5} \Rightarrow x = -2\ln(0.5) \approx 1.39$ min.

129. $N(t) = 100{,}000e^{rt} \Rightarrow 200{,}000 = 100{,}000e^{r(2)} \Rightarrow 2 = e^{2r} \Rightarrow \ln 2 = \ln e^{2r} \Rightarrow \ln 2 = 2r \Rightarrow$

$$\frac{\ln 2}{2} = r \Rightarrow r = 0.3466 \, . \text{ Now } 350{,}000 = 100{,}000e^{0.3466t} \Rightarrow 3.5 = e^{0.3466t} \Rightarrow \ln 3.5 = \ln e^{0.3466t} \Rightarrow$$

$$\ln 3.5 = 0.3466t \Rightarrow t = \frac{\ln 3.5}{0.3466} \Rightarrow t \approx 3.6 \text{ hrs.}$$

131. $A_n = A_0 e^{rn} \Rightarrow 2300 = 2000e^{r(4)} \Rightarrow 1.15 = e^{4r} \Rightarrow \ln 1.15 = \ln e^{4r} \Rightarrow \ln 1.15 = 4r \Rightarrow r = \dfrac{\ln 1.15}{4} \Rightarrow r = 0.03494 \, .$

Now $3200 = 2000e^{0.03494t} \Rightarrow 1.6 = e^{0.03494t} \Rightarrow \ln 1.6 = \ln e^{0.03494t} \Rightarrow$

$$\ln 1.6 = 0.03494t \Rightarrow t = \frac{\ln 1.6}{0.03494} \Rightarrow t \approx 13.5 \text{ years.}$$

133.   (a)   The initial concentration of the drug is 11 milligrams per liter since $C(0) = 11(0.72)^0 = 11$ milligrams per liter.

(b)   50% of $11 = 5.5$ milligrams per liter.

$$11(0.72)^t = 5.5 \Rightarrow 0.72^t = 0.5 \Rightarrow \ln 0.72^t = \ln 0.5 \Rightarrow t \ln 0.72 = \ln 0.5 \Rightarrow t = \frac{\ln 0.5}{\ln 0.72} \approx 2.11 \text{ The drug}$$

concentration decreases to 50% of its initial level after about 2 hours.

135. $P\left(1 + \dfrac{r}{n}\right)^{nt} = A \Rightarrow \left(1 + \dfrac{r}{n}\right)^{nt} = \dfrac{A}{P} \Rightarrow \log\left(1 + \dfrac{r}{n}\right)^{nt} = \log\left(\dfrac{A}{P}\right) \Rightarrow nt \, \log\left(1 + \dfrac{r}{n}\right) = \log\left(\dfrac{A}{P}\right) \Rightarrow t = \dfrac{\log(A/P)}{n\log(1 + r/n)}$

Extended and Discovery Exercises for Section 5.6

1. $f(x) = Ca^x = Ce^{\ln(a^x)} = Ce^{x\ln(a)}$, that is $k = \ln a$; $g(x) = 2^x = e^{\ln(2^x)} = e^{x\ln(2)} \Rightarrow k = \ln(2)$

Checking Basic Concepts for Sections 5.5 and 5.6

1. $\log\dfrac{x^2 y^3}{\sqrt[3]{z}} = \log x^2 + \log y^3 - \log z^{1/3} = 2\log x + 3\log y - \dfrac{1}{3}\log z$

3. (a) $5(1.4)^x - 4 = 25 \Rightarrow 5(1.4)^x = 29 \Rightarrow 1.4^x = \dfrac{29}{5} \Rightarrow \ln 1.4^x = \ln\dfrac{29}{5} \Rightarrow x\ln 1.4 = \ln\dfrac{29}{5} \Rightarrow x = \dfrac{\ln\frac{29}{5}}{\ln 1.4} \approx 5.224$

    (b) $4^{2-x} = 4^{2x+1} \Rightarrow \log_4 4^{2-x} = \log_4 4^{2x+1} \Rightarrow 2 - x = 2x + 1 \Rightarrow 3x = 1 \Rightarrow x = \dfrac{1}{3}$

5. (a) Looking at the graph of $y = 80 + 120(0.9)^x$ we see that eventually the values stay around 80. After a long time, the temperature of the object stays around 80°F

    (b) $80 + 120(0.9)^x = 100 \Rightarrow 120(0.9)^x = 20 \Rightarrow 0.9^x = \dfrac{1}{6} \Rightarrow \ln 0.9^x = \ln\dfrac{1}{6} \Rightarrow x\ln 0.9 = \ln\dfrac{1}{6} \Rightarrow$

    $x = \dfrac{\ln\frac{1}{6}}{\ln 0.9} \approx 17$; the object's temperature is 100°F after about 17 minutes.

5.7: Constructing Nonlinear Models

1. Growth slows down as x increases; logarithmic.

3. Growth rate increases as x increases; exponential.

5. Exponential; least-squares regression gives $f(x) = 1.2(1.7)^x$.

7. Logarithmic; least-squares regression gives $f(x) = 1.088 + 2.937\ln x$.

9. Logistic; least-squares regression gives $f(x) = \dfrac{9.96}{1 + 30.6e^{-1.51x}}$.

11. (a) Using the Logistic calculation, $C(x) = \dfrac{44.82}{1 + 973.66e^{-0.9686x}}$ thousand new AIDS cases per year after 1980.

    (b) See Figure 11.

    (c) $C(9) \approx 38.7$; or about 38,700 new cases.

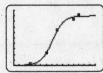
Figure 11

13. (a) Using the Quadratic Regression calculation, $R(x) = 105.732x^2 + 245.339x + 46.107$ \$ millions per year after 2015.

    (b) See Figure 13.

    (c) $R(7) \approx 6944$; or about \$6.944 billion.

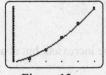
Figure 13

15. (a) The best fit model for the data set is a quadratic function. Using the regression function on your calculator yields $V(x) = 3.55x^2 + 0.21x + 1.9$.

   (b) Since x represents the number of years after 2007 we will let $x = 5$.

   $V(5) = 3.55(5)^2 + 0.21(5) + 1.9 = 91.7$ million

17. $a \approx 9.02$, $b \approx 1.03$ or $f(x) = 9.02 + 1.03 \ln x$.

19. (a) $a \approx 1.4734$, $b \approx 0.99986$, or $f(x) = 1.4734(0.99986)^x$.

   (b) $f(7000) = 1.4734(0.99986)^{7000} \approx 0.55$; approximately 0.55 kg/m^3.

21. (a) $f(x) = \dfrac{4.9955}{1 + 49.7081e^{-0.6998x}}$

   (b) For large x, $f(x) \approx \dfrac{4.9955}{1} \approx 5$; the density after a long time is about 5 thousand per acre.

23. (a) Using the logistic regression function on your calculator we have $C(x) = \dfrac{0.94}{1 + 17.97e^{-0.862x}}$.

   (b) Since x represents the number of years after 2005 we will let $x = 9$.

   $C(9) = \dfrac{0.94}{1 + 17.97e^{-0.862(9)}} \approx 0.93$ billion.

25. (a) Power: least-squares regression gives $A(w) \approx 101x^{0.662}$.

   (b) See Figure 25.

   (c) $500 = 101w^{0.662} \Rightarrow \dfrac{500}{101} = w^{0.662} \Rightarrow \left(\dfrac{500}{101}\right)^{1/0.662} = (w^{0.662})^{1/0.662} \Rightarrow w \approx 11.2$. lbs. A weight of about

   11.2 lbs. corresponds to a bird with a wing area of 500 in^2.

Figure 25

27. (a) From the table we see that $H(5) = 3$. After 5 years the tree is 3 feet tall.

   (b) Logistic: least-squares regressions gives $H(x) \approx \dfrac{50.1}{1 + 47.4e^{-0.221x}}$.

   (c) See Figure 27.

   (d) $25 = \dfrac{50.1}{1 + 47.4e^{-0.221x}} \Rightarrow \dfrac{50.1}{25} = 1 + 47.4e^{-0.221x} \Rightarrow \dfrac{50.1}{25} - 1 = 47.4e^{-0.221x} \Rightarrow$

   $\dfrac{\frac{50.1}{25} - 1}{47.4} = e^{-0.221x} \Rightarrow \ln\left(\dfrac{\frac{50.1}{25} - 1}{47.4}\right) = -0.221x \Rightarrow x = \dfrac{\ln\left(\dfrac{\frac{50.1}{25} - 1}{47.4}\right)}{-0.221} \approx 17.4$ years. After about 17.4 years the

   tree is 25 feet tall.

   (e) The answer involved interpolation.

29. (a) The data is not linear. See Figure 29.

   (b) $a \approx 12.42$, $b \approx 1.066$, or $f(x) = 12.42(1.066)^x$.

   (c) $f(39) = 12.42(1.066)^{39} \approx 150$; 150 kilograms per hectare. Chemical fertilizer use increased, but at a slower rate than predicted by f.

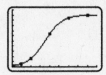

Figure 27   Figure 29

Extended and Discovery Exercises for Section 5.7

1. By taking a linear least-square regression of the logarithm of the distance values for x and the period values for y, the equation $y = 1.5x - 8.5$. This can be used to find the exponential form:

    $z = bw + d \Rightarrow z = 1.5w - 8.5 \Rightarrow b = 1.5,\ d = -8.5;\ a = e^d = e^{-8.5} \approx 0.0002 \Rightarrow y = ax^b = 0.0002x^{1.5}$

3. (a) A best fit for the exponential equation is found to be : $R(x) = 0.2e^{0.01214x}$

    (b) $3 = 0.2e^{0.01214x} \Rightarrow x = \ln\dfrac{15}{0.91214} + 1899 = 223 + 1800 = 2023$

Checking Basic Concepts for Section 5.7

1. Exponential; least-squares regression gives $f(x) \approx 0.5(1.2)^x$.

3. Logistic; least-squares regression gives $f(x) = \dfrac{4.5}{1 + 277e^{-1.4x}}$.

Chapter 5 Review Exercises

1. (a) $(f + g)(1) = f(1) + g(1) = 7 + 1 = 8$

    (a) $(f - g)(3) = f(3) - g(3) = 9 - 9 = 0$

    (a) $(fg)(-1) = f(-1)g(-1) = 3(-2) = -6$

    (a) $(f/g)(0) = \dfrac{f(0)}{g(0)} = \dfrac{5}{0}$; undefined.

3. (a) $f(x) = x^2 \Rightarrow f(3) = 9$ and $g(x) = 1 - x \Rightarrow g(3) = 1 - 3 = -2$. Thus,
        $(f + g)(3) = f(3) + g(3) = 9 + (-2) = 7$.

    (b) $f(-2) = 4$ and $g(-2) = 1 - (-2) = 3$. Thus, $(f - g)(-2) = f(-2) - g(-2) = 4 - 3 = 1$.

    (c) $f(1) = 1$ and $g(1) = 1 - 1 = 0$. Thus, $(fg)(1) = f(1)g(1) = (1)(0) = 0$.

    (d) $f(3) = 9$ and $g(3) = 1 - 3 = -2$. Thus, $(f/g)(3) = \dfrac{f(3)}{g(3)} = \dfrac{9}{-2} = -\dfrac{9}{2}$.

5. (a) $(g \circ f)(-2) = g(f(-2)) = g(1) = 2$

    (b) $(f \circ g)(3) = f(g(3)) = f(-2) = 1$

    (c) $f^{-1}(3) = 2$ since $f(2) = 3$

7. (a) $f(x) = \sqrt{x}$ and $g(x) = x^2 + x \Rightarrow (f \circ g)(2) = f(g(2)) = f(6) = \sqrt{6}$

(b) $(g \circ f)(9) = g(f(9)) = g(\sqrt{9}) = g(3) = 3^2 + 3 = 12$

9. $(f \circ g) = f(g(x)) = f\left(\dfrac{1}{x}\right) = \left(\dfrac{1}{x}\right)^3 - \left(\dfrac{1}{x}\right)^2 + 3\left(\dfrac{1}{x}\right) - 2$; $D = \{x \mid x \neq 0\}$

11. $(f \circ g) = f(g(x)) = f\left(\dfrac{1}{2}x^3 + \dfrac{1}{2}\right) = \sqrt[3]{2\left(\dfrac{1}{2}x^3 + \dfrac{1}{2}\right) - 1} = \sqrt[3]{x^3 + 1 - 1} = \sqrt[3]{x^3} = x$; $D =$ all real numbers

13. $h(x) = (g \circ f)(x) \Rightarrow h(x) = \sqrt{x^2 + 3} \Rightarrow f(x) = x^2 + 3$, $g(x) = \sqrt{x}$. *Answers may vary.*

15.   Subtract 6 from x and then multiply the results by 10. $\dfrac{x}{10} + 6$ and $10(x - 6)$.

17.   Since the graph of $f(x) = 3x - 1$ is a line sloping upward from left to right, a horizontal line can intersect it at most once. Therefore, f is one-to-one.

19. f is not one-to-one. It does not pass the Horizontal line test.

21.   See Figure 21. The domain of f is $D = \{-1, 0, 4, 6\}$ and its range is $R = \{1, 3, 4, 6\}$. The domain and range of f^{-1} are $D = \{1, 3, 4, 6\}$ and $R = \{-1, 0, 4, 6\}$, respectively.

x	6	4	3	1
$f^{-1}(x)$	−1	0	4	6

Figure 21

23.   The inverse operations are add 5 to x and divide by 3. Therefore, $f^{-1}(x) = \dfrac{x + 5}{3}$.

25. $(f \circ f^{-1})(x) = f(f^{-1}(x)) = f\left(\dfrac{x+1}{2}\right) = 2\left(\dfrac{x+1}{2}\right) - 1 = (x+1) - 1 = x$

      $(f^{-1} \circ f)(x) = f^{-1}(f(x)) = f^{-1}(2x - 1) = \dfrac{(2x-1)+1}{2} = \dfrac{2x}{2} = x$

27. $(f \circ g^{-1})(4) = f(g^{-1}(4)) = f(3) = 1$

29. $y = \sqrt{x+1} \Rightarrow y^2 = x + 1 \Rightarrow y^2 - 1 = x$; $\Rightarrow f^{-1}(x) = x^2 - 1$, $x \geq 0$ For $f : D = \{x \mid x \geq -1\}$ and $R = \{y \mid y \geq 0\}$; for $f^{-1} : D = \{x \mid x \geq 0\}$ and $R = \{y \mid y \geq -1\}$

31.   If $f(0) = 3 \Rightarrow Ca^x = 3 \Rightarrow Ca^0 = 3 \Rightarrow a^0 = 1$; $C(1) = 3 \Rightarrow C = 3$; then $f(3) = 24 \Rightarrow$ $Ca^x = 24 \Rightarrow 3a^3 = 24 \Rightarrow a^3 = 8 \Rightarrow a = 2$. Therefore $C = 3$ and $a = 2$.

33.   See Figure 33. $D =$ all real numbers.

35.   The graph of $f(x) = \log(-x)$ is similar to the graph of $f(x) = \log x$, except it is reflected through the y -axis. The domain is all real numbers less than 0. See Figure 35.

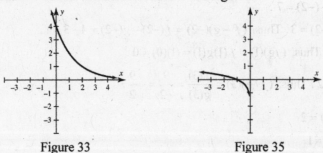

Figure 33   Figure 35

37. $y = Ca^x \Rightarrow y = 2(2)^x \Rightarrow C = 2; a = 2$

39. $A = 1200\left(1 + \dfrac{0.09}{2}\right)^{2(3)} \approx \$1562.71$

41. $e^x = 19 \Rightarrow \ln e^x = \ln 19 \Rightarrow x = \ln 19 \approx 2.9444$; the result can be supported graphically using the intersection method with the graphs $Y_1 = e^\wedge X$ and $Y_2 = 19$. The intersection point is near (2.9444, 19). The result may also be supported numerically with a table of $Y_1 = e^\wedge X$ starting at 2.9 and incrementing by 0.01. Here $Y_1 = 19$ when $x \approx 2.94$.

43. $\log 1000 = \log 10^3 = 3$

45. $10 \log 0.01 + \log \dfrac{1}{10} = 10 \log 10^{-2} + \log 10^{-1} = 10(-2) + (-1) = -21$

47. $\log_3 9 = \log_3 3^2 = 2$

49. $\ln e = \ln e^1 = 1$

51. $\log_3 18 = \dfrac{\log 18}{\log 3} \approx 2.631$

53. $10^x = 125 \Rightarrow \log 10^x = \log 125 \Rightarrow x = \log 125 \approx 2.097$

55. $e^{0.1x} = 5.2 \Rightarrow \ln e^{0.1x} = \ln 5.2 \Rightarrow 0.1x = \ln 5.2 \Rightarrow x = 10 \ln 5.2 \approx 16.49$

57. $5^{-x} = 10 \Rightarrow \log 5^{-x} = \log 10 \Rightarrow -x \log 5 = 1 \Rightarrow -x = \dfrac{1}{\log 5} \Rightarrow x = -\dfrac{1}{\log 5} \approx -1.431$

59. $50 - 3(0.78)^{x-10} = 21 \Rightarrow -3(0.78)^{x-10} = -29 \Rightarrow (0.78)^{x-10} = \dfrac{29}{3} \Rightarrow$

      $(x-10) \log (0.78) = \log \left(\dfrac{29}{3}\right) \Rightarrow x - 10 = \dfrac{\log \left(\frac{29}{3}\right)}{\log (0.78)} \Rightarrow x = 10 + \dfrac{\log \left(\frac{29}{3}\right)}{\log (0.78)} \approx 0.869$

61.   For each unit increase in x , the y -values are multiplied by 2, so the data are exponential. Since $y = 1.5$ when $x = 0$, the initial value is 1.5, so $f(x) = Ca^x \Rightarrow f(x) = 1.5a^x$. Since $y = 3$ when $x = 1$, $f(1) = 1.5a^1 \Rightarrow 3 = 1.5a \Rightarrow a = 2$. Thus the function $f(x) = 1.5(2^x)$ can model the data.

63. $\log x = 1.5 \Rightarrow 10^{\log x} = 10^{1.5} \Rightarrow x = 10^{1.5} \approx 31.62$

65. $\ln x = 3.4 \Rightarrow e^{\ln x} = e^{3.4} \Rightarrow x = e^{3.4} \approx 29.96$

67. $\log 6 + \log 5x = \log(6 \cdot 5x) = \log 30x$

69. $\ln \dfrac{y}{x^2} = \ln (y) - \ln (x^2) = \ln (y) - 2 \ln (x)$

71. $8 \log x = 2 \Rightarrow \log x = \dfrac{1}{4} \Rightarrow 10^{\log x} = 10^{1/4} \Rightarrow x = 10^{1/4} \Rightarrow x = \sqrt[4]{10} \approx 1.778$

73. $2 \log 3x + 5 = 15 \Rightarrow 2 \log 3x = 10 \Rightarrow \log 3x = 5 \Rightarrow 10^{\log 3x} = 10^5 \Rightarrow 3x = 100,000 \Rightarrow$

      $x = \dfrac{100,000}{3} \approx 33,333$

75. $2 \log_5 (x+2) = \log_5 (x+8) \Rightarrow \log_5 (x+2)^2 = \log_5 (x+8) \Rightarrow 10^{\log_5 (x+2)^2} = 10^{\log_5 (x+8)} \Rightarrow$

      $(x+2)^2 = (x+8) \Rightarrow x^2 + 4x + 4 = x + 8 \Rightarrow x^2 + 3x - 4 = 0 \Rightarrow (x+4)(x-1) = 0 \Rightarrow x = -4$ or $x = 1$; since $x = -4$ is undefined in $\log (x+2)$, the only solution is $x = 1$.

77. If b is the y-intercept, then the point $(0, b)$ is on the graph of f. Therefore, the point $(b, 0)$ is on the graph
     of f^{-1}. The point $(b, 0)$ lies on the x-axis and is the x-intercept of the graph of f^{-1}.

79. (a) $N(t) = N_0 e^{rt} \Rightarrow 6000 = 4000 e^{r(1)} \Rightarrow 1.5 = e^r \Rightarrow \ln 1.5 = \ln e^r \Rightarrow \ln 1.5 = r \Rightarrow r = 0.4055$

         Now $N(2.5) = 4000 e^{0.4055(2.5)} \Rightarrow N(2.5) \approx 11{,}022$

     (b) $8500 = 4000 e^{0.4055t} \Rightarrow 2.125 = e^{0.4055t} \Rightarrow \ln 2.125 = \ln e^{0.4055t} \Rightarrow \ln 2.125 = 0.4055t \Rightarrow$

         $\dfrac{\ln 2.125}{0.4055} = t \Rightarrow t \approx 1.86$ hours.

81. $h(x) = f(x) + g(x) = 10x + 5x = 15x$

83. Use the intersection method by graphing $Y_1 = 175.6(1 - 0.66e^{\wedge}(-0.24X))^{\wedge}3$ and $Y_2 = 50$ in the viewing
     rectangle [0, 14, 1] by [0, 175, 25]. The intersection is near the point (2.74, 50). See Figure 83. This means
     that at approximately 3 weeks the fish weighed 50 milligrams.
     Figure 83

85. $15 = 32 e^{-0.2t} \Rightarrow \dfrac{15}{32} = e^{-0.2t} \Rightarrow -0.2t = \ln\left(\dfrac{15}{32}\right) \Rightarrow t = \left(-\dfrac{1}{0.2}\right)\ln\left(\dfrac{15}{32}\right) \approx (-5)(-0.7577) \approx 3.8$; after about 3.8

     minutes.

87. Logistic: least-squares regression gives $f(x) \approx \dfrac{171.4}{1 + 18.4 e^{-0.0744x}}$

89. See Figure 89. $a \approx 3.50$, $b \approx 0.74$, or $f(x) = 3.50(0.74)^x$

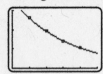

     Figure 89

Chapter 6: Systems of Equations and Inequalities

6.1: Functions and Systems of Equations in Two Variables

1. $A(5, 8) = \dfrac{1}{2}(5)(8) = 20$. The area of a triangle with a base of 5 and height of 8 is 20 square units.

3. $f(2, -3)$ if $f(x, y) = x^2 + y^2 \Rightarrow f(2, -3) = 2^2 + (-3)^2 = 4 + 9 = 13$

5. $f(-2, 3)$ if $f(x, y) = 3x - 4y \Rightarrow f(-2, 3) = 3(-2) - 4(3) = -6 - 12 = -18$

7. $f\left(\dfrac{1}{2}, -\dfrac{7}{4}\right)$ if $f(x, y) = \dfrac{2x}{y+3} \Rightarrow f\left(\dfrac{1}{2}, -\dfrac{7}{4}\right) = \dfrac{2(\frac{1}{2})}{(-\frac{7}{4})+3} = \dfrac{1}{\frac{5}{4}} = \dfrac{4}{5}$

9. The sum of y and twice x is computed by $f(x, y) = y + 2x$.

11. The product of x and y divided by $1 + x$ is computed by $f(x, y) = \dfrac{xy}{1+x}$.

13. $3x - 4y = 7 \Rightarrow 3x = 4y + 7 \Rightarrow x = \dfrac{4y+7}{3}$; $3x - 4y = 7 \Rightarrow -4y = -3x + 7 \Rightarrow y = \dfrac{3x-7}{4}$

15. $x - y^2 = 5 \Rightarrow x = y^2 + 5$; $x - y^2 = 5 \Rightarrow -y^2 = -x + 5 \Rightarrow y^2 = x - 5 \Rightarrow y = \pm\sqrt{x-5}$

17. $\dfrac{2x-y}{3y} = 1 \Rightarrow 2x - y = 3y \Rightarrow 2x = 4y \Rightarrow x = 2y$; $\dfrac{2x-y}{3y} = 1 \Rightarrow 2x - y = 3y \Rightarrow 2x = 4y \Rightarrow y = \dfrac{x}{2}$

19. The only ordered pair that satisfies both equations is $(2, 1)$. The system is linear.
    $2(2) + 1 = 5 \star$ $2(-2) + 1 = -3$ $2(1) + 0 = 2$
    $2 + 1 = 3 \star$ $-2 + 1 = -1$ $1 + 0 = 1$

21. The only ordered pair that satisfies both equations is $(4, -3)$. The system is non-linear.
    $4^2 + (-3)^2 = 25 \star$ $0^2 + 5^2 = 25 \star$ $4^2 + 3^2 = 25 \star$
    $2(4) + 3(-3) = -1 \star$ $2(0) + 3(5) = 15$ $2(4) + 3(3) = 17$

23. From the graph the solution is $(2, 2)$. The solution satisfies both $x - y = 0$ and $x + y = 4$.

25. From the graph the solution is $\left(\dfrac{1}{2}, -2\right)$. The solution satisfies both $6x + 4y = -5$ and $2x - 3y = 7$.

27. Since the lines intersect, the system is consistent with a unique solution at $(2, 2)$.
    $x + y = 4 \Rightarrow y = 4 - x$; We substitute $4 - x$ for y in the other equation.
    $2x - (4 - x) = 2 \Rightarrow 3x = 6 \Rightarrow x = 2$, then $2 + y = 4 \Rightarrow y = 2$. The solution is $(2, 2)$.

29. Since the lines are parallel, the system is inconsistent with no solutions.

31. The lines intersect at $(0, 1)$. The system is consistent and independent. See Figure 31.

33. The lines intersect at $(0, -2)$. The system is consistent and independent. See Figure 33.

35. Parallel lines $\Rightarrow$ inconsistent system. See Figure 35.

37. The lines intersect at $(2, -1)$. The system is consistent and independent. See Figure 37.

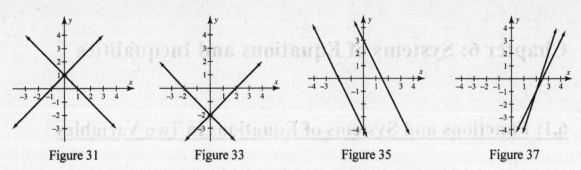

Figure 31   Figure 33   Figure 35   Figure 37

39. The lines intersect at $(-2, 2)$. The system is consistent and independent. See Figure 39.

41. The system has an infinite number of solutions. $\{(x, y)\,|\,2x - y = -4\}$. The system is consistent and dependent. See Figure 41.

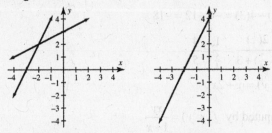

Figure 39   Figure 41

43. Solve the first equation for y: $x + y = 1 \Rightarrow y = 1 - x$. Substitute this in to the second equation.

$-2x - (1 - x) = 0 \Rightarrow -2x - 1 + x = 0 \Rightarrow -x - 1 = 0 \Rightarrow x = -1$.

If $x = -1$, then $y = 1 - (-1) = 2$. The solution is $(-1, 2)$.

Check: $(-1) + (2) = 1 \Rightarrow 1 = 1$; $-2(-1) - (2) = 0 \Rightarrow 2 - 2 = 0 \Rightarrow 0 = 0$.

45. Solve the first equation for x: $x + 2y = 0 \Rightarrow x = -2y$. Substitute this into the second equation.

$3x + 7y = 1 \Rightarrow 3(-2y) + 7y = 1 \Rightarrow -6y + 7y = 1 \Rightarrow y = 1$.

If $y = 1$, then $x = -2(1) = -2$. The solution is $(-2, 1)$.

Check: $x + 2y = 0 \Rightarrow (-2) + 2(1) = 0 \Rightarrow 0 = 0$; $x = -2y \Rightarrow -2 = -2(1) \Rightarrow -2 = -2$

47. Solve the first equation for x: $2x - 9y = -17 \Rightarrow 2x = 9y - 17 \Rightarrow x = \dfrac{9y - 17}{2}$. Substitute this into the second

equation. $8x + 5y = 14 \Rightarrow 8\left(\dfrac{9y - 17}{2}\right) + 5y = 14 \Rightarrow 4(9y - 17) + 5y = 14 \Rightarrow$

$36y - 68 + 5y = 14 \Rightarrow 41y = 82 \Rightarrow y = 2$. If $y = 2$, then $x = \dfrac{9(2) - 17}{2} = \dfrac{1}{2}$. The solution is $(0.5, 2)$.

Check: $2x - 9y = -17 \Rightarrow 2(0.5) - 9(2) = -17 \Rightarrow 1 - 18 = -17 \Rightarrow -17 = -17$;

$8x + 5y = 14 \Rightarrow 8(0.5) + 5(2) = 14 \Rightarrow 4 + 10 = 14 \Rightarrow 14 = 14$

49. Solve the second equation for x: $x + \dfrac{1}{2}y = 10 \Rightarrow x = -\dfrac{1}{2}y + 10$. Substitute this into the first equation.

$\dfrac{1}{2}x - y = -5 \Rightarrow \dfrac{1}{2}\left(-\dfrac{1}{2}y + 10\right) - y = -5 \Rightarrow -\dfrac{1}{4}y + 5 - y = -5 \Rightarrow -\dfrac{5}{4}y = -10 \Rightarrow y = 8$.

If $y = 8$, then $x = -\dfrac{1}{2}(8) + 10 = 6$. The solution is $(6, 8)$.

Check: $x + \frac{1}{2}y = 10 \Rightarrow 6 + \frac{1}{2}(8) = 10 \Rightarrow 6 + 4 = 10 \Rightarrow 10 = 10$

$\frac{1}{2}x - y = -5 \Rightarrow \frac{1}{2}(6) - 8 = -5 \Rightarrow 3 - 8 = -5 \Rightarrow -5 = -5$

51. Solve the first equation for y: $3x - 2y = 5 \Rightarrow -2y = -3x + 5 \Rightarrow y = \frac{3}{2}x - \frac{5}{2}$. Substitute this into the second

    equation. $-6x + 4\left(\frac{3}{2}x - \frac{5}{2}\right) = -10 \Rightarrow -6x + 6x - 10 = -10 \Rightarrow -10 = -10 \Rightarrow$ there are infinitely many solutions.

    $\{(x, y) | 3x - 2y = 5\}$.

53. Solve the first equation for x: $2x - 7y = 8 \Rightarrow 2x = 7y + 8 \Rightarrow x = \frac{7}{2}y + 4$. Substitute this into the second

    equation. $-3\left(\frac{7}{2}y + 4\right) + \frac{21}{2}y = 5 \Rightarrow -\frac{21}{2}y - 12 + \frac{21}{2}y = 5 \Rightarrow -12 = 5 \Rightarrow$ there are no real solutions.

55. $0.2x - 0.1y = 0.5 \Rightarrow 2x - y = 5$ and $0.4x + 0.3y = 2.5 \Rightarrow 4x + 3y = 25$

    Solve the first equation for y: $2x - y = 5 \Rightarrow y = 2x - 5$. Substitute this into the second equation.

    $4x + 3y = 25 \Rightarrow 4x + 3(2x - 5) = 25 \Rightarrow 4x + 6x - 15 = 25 \Rightarrow 10x = 40 \Rightarrow x = 4$.

    If $x = 4$, then $y = 2(4) - 5 \Rightarrow y = 3$. The solution is (4, 3).

57. Solve the second equation for y: $2x + y = 0 \Rightarrow y = -2x$. Substitute this into the first equation.

    $x^2 - y = 0 \Rightarrow x^2 - (-2x) = 0 \Rightarrow x^2 + 2x = 0 \Rightarrow x(x + 2) = 0 \Rightarrow x = 0$. or $x = -2$.

    When $x = 0$, $y = -2(0) = 0$ and when $x = -2$, $y = -2(-2) = 4$. The solutions are (0, 0) and $(-2, 4)$.

59. Solve the second equation for y: $x + y = 6 \Rightarrow y = 6 - x$. Substitute this into the first equation.

    $xy = 8 \Rightarrow x(6 - x) = 8 \Rightarrow 6x - x^2 = 8 \Rightarrow x^2 - 6x + 8 = 0 \Rightarrow (x - 4)(x - 2) = 0 \Rightarrow x = 4$ or $x = 2$.

    When $x = 4$, $y = 6 - 4 = 2$ and when $x = 2$, $y = 6 - 2 = 4$. The solutions are (4, 2) and (2, 4).

61. Substitute the second equation, $y = 2x$, into the first equation.

    $x^2 + y^2 = 20 \Rightarrow x^2 + (2x)^2 = 20 \Rightarrow x^2 + 4x^2 = 20 \Rightarrow 5x^2 = 20 \Rightarrow x^2 = 4 \Rightarrow x = \pm 2$.

    When $x = -2$, $y = 2(-2) = -4$ and when $x = 2$, $y = 2(2) = 4$. The solutions are $(-2, -4)$ and (2, 4).

63. Solve the second equation for x: $x - y = -2 \Rightarrow x = y - 2$. Substitute this into the first equation.

    $\sqrt{y - 2} - 2y = 0 \Rightarrow \sqrt{y - 2} = 2y \Rightarrow y - 2 = 4y^2 \Rightarrow 4y^2 - y + 2 = 0$. Using the quadratic formula to solve we

    get: $\frac{1 \pm \sqrt{1 - 4(4)(2)}}{2(4)} = \frac{1 \pm \sqrt{-31}}{8} \Rightarrow$ no real solutions.

65. Solve the first equation for y: $2x^2 - y = 5 \Rightarrow -y = -2x^2 + 5 \Rightarrow y = 2x^2 - 5$. Substitute this into the second

    equation. $-4x^2 + 2(2x^2 - 5) = -10 \Rightarrow -4x^2 + 4x^2 - 10 = -10 \Rightarrow -10 = -10 \Rightarrow$ there are infinitely many

    solutions, $\{(x, y) | 2x^2 - y = 5\}$.

67. Solve the second equation for y: $x^2 + y = 4 \Rightarrow y = 4 - x^2$. Substitute this into the first equation.

    $x^2 - y = 4 \Rightarrow x^2 - (4 - x^2) = 4 \Rightarrow 2x^2 = 8 \Rightarrow x^2 = 4 \Rightarrow x = \pm 2$.

    When $x = -2$, $y = 4 - (-2)^2 = 0$ and when $x = 2$, $y = 4 - (2)^2 = 0$.

    The solutions are $(-2, 0)$ and (2, 0).

69. Solve the second equation for y: $x - y = 0 \Rightarrow y = x$. Substitute this into the first equation.

    $x^3 - x = 3y \Rightarrow x^3 - x = 3x \Rightarrow x^3 - 4x = 0 \Rightarrow x(x + 2)(x - 2) = 0 \Rightarrow x = 0$, $x = -2$, or $x = 2$.

When $x = 0$, $y = 0$, when $x = -2$, $y = -2$, and when $x = 2$, $y = 2$.

The solutions are $(-2, -2)$, $(0, 0)$, and $(2, 2)$.

71. $x - y = 2$ and $2x + 2y = 38$. Multiply the second equation by $\dfrac{1}{2}$ and add to eliminate the y-variable.

$$\begin{array}{r} x - y = 2 \\ x + y = 19 \\ \hline 2x \quad\quad = 21 \end{array} \Rightarrow x = 10.5$$

Since, $x - y = 2$, $y = 8.5$. The solution is $(10.5, 8.5)$.

73. $x + y = 75$ and $4x + 7y = 456$. Multiply the first equation by 7 and subtract to eliminate the y-variable.

$$\begin{array}{r} 7x + 7y = 525 \\ 4x + 7y = 456 \\ \hline 3x \quad\quad = 69 \end{array} \Rightarrow x = 23$$

Since, $x + y = 75$, $y = 52$. The solution is $(23, 52)$.

75.   The given equations result in the following nonlinear system of equations.

$A(l, w) = 35 \Rightarrow lw = 35$ and $P(l, w) = 24 \Rightarrow 2l + 2w = 24$

Begin solving the second equation for l. $2l + 2w = 24 \Rightarrow l + w = 12 \Rightarrow l = 12 - w$. Substitute this into the first equation. $lw = 35 \Rightarrow (12 - w)w = 35 \Rightarrow 12w - w^2 = 35 \Rightarrow w^2 - 12w + 35 = 0$. This is a quadratic equation that can be solved by factoring, graphing, or the quadratic formula. The solutions to this quadratic are found using factoring. $w^2 - 12w + 35 = 0 \Rightarrow (w - 5)(w - 7) = 0 \Rightarrow w = 5$ or 7.

Since $l = 12 - w$, if $w = 5$, then $l = 7$, and if $w = 7$, then $l = 5$. If the length is greater than the width, the solution is $l = 7$ and $w = 5$. A rectangle with length 7 and width 5 has an area of 35 and a perimeter of 24.

77.   Add the two equations together to eliminate the y-variable.

$$\begin{array}{r} x + y = 20 \\ x - y = 8 \\ \hline 2x \quad\quad = 28 \end{array} \Rightarrow x = 14$$

Since $x + y = 20$, it follows that $y = 6$. The unique solution is $(14, 6)$. The system is consistent and independent. Graphical and numerical support are shown in Figures 77a and 77b, where $Y_1 = 20 - X$ and $Y_2 = X - 8$.

[0, 24, 4] by [0, 16, 4]

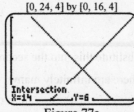
Figure 77a

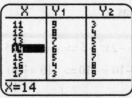
Figure 77b

79.   Subtract the two equations to eliminate the x-variable.

$$\begin{array}{r} x + 3y = 10 \\ x - 2y = -5 \\ \hline 5y = 15 \end{array} \Rightarrow y = 3$$

Since $x + 3y = 10$, it follows that $x = 1$. The solution is $(1, 3)$. The system is consistent and independent.

81.   Multiply the second equation by -1 and subtract to eliminate both variables.

$$x + y = 500$$
$$x + y = 500$$
$$\overline{0 = 0} \Rightarrow \text{infinite number of solutions}$$

The solution is $\{(x, y)|x + y = 500\}$. The system is consistent but dependent.

83. Multiply the second equation by 2 and add. This eliminates both variables.

$$2x + 4y = 7$$
$$-2x - 4y = 10$$
$$\overline{0 = 17} \Rightarrow \text{no solution}$$

Since $0 \neq 17$, the system is inconsistent.

85. Multiply the second equation by 2 and subtract to eliminate the x-variable.

$$2x + 3y = 2$$
$$2x - 4y = -10$$
$$\overline{7y = 12} \Rightarrow y = \frac{12}{7}$$

Since $2x + 3y = 2$, it follows that $x = \frac{2 - 3y}{2} \Rightarrow x = -\frac{11}{7}$. The solution is $\left(-\frac{11}{7}, \frac{12}{7}\right)$. The system is

consistent and independent.

87. Multiply the second equation by $-\frac{1}{2}$ and add to eliminate the x-variable.

$$\frac{1}{2}x - \phantom{\frac{1}{4}}y = 5$$
$$-\frac{1}{2}x + \frac{1}{4}y = -2$$
$$\overline{\phantom{-\frac{1}{2}x}-\frac{3}{4}y = 3} \Rightarrow y = -4$$

Then, $x - \frac{1}{2}y = 4 \Rightarrow x = 4 + \frac{1}{2}y \Rightarrow x = 2$. The solution is $(2, -4)$.

89. Multiply the first equation by 3 and add to eliminate both variables.

$$21x - 9y = -51$$
$$-21x + 9y = 51$$
$$\overline{0 = 0} \Rightarrow \text{infinite number of solutions}$$

There are infinitely many solutions of the form $\{(x, y)|7x - 3y = -17\}$.

91. Multiply the first equation by 3 and add to eliminate both variables.

$$2x + 4y = 1$$
$$-2x - 4y = 5$$
$$\overline{0 = 6} \Rightarrow \text{no solutions}$$

93. Clear decimals: $0.2x + 0.3y = 8 \Rightarrow 2x + 3y = 80$ and $-0.4x + 0.2y = 0 \Rightarrow -4x + 2y = 0$.

Multiply the first equation by 2 and add to eliminate the x-variable.

$$4x + 6y = 160$$
$$-4x + 2y = 0$$
$$\overline{8y = 160} \Rightarrow y = 20$$

Then, $-4x + 2y = 0 \Rightarrow 4x = 2y \Rightarrow x = \frac{1}{2}y \Rightarrow x = \frac{1}{2}(20) = 10$. The solution is $(10, 20)$.

95. Multiply the first equation by 3 and the second equation by 2. Add to eliminate the x-variable.

$$\begin{array}{r} 6x + 9y = 21 \\ -6x + 4y = -8 \\ \hline 13y = 13 \Rightarrow y = 1 \end{array}$$

Then, $2x + 3y = 7 \Rightarrow 2x + 3 = 7 \Rightarrow 2x = 4 \Rightarrow x = 2$. The solution is $(2, 1)$.

97. Multiply the first equation by 3, and the second equation by 5. Add to eliminate the y-variable.

$$\begin{array}{r} 21x - 15y = -45 \\ -10x + 15y = -10 \\ \hline 11x = -55 \Rightarrow x = -5 \end{array}$$

Then, $-10x + 15y = -10 \Rightarrow 50 + 15y = -10 \Rightarrow 15y = -60 \Rightarrow y = -4$. The solution is $(-5, -4)$.

99. Add the two equations:

$$\begin{array}{r} x^2 + y = 12 \\ x^2 - y = 6 \\ \hline 2x^2 = 18 \Rightarrow x^2 = 9 \Rightarrow x = \pm 3. \end{array}$$

If $x = 3$, then $3^2 + y = 12 \Rightarrow y = 3$, and if $x = -3$ then $(-3)^2 + y = 12 \Rightarrow y = 3$. Therefore the solutions are:
$(3, 3)$ and $(-3, 3)$.

101. Subtract the two equations:

$$\begin{array}{r} x^2 + y^2 = 25 \\ x^2 + 7y = 37 \\ \hline y^2 - 7y = -12 \Rightarrow y^2 - 7y + 12 = 0 \Rightarrow (y - 3)(y - 4) = 0 \Rightarrow y = 3, 4. \end{array}$$

If $y = 3$, then $x^2 + 7(3) = 37 x^2 = 16 \Rightarrow x = \pm 4$, and if $y = 4$ then $x^2 + 7(4) = 37 \Rightarrow x^2 = 9 \Rightarrow x = \pm 3$.
Therefore the solutions are: $(4, 3)$, $(-4, 3)$, $(3, 4)$, and $(-3, 4)$.

103. Subtract the two equations:

$$\begin{array}{r} x^2 + y^2 = 4 \\ 2x^2 + y^2 = 8 \\ \hline -x^2 = -4 \Rightarrow x^2 = 4 \Rightarrow x = \pm 2. \end{array}$$

If $x = -2$, then $(-2)^2 + y^2 = 4 \Rightarrow 4 + y^2 = 4 \Rightarrow y^2 = 0 \Rightarrow y = 0$, and if $x = 2$ then
$(2)^2 + y^2 = 4 \Rightarrow y^2 = 0 \Rightarrow y = 0$. Therefore the solutions are: $(-2, 0)$ and $(2, 0)$.

105. $x^2 + y^2 = 16 \Rightarrow y = \pm\sqrt{16 - x^2}$ and $x - y = 0 \Rightarrow y = x$

Graph $Y_1 = \sqrt{(16 - X^2)}$, $Y_2 = -\sqrt{(16 - X^2)}$ and $Y_3 = X$. Their graphs intersect near the points
$(-2.828, -2.828)$ and $(2.828, 2.828)$. See Figures 105a and 105b.

Substituting $y = x$ into the first equation gives $x^2 + x^2 = 16 \Rightarrow 2x^2 = 16 \Rightarrow x^2 = 8 \Rightarrow x = \pm\sqrt{8}$

Since $y = x$, the solutions are $(-\sqrt{8}, -\sqrt{8})$ and $(\sqrt{8}, \sqrt{8})$

107. $xy = 12 \Rightarrow y = \frac{12}{x}$ and $x - y = 4 \Rightarrow y = x - 4$. Graph $Y_1 = 12/X$, $Y_2 = X - 4$.

Their graphs intersect near the points $(6, 2)$ and $(-2, -6)$. See Figures 107a & 107b.

Substituting $y = x - 4$ into the first equation gives

$x(x-4) = 12 \Rightarrow x^2 - 4x - 12 = 0 \Rightarrow (x+2)(x-6) = 0 \Rightarrow x = -2$ or 6. Since $y = x - 4$, the solutions are $(-2, -6)$ and $(6, 2)$.

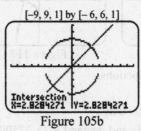
[−9, 9, 1] by [−6, 6, 1]
Figure 105a

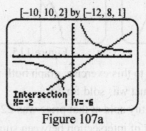
[−9, 9, 1] by [−6, 6, 1]
Figure 105b

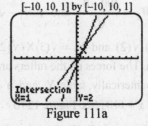
[−10, 10, 2] by [−12, 8, 1]
Figure 107a

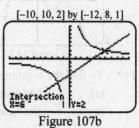
[−10, 10, 2] by [−12, 8, 1]
Figure 107b

109. (a) $2x + y = 1 \Rightarrow y = 1 - 2x$ and $x - 2y = 3 \Rightarrow y = \frac{1}{2}(x-3)$

Graph $Y_1 = 1 - 2X$ and $Y_2 = 0.5(X - 3)$. Their graphs intersect at the point $(1, -1)$, which is the solution. See Figure 109a.

(b)   Table $Y_1 = 1 - 2X$ and $Y_2 = 0.5(X - 3)$ starting at 0 and incrementing by 0.5. See Figure 109b. Here $Y_1 = Y_2 = -1$ when $x = 1$. The solution is $(1, -1)$.

(c)   Substituting $y = 1 - 2x$ into the equation $x - 2y = 3$ gives

$x - 2(1 - 2x) = 3 \Rightarrow x - 2 + 4x = 3 \Rightarrow 5x = 5 \Rightarrow x = 1$. If $x = 1$, then $y = 1 - 2(1) = -1$. The solution is $(1, -1)$.

111. (a) $-2x + y = 0 \Rightarrow y = 2x$ and $7x - 2y = 3 \Rightarrow 7x - 3 = 2y \Rightarrow y = \frac{7x - 3}{2}$

Graph $Y_1 = 2X$ and $Y_2 = (7X - 3)/2$. Their graphs intersect at the point $(1, 2)$, which is the solution. See Figure 111a.

(b)   Table $Y_1 = 2X$ and $Y_2 = (7X - 3)/2$ starting at 0 and incrementing by 0.5. See Figure 111b.

Here $Y_1 = Y_2 = 2$ when $x = 1$. The solution is $(1, 2)$.

(c)   Substituting $y = 2x$ into the second equation gives $7x - 2(2x) = 3 \Rightarrow 3x = 3 \Rightarrow x = 1$.

If $x = 1$, then $y = 2(1) = 2$. The solution is $(1, 2)$.

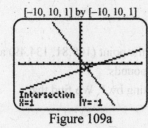
[−10, 10, 1] by [−10, 10, 1]
Figure 109a

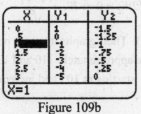
Figure 109b

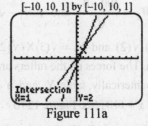
[−10, 10, 1] by [−10, 10, 1]
Figure 111a

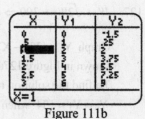
Figure 111b

113. $x^3 - 3x + y = 1 \Rightarrow y = 1 + 3x - x^3$ and $x^2 + 2y = 3 \Rightarrow y = \frac{3 - x^2}{2}$. Graph $Y_1 = 1 + 3X - X^3$ and

$Y_2 = (3 - X^2)/2$. See Figure 113. There are three points of intersection. The coordinates of these points are near $(-1.588, 0.239)$, $(0.164, 1.487)$, and $(1.924, -0.351)$.

115. $2x^3 - x^2 = 5y \Rightarrow y = \frac{2x^3 - x^2}{5}$ and $2^{-x} - y = 0 \Rightarrow y = 2^{-x}$. Graph $Y_1 = (2X^3 - X^2)/5$ and $Y_2 = 2^{\wedge}(-X)$.

See Figure 115. There is one point of intersection. The coordinates of this point are near $(1.220, 0.429)$.

117. $e^{2x} + y = 4 \Rightarrow y = 4 - e^{2x}$ and $\ln x - 2y = 0 \Rightarrow y = \frac{\ln x}{2}$. Graph $Y_1 = 4 - e^{\wedge}(2X)$ and $Y_2 = \ln(X)/2$. See

Figure 117. There is one point of intersection. The coordinates of this point are near $(0.714, -0.169)$.

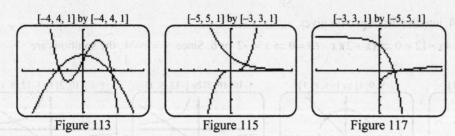

Figure 113   Figure 115   Figure 117

119. To find the solutions to this exercise, graph both functions.

    (a)   The product was sold for \$5.

    (b)   The quantity sold at that price was 22500 units.

    (c)   The point of intersection between supply and demand is $(5, 22500)$.

    (d)   If the product was sold for less than \$5, the demand would be greater than the supply.

121. (a)   The fixed costs are given by the constant, \$12000.

    (b)   The cost to make one unit is given by the slope, \$10.

    (c)   The breakeven point is the point of intersection, given by

$$C = R \Rightarrow 10x + 12000 = 20x \Rightarrow 12000 = 10x \Rightarrow 1200 = x. \quad \text{So} \quad C = 10(1200) + 12000 = 24000$$

and $R = 20(1200) = 24000$. The breakeven point is $(1200, 24000)$ where $x = 1200$ is the number

of items it takes for the cost to equal the revenue, which occurs at \$24000.

123. (a) $C(x) = 2.50x + 1000; \quad R(x) = 7.50x$

    (b)   The intersection of the functions above occurs at $(200, 1500)$. Therefore, the company needs to

sell 200 items to break even.

125. (a)   Let $x = $ population of Minneapolis and $y = $ population of St. Paul. Then $x + y = 695$ and $x - y = 105$.

$$x + y = 695$$

    (b) $\dfrac{x - y = 105}{2x \quad\quad = 800} \Rightarrow x = 400.$

Then, $x - y = 105 \Rightarrow 400 - y = 105 \Rightarrow y = 295$. The solution is $(400, 295)$.

    (c)   The system is consistent and independent.

127. $W_1 + \sqrt{2} W_2 = 300 \Rightarrow W_2 = \dfrac{300 - W_1}{\sqrt{2}}$ and $\sqrt{3} W_1 - \sqrt{2} W_2 = 0 \Rightarrow W_2 = \dfrac{\sqrt{3} W_1}{\sqrt{2}}$

Graph $Y_1 = (300 - X)/\sqrt{(2)}$ and $Y_2 = \sqrt{(3)}X/\sqrt{(2)}$. Their graphs intersect near the point $(109.81, 134.49)$ as shown in Figure 127a. The forces on the rafters are approximately 110 and 134 pounds.

To find the solution numerically, table Y_1 and Y_2 starting at 107 and incrementing by 1. We find that $Y_1 = Y_2 \approx 134$ when $x \approx 110$. See Figure 127b.

We can find the solution symbolically by using the substitution method. Since $W_1 + \sqrt{2} W_2 = 300 \Rightarrow$

$W_1 = 300 - \sqrt{2} W_2$, we will substitute into the other equation. $\sqrt{3}(300 - \sqrt{2} W_2) - \sqrt{2} W_2 = 0 \Rightarrow$

$300\sqrt{3} - \sqrt{6} W_2 - \sqrt{2} W_2 = 0 \Rightarrow 300\sqrt{3} = (\sqrt{6} + \sqrt{2}) W_2 \Rightarrow W_2 = \dfrac{300\sqrt{3}}{\sqrt{6} + \sqrt{2}}$ and

$W_1 = 300 - \sqrt{2} \left[\dfrac{300\sqrt{3}}{\sqrt{6} + \sqrt{2}} \right] \Rightarrow W_1 = 300 - \dfrac{300\sqrt{3}}{1 + \sqrt{3}} \Rightarrow W_1 = \dfrac{300}{1 + \sqrt{3}}$

$W_1 = \dfrac{300}{1 + \sqrt{3}} \approx 109.8 \text{ lbs}, \quad W_2 = \dfrac{300\sqrt{3}}{\sqrt{6} + \sqrt{2}} \approx 134.5 \text{ lbs}$

129. We must solve the system of nonlinear equations: $\pi r^2 h = 50$ and $2\pi r h = 65$.

Solving each equation for h results in the following: $\pi r^2 h = 50 \Rightarrow h = \dfrac{50}{\pi r^2}$ and $2\pi rh = 65 \Rightarrow h = \dfrac{65}{2\pi r}$.

Graph $Y_1 = 50/(\pi X^{\wedge}2)$ and $Y_2 = 65/(2\pi X)$. Their graphs intersect near $(1.538, 6.724)$. See Figure 129.

A cylinder with approximate measurements of $r \approx 1.538$ inches and $h \approx 6.724$ inches has a volume of 50 cubic inches and a lateral surface area of 65 square inches.

[0, 200, 50] by [0, 200, 50]

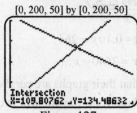

Figure 127a

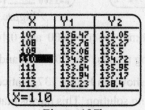

Figure 127b

[0, 4, 1] by [0, 20, 1]

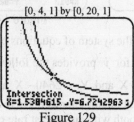

Figure 129

131. Let $x =$ the length of each side of the base and let $y =$ the height of the box. Since the volume $= 576$ in^3,

$x^2 y = 576$, and so $y = \dfrac{576}{x^2}$. Since the surface are is 336 in^2, $x^2 + 4xy = 336$. Substituting for y in this

equation yields $x^2 + 4x \cdot \dfrac{576}{x^2} = 336$. Simplifying we get: $x^3 - 336x + 2304 = 0$. By graphing the left side of

the equation, the x-intercepts are approximately 9.1 and 12. See Figure 131. When $x \approx 9.1$, the dimensions

are: 9.1 by 9.1 by $\dfrac{576}{(9.1)^2} \approx 6.96$ inches. When

$x = 12$, the dimensions are: 12 by 12 by $\dfrac{576}{(12)^2} = 4$ inches.

133. (a) Let x represent the number of global bank thefts in 2014 and let y represent the number of global bank thefts in 2015. Then $x + y = 1739$ and $x - y = 95$.

$x + y = 1739$

(b) $\dfrac{x - y = 95}{2x \quad\ = 1834} \Rightarrow x = 917$.

Then, $x - y = 95 \Rightarrow 917 - y = 95 \Rightarrow y = 822$. The solution is $(917, 822)$.

(c) Solve each equation for y and graph $Y_1 = 1739 - x$ and $Y_2 = x - 95$ as shown in Figure 133. The solution is the intersection point $(917, 822)$.

135. (a) To solve this problem start by letting x represent the amount of the 8% loan and y the 10% loan. Since the total of both loans is 3000, the equation $x + y = 3000$ must be satisfied. The annual interest rate for the 8% loan is given by $0.08x$, while the annual interest rate for the 10% loan is expressed by $0.10y$. the total interest for both loans is their sum $0.08x + 0.10y = 264$.

(b) Thus, to determine a solution, the following linear system of equations could be solved.

$x + y = 3000$ and $0.08x + 0.10y = 264$. Start by solving each for y.

$x + y = 3000 \Rightarrow y = 3000 - x$ and

$0.08x + 0.10y = 264 \Rightarrow 0.10y = 264 - 0.08x \Rightarrow y = \dfrac{264 - 0.08x}{0.10} \Rightarrow y = 2640 - 0.8x$. The two equations

can be solved using the intersection-of-graphs method. Let $Y_1 = 3000 - X$ and $Y_2 = 2640 - 0.8X$.

Since the system of equations is linear, each graph is a line. The lines are not parallel and intersect at the point $(1800, 1200)$, as shown in Figure 135. Thus, the 8% loan is for 1800 and the 10% loan is for 1200.

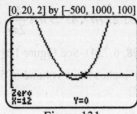
[0, 20, 2] by [-500, 1000, 100]

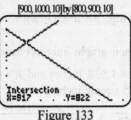
[900, 1000, 10] by [800, 900, 10]

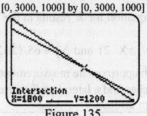
[0, 3000, 1000] by [0, 3000, 1000]

| Figure 131 | Figure 133 | Figure 135 |

137. With these conditions the system of equations becomes $x + y = 3000$ and $0.10x + 0.10y = 264$.

Solving each equation for y provides the following results, $y = 3000 - x$ and $y = 2640 - x$.

Graphs of $Y_1 = 3000 - X$ and $Y_2 = 2640 - X$ are shown in Figure 137. Notice that their graphs are parallel lines with slope -1 that do not intersect. There is no solution. This means that there is no way to have two loans totaling $3000, both with an interest rate of 10%, and only pay $264 in interest each year. The interest must be 10% of $3000 or $300. This system of equations is inconsistent - there is no solution. Graphs of inconsistent systems in two variables consist of parallel lines.

139. Let x represent the air speed of the plane and y the wind speed. Traveling with the wind, the average ground speed of the plane is $\dfrac{1680}{3} = 560$ mph, while its ground speed against the wind was $\dfrac{1680}{3.5} = 480$ mph. Thus,

$$\begin{array}{r} x + y = 560 \\ x - y = 480 \\ \hline 2x \quad\quad = 1040 \Rightarrow x = 520 \end{array}$$

Thus, $y = 560 - x = 40$. The air speed of the plane is 520 mph and the wind speed is 40 mph.

141. (a) The perimeter is 40, thus $2l + 2w = 40 \Rightarrow l + w = 20 \Rightarrow l = 20 - w$. Since the area is 91, we have that $lw = 91$. Use the substitution method to solve this system.

$(20 - w)w = 91 \Rightarrow 20w - w^2 = 91 \Rightarrow w^2 - 20w + 91 = 0 \Rightarrow (w - 13)(w - 7) = 0 \Rightarrow$

$w = 7$ or $w = 13$. When $w = 7$, $l = 13$ and when $w = 13$, $l = 7$. Since length is longer than width, the solution is $l = 13$ feet and $w = 7$ feet.

(b) $P = 2l + 2w = 40 \Rightarrow l + w = 20 \Rightarrow l = 20 - w$. Then, $A = lw = (20 - w)w = 20w - w^2$.

Graph $Y_1 = 20X - X\text{^}2$ in [0,25,5] by [0,150,25].

(c) The area can be any positive number less than or equal to 100 square feet. The maximum area of 100 square feet occurs when $w = 10$ and $l = 10$. See Figure 141. Since all sides are equal to 10, the shape of the rectangle is a square. That is, a square pen will provide the largest area.

143. (a) A 6' 11". person is 83 inches tall.

$w = 7.46(83) - 374 = 245.18 \approx 245$ lbs; $w = 7.93(83) - 405 = 253.19 \approx 253$ lbs.

(b) Graph $Y_1 = 7.46X - 374$ and $Y_2 = 7.93X - 405$. Their graphs intersect near (65.96, 118.04).

See Figure 143. The models agree when $h \approx 65.96$ inches and $w \approx 118$ pounds.

(c) The first model's coefficient for h is 7.46. Thus, for each increase in height of 1 inch, the weight increases by 7.46 lbs. Similarly, for the second equation the increase is 7.93 lbs.

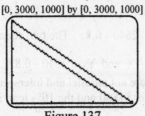
[0, 3000, 1000] by [0, 3000, 1000]

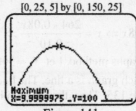
[0, 25, 5] by [0, 150, 25]

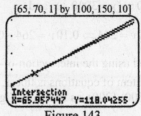
[65, 70, 1] by [100, 150, 10]

| Figure 137 | Figure 141 | Figure 143 |

145. Substitute 165.1 for h and 70 for w: $S(70,165.1) = 0.007184(70)^{0.425}(165.1)^{0.725} \approx 1.77 \text{m}^2$

147. $w = 132 \text{ lb} \approx \dfrac{132}{2.2} \text{ kg} = 60 \text{ kg}$; $h = 62 \text{ inches} \approx 62 \cdot 54 \text{ cm} = 157.48 \text{ cm}$.

$$S(w, h) = 0.007184(w^{0.425})(h^{0.725}) \Rightarrow S(60, 157.48) = 0.007184(60^{0.425})(157.48^{0.725}) \approx 1.6 \text{ m}^2$$

149. Since $z = kx^2y^3$ and $z = 31.9$ when $x = 2$ and $y = 2.5$.

$$31.9 = k(2)^2(2.5)^3 \Rightarrow 31.9 = 62.5k \Rightarrow k = \dfrac{31.9}{625} \approx 0.51$$

151. $z = k\sqrt{x} \cdot \sqrt[3]{y} \Rightarrow 10.8 = k\sqrt{4} \cdot \sqrt[3]{8} \Rightarrow 10.8 = k \cdot 2 \cdot 2 \Rightarrow 10.8 = 4k \Rightarrow k = 2.7$

Therefore: $z = 2.7\sqrt{x} \cdot \sqrt[3]{y}$ and now $z = 2.7\sqrt{16} \cdot \sqrt[3]{27} \Rightarrow z = 2.7(4)(3) \Rightarrow z = 32.4$.

153. Let d represent the diameter of the blades, v represent the wind velocity, and w represent the watts of power generated by the windmill. Then $w = kd^2v^3$ and $w = 2405$ when $d = 8$ and $v = 10$.

$$2405 = k(8)^2(10)^3 \Rightarrow 2405 = 64,000k \Rightarrow k = \dfrac{481}{12,800}. \text{ The variation equation becomes } w = \dfrac{481}{12,800}d^2v^3.$$

Thus, when $d = 6$ and $v = 20$; $w = \dfrac{481}{12,800}(6)^2(20)^3 = 10,822.5$.

With six-foot blades and a 20 mile-per-hour wind, the windmill will produce about 10,823 watts.

155. From the example $V = 0.00132h^{1.12}d^{1.98}$. When $h = 105$ and $d = 38$ we have $V = 0.00132(105)^{1.12}(38)^{1.98} \approx 325.295$. A tree which is 105 feet tall with a diameter of 38 inches contains approximately 325.295 cubic feet of wood. To find the number of cords, divide this result by 128:

$\dfrac{325.295}{128} \approx 2.54 \text{ cords}$.

157. From exercise 133, $S(w, h) = 0.007184(w^{0.425})(h^{0.725})$ where w is weight in kilograms and h is height in centimeters. Let $S = 1.77$, $w = 154$, and $h = 65$. Then, $1.77 = k(154^{0.425})(65^{0.725})$, which results in $k \approx 0.0101$. Thus, $S(w, h) = 0.0101(w^{0.425})(h^{0.725})$, where w is in pounds, h is in inches, and S is in square meters.

6.2: Systems of Inequalities in Two Variables

1. Graph the boundary line $y = x + 1$. Choose $(0, 2)$ as a test point. Since $2 \geq 1$, the region containing $(0, 2)$ is part of the solution set. The inequality $y \geq x + 1$ includes this region and the line $y = x + 1$. See Figure 1.

3. Graph the boundary line $x = y + 2 \Rightarrow y = x - 2$. Choose $(4, 0)$ as a test point. Since $4 \geq 2$, the region containing $(4, 0)$ is part of the solution set. The inequality $x \geq y + 2$ includes this region and the line $y = x - 2$. See Figure 3.

5. Graph the boundary line $y = x$. Choose $(1, 0)$ as a test point. Since $1 \geq 0$, the region containing $(1, 0)$ is part of the solution set. The inequality $x \geq y$ includes this region and the line $y = x$. See Figure 5.

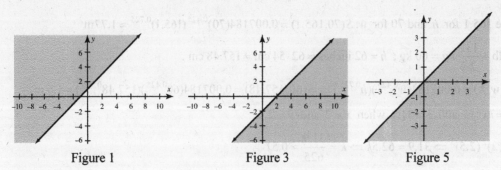

Figure 1   Figure 3   Figure 5

7. Graph the boundary line $x = 1$. Choose $(0, 0)$ as a test point. Since $0 < 1$, the region containing $(0, 0)$ is the solution set. The boundary line $x = 1$ is not part of the solution set. See Figure 7.

9. Graph the boundary line $x + y = 2 \Rightarrow y = 2 - x$. Choose $(0, 0)$ as a test point. Since $0 + 0 \le 2$, the region containing $(0, 0)$ is part of the solution set. The boundary line $y = 2 - x$ is also part of the solution set. See Figure 9.

11. Graph the boundary line $2x + y = 4 \Rightarrow y = 4 - 2x$. Choose $(3, 0)$ as a test point. Since $0 > 4 - 2(3)$, the region containing $(3, 0)$ is the solution set. The boundary line $y = 4 - 2x$ is not part of the solution set. See Figure 11.

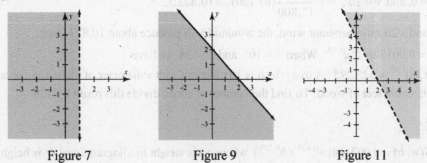

Figure 7   Figure 9   Figure 11

13. Graph the circle determined by $x^2 + y^2 = 4$. Choose $(0, 0)$ as a test point. Since $0 > 4$, the region inside the circle is not part of the solution $\Rightarrow$ shade the region outside the circle. Note, the circle is not part of the solution. See Figure 13.

15. Graph the boundary parabola determined by $x^2 + y = 2 \Rightarrow y = -x^2 + 2$. Choose $(0, 0)$ as a test point. Since $0 \le 2$, the region inside the parabola is part of the solution $\Rightarrow$ shade the region inside of the parabola. Note, the parabola is part of the solution. See Figure 15.

17. $x + y \ge 2 \Rightarrow y \ge 2 - x$ is above the line $y = 2 - x$. It includes the line.

    $x - y \le 1 \Rightarrow y \ge x - 1$ is above the line $y = x - 1$. It includes the line.

    The solution is above two lines, which matches Figure c. One solution is $(2, 3)$. *Answers may vary.*

19. $\frac{1}{2}x^3 - y > 0 \Rightarrow y < \frac{1}{2}x^3$ is below the curve $y = \frac{1}{2}x^3$. It does not include the curve.

    $2x - y \le 1 \Rightarrow y \ge 2x - 1$ is above the line $y = 2x - 1$. It includes the line.

    The solution is below a dotted curve and above a solid line, which matches Figure d. One solution is $(-1, -1)$. *Answers may vary.*

21. The solution region is above the line $y = 2x$ including the boundary, and below the line $y = 3x$, not including the boundary. See Figure 21. One solution is $(2, 5)$. *Answers may vary.*

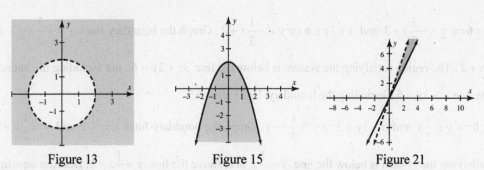

Figure 13   Figure 15   Figure 21

23. The solution region is below the line $y = 1 - 2x$ not including the boundary, and below the line $y = 1 - x$, not including the boundary. See Figure 23. One solution is $(0, 0)$. *Answers may vary.*

25. The solution region is above the parabola $y = x^2$ and below the line $y = 6 - x$. It includes the boundary. See Figure 25. One solution is $(0, 2)$. *Answers may vary.*

27. The solution region lies between the parallel lines $y = -\frac{1}{2}x - 1$ and $y = -\frac{1}{2}x + \frac{5}{2}$. It does not include the boundary. See Figure 27. One solution is $(0, 0)$. *Answers may vary.*

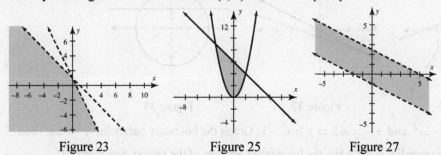

Figure 23   Figure 25   Figure 27

29. The solution region lies inside the circle centered at the origin with radius 4 and below the line $y = -x + 2$. It does not include the boundary determined by the line. See Figure 29. One solution is $(-1, 1)$. *Answers may vary.*

31. The solution region lies inside the circle centered at the origin with radius 3 and above the parabola $y = 2 - x^2$. It does not include the parabola as a boundary line. See Figure 31. One solution is $(2, 1)$. *Answers may vary.*

33. $x + 2y \le 4 \Rightarrow y \le -\frac{1}{2}x + 2$ and $2x - y \ge 6 \Rightarrow y \le 2x - 6$. Graph the boundary lines $y = -\frac{1}{2}x + 2$ and $y = 2x - 6$. The region satisfying the system is below the line $x + 2y = 4$ and below the line $2x - y = 6$. Because equality is included, the boundaries are part of the region. See Figure 33.

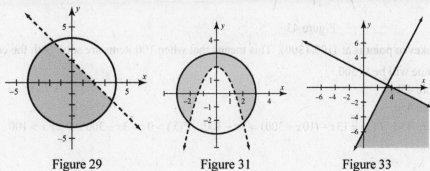

Figure 29   Figure 31   Figure 33

35. $3x + 2y < 6 \Rightarrow y < -\dfrac{3}{2}x + 3$ and $x + 3y \le 6 \Rightarrow y \le -\dfrac{1}{3}x + 2$. Graph the boundary lines $y = -\dfrac{3}{2}x + 3$ and

$y = -\dfrac{1}{3}x + 2$. The region satisfying the system is below the line $3x + 2y = 6$, not including the boundary and

below the line $x + 3y = 6$, including the boundary. See Figure 35.

37. $x - 2y \ge 0 \Rightarrow y \le \dfrac{1}{2}x$ and $x - 3y \le 3 \Rightarrow y \ge \dfrac{1}{3}x - 1$. Graph the boundary lines $y = \dfrac{1}{2}x$ and $y = \dfrac{1}{3}x - 1$. The

region satisfying the system is below the line $y = \dfrac{1}{2}x$ and above the line $y = \dfrac{1}{3}x - 1$. Because equality is

included, the boundaries are part of the region. See Figure 37.

39. $x^2 + y^2 \le 4$ and $y \ge 1$. Graph the boundary circle $x^2 + y^2 = 4$ and the boundary line $y = 1$. The region

satisfying the system is inside the circle $x^2 + y^2 = 4$ and above the line $y = 1$. Because equality is included,

the boundaries are part of the region. See Figure 39.

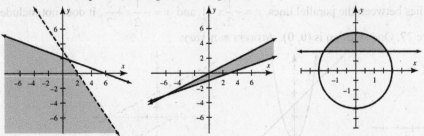

Figure 35   Figure 37   Figure 39

41. $2x^2 + y \le 0 \Rightarrow y \le -2x^2$ and $x^2 - y \le 3 \Rightarrow y \ge x^2 - 3$. Graph the boundary parabolas $y = -2x^2$ and

$y = x^2 - 3$. Because equality is included, the boundaries are part of the region. See Figure 41.

43. $x^2 + 2y \le 2 \Rightarrow y \le 1 - \dfrac{x^2}{2}$ and $x^2 + y^2 \le 4$. Graph the boundary parabola $y = 1 - \dfrac{x^2}{2}$ and the boundary circle

$x^2 + y^2 = 4$. Because equality is included, the boundaries are part of the region. See Figure 43.

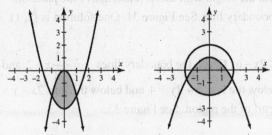

Figure 41   Figure 43

45.   (a)   The breakeven point is at $(100, 1300)$. This means that when 100 items are sold, both the cost and
                  the revenue will be $1300.

      (b)   See Figure 45b.

      (c)   See Figure 45c.

      (d)   See Figure 45d. $P(x) = 13x - (10x + 300) = 3x - 300$; $P(x) > 0 \Rightarrow 3x - 300 > 0 \Rightarrow x > 100$

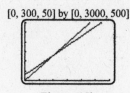

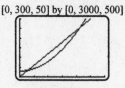
Figure 45b

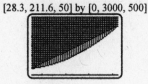
Figure 45c

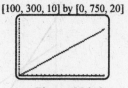
Figure 45d

47. (a)   The breakeven points are at $(28.35, 340.18)$ and $(211.65, 2539.82)$.

     (b)   See Figure 47b.

     (c)   See Figure 47c.

     (d) $P(x) = 12x - (0.05x^2 + 300) = -0.05x^2 + 12x - 300;$

     (e)   To maximize profit, 120 units must be sold. The maximum profit is $420.

[0, 300, 50] by [0, 3000, 500]   [28.3, 211.6, 50] by [0, 3000, 500]

Figure 47b   Figure 47c

49. The person will probably experience jitters.

51. The selected person could weigh 180 pounds or more.

53. The total number of vehicles entering intersection A is $500 + 150 = 650$ vehicles per hour. The expression $x + y$ represents the number of vehicles leaving intersection A each hour. Therefore, we have $x + y = 650$.

The total number of vehicles leaving intersection B is $50 + 400 = 450$. There are 100 vehicles entering intersection B from the south and y vehicles entering intersection B from the west. Thus, $y + 100 = 450$. We must solve the system;

$$
\begin{aligned}
x + y &= 650 \\
y + 100 &= 450 \\
\hline
x \qquad -100 &= 200 \quad \Rightarrow x = 300
\end{aligned}
$$

Thus, $y = 350$ and $x = 300$. At intersection A, a stoplight should allow for 300 vehicles per hour to travel south and 350 vehicles per hour to continue traveling east.

55. This region corresponds to weights that are less and heights that are greater than recommended. This individual has weight that is less than recommended for his or her height.

57. The upper left boundary is given by $25h - 7w = 800$ or $h = \dfrac{7w + 800}{25}$. The region is below and includes this line, which is described by $h \leq \dfrac{7w + 800}{25}$ or $25h - 7w \leq 800$. The lower right boundary of this region is given by $5h - w = 170$. The region is above and includes this line, which is described by $5h - w \geq 170$. Thus, the region can be described by the system of inequalities: $25h - 7w \leq 800$, $5h - w \geq 170$.

59. See Figure 59.

61. See Figure 61.

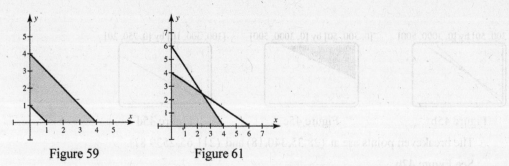

Figure 59   Figure 61

63. To find the maximum and minimum values of P in the region, we must evaluate P at each of the vertices. These values are shown in the table. See Figure 63. From the table the maximum is 65 and the minimum is 8.

65. To find the maximum and minimum values of C in the region, we must evaluate C at each of the vertices. These values are shown in the table. See Figure 65. From the table the maximum is 66 and the minimum is 3.

Vertex	$P = 3x + 5y$
(1, 1)	$3(1) + 5(1) = 8$
(6, 3)	$3(6) + 5(3) = 33$
(5, 10)	$3(5) + 5(10) = 65$
(2, 7)	$3(2) + 5(7) = 41$

Figure 63

Vertex	$C = 3x + 5y$
(1, 0)	$3(1) + 5(0) = 3$
(7, 6)	$3(7) + 5(6) = 51$
(7, 9)	$3(7) + 5(9) = 66$
(1, 10)	$3(1) + 5(10) = 53$

Figure 65

67. To find the maximum and minimum values of C in the region, we must evaluate C at each of the vertices. These values are shown in the table. See Figure 67. From the table the maximum is 100 and the minimum is 0.

Vertex	$C = 10y$
(1, 0)	$10(0) = 0$
(7, 6)	$10(6) = 60$
(7, 9)	$10(9) = 90$
(1, 10)	$10(10) = 100$

Figure 67

69. The line that goes through the points (0, 4) and (4, 0) has the equation $x + y = 4$. The shaded region is also bounded by the line $x = 0$ and the line $y = 0$. Thus, the shaded region is described by the system: $x + y \leq 4$, $x \geq 0$, and $y \geq 0$.

71. The region of feasible solutions is shown in Figure 71a. The vertices of this region are (3, 0), (6, 0), (0, 4), and (0, 3). To find the minimum value of C in the region, we must evaluate C. at each of the vertices. These values are shown in Figure 71b. From the table the minimum is 6 at the point (0, 3).

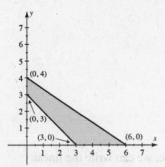

Figure 71a

Vertex	$C = 4x + 2y$
(3, 0)	$4(3) + 2(0) = 12$
(6, 0)	$4(6) + 2(0) = 24$
(0, 4)	$4(0) + 2(4) = 8$
(0, 3)	$4(0) + 2(3) = 6$

Figure 71b

73. The region of the feasible solutions is shown in Figure 73a. The vertices of this region are (0, 4), (0, 8), (4, 0), and (8, 0). To find the maximum and minimum values of $z = 7x + 6y$ in the region, we must evaluate z at each of the vertices. These values are shown in Figure 73b. From the table the maximum is 56 and the minimum is 24.

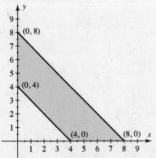

Figure 73a

Vertex	$z = 7x + 6y$
(0, 4)	$7(0) + 6(4) = 24$
(0, 8)	$7(0) + 6(8) = 48$
(4, 0)	$7(4) + 6(0) = 28$
(8, 0)	$7(8) + 6(0) = 56$

Figure 73b

75. The new objective equation is $P = 20x + 15y$. To find the maximum profit we must evaluate P at each of the vertices. These values are shown in Figure 75. From the table the maximum is 950 at the vertex (25, 30). The maximum profit will be 950 when 25 radios and 30 CD players are manufactured.

Vertex	$P + 20x = 15y$
(5, 5)	$20(5) + 15(5) = 175$
(25, 25)	$20(25) + 15(25) = 875$
(25, 30)	$20(25) + 15(30) = 950$
(5, 30)	$20(5) + 15(30) = 550$

Figure 75

77.   Make a table to list the information given. See Figure 77a. Using the table, we can write the linear programming problem as follows;

Cost: $C = 80x + 50y$, Protein: $15x + 20y \geq 60$, Fat: $10x + 5y \geq 30$, $x \geq 0$, and $y \geq 0$.

Brand	Units	Protein	Fat	Cost
A	x	15	10	80¢
B	y	20	5	50¢
Minimum		60	30	

Figure 77a

The region of feasible solutions is shown in Figure 77b. The vertices of this region are (0, 6), (2.4, 1.2), and (4, 0). To find the minimum value of C in the region, we must evaluate C at each of the vertices. These values are shown in Figure 77c. The minimum cost occurs when 2.4 units of Brand A and 1.2 units of Brand B are mixed, to give a cost of $2.52 per serving.

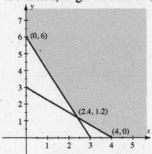

Figure 77b

Vertex	$C = 80x + 50y$
(0, 6)	$80(0) + 50(6) = 300$
(2.4, 1.2)	$80(2.4) + 50(1.2) = 252$
(4, 0)	$80(4) + 50(0) = 320$

Figure 77c

79.   Let x and y represent the number of hamsters and mice respectively. Since the total number of animals cannot exceed 50, $x + y \leq 50$. Because no more than 20 hamsters can be raised, $x \leq 20$. Here the revenue function is $R = 15x + 10y$. From the graph of the region of feasible solutions (not shown), the vertices are (0, 0), (0, 50), (20, 30), and (20, 0). To find (20, 30) solve the equations $x + y = 50$ and $x = 20$. The maximum value of R occurs at one of the vertices. For (0, 50), $R = 15(0) + 10(50) = 500$.

For (20, 30), $R = 15(20) + 10(30) = 600$. For (20, 0), $R = 15(20) + 10(0) = 300$.

The maximum revenue is $600.

81.   The number of hours on machine A needed to manufacture x units of part X is $4x$ while the number of hours on machine A needed to manufacture y units of part Y is $1y$. Since machine A is only available for 40 hours each week we have the constraint $4x + y \leq 40$. Similarly, the number of hours on machine B needed to manufacture x units of part X is $2x$ while the number of hours on machine B to make y units of part Y is $3y$. Since machine B is only available for 30 hours each week we have the constraint $2x + 3y \leq 30$. It should be

noted that the number of parts of each type cannot be negative. This gives the constraint $x \geq 0$ and $y \geq 0$.
The profit earned on x units of part X is $500x$ while the profit earned on y units of part Y is $600y$. Thus, the
total weekly profit is $P = 500x + 600y$. This is our objective function. Graph the constraints and shade the
region. The region of feasible solutions is shown in Figure 81a. The vertices of this region are (0, 0), (10, 0),
(9, 4), and (0, 10). To find the maximum value of P in the region, we must evaluate P at each of the vertices.
These values are shown in Figure 81b. From the table the maximum is 6900 at the point (9, 4). The maximum
profit is $6900 when there are 9 parts of type X and 4 parts of type Y manufactured.

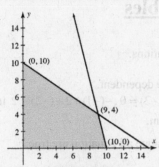

Figure 81a

Vertex	$P = 500x + 600y$
(0, 0)	$500(0) + 600(0) = 0$
(10, 0)	$500(10) + 600(0) = 5000$
(9, 4)	$500(9) + 600(4) = 6900$
(0, 10)	$500(0) + 600(10) = 6000$

Figure 81b

Checking Basic Concepts for Sections 6.1 and 6.2

1. $d(13, 18) = \sqrt{(13-1)^2 + (18-2)^2} = \sqrt{12^2 + 16^2} = \sqrt{400} = 20$

3. $z = x^2 + y^2 \Rightarrow y^2 = z - x^2 \Rightarrow y = \pm\sqrt{z - x^2}$

5. Graph the boundary line $3x - 2y = 6 \Rightarrow y = \dfrac{3}{2}x - 3$. Choose (0, 0) as a test point. Since $0 \leq 6$, the region

    containing (0, 0) is the solution set. The boundary line $y = \dfrac{3}{2}x - 3$ is part of the solution set. See Figure 5.

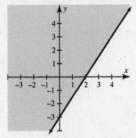

Figure 5

7. Let x represent the time watched in 2008 and y represent the time in 2018.

    (a) $x - y = 22; 0.8x - y = 0$

(b)   Substitute the second equation into the first: $x - (0.8x) = 22 \Rightarrow 0.2x = 22 \Rightarrow x = 110$;

Next, find y: $y = 0.8(110) = 88$.

The solution is $(110, 88)$. In 2008, the average daily time spent watching broadcast television was 110 minutes, whereas it was 88 minutes in 2018.

6.3 Systems of Linear Equations in Three Variables

1.   No, systems of linear equations can have zero, one, or infinitely many solutions.
3.   2, the same as the number of variables.
5.   A system with an infinite number of solutions means that the equations are dependent.
7.   Testing $(0, 2, -2)$: $0 + 2 - (-2) = 4$, true; $-0 + 2 + (-2) = 2$, false; $0 + 2 + (-2) = 0$, $-(-1) + 3 + (-2) = 2$ true; since the second equation is false, $(0, 2, -2)$ is not a solution for the system.

Testing $(-1, 3, -2)$: $-1 + 3 - (-2) = 4$, true; , true; $-1 + 3 + (-2) = 0$, true; since all equation are true, $(-1, 3, -2)$ is a solution for the system.

9. Testing $\left(-\frac{5}{11},\frac{20}{11},-2\right)$: $-\frac{5}{11}+3\left(\frac{20}{11}\right)-2(-2)=9\Rightarrow-\frac{5}{11}+\frac{60}{11}+\frac{44}{11}=9\Rightarrow\frac{99}{11}=9$, true;

$-3\left(-\frac{5}{11}\right)+2\left(\frac{20}{11}\right)+4(-2)=-3\Rightarrow\frac{15}{11}+\frac{40}{11}-\frac{88}{11}=-3\Rightarrow-\frac{33}{11}=-3$, true;

$-2\left(-\frac{5}{11}\right)+5\left(\frac{20}{11}\right)+2(-2)=6\Rightarrow\frac{10}{11}+\frac{100}{11}-\frac{44}{11}=6\Rightarrow\frac{66}{11}=6$, true; since all of the equations are true,

$\left(-\frac{5}{11},\frac{20}{11},-2\right)$ is a solution for the system.

Testing $(1,2,-1)$: $1+3(2)-2(-1)=9\Rightarrow1+6+2=9\Rightarrow9=9$, true;

$-3(1)+2(2)+4(-1)=-3\Rightarrow-3+4+(-4)=-3\Rightarrow-3=-3$, true;

$-2(1)+5(2)+2(-1)=6\Rightarrow(-2)+10+(-2)=6\Rightarrow6=6$, true; since all equations are true, $(1,2,-1)$ is a

solution for the system. Therefore they are both solutions.

11. Add all of the equations:

$1x+0y+1z=4$
$0x-1y+1z=2$
$-1x+1y-1z=-3$
$z=3$

Substitute $z=3$ into the equation $x+z=4\Rightarrow x+(3)=4\Rightarrow x=1$. Finally, substitute $z=3$ into the equation $-y+z=2\Rightarrow-y+(3)=2\Rightarrow y=1$. The solution is $(1,1,3)$.

13. Subtract the first two equations:

$1x+0y-1z=-3$
$0x-1y+1z=2$
$x-y=-1$

Add the new equation to the third equation:

$1x-1y=-1$
$2x+1y=-5$
$3x=-6$

$3x=-6\Rightarrow x=-2$. Substitute $x=-2$ into the equation: $x-z=-3\Rightarrow(-2)-z=-3\Rightarrow z=1$. Finally, substitute $z=1$ into the equation: $y-z=-2\Rightarrow y-(1)=-2\Rightarrow y=-1$. The solution is $(-2,-1,1)$.

15. Add the first two equations:

$x+y+z=6$
$-x+2y+z=6$
$3y+2z=12$.

Subtract 2 times the third equation from this equation:

$3y+2z=12$
$2y+2z=10$
$y=2$.

Substitute $y=2$ into the equation: $3y+2z=12\Rightarrow3(2)+2z=12\Rightarrow6+2z=12\Rightarrow2z=6\Rightarrow z=3$. Finally, substitute $y=2$ and $z=3$ into one of the original equations: $x+2+3=6\Rightarrow x=1$. The solution is $(1,2,3)$.

17. Multiply the first equation by 2 and subtract the second equation:

$$2x + 4y + 6z = 8$$
$$\underline{2x + y + 3z = 5}$$
$$3y + 3z = 3.$$

Subtract the third equation from the first:

$$x + 2y + 3z = 4$$
$$\underline{x - y + z = 2}$$
$$3y + 2z = 2.$$

Now subtract these two equations:

$$3y + 3z = 3$$
$$\underline{3y + 2z = 2}$$
$$z = 1.$$

Substitute $z = 1$ into $3y + 2z = 2 : 3y + 2(1) = 2 \Rightarrow 3y = 0 \Rightarrow y = 0$.

Finally substitute $y = 0$ and $z = 1$ into the original equation $x - y + z = 2 : x - 0 + 1 = 2 \Rightarrow x = 1$.

The solution is: $(1, 0, 1)$.

19.   Add the second and third equations:

$$4x + 2y + z = 1$$
$$\underline{2x - 2y - z = 2}$$
$$6x = 3 \quad \Rightarrow x = \frac{1}{2}.$$

Subtract the first two equations:

$$3x + y + z = 0$$
$$\underline{4x + 2y + z = 1}$$
$$-x - y = -1.$$

Substitute $x = \frac{1}{2}$ into $-x - y = -1: -\frac{1}{2} - y = -1 \Rightarrow -y = -\frac{1}{2} \Rightarrow y = \frac{1}{2}$.

Substitute $x = \frac{1}{2}$ and $y = \frac{1}{2}$ into $3x + y + z = 0: 3\left(\frac{1}{2}\right) + \frac{1}{2} + z = 0 \Rightarrow \frac{3}{2} + \frac{1}{2} + z = 0 \Rightarrow$

$2 + z = 0 \Rightarrow z = -2$.

The solution is: $\left(\frac{1}{2}, \frac{1}{2}, -2\right)$.

21.   Subtract the third equation from the first:

$$x + 3y + z = 6$$
$$\underline{x - y - z = 0}$$
$$4y + 2z = 6.$$

Multiply the third equation by 3 and subtract it from the second equation:

$$3x + y - z = 6$$
$$\underline{3x - 3y - 3z = 0}$$
$$4y + 2z = 6.$$

Subtracting these two equations we get:

$$4y + 2z = 6$$
$$\underline{4y + 2z = 6}$$
$$0 = 0.$$

Therefore infinitely many solutions and $4y + 2z = 6 \Rightarrow 4y = -2z + 6 \Rightarrow y = \dfrac{-z+3}{2}$.

Adding the last two original equations we get:

$$\begin{array}{r} 3x + y - z = 6 \\ x - y - z = 0 \\ \hline 4x \quad\ - 2z = 6 \end{array} \Rightarrow 4x = 2z + 6 \Rightarrow x = \dfrac{z+3}{2}.$$

We have infinitely many solutions: $\left(\dfrac{z+3}{2}, \dfrac{-z+3}{2}, z \right)$.

23. Add the first two equations:

$$\begin{array}{r} x - 4y + 2z = -2 \\ x + 2y - 2z = -3 \\ \hline 2x - 2y \quad\quad = -5. \end{array}$$

Multiply the third equation by 2 and subtract from $2x - 2y = -5$:

$$\begin{array}{r} 2x - 2y = -5 \\ 2x - 2y = \ \ 8 \\ \hline 0 = -13. \end{array}$$

Therefore we have no solutions.

25. Add the last two equations:

$$\begin{array}{r} 2a + b - c = -11 \\ 2a - 2b + c = \ \ 3 \\ \hline 4a - b \quad\ = -8. \end{array}$$

Multiply the second equation by 2 and add to the first equation:

$$\begin{array}{r} 4a - b + 2c = \ \ 0 \\ 4a + 2b - 2c = -22 \\ \hline 8a + b \quad\quad = -22. \end{array}$$

Now add the equation $4a - b = -8$ to $8a + b = -8$:

$$\begin{array}{r} 4a - b = -8 \\ 8a + b = -22 \\ \hline 12a \quad\ = -30 \end{array} \Rightarrow a = \dfrac{-30}{12} \Rightarrow a = \dfrac{-5}{2}.$$

Substitute $a = \dfrac{-5}{2}$: $8\left(\dfrac{-5}{2}\right) + b = -22 \Rightarrow -20 + b = -22 \Rightarrow b = -2$.

Substitute $a = \dfrac{-5}{2}$ and $b = -2$ into $4\left(\dfrac{-5}{2}\right) - (-2) + 2c = 0 \Rightarrow -10 + 2 + 2c = 0 \Rightarrow 2c = 8 \Rightarrow c = 4$. The solution

is: $\left(\dfrac{-5}{2}, -2, 4 \right)$.

27. Subtract the first and third equations:

$$\begin{array}{r} a + b + c = \ \ 0 \\ a + 3b + 3c = \ \ 5 \\ \hline -2b - 2c = -5. \end{array}$$

Subtract the second and third equations:

$$a - b - c = 3$$
$$a + 3b + 3c = 5$$
$$\overline{ -4b - 4c = -2.}$$

Multiply $-2b - 2c = -5$ by 2 and subtract $-4b - 4c = -2$:

$$-4b - 4c = -10$$
$$-4b - 4c = -2$$
$$\overline{ 0 = -8.}$$

Therefore, we have no solution.

29. Add the first two equations:

$$3x + 2y + z = -1$$
$$3x + 4y - z = 1$$
$$\overline{6x + 6y = 0.}$$

Add the last two equations:

$$3x + 4y - z = 1$$
$$x + 2y + z = 0$$
$$\overline{4x + 6y = 1.}$$

Now, subtract $6x + 6y = 0$ and $4x + 6y = 1$:

$$6x + 6y = 0$$
$$4x + 6y = 1$$
$$\overline{2x = -1} \Rightarrow x = -\frac{1}{2}.$$

Substitute $x = -\frac{1}{2}$ into $6x + 6y = 0$: $6\left(-\frac{1}{2}\right) + 6y = 0 \Rightarrow 6y = 3 \Rightarrow y = \frac{1}{2}.$

Substitute $x = -\frac{1}{2}$ and $y = \frac{1}{2}$ into $x + 2y + x = 0 \Rightarrow -\frac{1}{2} + 2\left(\frac{1}{2}\right) + z = 0 \Rightarrow z = -\frac{1}{2}.$

The solution is: $\left(-\frac{1}{2}, \frac{1}{2}, -\frac{1}{2}\right).$

31. Multiply the first equation by 2 and add the second equation:

$$-2x + 6y + 2z = 6$$
$$2x + 7y + 4z = 13$$
$$\overline{ 13y + 6z = 19.}$$

Multiply the second equation by 2 and subtract the third equation:

$$4x + 14y + 8z = 26$$
$$4x + y + 2z = 7$$
$$\overline{ 13y + 6z = 19.}$$

Subtracting the two new equations we get:

$$13y + 6z = 19$$
$$13y + 6z = 19$$
$$\overline{ 0 = 0.}$$

Therefore, we have infinitely many solutions.

$$13y + 6z = 19 \Rightarrow 13y = -6z + 19 \Rightarrow y = \frac{-6z + 19}{13}.$$

Multiply the third equation by 3 and subtract from the first equation:

$$-x + 3y + z = 3$$
$$12x + 3y + 6z = 21$$
$$\overline{-13x \quad\quad -5z = -18} \Rightarrow x = \frac{-5z+18}{13}.$$

We have infinitely many solutions: $\left(\dfrac{-5z+18}{13}, \dfrac{-6z+19}{13}, z \right)$.

33. Subtract the second and third equations:

$$y + 4z = -13$$
$$3x + y \quad\quad = 13$$
$$\overline{-3x \quad\quad +4z = -26.}$$

Multiply the first equation by 3 and subtract $-3x + 4z = -26$ from it:

$$-3x \quad +6z = -27$$
$$-3x \quad +4z = -26$$
$$\overline{\quad\quad 2z = -1} \Rightarrow z = -\frac{1}{2}.$$

Substitute $z = -\dfrac{1}{2}$ into $y + 4z = -13$: $y + 4\left(-\dfrac{1}{2}\right) = -13 \Rightarrow y = -11$.

Substitute $z = -\dfrac{1}{2}$ into $-x + 2z = -9 \Rightarrow -x + 2\left(-\dfrac{1}{2}\right) = -9 \Rightarrow -x = -8 \Rightarrow x = 8$.

The solution is: $\left(8, -11, -\dfrac{1}{2} \right)$.

35. Multiply the first equation by 2 and subtract the second equation:

$$x - 2y + z = -8$$
$$x + 2y - 3z = 20$$
$$\overline{\quad -4y + 4z = -28.}$$

Add the first and third equations:

$$\tfrac{1}{2}x - y + \tfrac{1}{2}z = -4$$
$$-\tfrac{1}{2}x + 3y + 2z = 0$$
$$\overline{\quad\quad 2y + \tfrac{5}{2}z = -4.}$$

Multiply $2y + \dfrac{5}{2}z = -4$ by 2 and add $-4y + 4z = -28$:

$$4y + 5z = -8$$
$$-4y + 4z = -28$$
$$\overline{\quad\quad 9z = -36} \Rightarrow z = -4.$$

Substitute $z = -4$ into $-4y + 4z = -28$: $-4y + 4(-4) = -28 \Rightarrow -4y - 16 = -28 \Rightarrow -4y = -12 \Rightarrow y = 3$.

Substitute $y = 3$ and $z = -4$ into $x + 2y - 3z = 20$: $x + 2(3) - 3(-4) = 20 \Rightarrow x + 18 = 20 \Rightarrow x = 2$. The solution is: $(2, 3, -4)$.

37. Let x = children tickets sold, y = student tickets sold, and z = adult tickets sold. Then:
$x + y + z = 500$, $5x + 7y + 10z = 3560$, and $y = z + 180$ or $y - z = 180$.

Multiply the first equation by 5 and subtract the second equation:

$$\begin{array}{rcl}5x+5y+ 5z &=& 2500\\ 5x+7y+10z &=& 3560\\ \hline -2y- 5z &=& -1060\end{array}$$

Now, multiply $y-z=180$ by 2 and add $-2y-5z=-1060$:

$$\begin{array}{rcl}2y-2z &=& 360\\ -2y-5z &=& -1060\\ \hline -7z &=& -700 \quad \Rightarrow z=100.\end{array}$$

Substitute $z=100$ into $y-z=180$: $y-100=180 \Rightarrow y=280$.

Substitute $y=280$ and $z=100$ into $x+y+z=500$: $x+280+100=500 \Rightarrow x=120$.

There were 120 children tickets, 280 student tickets, and 100 adult tickets sold.

39.   Let $x=$ cost of a hamburger, $y=$ cost of fries, and $z=$ cost of a soda. Then $2x+2y+z=9$; $x+y+z=5$; and $x+y=5$.

Multiply the equation $x+y+z=5$ by 2 and subtract from the equation $2x+2y+z=9$:

$$\begin{array}{rcl}2x+2y+ z &=& 9\\ 2x+2y+2z &=& 10\\ \hline -z &=& -1 \Rightarrow z=1.\end{array}$$

Subtract $x+y+z=5$ and $x+y=5$:

$$\begin{array}{rcl}x+y+z &=& 5\\ x+y &=& 5\\ \hline z &=& 0.\end{array}$$

z cannot equal both 0 and 1 therefore there is no solution, at least one student was charged incorrectly.

41.   (a) $x+y+z=180$; $x=z+25 \Rightarrow x-z=25$; and $y+z=x+30 \Rightarrow -x+y+z=30$.

       (b)   Add $x-z=25$ to $-x+y+z=30$:

$$\begin{array}{rcl}x \quad -z &=& 25\\ -x+y+z &=& 30\\ \hline y &=& 55.\end{array}$$

Add $x+y+z=180$ to $x-z=25$:

$$\begin{array}{rcl}x+y+z &=& 180\\ x \quad -z &=& 25\\ \hline 2x+y &=& 205.\end{array}$$

Substitute $y=55$ into $2x+y=205$: $2x+55=205 \Rightarrow 2x=150 \Rightarrow x=75$.

Substitute $x=75$ and $y=55$ into $x+y+z=180$: $75+55+z=180 \Rightarrow z=50$.

The angles are: 75°, 55°, and 50°.

Check: $75+55+50=180 \Rightarrow 180=180$; $75-50=25 \Rightarrow 25=25$; $-75+55+50=30 \Rightarrow 30=30$.

43.   Let x , y , and z equal the amounts invested in the three mutual funds.

Then $x+y+z=20,000$; $0.05x+0.07y+0.10z=1650$; and $4x=z \Rightarrow 4x-z=0$.

Multiply $x+y+z=20,000$ by 0.07 and subtract $0.05x+0.07y+0.10z=1650$:

$$\begin{array}{rcl}0.07x+0.07y+0.07z &=& 1400\\ 0.05x+0.07y+0.10z &=& 1650\\ \hline 0.02x \quad\quad -0.03z &=& -250.\end{array}$$

Now multiply $0.02x - 0.03z = -250$ by 200 and subtract $4x - z = 0$:

$$4x - 6z = -50,000$$
$$\underline{4x - z = 0}$$
$$-5z = -50,000 \Rightarrow z = 10,000.$$

Substitute $z = 10,000$ into $4x - z = 0$: $4x - 10,000 = 0 \Rightarrow 4x = 10,000 \Rightarrow x = 2500$.

Substitute $x = 2500$ and $z = 10,000$ into $x + y + z = 20,000$: $2500 + y + 10,000 = 20,000 \Rightarrow y = 7500$.

The fund amounts are: \$2500 at 5%, \$7500 at 7%, and \$10,000 at 10%.

45. (a) $\begin{aligned} N + P + K &= 80 \\ N + P - K &= 8 \\ 9P - K &= 0 \end{aligned}$

    (b) Using technology to solve the system, the solution is (40, 4, 36).
    The sample contains 40 pounds of nitrogen, 4 pounds of phosphorus and 36 pounds of potassium.

6.4 Solutions to Linear Systems Using Matrices

1. Since there are three rows and one column, its dimension is 3×1.
3. Since there are two rows and two columns, its dimension is 2×2.
5. 3×2
7. This system can be written using a 2×3 matrix:
$$\begin{bmatrix} 5 & -2 & | & 3 \\ -1 & 3 & | & -1 \end{bmatrix}$$
9. This system can be written using a 3×4 matrix:
$$\begin{bmatrix} -3 & 2 & 1 & | & -4 \\ 5 & 0 & -1 & | & 9 \\ 1 & -3 & -6 & | & -9 \end{bmatrix}$$
11. $3x + 2y = 4$ and $y = 5$
13. $3x + y + 4z = 0$, $5y + 8z = -1$, and $-7z = 1$
15. (a) Yes
    (b) No, since $a_{22} = -1$ and $a_{32} \neq 0$. The diagonal is not all 1's and there are not all 0's below the diagonal.
    (c) Yes
17. The system can be written as $x + y = 4$ and $y = 3$. Substituting $y = 3$ into the first equation gives $x + (3) = 4 \Rightarrow x = 1$. The solution is $(1, 3)$.
19. The system can be written as $x - 2y = 1$ and $y = -1$. Substituting $y = -1$ into the first equation gives $x - 2(-1) = 1 \Rightarrow x = -1$. The solution is $(-1, -1)$.
21. The system can be written as $x + 2y = 3$ and $y = -1$. Substituting $y = -1$ into the first equation gives $x + 2(-1) = 3 \Rightarrow x = 5$. The solution is $(5, -1)$.
23. The system can be written as $x - y = 2$ and $y = 0$. Substituting $y = 0$ into the first equation gives $x - 0 = 2 \Rightarrow x = 2$. The solution is $(2, 0)$.
25. The system can be written as $x + y - z = 4$, $y - z = 2$, and $z = 1$. Substituting $z = 1$ into the second equation gives $y - (1) = 2 \Rightarrow y = 3$. Substituting $y = 3$ and $z = 1$ into the first equation gives $x + (3) - (1) = 4 \Rightarrow x = 2$. The solution is $(2, 3, 1)$.

27. The system can be written as $x+2y-z=5$, $y-2z=1$, and $0=0$. Since $0=0$, there are an infinite number of solutions. The second equation gives $y=1+2z$. Substituting this into the first equation gives $x+2(1+2z)-z=5 \Rightarrow x=3-3z$. The solution can be written as $\{(3-3z,1+2z,z)\,|\,z \text{ is a real number}\}$.

29. The system can be written as $x+2y+z=-3$, $y-3z=\dfrac{1}{2}$, and $0=4$. Since $0=4$ is false, there are no solutions.

31. $\begin{array}{c}(1/2)R_1 \rightarrow \\ \\ (1/4)R_3 \rightarrow\end{array}\left[\begin{array}{ccc|c} 1 & -2 & 3 & 5 \\ -3 & 5 & 3 & 2 \\ 1 & 2 & 1 & -2 \end{array}\right]$

33. $\begin{array}{c}R_2+R_1 \rightarrow \\ R_3-R_1 \rightarrow\end{array}\left[\begin{array}{ccc|c} 1 & -1 & 1 & 2 \\ 0 & 1 & -1 & 2 \\ 0 & 8 & -1 & 3 \end{array}\right]$

35. The system can be written as follows:

$\left[\begin{array}{cc|c} 1 & 2 & 3 \\ -1 & -1 & 7 \end{array}\right] R_2+R_1 \rightarrow \left[\begin{array}{cc|c} 1 & 2 & 3 \\ 0 & 1 & 10 \end{array}\right]$

The solution is $y=10$ and $x+2y=3 \Rightarrow x+2(10)=3 \Rightarrow x=-17$. The solution is $(-17,10)$.

37. The system can be written as follows:

$\left[\begin{array}{cc|c} 2 & 3 & 6 \\ 1 & -2 & -4 \end{array}\right](1/2)R_1 \rightarrow \left[\begin{array}{cc|c} 1 & \frac{3}{2} & 3 \\ 1 & -2 & -4 \end{array}\right] R_2-R_1 \rightarrow \left[\begin{array}{cc|c} 1 & \frac{3}{2} & 3 \\ 0 & -\frac{7}{2} & 7 \end{array}\right](\frac{2}{7})R_2 \rightarrow \left[\begin{array}{cc|c} 1 & \frac{3}{2} & 3 \\ 0 & 1 & 2 \end{array}\right]$

The solution is $y=2$ and $x-2y=-4 \Rightarrow x-4=-4 \Rightarrow x=0$. The solution is $(0,2)$.

39. The system can be written as follows:

$\left[\begin{array}{cc|c} 1 & -3 & -2 \\ 2 & -1 & -4 \end{array}\right]2R_1-R_2 \rightarrow \left[\begin{array}{cc|c} 1 & -3 & -2 \\ 0 & -5 & 0 \end{array}\right](-1/5)R_2 \rightarrow \left[\begin{array}{cc|c} 1 & -3 & -2 \\ 0 & 1 & 0 \end{array}\right]$

The solution is $y=0$ and $x-3y=-2 \Rightarrow x-3(0)=-2 \Rightarrow x=-2$. The solution is $(-2,0)$.

41. The system can be written as follows:

$\left[\begin{array}{ccc|c} 1 & 2 & 1 & 3 \\ 1 & 1 & -1 & 3 \\ -1 & -2 & 1 & -5 \end{array}\right] \begin{array}{c} R_2-R_1 \rightarrow \\ R_3+R_1 \rightarrow \end{array} \left[\begin{array}{ccc|c} 1 & 2 & 1 & 3 \\ 0 & -1 & -2 & 0 \\ 0 & 0 & 2 & -2 \end{array}\right] \begin{array}{c} (-1)R_2 \rightarrow \\ (1/2)R_3 \rightarrow \end{array} \left[\begin{array}{ccc|c} 1 & 2 & 1 & 3 \\ 0 & 1 & 2 & 0 \\ 0 & 0 & 1 & -1 \end{array}\right]$

Back substitution produces $z=-1$; $y+2z=0 \Rightarrow y=2$; $x+2y+z=3 \Rightarrow x=0$.
The solution is $(0,2,-1)$.

43. The system can be written as follows:

$\left[\begin{array}{ccc|c} 1 & 2 & -1 & -1 \\ 2 & -1 & 1 & 0 \\ -1 & -1 & 2 & 7 \end{array}\right] \begin{array}{c} -2R_1+R_2 \rightarrow \\ R_1+R_3 \rightarrow \end{array} \left[\begin{array}{ccc|c} 1 & 2 & -1 & -1 \\ 0 & -5 & 3 & 2 \\ 0 & 1 & 1 & 6 \end{array}\right] R_3 \Leftrightarrow R_2 \rightarrow \left[\begin{array}{ccc|c} 1 & 2 & -1 & -1 \\ 0 & 1 & 1 & 6 \\ 0 & -5 & 3 & 2 \end{array}\right]$

$5R_2+R_3 \rightarrow \left[\begin{array}{ccc|c} 1 & 2 & -1 & -1 \\ 0 & 1 & 1 & 6 \\ 0 & 0 & 8 & 32 \end{array}\right] \frac{1}{8}R_3 \rightarrow \left[\begin{array}{ccc|c} 1 & 2 & -1 & -1 \\ 0 & 1 & 1 & 6 \\ 0 & 0 & 1 & 4 \end{array}\right]$

Back substitution produces $z=4$; $y+z=6 \Rightarrow y=2$; $x+2y-z=-1 \Rightarrow x=-1$.

The solution is $(-1, 2, 4)$.

45. The system can be written as follows:

$$\begin{bmatrix} 3 & 1 & 3 & | & 14 \\ 1 & 1 & 1 & | & 6 \\ -2 & -2 & 3 & | & -7 \end{bmatrix} \begin{matrix} R_2 \to \\ R_1 \to \\ \ \end{matrix} \begin{bmatrix} 1 & 1 & 1 & | & 6 \\ 3 & 1 & 3 & | & 14 \\ -2 & -2 & 3 & | & -7 \end{bmatrix} \begin{matrix} \\ R_2 - 3R_1 \to \\ R_3 + 2R_1 \to \end{matrix} \begin{bmatrix} 1 & 1 & 1 & | & 6 \\ 0 & -2 & 0 & | & -4 \\ 0 & 0 & 5 & | & 5 \end{bmatrix}$$

$$\begin{matrix} \\ (-1/2)R_2 \to \\ (1/5)R_3 \to \end{matrix} \begin{bmatrix} 1 & 1 & 1 & | & 6 \\ 0 & 1 & 0 & | & 2 \\ 0 & 0 & 1 & | & 1 \end{bmatrix}$$

Back substitution produces $z = 1$; $y = 2$; $x + y + z = 6 \Rightarrow x = 3$. The solution is $(3, 2, 1)$.

47. The system can be written as follows:

$$\begin{bmatrix} 1 & 2 & -1 & | & 2 \\ 2 & 5 & 1 & | & 8 \\ 3 & 7 & 0 & | & 5 \end{bmatrix} \begin{matrix} \\ R_2 - 2R_1 \to \\ R_3 - 3R_1 \to \end{matrix} \begin{bmatrix} 1 & 2 & -1 & | & 2 \\ 0 & 1 & 3 & | & 4 \\ 0 & 1 & 3 & | & -1 \end{bmatrix} \begin{matrix} \\ \\ R_3 - R_2 \to \end{matrix} \begin{bmatrix} 1 & 2 & -1 & | & 2 \\ 0 & 1 & 3 & | & 4 \\ 0 & 0 & 0 & | & -5 \end{bmatrix}$$

The last equation indicates that $0 = -5$, which is false. Therefore, there are no solutions.

49. The system can be written as follows:

$$\begin{bmatrix} -1 & 2 & 4 & | & 10 \\ 3 & -2 & -2 & | & -12 \\ 1 & 2 & 6 & | & 8 \end{bmatrix} \begin{matrix} (-1)R_1 \to \\ R_2 + 3R_1 \to \\ R_1 + R_3 \to \end{matrix} \begin{bmatrix} 1 & -2 & -4 & | & -10 \\ 0 & 4 & 10 & | & 18 \\ 0 & 4 & 10 & | & 18 \end{bmatrix} \begin{matrix} R_1 + (1/2)R_2 \to \\ (1/4)R_2 \to \\ R_2 - R_3 \to \end{matrix} \begin{bmatrix} 1 & 0 & 1 & | & -1 \\ 0 & 1 & \frac{5}{2} & | & \frac{9}{2} \\ 0 & 0 & 0 & | & 0 \end{bmatrix}$$

The last equation indicates that $0 = 0$, which is true. Therefore, there is an infinite number of solutions. The second equation gives: $y + \frac{5}{2}z = \frac{9}{2} \Rightarrow y = \frac{-5z + 9}{2}$. The first equation gives: $x + z = -1 \Rightarrow x = -1 - z$.

Therefore there are infinitely many solutions which can be written as $\left(-1 - z, \frac{-5z + 9}{2}, z \right)$.

51. $$\begin{bmatrix} 1 & -1 & 1 & | & 1 \\ 1 & 2 & -1 & | & 2 \\ 0 & 1 & -1 & | & 0 \end{bmatrix} \begin{matrix} \\ R_2 - R_1 \to \\ \ \end{matrix} \begin{bmatrix} 1 & -1 & 1 & | & 1 \\ 0 & 3 & -2 & | & 1 \\ 0 & 1 & -1 & | & 0 \end{bmatrix} \begin{matrix} \\ (1/3)R_2 \to \\ \ \end{matrix} \begin{bmatrix} 1 & -1 & 1 & | & 1 \\ 0 & 1 & -\frac{2}{3} & | & \frac{1}{3} \\ 0 & 1 & -1 & | & 0 \end{bmatrix}$$

$$\begin{matrix} \\ \\ R_3 - R_2 \to \end{matrix} \begin{bmatrix} 1 & -1 & 1 & | & 1 \\ 0 & 1 & -\frac{2}{3} & | & \frac{1}{3} \\ 0 & 0 & -\frac{1}{3} & | & -\frac{1}{3} \end{bmatrix} \begin{matrix} \\ \\ -3R_3 \to \end{matrix} \begin{bmatrix} 1 & -1 & 1 & | & 1 \\ 0 & 1 & -\frac{2}{3} & | & \frac{1}{3} \\ 0 & 0 & 1 & | & 1 \end{bmatrix}$$

The matrix is now in row-echelon form. We see that $z = 1$. Thus, $y - \frac{2}{3}z = \frac{1}{3} \Rightarrow y - \frac{2}{3}(1) = \frac{1}{3} \Rightarrow y = 1$ and

$x - y + z = 1 \Rightarrow x - 1 + 1 = 1 \Rightarrow x = 1$. The solution is $(1, 1, 1)$.

53. $$\begin{bmatrix} 2 & -4 & 2 & | & 11 \\ 1 & 3 & -2 & | & -9 \\ 4 & -2 & 1 & | & 7 \end{bmatrix} \begin{matrix} (1/2)R_1 \to \\ \\ \ \end{matrix} \begin{bmatrix} 1 & -2 & 1 & | & \frac{11}{2} \\ 1 & 3 & -2 & | & -9 \\ 4 & -2 & 1 & | & 7 \end{bmatrix} \begin{matrix} \\ R_2 - R_3 \to \\ R_3 - 4R_1 \to \end{matrix} \begin{bmatrix} 1 & -2 & 1 & | & \frac{11}{2} \\ 0 & 5 & -3 & | & -\frac{29}{2} \\ 0 & 6 & -3 & | & -15 \end{bmatrix}$$

$$\begin{matrix} \\ (1/5)R_2 \to \\ \ \end{matrix} \begin{bmatrix} 1 & -2 & 1 & | & \frac{11}{2} \\ 0 & 1 & -\frac{3}{5} & | & -\frac{29}{10} \\ 0 & 6 & -3 & | & -15 \end{bmatrix} \begin{matrix} \\ \\ R_3 - 6R_2 \to \end{matrix} \begin{bmatrix} 1 & -2 & 1 & | & \frac{11}{2} \\ 0 & 1 & -\frac{3}{5} & | & -\frac{29}{10} \\ 0 & 0 & \frac{3}{5} & | & \frac{24}{10} \end{bmatrix} \begin{matrix} \\ \\ (5/3)R_3 \to \end{matrix} \begin{bmatrix} 1 & -2 & 1 & | & \frac{11}{2} \\ 0 & 0 & -\frac{3}{5} & | & -\frac{29}{10} \\ 0 & 0 & 1 & | & 4 \end{bmatrix}$$

The matrix is now in row-echelon form. We see that $z = 4$. Thus, $y - \dfrac{3}{5}z = -\dfrac{29}{10} \Rightarrow$

$y - \dfrac{3}{5}(4) = -\dfrac{29}{10} \Rightarrow y = -\dfrac{1}{2}$ and $x - 2y + z = \dfrac{11}{2} \Rightarrow x - 2\left(-\dfrac{1}{2}\right) + 4 = \dfrac{11}{2} \Rightarrow x = \dfrac{1}{2}$. The solution is $\left(\dfrac{1}{2}, -\dfrac{1}{2}, 4\right)$.

55. $\begin{bmatrix} 3 & -2 & 2 & | & -18 \\ -1 & 2 & -4 & | & 16 \\ 4 & -3 & -2 & | & -21 \end{bmatrix} (1/3)R_1 \rightarrow \begin{bmatrix} 1 & -\frac{2}{3} & \frac{2}{3} & | & -6 \\ -1 & 2 & -4 & | & 16 \\ 4 & -3 & -2 & | & -21 \end{bmatrix} \begin{matrix} \\ R_2 + R_1 \rightarrow \\ R_3 - 4R_1 \rightarrow \end{matrix} \begin{bmatrix} 1 & -\frac{2}{3} & \frac{2}{3} & | & -6 \\ 0 & \frac{4}{3} & -\frac{10}{3} & | & 10 \\ 0 & -\frac{1}{3} & -\frac{14}{3} & | & 3 \end{bmatrix}$

$\begin{matrix} \\ (3/4)R_2 \rightarrow \\ 4R_3 + R_2 \rightarrow \end{matrix} \begin{bmatrix} 1 & -\frac{2}{3} & \frac{2}{3} & | & -6 \\ 0 & 1 & -\frac{5}{2} & | & \frac{15}{2} \\ 0 & 0 & -22 & | & 22 \end{bmatrix} -(1/22)R_3 \rightarrow \begin{bmatrix} 1 & -\frac{2}{3} & \frac{2}{3} & | & -6 \\ 0 & 1 & -\frac{5}{2} & | & \frac{15}{2} \\ 0 & 0 & 1 & | & -1 \end{bmatrix}$

The matrix is now in row-echelon form. We see that $z = -1$. Thus, $y - \dfrac{5}{2}z = \dfrac{15}{2} \Rightarrow y - \dfrac{5}{2}(-1) = \dfrac{15}{2} \Rightarrow y = 5$

and $x - \dfrac{2}{3}y + \dfrac{2}{3}z = -6 \Rightarrow x - \dfrac{2}{3}(5) + \dfrac{2}{3}(-1) = -6 \Rightarrow x = -2$.

The solution is $(-2, 5, -1)$.

57. $\begin{bmatrix} 1 & -4 & 3 & | & 26 \\ -1 & 3 & -2 & | & -19 \\ 0 & -1 & 1 & | & 10 \end{bmatrix} R_2 + R_1 \rightarrow \begin{bmatrix} 1 & -4 & 3 & | & 26 \\ 0 & -1 & 1 & | & 7 \\ 0 & -1 & 1 & | & 10 \end{bmatrix} \begin{matrix} -R_2 \rightarrow \\ R_3 - R_2 \rightarrow \end{matrix} \begin{bmatrix} 1 & -4 & 3 & | & 26 \\ 0 & 1 & -1 & | & -7 \\ 0 & 0 & 0 & | & 3 \end{bmatrix}$

The last equation indicates that $0 = 3$, which is always false. Therefore, the system has no solutions.

59. $\begin{bmatrix} 5 & 0 & 4 & | & 7 \\ 2 & -4 & 0 & | & 6 \\ 0 & 3 & 3 & | & 3 \end{bmatrix} \begin{matrix} (1/5)R_1 \rightarrow \\ 2R_1 - 5R_2 \rightarrow \\ (1/3)R_3 \rightarrow \end{matrix} \begin{bmatrix} 1 & 0 & \frac{4}{5} & | & \frac{7}{5} \\ 0 & 20 & 8 & | & -16 \\ 0 & 1 & 1 & | & 1 \end{bmatrix} \begin{matrix} \\ (1/20)R_2 \rightarrow \\ R_2 - 20R_3 \rightarrow \end{matrix} \begin{bmatrix} 1 & 0 & \frac{4}{5} & | & \frac{7}{5} \\ 0 & 1 & \frac{2}{5} & | & -\frac{4}{5} \\ 0 & 0 & -12 & | & -36 \end{bmatrix}$

$\begin{matrix} \\ \\ (-1/12)R_3 \rightarrow \end{matrix} \begin{bmatrix} 1 & 0 & \frac{4}{5} & | & \frac{7}{5} \\ 0 & 1 & \frac{2}{5} & | & -\frac{4}{5} \\ 0 & 0 & 1 & | & 3 \end{bmatrix} \begin{matrix} (-4/5)R_3 + R_1 \rightarrow \\ (-2/5)R_3 + R_2 \rightarrow \\ \end{matrix} \begin{bmatrix} 1 & 0 & 0 & | & -1 \\ 0 & 1 & 0 & | & -2 \\ 0 & 0 & 1 & | & 3 \end{bmatrix}$ The solution is $(-1, -2, 3)$.

61. $\begin{bmatrix} 5 & -2 & 1 & | & 5 \\ 1 & 1 & -2 & | & -2 \\ 4 & -3 & 3 & | & 7 \end{bmatrix} \begin{matrix} R_2 \rightarrow \\ 5R_2 - R_1 \rightarrow \\ 4R_2 - R_3 \rightarrow \end{matrix} \begin{bmatrix} 1 & 1 & -2 & | & -2 \\ 0 & 7 & -11 & | & -15 \\ 0 & 7 & -11 & | & -15 \end{bmatrix} \begin{matrix} \\ (1/7)R_2 \rightarrow \\ R_2 - R_3 \rightarrow \end{matrix} \begin{bmatrix} 1 & 1 & -2 & | & -2 \\ 0 & 1 & -\frac{11}{7} & | & -\frac{15}{7} \\ 0 & 0 & 0 & | & 0 \end{bmatrix}$

$(-1)R_2 + R_1 \rightarrow \begin{bmatrix} 1 & 0 & -\frac{3}{7} & | & \frac{1}{7} \\ 0 & 1 & -\frac{11}{7} & | & -\frac{15}{7} \\ 0 & 0 & 0 & | & 0 \end{bmatrix}$

The last equation indicates that $0 = 0$, which is true. Therefore, there is an infinite number of solutions. The

second equation gives: $y - \dfrac{11}{7}z = -\dfrac{15}{7} \Rightarrow y = \dfrac{11z - 15}{7}$. The first equation gives: $x - \dfrac{3}{7}z = \dfrac{1}{7} \Rightarrow x = \dfrac{3z + 1}{7}$.

Therefore, there are infinitely many solutions which can be written as: $\left(\dfrac{3z + 1}{7}, \dfrac{11z - 15}{7}, z\right)$.

63. The equations are: $x = 12$ and $y = 3 \Rightarrow (12, 3)$.

65. The equations are: $x = -2$, $y = 4$, and $z = \dfrac{1}{2} \Rightarrow \left(-2, 4, \dfrac{1}{2}\right)$.

67. The last equation indicates that $0 = 0$, which is true. Therefore, there is an infinite number of solutions. The first equation gives: $x + 2z = 4 \Rightarrow x = -2z + 4$. The second equation gives: $y - z = -3 \Rightarrow y = z - 3$. The system has infinitely many solutions which can be written as: $(-2z + 4, z - 3, z)$.

69. The last equation indicates that $0 = \dfrac{2}{3}$, which is always false. Therefore, there are no solutions.

71. $\begin{bmatrix} 1 & -1 & | & 1 \\ 1 & 1 & | & 5 \end{bmatrix} R_2 - R_1 \rightarrow \begin{bmatrix} 1 & -1 & | & 1 \\ 0 & 2 & | & 4 \end{bmatrix} (1/2)R_2 \rightarrow \begin{bmatrix} 1 & -1 & | & 1 \\ 0 & 1 & | & 2 \end{bmatrix} R_1 + R_2 \rightarrow \begin{bmatrix} 1 & 0 & | & 3 \\ 0 & 1 & | & 2 \end{bmatrix}$

The solution is $(3, 2)$

73. $\begin{bmatrix} 1 & 2 & 1 & | & 3 \\ 0 & 1 & -1 & | & -2 \\ -1 & -2 & 2 & | & 6 \end{bmatrix} \begin{matrix} \\ \\ R_1 + R_3 \rightarrow \end{matrix} \begin{bmatrix} 1 & 2 & 1 & | & 3 \\ 0 & 1 & -1 & | & -2 \\ 0 & 0 & 3 & | & 9 \end{bmatrix} \begin{matrix} R_1 - 2R_2 \rightarrow \\ \\ (1/3)R_3 \rightarrow \end{matrix} \begin{bmatrix} 1 & 0 & 3 & | & 7 \\ 0 & 1 & -1 & | & -2 \\ 0 & 0 & 1 & | & 3 \end{bmatrix} \begin{matrix} R_1 - 3R_3 \rightarrow \\ R_2 + R_3 \rightarrow \\ \end{matrix} \begin{bmatrix} 1 & 0 & 0 & | & -2 \\ 0 & 1 & 0 & | & 1 \\ 0 & 0 & 1 & | & 3 \end{bmatrix}$

The solution is $(-2, 1, 3)$.

75. $\begin{bmatrix} 1 & -1 & 2 & | & 7 \\ 2 & 1 & -4 & | & -27 \\ -1 & 1 & -1 & | & 0 \end{bmatrix} \begin{matrix} \\ R_2 - 2R_1 \rightarrow \\ R_1 + R_3 \rightarrow \end{matrix} \begin{bmatrix} 1 & -1 & 2 & | & 7 \\ 0 & 3 & -8 & | & -41 \\ 0 & 0 & 1 & | & 7 \end{bmatrix} (1/3)R_2 \rightarrow \begin{bmatrix} 1 & -1 & 2 & | & 7 \\ 0 & 1 & -\frac{8}{3} & | & -\frac{41}{3} \\ 0 & 0 & 1 & | & 7 \end{bmatrix}$

$R_2 + R_1 \rightarrow \begin{bmatrix} 1 & 0 & -\frac{2}{3} & | & -\frac{20}{3} \\ 0 & 1 & -\frac{8}{3} & | & -\frac{41}{3} \\ 0 & 0 & 1 & | & 7 \end{bmatrix} \begin{matrix} (2/3)R_3 + R_1 \rightarrow \\ (8/3)R_3 + R_2 \rightarrow \\ \end{matrix} \begin{bmatrix} 1 & 0 & 0 & | & -2 \\ 0 & 1 & 0 & | & 5 \\ 0 & 0 & 1 & | & 7 \end{bmatrix}$

The solution is $(-2, 5, 7)$.

77. $\begin{bmatrix} 2 & 1 & -1 & | & 2 \\ 1 & -2 & 1 & | & 0 \\ 1 & 3 & -2 & | & 4 \end{bmatrix} \begin{matrix} R_2 \rightarrow \\ 2R_2 - R_1 \rightarrow \\ R_2 - R_3 \rightarrow \end{matrix} \begin{bmatrix} 1 & -2 & 1 & | & 0 \\ 0 & -5 & 3 & | & -2 \\ 0 & -5 & 3 & | & -4 \end{bmatrix} \begin{matrix} \\ (-1/5)R_2 \rightarrow \\ R_2 - R_3 \rightarrow \end{matrix} \begin{bmatrix} 1 & -2 & 1 & | & 0 \\ 0 & 1 & -\frac{3}{5} & | & \frac{2}{5} \\ 0 & 0 & 0 & | & 2 \end{bmatrix}$

The last equation indicates that $0 = 2$, which is always false. Therefore, there are no solutions.

79. Enter the coefficients of the linear system into a 3×4 matrix, as shown in Figure 79a. Reduce the matrix to reduced row-echelon form, as shown in Figure 79b. The solution is the ordered triple $(-9.266, -9.167, 2.440)$.

81. Enter the coefficients of the linear system into a 3×4 matrix, as shown in Figure 81a. Reduce the matrix to reduced row-echelon form, as shown in Figure 81b. The solution is the ordered triple $(5.211, 3.739, -4.655)$.

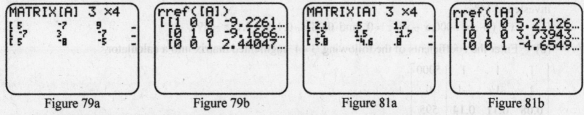

Figure 79a　　　　　Figure 79b　　　　　Figure 81a　　　　　Figure 81b

83. Enter the coefficients of the linear system into a 3×4 matrix, as shown in Figure 83a. Reduce the matrix to reduced row-echelon form, as shown in Figure 83b. The solution is the ordered triple $(7.993, 1.609, -0.401)$.

85. Let x represent the fraction of the pool that the first pump can empty each hour, and y the fraction of the pool that the second pump can empty each hour, and z the fraction for the third pump. Since the first pump is twice as fast we have $x = 2y$ or $x - 2y = 0$. Since the first two pumps can empty the pool in 8 hours, it

follows that they can empty $\frac{1}{8}$ of the pool in an hour. Thus, $x + y = \frac{1}{8}$. Similarly, all three pumps can empty

the pool in 6 hours, so $x + y + z = \frac{1}{6}$. Putting these equations in an augmented matrix results in the following:

$$\begin{bmatrix} 1 & -2 & 0 & | & 0 \\ 1 & 1 & 0 & | & \frac{1}{8} \\ 1 & 1 & 1 & | & \frac{1}{6} \end{bmatrix} \begin{bmatrix} 1 & -2 & 0 & | & 0 \\ 0 & 3 & 0 & | & \frac{1}{8} \\ 0 & 3 & 1 & | & \frac{1}{6} \end{bmatrix} \begin{bmatrix} 1 & -2 & 0 & | & 0 \\ 0 & 1 & 0 & | & \frac{1}{24} \\ 0 & 0 & 1 & | & \frac{1}{24} \end{bmatrix} \begin{bmatrix} 1 & 0 & 0 & | & \frac{1}{12} \\ 0 & 1 & 0 & | & \frac{1}{24} \\ 0 & 0 & 1 & | & \frac{1}{24} \end{bmatrix}$$

Thus, $x = \frac{1}{12}$, $y = z = \frac{1}{24}$. The first pump could empty $\frac{1}{12}$ of the pool in one hour or the entire pool in 12

hours, while the second and third pumps individually could empty the pool in 24 hours. Using technology, this
solution can be obtained as shown in Figure 85.

87. (a) Using technology, find a solution for the augmented matrix.

$$\begin{bmatrix} 1800 & 5000 & 1 & | & 1300 \\ 3200 & 12000 & 1 & | & 5300 \\ 4500 & 13000 & 1 & | & 6500 \end{bmatrix} \rightarrow \begin{bmatrix} 1 & 0 & 0 & | & 0.5714 \\ 0 & 1 & 0 & | & 0.4571 \\ 0 & 0 & 1 & | & -2014 \end{bmatrix}$$

Then $F = 0.5714N + 0.4571R - 2014$

(b) $F \approx 0.5714(3500) + 0.4571(12500) - 2014 \approx \$5700$

89. $I_1 = I_2 + I_3 \Rightarrow I_1 - I_2 - I_3 = 0, 15 + 4I_3 = 14I_2 \Rightarrow -14I_2 + 4I_3 = -15$, and $10 + 4I_3 = 5I_1 \Rightarrow -5I_1 + 4I_3 = -10$

Therefore the matrix is:

$$\begin{bmatrix} 1 & -1 & -1 & | & 0 \\ 0 & -14 & 4 & | & -15 \\ -5 & 0 & 4 & | & -10 \end{bmatrix} \begin{matrix} \\ (-1/14)R_2 \rightarrow \\ 5R_1 + R_3 \rightarrow \end{matrix} \begin{bmatrix} 1 & -1 & -1 & | & 0 \\ 0 & 1 & -\frac{2}{7} & | & \frac{15}{14} \\ 0 & -5 & -1 & | & -10 \end{bmatrix} 5R_2 + R_3 \rightarrow \begin{bmatrix} 1 & -1 & -1 & | & 0 \\ 0 & 1 & -\frac{2}{7} & | & \frac{15}{14} \\ 0 & 0 & -\frac{17}{7} & | & -\frac{65}{14} \end{bmatrix}$$

$$\begin{matrix} \\ \\ (-7/17)R_3 \rightarrow \end{matrix} \begin{bmatrix} 1 & -1 & -1 & | & 0 \\ 0 & 1 & -\frac{2}{7} & | & \frac{15}{14} \\ 0 & 0 & 1 & | & \frac{65}{34} \end{bmatrix} \begin{matrix} R_1 + R_3 \rightarrow \\ (2/7)R_3 + R_2 \rightarrow \\ \\ \end{matrix} \begin{bmatrix} 1 & -1 & 0 & | & \frac{65}{34} \\ 0 & 1 & 0 & | & \frac{385}{238} \\ 0 & 0 & 1 & | & \frac{65}{34} \end{bmatrix} R_2 + R_1 \rightarrow \begin{bmatrix} 1 & 0 & 0 & | & \frac{840}{238} \\ 0 & 1 & 0 & | & \frac{385}{238} \\ 0 & 0 & 1 & | & \frac{65}{34} \end{bmatrix}$$

The solution is: $(3.53, 1.62, 1.91)$.

91. Let x equal the amount invested at 8%, y equal the amount invested at 11% and z equal the amount
invested at 14%.

(a) $x + y + z = 5000, x + y - z = 0$, and $0.08x + 0.11y + 0.14z = 595$

(b) Enter the coefficients of the following 3×4 augmented matrix into a calculator.

$$\begin{bmatrix} 1 & 1 & 1 & | & 5000 \\ 1 & 1 & -1 & | & 0 \\ 0.08 & 0.11 & 0.14 & | & 595 \end{bmatrix}$$

Using technology, the solution is $x = 1000$, $y = 1500$, and $z = 2500$. See Figure 91. Thus, $1000 needs to be
invested at 8%, $1500 at 11%, and $2500 at 14% in order to earn the total amount interest of $595.

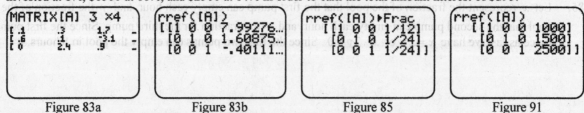

Figure 83a　　　　　Figure 83b　　　　　Figure 85　　　　　Figure 91

93. (a) At intersection A incoming traffic is equal to $x+5$. The outgoing traffic is given by $y+7$. Therefore, $x+5=y+7$, which is the first equation. The incoming traffic at intersection B is $z+6$ and the outgoing traffic is $x+3$, so $z+6=x+3$. Finally at intersection C, the incoming flow is $y+3$ and the outgoing flow is $z+4$, so $y+3=z+4$.

     (b) These three equations can be written as: $x-y=2$, $x-z=3$, and $y-z=1$.

          The system of linear equations can be represented by the following augmented matrix:

          $$\begin{bmatrix} 1 & -1 & 0 & | & 2 \\ 1 & 0 & -1 & | & 3 \\ 0 & 1 & -1 & | & 1 \end{bmatrix}$$

          Begin by subtracting the first row from the second, followed by subtracting the second row from the third. Gaussian elimination results in the following augmented matrix:

          $$\begin{bmatrix} 1 & -1 & 0 & | & 2 \\ 0 & 1 & -1 & | & 1 \\ 0 & 0 & 0 & | & 0 \end{bmatrix}$$

          The last row of zeros indicates that the linear system is dependent and has an infinite number of solutions. Back solving produces $y-z=1 \Rightarrow y=z+1$. Substituting into the first equation gives $x-(z+1)=2 \Rightarrow x=z+3$. Thus, the solution can be written $\{(z+3, z+1, z)|z$ is any nonnegative real number$\}$.

     (c) There are an infinite number of solutions to the system. However, solutions such as $z=1000$, $x=1003$ and $y=1001$ are likely, unless a large number of people are simply driving around the block. In reality there is an average traffic flow rate for z that could be measured. From this, values for both x and y could be determined.

95. (a) The three equations can be written as: $3=1^2 a+1b+c$, $29=5^2 a+5b+c$, and $40=6^2 a+6b+c \Rightarrow 36a+6b+c=40$, $25a+5b+c=29$, $a+b+c=3$.

          Therefore, the matrix is:

          $$\begin{bmatrix} 36 & 6 & 1 & | & 40 \\ 25 & 5 & 1 & | & 29 \\ 1 & 1 & 1 & | & 3 \end{bmatrix}$$

     (b) Using technology, $a=\dfrac{9}{10}$, $b=\dfrac{11}{10}$, $c=1 \Rightarrow f(x)=\dfrac{9}{10}x^2+\dfrac{11}{10}x+1$

     (c) See Figure 95.

     (d) For example, 6 quarters after its release the sales was $f(6)=\dfrac{9}{10}(6)^2+\dfrac{11}{10}(6)+1=40$

          [-0.5, 10, 1] by [-5, 90, 10]
          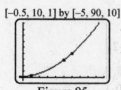
          Figure 95

97. (a) Let $f(x)=ax^2+bx+c$. The constants a, b, and c must satisfy the following equations:

          $f(1990)=a(1990)^2+b(1990)+c=11$, $f(2010)=a(2010)^2+b(2010)+c=10$, and

          $f(2030)=a(2030)^2+b(2030)+c=6$. This system of equations can be represented by the following augmented matrix:

$$\begin{bmatrix} 1990^2 & 1990 & 1 & | & 11 \\ 2010^2 & 2010 & 1 & | & 10 \\ 2030^2 & 2030 & 1 & | & 6 \end{bmatrix}$$

(b)   Using technology to solve the system results in the reduced row-echelon form. The solution is $a = -0.00375$, $b = 14.95$, and $c = -14,889.125$. Therefore, the symbolic representation of f is

$$f(x) = -0.00375x^2 + 14.95x - 14,889.125.$$

(c)   The data and f are graphed in Figure 97. Notice that the graph of f passes through each data point.

(d)   *Answers may vary*. For example, in 2015 the ratio could be $f(2015) \approx 9.3$.

[1985, 2035, 5] by [5, 12, 1]

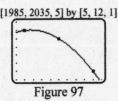

Figure 97

Extended and Discovery Exercise for Section 6.4

1.   Using technology: $w = 1$, $x = -1$, $y = 2$, $z = 0$. The solution is $(1, -1, 2, 0)$.

Checking Basic Concepts for Sections 6.3 and 6.4

1.   (a) $\begin{bmatrix} 1 & -2 & 1 & | & -2 \\ 1 & 1 & 2 & | & 3 \\ 2 & -1 & -1 & | & 5 \end{bmatrix} \begin{matrix} R_1 - R_2 \to \\ 2R_2 - R_3 \to \end{matrix} \begin{bmatrix} 1 & -2 & 1 & | & -2 \\ 0 & -3 & -1 & | & -5 \\ 0 & 3 & 5 & | & 1 \end{bmatrix} \begin{matrix} (-1/3)R_2 \to \\ R_2 + R_3 \to \end{matrix} \begin{bmatrix} 1 & -2 & 1 & | & -2 \\ 0 & 1 & \frac{1}{3} & | & \frac{5}{3} \\ 0 & 0 & 4 & | & -4 \end{bmatrix}$

$(1/4)R_3 \to \begin{bmatrix} 1 & -2 & 1 & | & -2 \\ 0 & 1 & \frac{1}{3} & | & \frac{5}{3} \\ 0 & 0 & 1 & | & -1 \end{bmatrix} \begin{matrix} (-1)R_3 + R_1 \to \\ (-1/3)R_3 + R_2 \to \end{matrix} \begin{bmatrix} 1 & -2 & 0 & | & -1 \\ 0 & 1 & 0 & | & 2 \\ 0 & 0 & 1 & | & -1 \end{bmatrix}$

$2R_2 + R_1 \to \begin{bmatrix} 1 & 0 & 0 & | & 3 \\ 0 & 1 & 0 & | & 2 \\ 0 & 0 & 1 & | & -1 \end{bmatrix}$

The solution is: $(3, 2, -1)$.

(b) $\begin{bmatrix} 1 & -2 & 1 & | & -2 \\ 1 & 1 & 2 & | & 3 \\ 2 & -1 & 3 & | & 1 \end{bmatrix} \begin{matrix} R_1 - R_2 \to \\ 2R_2 - R_3 \to \end{matrix} \begin{bmatrix} 1 & -2 & 1 & | & -2 \\ 0 & -3 & -1 & | & -5 \\ 0 & 3 & 1 & | & 5 \end{bmatrix} \begin{matrix} (-1/3)R_2 \to \\ R_2 + R_3 \to \end{matrix} \begin{bmatrix} 1 & -2 & 1 & | & -2 \\ 0 & 1 & \frac{1}{3} & | & \frac{5}{3} \\ 0 & 0 & 0 & | & 0 \end{bmatrix}$

$2R_2 + R_1 \to \begin{bmatrix} 1 & 0 & \frac{5}{3} & | & \frac{4}{3} \\ 0 & 1 & \frac{1}{3} & | & \frac{5}{3} \\ 0 & 0 & 0 & | & 0 \end{bmatrix}$

The last equation indicates that $0 = 0$, which is true. Therefore, there is an infinite number of solutions.

The second equation gives: $y + \frac{1}{3}z = \frac{5}{3} \Rightarrow y = \frac{5-z}{3}$. The first equation gives:

$x + \frac{5}{3}z = \frac{4}{3} \Rightarrow x = \frac{4-5z}{3}$. Therefore, there are infinitely many solutions which can be written as:

$\left(\frac{4-5z}{3}, \frac{-z+5}{3}, z \right)$.

(c) $\begin{bmatrix} 1 & -2 & 1 & | & -2 \\ 1 & 1 & 2 & | & 3 \\ 2 & -1 & 3 & | & 5 \end{bmatrix} \begin{matrix} R_1 - R_2 \rightarrow \\ 2R_2 - R_3 \rightarrow \end{matrix} \begin{bmatrix} 1 & -2 & 1 & | & -2 \\ 0 & -3 & -1 & | & -5 \\ 0 & 3 & 1 & | & 1 \end{bmatrix} \begin{matrix} (-1/3)R_2 \rightarrow \\ R_2 + R_3 \rightarrow \end{matrix} \begin{bmatrix} 1 & -2 & 1 & | & -2 \\ 0 & 1 & \frac{1}{3} & | & \frac{5}{3} \\ 0 & 0 & 0 & | & -4 \end{bmatrix}$

The last equation gives: $0 = -4$, which is false. Therefore there is no solution.

3. $\begin{bmatrix} 1 & 0 & 1 & | & 2 \\ 1 & 1 & -1 & | & 1 \\ -1 & -2 & -1 & | & 0 \end{bmatrix} \begin{matrix} R_2 - R_1 \rightarrow \\ R_3 + R_1 \rightarrow \end{matrix} \begin{bmatrix} 1 & 0 & 1 & | & 2 \\ 0 & 1 & -2 & | & -1 \\ 0 & -2 & 0 & | & 2 \end{bmatrix} (-1/2)R_3 \rightarrow \begin{bmatrix} 1 & 0 & 1 & | & 2 \\ 0 & 1 & -2 & | & -1 \\ 0 & 1 & 0 & | & -1 \end{bmatrix}$

$R_3 - R_2 \rightarrow \begin{bmatrix} 1 & 0 & 0 & | & -2 \\ 0 & 1 & -2 & | & -1 \\ 0 & 0 & 2 & | & 0 \end{bmatrix}$

Backward substitution produces $z = 0$, $y - 2z = -1 \Rightarrow y - 0 = -1 \Rightarrow y = -1$. Substituting $z = 0$ and $y = -1$ into the first equation results in $x + 0(-1) + (0) = 2 \Rightarrow x = 2$. The solution is $(2, -1, 0)$.

6.5 Properties and Applications of Matrices

1. (a) $a_{12} = 3$, $b_{32} = 1$, $b_{22} = 0$

   (b) $a_{11} = 1$, $b_{11} = 1 \Rightarrow a_{11}b_{11} = 1$; $a_{12} = 3$, $b_{21} = 3 \Rightarrow a_{12}b_{21} = 9$; $a_{13} = -4$,
   $b_{31} = 3 \Rightarrow a_{13}b_{31} = -12 \Rightarrow a_{11}b_{11} + a_{12}b_{21} + a_{13}b_{31} = -2$

   (c) If $A = B$, then all elements of A must be equal to all elements of B. Then $x = 3$.

3. $x = 1$ and $y = 1$

5. The matrices are not the same size, therefore not possible.

7. (a) $\begin{bmatrix} 4 & -1 \\ -1 & 4 \end{bmatrix} + \begin{bmatrix} -1 & 4 \\ 4 & -1 \end{bmatrix} = \begin{bmatrix} 3 & 3 \\ 3 & 3 \end{bmatrix}$

   (b) $\begin{bmatrix} -1 & 4 \\ 4 & -1 \end{bmatrix} + \begin{bmatrix} 4 & -1 \\ -1 & 4 \end{bmatrix} = \begin{bmatrix} 3 & 3 \\ 3 & 3 \end{bmatrix}$

   (c) $\begin{bmatrix} 4 & -1 \\ -1 & 4 \end{bmatrix} - \begin{bmatrix} -1 & 4 \\ 4 & -1 \end{bmatrix} = \begin{bmatrix} 5 & -5 \\ -5 & 5 \end{bmatrix}$

9. (a) $\begin{bmatrix} 3 & 4 & -1 \\ 0 & -3 & 2 \\ -2 & 5 & 10 \end{bmatrix} + \begin{bmatrix} 11 & 5 & -2 \\ 4 & -7 & 12 \\ 6 & 6 & 6 \end{bmatrix} = \begin{bmatrix} 14 & 9 & -3 \\ 4 & -10 & 14 \\ 4 & 11 & 16 \end{bmatrix}$

   (b) $\begin{bmatrix} 11 & 5 & -2 \\ 4 & -7 & 12 \\ 6 & 6 & 6 \end{bmatrix} + \begin{bmatrix} 3 & 4 & -1 \\ 0 & -3 & 2 \\ -2 & 5 & 10 \end{bmatrix} = \begin{bmatrix} 14 & 9 & -3 \\ 4 & -10 & 14 \\ 4 & 11 & 16 \end{bmatrix}$

(c) $\begin{bmatrix} 3 & 4 & -1 \\ 0 & -3 & 2 \\ -2 & 5 & 10 \end{bmatrix} - \begin{bmatrix} 11 & 5 & -2 \\ 4 & -7 & 12 \\ 6 & 6 & 6 \end{bmatrix} = \begin{bmatrix} -8 & -1 & 1 \\ -4 & 4 & -10 \\ -8 & -1 & 4 \end{bmatrix}$

11. (a) $A + B = \begin{bmatrix} 2 & -6 \\ 3 & 1 \end{bmatrix} + \begin{bmatrix} -1 & 0 \\ -2 & 3 \end{bmatrix} = \begin{bmatrix} 1 & -6 \\ 1 & 4 \end{bmatrix}$

   (b) $3A = 3 \begin{bmatrix} 2 & -6 \\ 3 & 1 \end{bmatrix} = \begin{bmatrix} 6 & -18 \\ 9 & 3 \end{bmatrix}$

   (c) $2A - 3B = 2 \begin{bmatrix} 2 & -6 \\ 3 & 1 \end{bmatrix} - 3 \begin{bmatrix} -1 & 0 \\ -2 & 3 \end{bmatrix} = \begin{bmatrix} 7 & -12 \\ 12 & -7 \end{bmatrix}$

13. (a) $A + B$ is undefined since A is 3×3 and B is 2×3. They do not have the same dimension.

   (b) $3A = 3 \begin{bmatrix} 1 & -1 & 0 \\ 1 & 5 & 9 \\ -4 & 8 & -5 \end{bmatrix} = \begin{bmatrix} 3 & -3 & 0 \\ 3 & 15 & 27 \\ -12 & 24 & -15 \end{bmatrix}$

   (c) $2A - 3B$ is undefined since A is 3×3 and B is 2×3. They do not have the same dimension.

15. (a) $A + B = \begin{bmatrix} -2 & -1 \\ -5 & 1 \\ 2 & -3 \end{bmatrix} + \begin{bmatrix} 2 & -1 \\ 3 & 1 \\ 7 & -5 \end{bmatrix} = \begin{bmatrix} 0 & -2 \\ -2 & 2 \\ 9 & -8 \end{bmatrix}$

   (b) $3A = 3 \begin{bmatrix} -2 & -1 \\ -5 & 1 \\ 2 & -3 \end{bmatrix} = \begin{bmatrix} -6 & -3 \\ -15 & 3 \\ 6 & -9 \end{bmatrix}$

   (c) $2A - 3B = 2 \begin{bmatrix} -2 & -1 \\ -5 & 1 \\ 2 & -3 \end{bmatrix} - 3 \begin{bmatrix} 2 & -1 \\ 3 & 1 \\ 7 & -5 \end{bmatrix} = \begin{bmatrix} -10 & 1 \\ -19 & -1 \\ -17 & 9 \end{bmatrix}$

17. $2 \begin{bmatrix} 2 & -1 \\ 5 & 1 \\ 0 & 3 \end{bmatrix} + \begin{bmatrix} 5 & 0 \\ 7 & -3 \\ 1 & 1 \end{bmatrix} - \begin{bmatrix} 9 & -4 \\ 4 & 4 \\ 1 & 6 \end{bmatrix} = \begin{bmatrix} 0 & 2 \\ 13 & -5 \\ 0 & 1 \end{bmatrix}$

19. $\begin{bmatrix} 4 & 6 \\ 3 & -7 \end{bmatrix} - 2 \begin{bmatrix} 1 & 0 \\ -4 & 1 \end{bmatrix} = \begin{bmatrix} 2 & 6 \\ 11 & -9 \end{bmatrix}$

21. $2 \begin{bmatrix} 2 & -1 & -1 \\ -1 & 2 & -1 \\ -1 & -1 & 2 \end{bmatrix} + 3 \begin{bmatrix} 1 & 2 & 3 \\ 2 & 1 & 3 \\ 2 & 3 & 1 \end{bmatrix} = \begin{bmatrix} 7 & 4 & 7 \\ 4 & 7 & 7 \\ 4 & 7 & 7 \end{bmatrix}$

23. The "1" is dark gray and the background is light gray.

$A = \begin{bmatrix} 1 & 2 & 1 \\ 1 & 2 & 1 \\ 1 & 2 & 1 \end{bmatrix}$

25. To enhance the contrast, change light gray to white and change dark gray to black. This could be accomplished by adding the 3×3 matrix B to A.

$B = \begin{bmatrix} -1 & 1 & -1 \\ -1 & 1 & -1 \\ -1 & 1 & -1 \end{bmatrix}$; $A + B = \begin{bmatrix} 1 & 2 & 1 \\ 1 & 2 & 1 \\ 1 & 2 & 1 \end{bmatrix} + \begin{bmatrix} -1 & 1 & -1 \\ -1 & 1 & -1 \\ -1 & 1 & -1 \end{bmatrix} = \begin{bmatrix} 0 & 3 & 0 \\ 0 & 3 & 0 \\ 0 & 3 & 0 \end{bmatrix}$

27. A and B are both 2×2 so AB and BA are also both 2×2.

$$AB = \begin{bmatrix} 1 & -1 \\ 2 & 0 \end{bmatrix}\begin{bmatrix} -2 & 3 \\ 1 & 2 \end{bmatrix} = \begin{bmatrix} -3 & 1 \\ -4 & 6 \end{bmatrix}; \quad BA = \begin{bmatrix} -2 & 3 \\ 1 & 2 \end{bmatrix}\begin{bmatrix} 1 & -1 \\ 2 & 0 \end{bmatrix} = \begin{bmatrix} 4 & 2 \\ 5 & -1 \end{bmatrix}$$

29.   Since both A and B are 2×3, the number of rows in B is not equal to the number of columns in A, so AB is undefined. Also, the number of rows in A is not equal to the number of columns in B so BA is undefined.

31. $AB = \begin{bmatrix} 3 & -1 \\ 1 & 0 \\ -2 & -4 \end{bmatrix}\begin{bmatrix} -2 & 5 & -3 \\ 9 & -7 & 0 \end{bmatrix} =$

$$\begin{bmatrix} 3(-2)+(-1)(9) & 3(5)+(-1)(-7) & 3(-3)+(-1)(0) \\ 1(-2)+0(9) & 1(5)+0(-7) & 1(-3)+0(0) \\ -2(-2)+(-4)(9) & -2(5)+(-4)(-7) & -2(-3)+(-4)(0) \end{bmatrix} \Rightarrow AB = \begin{bmatrix} -15 & 22 & -9 \\ -2 & 5 & -3 \\ -32 & 18 & 6 \end{bmatrix}$$

$$BA = \begin{bmatrix} -2 & 5 & -3 \\ 9 & -7 & 0 \end{bmatrix}\begin{bmatrix} 3 & -1 \\ 1 & 0 \\ -2 & -4 \end{bmatrix} =$$

$$\begin{bmatrix} -2(3)+5(1)+(-3)(-2) & -2(-1)+(5)(0)+(-3)(-4) \\ 9(3)+(-7)(1)+0(-2) & 9(-1)+(-7)(0)+0(-4) \end{bmatrix} \Rightarrow BA = \begin{bmatrix} 5 & 14 \\ 20 & -9 \end{bmatrix}$$

33. AB is undefined, we cannot multiply a 3×3 by a 2×3.

$$BA = \begin{bmatrix} -1 & 3 & -1 \\ 7 & -7 & 1 \end{bmatrix}\begin{bmatrix} 1 & -1 & 0 \\ 2 & -1 & 5 \\ 6 & 1 & -4 \end{bmatrix} =$$

$$\begin{bmatrix} -1(1)+3(2)+(-1)(6) & -1(-1)+3(-1)+(-1)(1) & -1(0)+3(5)+(-1)(-4) \\ 7(1)+(-7)(2)+1(6) & 7(-1)+(-7)(-1)+1(1) & 7(0)+(-7)(5)+1(-4) \end{bmatrix} \Rightarrow BA = \begin{bmatrix} -1 & -3 & 19 \\ -1 & 1 & -39 \end{bmatrix}$$

35.   Since A is 2×2 and B is 3×1, the number of rows in B is not equal to the number of columns in A, so AB is undefined. Also, the number of rows in A is not equal to the number of columns in B so BA is undefined.

37. A and B are both 3×3 so AB and BA are also both 3×3.

$$AB = \begin{bmatrix} 2 & -1 & 3 \\ 0 & 1 & 0 \\ 2 & -2 & 3 \end{bmatrix}\begin{bmatrix} 1 & 5 & -1 \\ 0 & 1 & 3 \\ -1 & 2 & 1 \end{bmatrix} = \begin{bmatrix} -1 & 15 & -2 \\ 0 & 1 & 3 \\ -1 & 14 & -5 \end{bmatrix};$$

$$BA = \begin{bmatrix} 1 & 5 & -1 \\ 0 & 1 & 3 \\ -1 & 2 & 1 \end{bmatrix}\begin{bmatrix} 2 & -1 & 3 \\ 0 & 1 & 0 \\ 2 & -2 & 3 \end{bmatrix} = \begin{bmatrix} 0 & 6 & 0 \\ 6 & -5 & 9 \\ 0 & 1 & 0 \end{bmatrix}$$

39. A is 2×2 and B is 2×1 so AB is 2×1. However BA is undefined since the number of rows in A is not equal to the number of columns in B.

$$AB = \begin{bmatrix} 2 & -1 \\ 3 & 1 \end{bmatrix}\begin{bmatrix} 1 \\ 3 \end{bmatrix} = \begin{bmatrix} -1 \\ 6 \end{bmatrix}$$

41. A is 2×2 and B is 2×3 so AB is 2×3. However BA is undefined since the number of rows in A is not equal to the number of columns in B.

$$AB = \begin{bmatrix} -3 & 1 \\ 2 & -4 \end{bmatrix}\begin{bmatrix} 1 & 0 & -2 \\ -4 & 8 & 1 \end{bmatrix} = \begin{bmatrix} -7 & 8 & 7 \\ 18 & -32 & -8 \end{bmatrix}$$

43. A is 3×3 and B is 3×1 so AB is 3×1. However BA is undefined since the number of rows in A is not equal to the number of columns in B.

$$AB = \begin{bmatrix} 1 & 0 & -2 \\ 3 & -4 & 1 \\ 2 & 0 & 5 \end{bmatrix} \begin{bmatrix} 1 \\ -1 \\ 3 \end{bmatrix} = \begin{bmatrix} -5 \\ 10 \\ 17 \end{bmatrix}$$

45. Using technology; $BA = \begin{bmatrix} 3 & -2 & 4 \\ 5 & 2 & 3 \\ 7 & 5 & 4 \end{bmatrix} \begin{bmatrix} 1 & 1 & -5 \\ -1 & 0 & -7 \\ -6 & 4 & 3 \end{bmatrix} = \begin{bmatrix} -19 & 19 & 11 \\ 21 & -7 & -48 \\ -22 & 23 & -58 \end{bmatrix}$

47. Using technology; $3A^2 + 2B = 3\begin{bmatrix} 3 & -2 & 4 \\ 5 & 2 & 3 \\ 7 & 5 & 4 \end{bmatrix}^2 + 2\begin{bmatrix} 1 & 1 & -5 \\ -1 & 0 & -7 \\ -6 & 4 & 3 \end{bmatrix} = \begin{bmatrix} 83 & 32 & 92 \\ 10 & -63 & -8 \\ 210 & 56 & 93 \end{bmatrix}$

49. (a) $B + C = \begin{bmatrix} 6 & 2 & 7 \\ 3 & -4 & -5 \\ 7 & 1 & 0 \end{bmatrix} + \begin{bmatrix} 1 & 4 & -3 \\ 8 & 1 & -1 \\ 4 & 6 & -2 \end{bmatrix} = \begin{bmatrix} 7 & 6 & 4 \\ 11 & -3 & -6 \\ 11 & 7 & -2 \end{bmatrix}$

   $A(B + C) = \begin{bmatrix} 2 & -1 & 3 \\ 1 & 3 & -5 \\ 0 & -2 & 1 \end{bmatrix} \begin{bmatrix} 7 & 6 & 4 \\ 11 & -3 & -6 \\ 11 & 7 & -2 \end{bmatrix} = \begin{bmatrix} 36 & 36 & 8 \\ -15 & -38 & -4 \\ -11 & 13 & 10 \end{bmatrix}$

   (b) $AB = \begin{bmatrix} 2 & -1 & 3 \\ 1 & 3 & -5 \\ 0 & -2 & 1 \end{bmatrix} \begin{bmatrix} 6 & 2 & 7 \\ 3 & -4 & -5 \\ 7 & 1 & 0 \end{bmatrix} = \begin{bmatrix} 30 & 11 & 19 \\ -20 & -15 & -8 \\ 1 & 9 & 10 \end{bmatrix}$

   $AC = \begin{bmatrix} 2 & -1 & 3 \\ 1 & 3 & -5 \\ 0 & -2 & 1 \end{bmatrix} \begin{bmatrix} 1 & 4 & -3 \\ 8 & 1 & -1 \\ 4 & 6 & -2 \end{bmatrix} = \begin{bmatrix} 6 & 25 & -11 \\ 5 & -23 & 4 \\ -12 & 4 & 0 \end{bmatrix}$; $AB + AC = \begin{bmatrix} 36 & 36 & 8 \\ -15 & -38 & -4 \\ -11 & 13 & 10 \end{bmatrix}$

   $A(B+C) = AB + AC$, which indicates that the distributive property holds for matrices.

51. (a) $(A - B)^2 = \begin{bmatrix} -4 & -3 & -4 \\ -2 & 7 & 0 \\ -7 & -3 & 1 \end{bmatrix} \begin{bmatrix} -4 & -3 & -4 \\ -2 & 7 & 0 \\ -7 & -3 & 1 \end{bmatrix} = \begin{bmatrix} 50 & 3 & 12 \\ -6 & 55 & 8 \\ 27 & -3 & 29 \end{bmatrix}$

   (b) $A^2 - AB - BA + B^2 =$

   $\begin{bmatrix} 3 & -11 & 14 \\ 5 & 18 & -17 \\ -2 & -8 & 11 \end{bmatrix} - \begin{bmatrix} 30 & 11 & 19 \\ -20 & -15 & -8 \\ 1 & 9 & 10 \end{bmatrix} - \begin{bmatrix} 14 & -14 & 15 \\ 2 & -5 & 24 \\ 15 & -4 & 16 \end{bmatrix} + \begin{bmatrix} 91 & 11 & 32 \\ -29 & 17 & 41 \\ 45 & 10 & 44 \end{bmatrix} = \begin{bmatrix} 50 & 3 & 12 \\ -6 & 55 & 8 \\ 27 & -3 & 29 \end{bmatrix}$

   $(A - B)^2 = A^2 - AB - BA + B^2$, which indicates that matrices seem to conform to common rules of algebra except for the commutative property since $AB \neq BA$, in general.

53. Because person 1, likes person 4, we put a 1 in row 1 column 4. Similarly, Person 4 likes person 2, so we put a 1 in row 4 column 2. When no arrow exists to indicate that one person likes another, we place a 0 in the appropriate row and column matrix. Using this process results in the following matrix.

$$\begin{bmatrix} 0 & 0 & 1 & 1 \\ 1 & 0 & 0 & 0 \\ 1 & 0 & 0 & 1 \\ 1 & 1 & 1 & 0 \end{bmatrix}$$

55. Since there is only one 1 in the second column then Person 2 is the least liked person in the network.

57.   Because person 1, likes person 3, there is a 1 in row 1 column 3. Similarly, Person 3 likes person 2, so there is
       a 1 in row 3 column 2. When no arrow exists to indicate that one person likes another, there is a 0 in the
       appropriate row and column matrix. Using this process results in the following diagram.

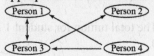

59.   No one likes Person 4.

61. $A^2 = \begin{bmatrix} 0 & 1 & 0 & 0 & 0 \\ 1 & 0 & 0 & 0 & 0 \\ 1 & 1 & 0 & 1 & 1 \\ 1 & 0 & 0 & 0 & 1 \\ 1 & 0 & 0 & 1 & 0 \end{bmatrix}^2 = \begin{bmatrix} 1 & 0 & 0 & 0 & 0 \\ 0 & 1 & 0 & 0 & 0 \\ 3 & 1 & 0 & 1 & 1 \\ 1 & 1 & 0 & 1 & 0 \\ 1 & 1 & 0 & 0 & 1 \end{bmatrix}$

63.   Person 3 likes three people that also like Person 1.

65.   To make a negative image, subtract the matrix A from a completely black image matrix B.

       $B = \begin{bmatrix} 3 & 3 & 3 \\ 3 & 3 & 3 \\ 3 & 3 & 3 \end{bmatrix}$; $B - A = \begin{bmatrix} 3 & 3 & 3 \\ 3 & 3 & 3 \\ 3 & 3 & 3 \end{bmatrix} - \begin{bmatrix} 0 & 3 & 0 \\ 0 & 3 & 0 \\ 0 & 3 & 0 \end{bmatrix} = \begin{bmatrix} 3 & 0 & 3 \\ 3 & 0 & 3 \\ 3 & 0 & 3 \end{bmatrix}$

67. $A = \begin{bmatrix} 3 & 3 & 3 & 3 \\ 3 & 0 & 0 & 0 \\ 3 & 3 & 3 & 0 \\ 3 & 0 & 0 & 0 \\ 3 & 0 & 0 & 0 \end{bmatrix}$

69.   (a)   One possible solution for a "Z" is $A = \begin{bmatrix} 3 & 3 & 3 & 3 \\ 0 & 0 & 3 & 0 \\ 0 & 3 & 0 & 0 \\ 3 & 3 & 3 & 3 \end{bmatrix}$

       (b)   If A is the matrix in part (a) then

              $B = \begin{bmatrix} 3 & 3 & 3 & 3 \\ 3 & 3 & 3 & 3 \\ 3 & 3 & 3 & 3 \\ 3 & 3 & 3 & 3 \end{bmatrix}$ and $B - A = \begin{bmatrix} 3 & 3 & 3 & 3 \\ 3 & 3 & 3 & 3 \\ 3 & 3 & 3 & 3 \\ 3 & 3 & 3 & 3 \end{bmatrix} - \begin{bmatrix} 3 & 3 & 3 & 3 \\ 0 & 0 & 3 & 0 \\ 0 & 3 & 0 & 0 \\ 3 & 3 & 3 & 3 \end{bmatrix} = \begin{bmatrix} 0 & 0 & 0 & 0 \\ 3 & 3 & 0 & 3 \\ 3 & 0 & 3 & 3 \\ 0 & 0 & 0 & 0 \end{bmatrix}$

71.   (a)   One possible solution for a "L" is $A = \begin{bmatrix} 3 & 0 & 0 & 0 \\ 3 & 0 & 0 & 0 \\ 3 & 0 & 0 & 0 \\ 3 & 3 & 3 & 3 \end{bmatrix}$

       (b)   If A is the matrix in part (a) then

              $B = \begin{bmatrix} 3 & 3 & 3 & 3 \\ 3 & 3 & 3 & 3 \\ 3 & 3 & 3 & 3 \\ 3 & 3 & 3 & 3 \end{bmatrix}$ and $B - A = \begin{bmatrix} 3 & 3 & 3 & 3 \\ 3 & 3 & 3 & 3 \\ 3 & 3 & 3 & 3 \\ 3 & 3 & 3 & 3 \end{bmatrix} - \begin{bmatrix} 3 & 0 & 0 & 0 \\ 3 & 0 & 0 & 0 \\ 3 & 0 & 0 & 0 \\ 3 & 3 & 3 & 3 \end{bmatrix} = \begin{bmatrix} 0 & 3 & 3 & 3 \\ 0 & 3 & 3 & 3 \\ 0 & 3 & 3 & 3 \\ 0 & 0 & 0 & 0 \end{bmatrix}$

73.   (a)   These tables can be represented by the matrices A and B where $A = \begin{bmatrix} 12 & 4 \\ 8 & 7 \end{bmatrix}$ and $B = \begin{bmatrix} 55 \\ 70 \end{bmatrix}$.

(b)   The product AB of these matrices calculates tuition cost for each student.

$$AB = \begin{bmatrix} 12 & 4 \\ 8 & 7 \end{bmatrix} \begin{bmatrix} 55 \\ 70 \end{bmatrix} = \begin{bmatrix} 12(55) + & 4(70) \\ 8(55) + & 7(70) \end{bmatrix} = \begin{bmatrix} 940 \\ 930 \end{bmatrix}$$

Student 1 is taking 12 credits at \$55 each and 4 credits at \$70 each. The total tuition for student 1 is $12(\$55) + 4(\$70) = \$940$. Similarly, the tuition for student 2 is \$930.

75. (a)   These tables can be represented by the matrices A and B where $A = \begin{bmatrix} 10 & 5 \\ 9 & 8 \\ 11 & 3 \end{bmatrix}$ and $B = \begin{bmatrix} 60 \\ 70 \end{bmatrix}$.

    (b)   The product AB of these matrices calculates tuition cost for each student.

$$AB = \begin{bmatrix} 10 & 5 \\ 9 & 8 \\ 11 & 3 \end{bmatrix} \begin{bmatrix} 60 \\ 70 \end{bmatrix} = \begin{bmatrix} 10(60) + 5(70) \\ 9(60) + 8(70) \\ 11(60) + 3(70) \end{bmatrix} = \begin{bmatrix} 950 \\ 1100 \\ 870 \end{bmatrix}$$

The total tuition for student 1 is $10(\$60) + 5(\$70) = \$950$. Similarly, the tuition for student 2 is \$1100 and the tuition for student 3 is \$870.

77. $AB = \begin{bmatrix} 3 & 4 & 8 \\ 5 & 6 & 2 \end{bmatrix} \begin{bmatrix} 10 \\ 20 \\ 30 \end{bmatrix} = \begin{bmatrix} 350 \\ 230 \end{bmatrix}$; The total cost of Order 1 is \$350, and the total cost of Order 2 is \$230.

79.   Because there is a link from Page 1 to Page 3 we put a 1 in row 1 column 3. Similarly, there is a link from Page 2 to Page 4, so we put a 1 in row 2 column 4. When no link exists from one web page to another, we place a 0 in the appropriate row and column. Using this process results in the following matrix.

$$\begin{bmatrix} 0 & 0 & 1 & 1 \\ 1 & 0 & 0 & 1 \\ 0 & 0 & 0 & 1 \\ 0 & 0 & 1 & 0 \end{bmatrix}$$

81. $\begin{bmatrix} 0 & 0 & 1 & 1 \\ 1 & 0 & 0 & 1 \\ 0 & 0 & 0 & 1 \\ 0 & 0 & 1 & 0 \end{bmatrix} \cdot \begin{bmatrix} 0 & 0 & 1 & 1 \\ 1 & 0 & 0 & 1 \\ 0 & 0 & 0 & 1 \\ 0 & 0 & 1 & 0 \end{bmatrix} = \begin{bmatrix} 0 & 0 & 1 & 1 \\ 0 & 0 & 2 & 1 \\ 0 & 0 & 1 & 0 \\ 0 & 0 & 0 & 1 \end{bmatrix}$

83.   There are two different 2-click paths from page 2 to page 3.

Extended and Discovery Exercises for Section 6.5

1. $\begin{bmatrix} C \\ M \\ Y \end{bmatrix} = \begin{bmatrix} 1 \\ 1 \\ 1 \end{bmatrix} - \begin{bmatrix} 0.631 \\ 1 \\ 0.933 \end{bmatrix} = \begin{bmatrix} 0.369 \\ 0 \\ 0.067 \end{bmatrix}$

Aquamarine is represented by $(0.369, 0, 0.067)$ in CMY.

3. $\begin{bmatrix} R \\ G \\ B \end{bmatrix} = \begin{bmatrix} 1 \\ 1 \\ 1 \end{bmatrix} - \begin{bmatrix} C \\ M \\ Y \end{bmatrix}$

6.6 Inverses of Matrices

1. B is the inverse of A.

$AB = \begin{bmatrix} 4 & 3 \\ 5 & 4 \end{bmatrix} \begin{bmatrix} 4 & -3 \\ -5 & 4 \end{bmatrix} = \begin{bmatrix} 1 & 0 \\ 0 & 1 \end{bmatrix}$ and $BA = \begin{bmatrix} 4 & -3 \\ -5 & 4 \end{bmatrix} \begin{bmatrix} 4 & 3 \\ 5 & 4 \end{bmatrix} = \begin{bmatrix} 1 & 0 \\ 0 & 1 \end{bmatrix}$

3. B is the inverse of A.

$AB = \begin{bmatrix} 1 & -1 & 2 \\ 0 & 1 & -1 \\ 1 & 0 & 2 \end{bmatrix} \begin{bmatrix} 2 & 2 & -1 \\ -1 & 0 & 1 \\ -1 & -1 & 1 \end{bmatrix} = \begin{bmatrix} 1 & 0 & 0 \\ 0 & 1 & 0 \\ 0 & 0 & 1 \end{bmatrix}$;

$BA = \begin{bmatrix} 2 & 2 & -1 \\ -1 & 0 & 1 \\ -1 & -1 & 1 \end{bmatrix} \begin{bmatrix} 1 & -1 & 2 \\ 0 & 1 & -1 \\ 1 & 0 & 2 \end{bmatrix} = \begin{bmatrix} 1 & 0 & 0 \\ 0 & 1 & 0 \\ 0 & 0 & 1 \end{bmatrix}$

5. B is not the inverse of A.

$AB = \begin{bmatrix} 2 & 1 & -1 \\ 3 & 0 & 2 \\ -1 & 0 & 1 \end{bmatrix} \begin{bmatrix} 0 & 1 & -2 \\ 1 & -3 & 7 \\ 0 & -1 & 3 \end{bmatrix} = \begin{bmatrix} 1 & 0 & 0 \\ 0 & 1 & 0 \\ 0 & -2 & 5 \end{bmatrix}$;

$BA = \begin{bmatrix} 0 & 1 & -2 \\ 1 & -3 & 7 \\ 0 & -1 & 3 \end{bmatrix} \begin{bmatrix} 2 & 1 & -1 \\ 3 & 0 & 2 \\ -1 & 0 & 1 \end{bmatrix} = \begin{bmatrix} 5 & 0 & 0 \\ -14 & 1 & 0 \\ -6 & 0 & 1 \end{bmatrix}$

7. $AA^{-1} = \begin{bmatrix} 1 & 1 \\ 1 & 2 \end{bmatrix} \begin{bmatrix} 2 & -1 \\ -1 & k \end{bmatrix} = \begin{bmatrix} 1 & -1+k \\ 0 & 2k-1 \end{bmatrix}$

We must have $-1+k=0$ and $2k-1=1$. The solution to both equations is $k=1$.

9. $AA^{-1} = \begin{bmatrix} 1 & 3 \\ -1 & -5 \end{bmatrix} \begin{bmatrix} k & 1.5 \\ -0.5 & -0.5 \end{bmatrix} = \begin{bmatrix} k-1.5 & 0 \\ -k+2.5 & 1 \end{bmatrix}$

We must have $k-1.5=1$ and $-k+2.5=0$. The solution to both equations is $k=2.5$.

11. I_2 multiplied by any 2×2 matrix A is equal to A, that is, $I_2 A = A I_2 = A$.

13. I_3 multiplied by any 3×3 matrix A is equal to A, that is, $I_3 A = A I_3 = A$.

15. $A|I_2 = \begin{bmatrix} 1 & 2 & 1 & 0 \\ 1 & 3 & 0 & 1 \end{bmatrix} \begin{matrix} \\ R_2 - R_1 \to \end{matrix} \begin{bmatrix} 1 & 2 & 1 & 0 \\ 0 & 1 & -1 & 1 \end{bmatrix} R_1 - 2R_2 \to \begin{bmatrix} 1 & 0 & 3 & -2 \\ 0 & 0 & -1 & 1 \end{bmatrix}; \; A^{-1} = \begin{bmatrix} 3 & -2 \\ -1 & 1 \end{bmatrix}$

17. $A|I_2 = \begin{bmatrix} -1 & 2 & 1 & 0 \\ 3 & -5 & 0 & 1 \end{bmatrix} -1R_1 \to \begin{bmatrix} 1 & -2 & -1 & 0 \\ 3 & -5 & 0 & 1 \end{bmatrix} \begin{matrix} \\ R_2 - 3R_1 \to \end{matrix} \begin{bmatrix} 1 & -2 & -1 & 0 \\ 0 & 1 & 3 & -1 \end{bmatrix} R_1 + 2R_2 \to \begin{bmatrix} 1 & 0 & 5 & 2 \\ 0 & 1 & 3 & 1 \end{bmatrix};$

$A^{-1} = \begin{bmatrix} 5 & 2 \\ 3 & 1 \end{bmatrix}$

19. $A|I_2 = \begin{bmatrix} 8 & 5 & 1 & 0 \\ 2 & 1 & 0 & 1 \end{bmatrix} \begin{matrix} (1/8)R_1 \to \\ R_1 - 4R_2 \to \end{matrix} \begin{bmatrix} 1 & \frac{5}{8} & \frac{1}{8} & 0 \\ 0 & 1 & 1 & -4 \end{bmatrix} (-5/8)R_2 + R_1 \to \begin{bmatrix} 1 & 0 & -\frac{1}{2} & \frac{5}{2} \\ 0 & 1 & 1 & -4 \end{bmatrix}; \; A^{-1} = \begin{bmatrix} -\frac{1}{2} & \frac{5}{2} \\ 1 & -4 \end{bmatrix}$

21. $A|I_3 = \begin{bmatrix} 0 & 0 & 1 & 1 & 0 & 0 \\ 1 & 0 & 0 & 0 & 1 & 0 \\ 0 & 1 & 0 & 0 & 0 & 1 \end{bmatrix} \begin{matrix} R_2 \to \\ R_3 \to \\ R_1 \to \end{matrix} \begin{bmatrix} 1 & 0 & 0 & 0 & 1 & 0 \\ 0 & 1 & 0 & 0 & 0 & 1 \\ 0 & 0 & 1 & 1 & 0 & 0 \end{bmatrix}; \; A^{-1} = \begin{bmatrix} 0 & 1 & 0 \\ 0 & 0 & 1 \\ 1 & 0 & 0 \end{bmatrix}$

23. $A|I_3 = \begin{bmatrix} 1 & 0 & 1 & 1 & 0 & 0 \\ 2 & 1 & 3 & 0 & 1 & 0 \\ -1 & 1 & 1 & 0 & 0 & 1 \end{bmatrix} \begin{matrix} \\ R_2 - 2R_1 \to \\ R_3 + R_1 \to \end{matrix} \begin{bmatrix} 1 & 0 & 1 & 1 & 0 & 0 \\ 0 & 1 & 1 & -2 & 1 & 0 \\ 0 & 1 & 2 & 1 & 0 & 1 \end{bmatrix} R_3 - R_2 \to$

$\begin{bmatrix} 1 & 0 & 1 & 1 & 0 & 0 \\ 0 & 1 & 1 & -2 & 1 & 0 \\ 0 & 0 & 1 & 3 & -1 & 1 \end{bmatrix} \begin{matrix} R_1 - R_3 \to \\ R_2 - R_3 \to \\ \\ \end{matrix} \begin{bmatrix} 1 & 0 & 0 & -2 & 1 & -1 \\ 0 & 1 & 0 & -5 & 2 & -1 \\ 0 & 0 & 1 & 3 & -1 & 1 \end{bmatrix}; \; A^{-1} = \begin{bmatrix} -2 & 1 & -1 \\ -5 & 2 & -1 \\ 3 & -1 & 1 \end{bmatrix}$

25. $\begin{bmatrix} 1 & 2 & -1 & 1 & 0 & 0 \\ 2 & 5 & 0 & 0 & 1 & 0 \\ -1 & -1 & 2 & 0 & 0 & 1 \end{bmatrix} \begin{matrix} \\ 2R_1 - R_2 \to \\ R_3 + R_1 \to \end{matrix} \begin{bmatrix} 1 & 2 & -1 & 1 & 0 & 0 \\ 0 & -1 & -2 & 2 & -1 & 0 \\ 0 & 1 & 1 & 1 & 0 & 1 \end{bmatrix}$

$\begin{matrix} (-1R_2) \to \\ R_2 + R_3 \to \end{matrix} \begin{bmatrix} 1 & 2 & -1 & 1 & 0 & 0 \\ 0 & 1 & 2 & -2 & 1 & 0 \\ 0 & 0 & -1 & 3 & -1 & 1 \end{bmatrix} \begin{matrix} R_1 - R_3 \to \\ 2R_3 - R_2 \to \\ (-1)R_3 \to \end{matrix} \begin{bmatrix} 1 & 2 & 0 & -2 & 1 & -1 \\ 0 & 1 & 0 & 4 & -1 & 2 \\ 0 & 0 & 1 & -3 & 1 & -1 \end{bmatrix} R_1 - 2R_2 \to \begin{bmatrix} 1 & 0 & 0 & -10 & 3 & -5 \\ 0 & 1 & 0 & 4 & -1 & 2 \\ 0 & 0 & 1 & -3 & 1 & -1 \end{bmatrix};$

$A^{-1} = \begin{bmatrix} -10 & 3 & -5 \\ 4 & -1 & 2 \\ -3 & 1 & -1 \end{bmatrix}$

27. $\begin{bmatrix} -2 & 1 & -3 & 1 & 0 & 0 \\ 0 & 1 & 2 & 0 & 1 & 0 \\ 1 & -2 & 1 & 0 & 0 & 1 \end{bmatrix} \begin{matrix} (-1/2)R_1 \to \\ \\ 2R_3 + R_1 \to \end{matrix} \begin{bmatrix} 1 & -\frac{1}{2} & \frac{3}{2} & -\frac{1}{2} & 0 & 0 \\ 0 & 1 & 2 & 0 & 1 & 0 \\ 0 & -3 & -1 & 1 & 0 & 2 \end{bmatrix}$

$3R_2 + R_3 \to \begin{bmatrix} 1 & -\frac{1}{2} & \frac{3}{2} & -\frac{1}{2} & 0 & 0 \\ 0 & 1 & 2 & 0 & 1 & 0 \\ 0 & 0 & 5 & 1 & 3 & 2 \end{bmatrix} (1/5)R_3 \to \begin{bmatrix} 1 & -\frac{1}{2} & \frac{3}{2} & -\frac{1}{2} & 0 & 0 \\ 0 & 1 & 2 & 0 & 1 & 0 \\ 0 & 0 & 1 & \frac{1}{5} & \frac{3}{5} & \frac{2}{5} \end{bmatrix}$

$\begin{matrix} R_1 - (3/2)R_3 \to \\ (-2)R_3 + R_2 \to \end{matrix} \begin{bmatrix} 1 & -\frac{1}{2} & 0 & -\frac{4}{5} & -\frac{9}{10} & -\frac{3}{5} \\ 0 & 1 & 0 & -\frac{2}{5} & -\frac{1}{5} & -\frac{4}{5} \\ 0 & 0 & 1 & \frac{1}{5} & \frac{3}{5} & \frac{2}{5} \end{bmatrix} R_1 + (1/2)R_2 \to \begin{bmatrix} 1 & 0 & 0 & -1 & -1 & -1 \\ 0 & 1 & 0 & -\frac{2}{5} & -\frac{1}{5} & -\frac{4}{5} \\ 0 & 0 & 1 & \frac{1}{5} & \frac{3}{5} & \frac{2}{5} \end{bmatrix}; \; A^{-1} = \begin{bmatrix} -1 & -1 & -1 \\ -\frac{2}{5} & -\frac{1}{5} & -\frac{4}{5} \\ \frac{1}{5} & \frac{3}{5} & \frac{2}{5} \end{bmatrix}$

29. $A = \begin{bmatrix} 0.5 & -1.5 \\ 0.2 & -0.5 \end{bmatrix} \Rightarrow A^{-1} = \begin{bmatrix} -10 & 30 \\ -4 & 10 \end{bmatrix}$ as shown in Figure 29.

31. $A=\begin{bmatrix}1 & 2 & 0\\ -1 & 4 & -1\\ 2 & -1 & 0\end{bmatrix} \Rightarrow A^{-1}=\begin{bmatrix}0.2 & 0 & 0.4\\ 0.4 & 0 & -0.2\\ 1.4 & -1 & -1.2\end{bmatrix}$ as shown in Figure 31.

33. $A=\begin{bmatrix}2 & -2 & 1\\ 0 & 5 & 8\\ 0 & 0 & -1\end{bmatrix} \Rightarrow A^{-1}=\begin{bmatrix}0.5 & 0.2 & 2.1\\ 0 & 0.2 & 1.6\\ 0 & 0 & -1\end{bmatrix}$ as shown in Figure 33.

35. $A=\begin{bmatrix}3 & -1 & -1\\ -1 & 3 & -1\\ -1 & -1 & 3\end{bmatrix} \Rightarrow A^{-1}=\begin{bmatrix}0.5 & 0.25 & 0.25\\ 0.25 & 0.5 & 0.25\\ 0.25 & 0.25 & 0.5\end{bmatrix}$ as shown in Figure 35.

```
[A]⁻¹
      [[-10 30]
       [-4  10]]
```

Figure 29

```
[A]⁻¹
 [[.2  0   .4  ]
  [.4  0   -.2 ]
  [1.4 -1  -1.2]]
```

Figure 31

```
[A]⁻¹
   [[.5 .2 2.1]
    [0  .2 1.6]
    [0  0  -1 ]]
```

Figure 33

```
[A]⁻¹
  [[.5  .25 .25]
   [.25 .5  .25]
   [.25 .25 .5 ]]
```

Figure 35

37. $A=\begin{bmatrix}1 & -1 & 0 & 0\\ -1 & 5 & -1 & 0\\ 0 & -1 & 5 & -1\\ 0 & 0 & -1 & 1\end{bmatrix} \Rightarrow A^{-1}=\begin{bmatrix}1.2\overline{6} & 0.2\overline{6} & 0.0\overline{6} & 0.0\overline{6}\\ 0.2\overline{6} & 0.2\overline{6} & 0.0\overline{6} & 0.0\overline{6}\\ 0.0\overline{6} & 0.0\overline{6} & 0.2\overline{6} & 0.2\overline{6}\\ 0.0\overline{6} & 0.0\overline{6} & 0.2\overline{6} & 1.2\overline{6}\end{bmatrix}$ as shown in Figure 37.

```
[A]⁻¹
[[1.266666667 ...
 [.2666666667 ...
 [.0666666667 ...
 [.0666666667 ...
```

Figure 37

39. $\begin{aligned}2x-3y&=7\\ -3x-4y&=9\end{aligned} \Rightarrow AX=\begin{bmatrix}2 & -3\\ -3 & -4\end{bmatrix}\begin{bmatrix}x\\ y\end{bmatrix}=\begin{bmatrix}7\\ 9\end{bmatrix}=B$

41. $\begin{aligned}\tfrac{1}{2}x-\tfrac{3}{2}y&=\tfrac{1}{4}\\ -x+2y&=5\end{aligned} \Rightarrow AX=\begin{bmatrix}\frac{1}{2} & -\frac{3}{2}\\ -1 & 2\end{bmatrix}\begin{bmatrix}x\\ y\end{bmatrix}=\begin{bmatrix}\frac{1}{4}\\ 5\end{bmatrix}=B$

43. $\begin{aligned}x-2y+z&=5\\ 3y-z&=6\\ 5x-4y-7z&=0\end{aligned} \Rightarrow AX=\begin{bmatrix}1 & -2 & 1\\ 0 & 3 & -1\\ 5 & -4 & -7\end{bmatrix}\begin{bmatrix}x\\ y\\ z\end{bmatrix}=\begin{bmatrix}5\\ 6\\ 0\end{bmatrix}=B$

45. $\begin{aligned}4x-y+3z&=-2\\ x+2y+5z&=11\\ 2x-3y&=-1\end{aligned} \Rightarrow AX=\begin{bmatrix}4 & -1 & 3\\ 1 & 2 & 5\\ 2 & -3 & 0\end{bmatrix}\begin{bmatrix}x\\ y\\ z\end{bmatrix}=\begin{bmatrix}-2\\ 11\\ -1\end{bmatrix}=B$

47. (a) $\begin{aligned}x+2y&=3\\ x+3y&=6\end{aligned} \Rightarrow AX=\begin{bmatrix}1 & 2\\ 1 & 3\end{bmatrix}\begin{bmatrix}x\\ y\end{bmatrix}=\begin{bmatrix}3\\ 6\end{bmatrix}=B$

(b)   If $AX = B \Rightarrow \begin{bmatrix} 1 & 2 \\ 1 & 3 \end{bmatrix} \begin{bmatrix} x \\ y \end{bmatrix} = \begin{bmatrix} 3 \\ 6 \end{bmatrix} \Rightarrow X = A^{-1}B \Rightarrow \begin{bmatrix} x \\ y \end{bmatrix} = \begin{bmatrix} 3 & -2 \\ -1 & 1 \end{bmatrix} \begin{bmatrix} 3 \\ 6 \end{bmatrix} = \begin{bmatrix} -3 \\ 3 \end{bmatrix}$

The solution to the system is $(-3, 3)$.

49.   (a) $\begin{aligned} -x + 2y &= 5 \\ 3x - 5y &= -2 \end{aligned} \Rightarrow AX = \begin{bmatrix} -1 & 2 \\ 3 & -5 \end{bmatrix} \begin{bmatrix} x \\ y \end{bmatrix} = \begin{bmatrix} 5 \\ -2 \end{bmatrix} = B$

      (b)   If $AX = B \Rightarrow \begin{bmatrix} -1 & 2 \\ 3 & -5 \end{bmatrix} \begin{bmatrix} x \\ y \end{bmatrix} = \begin{bmatrix} 5 \\ -2 \end{bmatrix} \Rightarrow X = A^{-1}B \Rightarrow \begin{bmatrix} x \\ y \end{bmatrix} = \begin{bmatrix} 5 & 2 \\ 3 & 1 \end{bmatrix} \begin{bmatrix} 5 \\ -2 \end{bmatrix} = \begin{bmatrix} 21 \\ 13 \end{bmatrix}$

The solution to the system is $(21, 13)$.

51.   (a) $\begin{aligned} x \quad\quad + z &= -7 \\ 2x + y + 3z &= -13 \\ -x + y + z &= -4 \end{aligned} \Rightarrow AX = \begin{bmatrix} 1 & 0 & 1 \\ 2 & 1 & 3 \\ -1 & 1 & 1 \end{bmatrix} \begin{bmatrix} x \\ y \\ z \end{bmatrix} = \begin{bmatrix} -7 \\ -13 \\ -4 \end{bmatrix} = B$

      (b)   If $AX = B \Rightarrow \begin{bmatrix} 1 & 0 & 1 \\ 2 & 1 & 3 \\ -1 & 1 & 1 \end{bmatrix} \begin{bmatrix} x \\ y \\ z \end{bmatrix} = \begin{bmatrix} -7 \\ -13 \\ -4 \end{bmatrix} \Rightarrow X = A^{-1}B = \begin{bmatrix} x \\ y \\ z \end{bmatrix} = \begin{bmatrix} -2 & 1 & -1 \\ -5 & 2 & -1 \\ 3 & -1 & 1 \end{bmatrix} \begin{bmatrix} -7 \\ -13 \\ -4 \end{bmatrix} =$

$\begin{bmatrix} 5 \\ 13 \\ -12 \end{bmatrix}$ The solution to the system is $(5, 13, -12)$.

53.   (a) $\begin{aligned} x + 2y - z &= 2 \\ 2x + 5y \quad\quad &= -1 \\ -x - y + 2z &= 0 \end{aligned} \Rightarrow AX = \begin{bmatrix} 1 & 2 & -1 \\ 2 & 5 & 0 \\ -1 & -1 & 2 \end{bmatrix} \begin{bmatrix} x \\ y \\ z \end{bmatrix} = \begin{bmatrix} 2 \\ -1 \\ 0 \end{bmatrix} = B$

      (b)   If $AX = B \Rightarrow \begin{bmatrix} 1 & 2 & -1 \\ 2 & 5 & 0 \\ -1 & -1 & 2 \end{bmatrix} \begin{bmatrix} x \\ y \\ z \end{bmatrix} = \begin{bmatrix} 2 \\ -1 \\ 0 \end{bmatrix} \Rightarrow X = A^{-1}B = \begin{bmatrix} x \\ y \\ z \end{bmatrix} = \begin{bmatrix} -10 & 3 & -5 \\ 4 & -1 & 2 \\ -3 & 1 & -1 \end{bmatrix} \begin{bmatrix} 2 \\ -1 \\ 0 \end{bmatrix} = \begin{bmatrix} -23 \\ 9 \\ -7 \end{bmatrix}$ The

solution to the system is $(-23, 9, -7)$.

55.   (a) $AX = B \Rightarrow \begin{bmatrix} 1.5 & 3.7 \\ -0.4 & -2.1 \end{bmatrix} \begin{bmatrix} x \\ y \end{bmatrix} = \begin{bmatrix} 0.32 \\ 0.36 \end{bmatrix}$

      (b)   See Figure 55. $X = A^{-1}B \Rightarrow X = \begin{bmatrix} 1.2 \\ -0.4 \end{bmatrix}$

57.   (a) $AX = B \Rightarrow \begin{bmatrix} 0.08 & -0.7 \\ 1.1 & -0.05 \end{bmatrix} \begin{bmatrix} x \\ y \end{bmatrix} = \begin{bmatrix} -0.504 \\ 0.73 \end{bmatrix}$

      (b)   See Figure 57. $X = A^{-1}B \Rightarrow X = \begin{bmatrix} 0.7 \\ 0.8 \end{bmatrix}$

59.   (a) $AX = B \Rightarrow \begin{bmatrix} 3.1 & 1.9 & -1 \\ 6.3 & 0 & -9.9 \\ -1 & 1.5 & 7 \end{bmatrix} \begin{bmatrix} x \\ y \\ z \end{bmatrix} = \begin{bmatrix} 1.99 \\ -3.78 \\ 5.3 \end{bmatrix}$

      (b)   See Figure 59. $X = A^{-1}B \Rightarrow X = \begin{bmatrix} 0.5 \\ 0.6 \\ 0.7 \end{bmatrix}$

61. (a) $AX = B \Rightarrow \begin{bmatrix} 3 & -1 & 1 \\ 5.8 & -2.1 & 0 \\ -1 & 0 & 2.9 \end{bmatrix} \begin{bmatrix} x \\ y \\ z \end{bmatrix} = \begin{bmatrix} 4.9 \\ -3.8 \\ 3.8 \end{bmatrix}$

     (b)   See Figure 61. $X = A^{-1}B \Rightarrow X \approx \begin{bmatrix} 9.26 \\ 27.39 \\ 4.50 \end{bmatrix}$

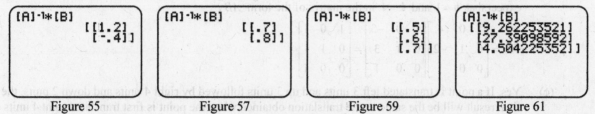

Figure 55   Figure 57   Figure 59   Figure 61

63. (a)   Since $h = 2$ and $k = 3$, the matrix A will translate the point $(0, 1)$ to the right 2 units and up 3 units. Its new location will be $(0 + 2, 1 + 3) = (2, 4)$. This is verified by the following computation:

$$AX = \begin{bmatrix} 1 & 0 & 2 \\ 0 & 1 & 3 \\ 0 & 0 & 1 \end{bmatrix} \begin{bmatrix} 0 \\ 1 \\ 1 \end{bmatrix} = \begin{bmatrix} 2 \\ 4 \\ 1 \end{bmatrix}$$

     (b) $A^{-1}Y = X$. That is, A^{-1} will translate $(2, 4)$ back to $(0, 1)$ by moving it left 2 units and down 3 units. Thus $h = -2$ and $k = -3$ in A^{-1}.

$$A^{-1}Y = \begin{bmatrix} 1 & 0 & -2 \\ 0 & 1 & -3 \\ 0 & 0 & 1 \end{bmatrix} \begin{bmatrix} 2 \\ 4 \\ 1 \end{bmatrix} = \begin{bmatrix} 0 \\ 1 \\ 1 \end{bmatrix}$$

     (c)   The product $AA^{-1} = I_3 = A^{-1}A$, since they are 3×3 inverse matrices.

65. Three units left implies that $h = -3$ and five units down implies that $k = -5$.

$$A = \begin{bmatrix} 1 & 0 & -3 \\ 0 & 1 & -5 \\ 0 & 0 & 1 \end{bmatrix} \text{ and } A^{-1} = \begin{bmatrix} 1 & 0 & 3 \\ 0 & 1 & 5 \\ 0 & 0 & 1 \end{bmatrix}$$

     A^{-1} will translate a point 3 units to the right and 5 units up.

67. (a) $BX = \begin{bmatrix} \dfrac{1}{\sqrt{2}} & \dfrac{1}{\sqrt{2}} & 0 \\[2mm] -\dfrac{1}{\sqrt{2}} & \dfrac{1}{\sqrt{2}} & 0 \\[2mm] 0 & 0 & 1 \end{bmatrix} \begin{bmatrix} -\sqrt{2} \\ -\sqrt{2} \\ 1 \end{bmatrix} = \begin{bmatrix} -2 \\ 0 \\ 1 \end{bmatrix} = Y$

     (b) $B^{-1}Y = \begin{bmatrix} \dfrac{1}{\sqrt{2}} & -\dfrac{1}{\sqrt{2}} & 0 \\[2mm] \dfrac{1}{\sqrt{2}} & \dfrac{1}{\sqrt{2}} & 0 \\[2mm] 0 & 0 & 1 \end{bmatrix} \begin{bmatrix} -2 \\ -0 \\ 1 \end{bmatrix} = \begin{bmatrix} -\sqrt{2} \\ -\sqrt{2} \\ 1 \end{bmatrix} = X$

     B^{-1} rotates the point represented by Y counterclockwise $45°$ about the origin.

69. (a) In the computation ABX, B translates $(1, 1)$ left 3 units and up 3 units to $(-2, 4)$. Then, A translates $(-2, 4)$ right 4 units and down 2 units to $(2, 2)$.

$$ABX = \begin{bmatrix} 1 & 0 & 4 \\ 0 & 1 & -2 \\ 0 & 0 & 1 \end{bmatrix} \begin{bmatrix} 1 & 0 & -3 \\ 0 & 1 & 3 \\ 0 & 0 & 1 \end{bmatrix} \begin{bmatrix} 1 \\ 1 \\ 1 \end{bmatrix} = \begin{bmatrix} 2 \\ 2 \\ 1 \end{bmatrix} = Y;$$ This represents the point $(2, 2)$ as expected.

(b) The net result of A and B is to translate a point 1 unit right and 1 unit up. Therefore, it is reasonable to expect that $h = 1$ and $k = 1$ in the matrix of the form AB.

$$AB = \begin{bmatrix} 1 & 0 & 4 \\ 0 & 1 & -2 \\ 0 & 0 & 1 \end{bmatrix} \begin{bmatrix} 1 & 0 & -3 \\ 0 & 1 & 3 \\ 0 & 0 & 1 \end{bmatrix} = \begin{bmatrix} 1 & 0 & 1 \\ 0 & 1 & 1 \\ 0 & 0 & 1 \end{bmatrix}$$

(c) Yes. If a point is translated left 3 units and up 3 units followed by right 4 units and down 2 units, the final result will be the same as the translation obtained when the point is first translated right 4 units and down 2 units followed by left 3 units and up 3. Therefore, we might expect that $AB = BA$.

$$BA = \begin{bmatrix} 1 & 0 & -3 \\ 0 & 1 & 3 \\ 0 & 0 & 1 \end{bmatrix} \begin{bmatrix} 1 & 0 & 4 \\ 0 & 1 & -2 \\ 0 & 0 & 1 \end{bmatrix} = \begin{bmatrix} 1 & 0 & 1 \\ 0 & 1 & 1 \\ 0 & 0 & 1 \end{bmatrix} = AB$$

(d) Since AB translates a point 1 unit right and 1 unit up, the inverse of AB would translate a point 1 unit left and 1 unit down. So $h = -1$ and $k = -1$ in $(AB)^{-1}$.

$$(AB)^{-1} = \begin{bmatrix} 1 & 0 & -1 \\ 0 & 1 & -1 \\ 0 & 0 & 1 \end{bmatrix};$$ Notice that $(AB)(AB)^{-1} = I_3$ as expected.

71. Let x be the number of CDs of type A purchased, let y be the CDs of type B and let z be the CDs of type C. The first row in the table implies that $2x + 3y + 4z = 120.91$. The other rows can be interpreted similarly. The system in matrix form is shown below.

$$AX = B \Rightarrow \begin{bmatrix} 2 & 3 & 4 \\ 1 & 4 & 0 \\ 2 & 1 & 3 \end{bmatrix} \begin{bmatrix} x \\ y \\ z \end{bmatrix} = \begin{bmatrix} 120.91 \\ 62.95 \\ 79.94 \end{bmatrix}$$

Use a graphing calculator to find the solution as shown in Figure 71. Type A CDs cost $ 10.99, type B cost $ 12.99, and type C cost $ 14.99.

73. (a) The following three equations must be solved, using the equation $P = a + bS + cC$.

$a + b(1500) + c(8) = 122$

$a + b(2000) + c(5) = 130$

$a + b(2200) + c(10) = 158$

These equations can be written in matrix form as follows:

$$AX = B \Rightarrow \begin{bmatrix} 1 & 1500 & 8 \\ 1 & 2000 & 5 \\ 1 & 2200 & 10 \end{bmatrix} \begin{bmatrix} a \\ b \\ c \end{bmatrix} = \begin{bmatrix} 122 \\ 130 \\ 158 \end{bmatrix}$$

The solution is given by $X = A^{-1}B$ as shown in Figure 73. It is $a = 30$, $b = 0.04$, and $c = 4$. That is, P is given by $P = 30 + 0.04S + 4C$

(b) $P = 30 + 0.04(1800) + 4(7) = 130$, or $130,000

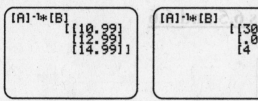

Figure 71   Figure 73

75.   (a)   Although this system of equations can be solved using matrices, we will use elimination to avoid performing row operations on variables. Subtracting the second equation from the third equations gives

$$0.4S + 0.4E + 0.1T = T$$
$$\underline{-0.4S + 0.2E + 0.1T = E}$$
$$0.2E \qquad = T - E \Rightarrow E = \frac{5}{6}T.$$

Substituting $E = \frac{5}{6}T$ in the first equation gives $0.2S + 0.4\left(\frac{5}{6}T\right) + 0.8T = S \Rightarrow .$

$\frac{17}{15}T = \frac{4}{5}S \Rightarrow S = \frac{17}{12}T$. The solution is $\left(\frac{17}{12}T, \frac{5}{6}T, T\right)$.

(b)   When $T = 60$ units, $S = \frac{17}{12}(60) = 85$ units and $E = \frac{5}{6}(60) = 50$ units. That is, service should produce 85 units and electrical should produce 50 units.

Extended and Discovery Exercises for 6.6

1.   (a) $A^{\mathrm{T}} = \begin{bmatrix} 3 & 2 & 4 \\ -3 & 6 & 2 \end{bmatrix}$. (b) $A^{\mathrm{T}} = \begin{bmatrix} 0 & 2 & -4 \\ 1 & 5 & 3 \\ -2 & 4 & 9 \end{bmatrix}$ (c) $A^{\mathrm{T}} = \begin{bmatrix} 5 & 1 & 6 & -9 \\ 7 & -7 & 3 & 2 \end{bmatrix}$

3.   Start by forming the following matrix equation:

$$AX = B \Rightarrow \begin{bmatrix} 0 & 1 \\ 1 & 1 \\ 2 & 1 \\ 3 & 1 \\ 4 & 1 \\ 5 & 1 \end{bmatrix} \begin{bmatrix} a \\ b \end{bmatrix} = \begin{bmatrix} 2.2 \\ 4.5 \\ 7.9 \\ 10.5 \\ 13 \\ 15 \end{bmatrix}$$

To find the least-squares solution to this system of linear equations, solve the matrix equation $A^{\mathrm{T}}AX = A^{\mathrm{T}}B$. for X. The solution is given by $X = (A^{\mathrm{T}}A)^{-1}A^{\mathrm{T}}B$. Enter the matrices A and B and compute the solution as shown in Figure 3a. Thus, $f(x) = 2.6314x + 2.2714$. The data and f are graphed in Figure 3b.

Figure 3a

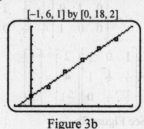

Figure 3b

Checking Basic Concepts for Sections 6.5 and 6.6

1. (a) $A + B = \begin{bmatrix} 1 & 0 & 1 \\ -1 & 1 & 2 \\ 1 & 3 & 0 \end{bmatrix} + \begin{bmatrix} -1 & 1 & 2 \\ 0 & 4 & 1 \\ 1 & -2 & 0 \end{bmatrix} = \begin{bmatrix} 0 & 1 & 3 \\ -1 & 5 & 3 \\ 2 & 1 & 0 \end{bmatrix}$

    (b) $2A - B = 2\begin{bmatrix} 1 & 0 & 1 \\ -1 & 1 & 2 \\ 1 & 3 & 0 \end{bmatrix} - \begin{bmatrix} -1 & 1 & 2 \\ 0 & 4 & 1 \\ 1 & -2 & 0 \end{bmatrix} = \begin{bmatrix} 3 & -1 & 0 \\ -2 & -2 & 3 \\ 1 & 8 & 0 \end{bmatrix}$

    (c) $AB = \begin{bmatrix} 1 & 0 & 1 \\ -1 & 1 & 2 \\ 1 & 3 & 0 \end{bmatrix}\begin{bmatrix} -1 & 1 & 2 \\ 0 & 4 & 1 \\ 1 & -2 & 0 \end{bmatrix} = \begin{bmatrix} 0 & -1 & 2 \\ 3 & -1 & -1 \\ -1 & 13 & 5 \end{bmatrix}$

3. (a) If $\begin{array}{r} x - 2y = 13 \\ 2x + 3y = 5 \end{array} \Rightarrow AX = \begin{bmatrix} 1 & -2 \\ 2 & 3 \end{bmatrix}\begin{bmatrix} x \\ y \end{bmatrix} = \begin{bmatrix} 13 \\ 5 \end{bmatrix}$

    $A^{-1} = \begin{bmatrix} 1 & -2 & | & 1 & 0 \\ 2 & 3 & | & 0 & 1 \end{bmatrix} 2R_1 - R_2 \rightarrow \begin{bmatrix} 1 & -2 & | & 1 & 0 \\ 0 & -7 & | & 2 & -1 \end{bmatrix} (-1/7)R_2 \rightarrow \begin{bmatrix} 1 & -2 & | & 1 & 0 \\ 0 & 1 & | & -\frac{2}{7} & \frac{1}{7} \end{bmatrix}$

    $2R_2 + R_1 \rightarrow \begin{bmatrix} 1 & 0 & | & \frac{3}{7} & \frac{2}{7} \\ 0 & 1 & | & -\frac{2}{7} & \frac{1}{7} \end{bmatrix} \Rightarrow A^{-1} = \begin{bmatrix} \frac{3}{7} & \frac{2}{7} \\ -\frac{2}{7} & \frac{1}{7} \end{bmatrix} \Rightarrow X = A^{-1}B \Rightarrow \begin{bmatrix} x \\ y \end{bmatrix} = \begin{bmatrix} \frac{3}{7} & \frac{2}{7} \\ -\frac{2}{7} & \frac{1}{7} \end{bmatrix}\begin{bmatrix} 13 \\ 5 \end{bmatrix} = \begin{bmatrix} 7 \\ -3 \end{bmatrix}$

    The solution to the system is $(7, -3)$. See Figure 3a.

    (b) If $\begin{array}{r} x - y + z = 2 \\ -x + y + z = 4 \\ y - z = -1 \end{array} \Rightarrow AX = \begin{bmatrix} 1 & -1 & 1 \\ -1 & 1 & 1 \\ 0 & 1 & -1 \end{bmatrix}\begin{bmatrix} x \\ y \\ z \end{bmatrix} = \begin{bmatrix} 2 \\ 4 \\ -1 \end{bmatrix} \Rightarrow$

    $A^{-1} = \begin{bmatrix} 1 & -1 & 1 & | & 1 & 0 & 0 \\ -1 & 1 & 1 & | & 0 & 1 & 0 \\ 0 & 1 & -1 & | & 0 & 0 & 1 \end{bmatrix}$

    $R_1 + R_2 \rightarrow \begin{bmatrix} 1 & -1 & 1 & | & 1 & 0 & 0 \\ 0 & 0 & 2 & | & 1 & 1 & 0 \\ 0 & 1 & -1 & | & 0 & 0 & 1 \end{bmatrix} \begin{array}{l} R_3 \rightarrow \\ R_2 \rightarrow \end{array} \begin{bmatrix} 1 & -1 & 1 & | & 1 & 0 & 0 \\ 0 & 1 & -1 & | & 0 & 0 & 1 \\ 0 & 0 & 2 & | & 1 & 1 & 0 \end{bmatrix}$

    $\begin{array}{l} R_2 + R_3 \rightarrow \\ (1/2)R_3 \rightarrow \end{array} \begin{bmatrix} 1 & -1 & 1 & | & 1 & 0 & 0 \\ 0 & 1 & 1 & | & 1 & 1 & 1 \\ 0 & 0 & 1 & | & \frac{1}{2} & \frac{1}{2} & 0 \end{bmatrix} R_1 - R_3 \rightarrow \begin{bmatrix} 1 & -1 & 0 & | & \frac{1}{2} & -\frac{1}{2} & 0 \\ 0 & 1 & 1 & | & 1 & 1 & 1 \\ 0 & 0 & 1 & | & \frac{1}{2} & \frac{1}{2} & 0 \end{bmatrix}$

    $R_2 - R_3 \rightarrow \begin{bmatrix} 1 & -1 & 0 & | & \frac{1}{2} & -\frac{1}{2} & 0 \\ 0 & 1 & 0 & | & \frac{1}{2} & \frac{1}{2} & 1 \\ 0 & 0 & 1 & | & \frac{1}{2} & \frac{1}{2} & 0 \end{bmatrix} R_1 + R_2 \rightarrow \begin{bmatrix} 1 & 0 & 0 & | & 1 & 0 & 1 \\ 0 & 1 & 0 & | & \frac{1}{2} & \frac{1}{2} & 1 \\ 0 & 0 & 1 & | & \frac{1}{2} & \frac{1}{2} & 0 \end{bmatrix} \Rightarrow$

    $A^{-1} = \begin{bmatrix} 1 & 0 & 1 \\ \frac{1}{2} & \frac{1}{2} & 1 \\ \frac{1}{2} & \frac{1}{2} & 0 \end{bmatrix} \Rightarrow X = A^{-1}B \Rightarrow \begin{bmatrix} x \\ y \\ z \end{bmatrix} = \begin{bmatrix} 1 & 0 & 1 \\ \frac{1}{2} & \frac{1}{2} & 1 \\ \frac{1}{2} & \frac{1}{2} & 0 \end{bmatrix}\begin{bmatrix} 2 \\ 4 \\ -1 \end{bmatrix} = \begin{bmatrix} 1 \\ 2 \\ 3 \end{bmatrix}$

    The solution to the system is $(1, 2, 3)$. See Figure 3b.

(c) $AX = B \Rightarrow \begin{bmatrix} 3.1 & -5.3 \\ -0.1 & 1.8 \end{bmatrix} \begin{bmatrix} x \\ y \end{bmatrix} = \begin{bmatrix} -2.682 \\ 0.787 \end{bmatrix}$. Then $X = A^{-1}B \Rightarrow X = \begin{bmatrix} -0.13 \\ 0.43 \end{bmatrix}$ as shown in Figure 3c.

```
[A]⁻¹*[B]              [A]⁻¹*[B]              [A]⁻¹*[B]
        [[7 ]                   [[1]                   [[-.13]
         [-3]]                   [2]                    [.43 ]]
                                 [3]]
```

| Figure 3a | Figure 3b | Figure 3c |

6.7 Determinants

1. $\det A = \det \begin{bmatrix} 4 & 3 \\ 5 & 4 \end{bmatrix} = (4)(4) - (5)(3) = 1 \neq 0$; A is invertible.

3. $\det A = \det \begin{bmatrix} -4 & 6 \\ -8 & 12 \end{bmatrix} = (-4)(12) - (-8)(6) = 0$; A is not invertible.

5. Deleting the first row and second column gives $M_{12} = \det A = \det \begin{bmatrix} 2 & -2 \\ 0 & 5 \end{bmatrix} = (2)(5) - (0)(-2) = 10$.

    The cofactor is $A_{12} = (-1)^{1+2} M_{12} = -1(10) = -10$.

7. Deleting the second row and second column gives $M_{22} = \det A = \det \begin{bmatrix} 7 & 1 \\ 1 & -2 \end{bmatrix} = (7)(-2) - (1)(1) = -15$.

    The cofactor is $A_{22} = (-1)^{2+2} M_{22} = 1(-15) = -15$.

9. $\det A = a_{11}A_{11} + a_{21}A_{21} + a_{31}A_{31} = a_{11}M_{11} - a_{21}M_{21} + a_{31}M_{31} \Rightarrow$

    $\det A = (1)\det \begin{bmatrix} 2 & -3 \\ -1 & 3 \end{bmatrix} - (0)\det \begin{bmatrix} 4 & -7 \\ -1 & 3 \end{bmatrix} + (0)\det \begin{bmatrix} 4 & -7 \\ 2 & -3 \end{bmatrix} = (1)(6-3) = 3$

    Since $\det A = 3 \neq 0$, A^{-1} exists.

11. $\det A = a_{11}A_{11} + a_{21}A_{21} + a_{31}A_{31} = a_{11}M_{11} - a_{21}M_{21} + a_{31}M_{31} \Rightarrow$

    $\det A = (5)\det \begin{bmatrix} -2 & 0 \\ 4 & 0 \end{bmatrix} - (0)\det \begin{bmatrix} 1 & 6 \\ 4 & 0 \end{bmatrix} + (0)\det \begin{bmatrix} 1 & 6 \\ -2 & 0 \end{bmatrix} = (5)(0-0) = 0$

    Since $\det A = 0$, A^{-1} does not exist.

13. Expanding about the first column results in $\det A = 2\det \begin{bmatrix} 3 & 0 \\ 0 & 5 \end{bmatrix} - 0 + 0 = (2)(15-0) = 30$.

15. Expanding about the first row results in $\det A = 0$.

17. Expanding about the first column results in

    $\det A = 3\det \begin{bmatrix} 5 & 7 \\ 0 & -1 \end{bmatrix} - 0 + 1\det \begin{bmatrix} -1 & 2 \\ 5 & 7 \end{bmatrix} = (3)(-5-0) + (1)(-7-10) = -32$.

19. Expanding about the first column results in

    $\det A = 1\det \begin{bmatrix} 1 & 3 \\ 4 & -2 \end{bmatrix} - (-7)\det \begin{bmatrix} -5 & 2 \\ 4 & -2 \end{bmatrix} + 0 = (1)(-2-12) - (-7)(10-8) = 0$.

21. $\det A = \det \begin{bmatrix} 11 & -32 \\ 1.2 & 55 \end{bmatrix} = 643.4$

23. $\det A = \det \begin{bmatrix} 2.3 & 5.1 & 2.8 \\ 1.2 & 4.5 & 8.8 \\ -0.4 & -0.8 & -1.2 \end{bmatrix} = -4.484$

25.   By Cramer's rule, the solution can be found as follows:

$E = \det \begin{bmatrix} 5 & 2 \\ 1 & 3 \end{bmatrix} = 13$; $F = \det \begin{bmatrix} -1 & 5 \\ 3 & 1 \end{bmatrix} = -16$; $D = \det \begin{bmatrix} -1 & 2 \\ 3 & 3 \end{bmatrix} = -9$

Thus $x = \dfrac{E}{D} = \dfrac{13}{-9}$ and $y = \dfrac{F}{D} = \dfrac{-16}{-9} = \dfrac{16}{9}$. The solution is $\left(-\dfrac{13}{9}, \dfrac{16}{9} \right)$.

27.   By Cramer's rule, the solution can be found as follows:

$E = \det \begin{bmatrix} 8 & 3 \\ 3 & -5 \end{bmatrix} = -49$, $F = \det \begin{bmatrix} -2 & 8 \\ 4 & 3 \end{bmatrix} = -38$; $D = \det \begin{bmatrix} -2 & 3 \\ 4 & -5 \end{bmatrix} = -2$

Thus $x = \dfrac{E}{D} = \dfrac{-49}{-2} = \dfrac{49}{2}$ and $y = \dfrac{F}{D} = \dfrac{-38}{-2} = 19$. The solution is $\left(\dfrac{49}{2}, 19 \right)$.

29.   By Cramer's rule, the solution can be found as follows:

$E = \det \begin{bmatrix} 23 & 4 \\ 70 & -5 \end{bmatrix} = -395$; $F = \det \begin{bmatrix} 7 & 23 \\ 11 & 70 \end{bmatrix} = 237$; $D = \det \begin{bmatrix} 7 & 4 \\ 11 & -5 \end{bmatrix} = -79$

Thus $x = \dfrac{E}{D} = \dfrac{-395}{-79} = 5$ and $y = \dfrac{F}{D} = \dfrac{237}{-79} = -3$. The solution is $(5, -3)$.

31.   By Cramer's rule, the solution can be found as follows:

$E = \det \begin{bmatrix} -0.91 & -2.5 \\ 0.423 & 0.9 \end{bmatrix} = 0.2385$; $F = \det \begin{bmatrix} 1.7 & -0.91 \\ -0.4 & 0.423 \end{bmatrix} = 0.3551$; $D = \det \begin{bmatrix} 1.7 & -2.5 \\ -0.4 & 0.9 \end{bmatrix} = 0.53$

Thus $x = \dfrac{E}{D} = \dfrac{0.2385}{0.53} = 0.45$ and $y = \dfrac{F}{D} = \dfrac{0.3551}{0.53} = 0.67$. The solution is $(0.45, 0.67)$.

33.   Enter the vertices as columns in a counterclockwise direction.

$D = \dfrac{1}{2} \det \begin{bmatrix} 0 & 4 & 1 \\ 0 & 2 & 4 \\ 1 & 1 & 1 \end{bmatrix} = 7$; The area of the triangle is 7 square units.

35.   A line segment between $(1, 3)$ and $(3, 2)$ divides the quadrangle into two triangles whose areas can be found using determinants. Enter the vertices of each triangle as columns in a counterclockwise direction.

$D = \dfrac{1}{2} \det \begin{bmatrix} 0 & 3 & 1 \\ 0 & 2 & 3 \\ 1 & 1 & 1 \end{bmatrix} + \dfrac{1}{2} \det \begin{bmatrix} 1 & 3 & 5 \\ 3 & 2 & 4 \\ 1 & 1 & 1 \end{bmatrix} = 6.5$; The area of the quadrangle is 6.5 square units.

37.   If the three points form a triangle with no area ($D = 0$ using determinants), then the points must be collinear.

$D = \dfrac{1}{2} \det \begin{bmatrix} 1 & -3 & 2 \\ 3 & 11 & 1 \\ 1 & 1 & 1 \end{bmatrix} = 0$; The points are collinear.

39.   If the three points form a triangle with no area ($D = 0$ using determinants), then the points must be collinear.

$$D = \frac{1}{2}\det\begin{bmatrix} -2 & 4 & 2 \\ -5 & 4 & 3 \\ 1 & 1 & 1 \end{bmatrix} = 6 \neq 0 \text{ ; The points are not collinear.}$$

41. Use cofactors to expand about row 1 of $\begin{bmatrix} x & y & 1 \\ 2 & 1 & 1 \\ -1 & 4 & 1 \end{bmatrix} = 0 \Rightarrow x(-3) - y(3) + 9 = 0 \Rightarrow$

    $-3x - 3y = -9 \Rightarrow x + y = 3$.

43. Use cofactors to expand about row 1 of $\begin{bmatrix} x & y & 1 \\ 6 & -7 & 1 \\ 4 & -3 & 1 \end{bmatrix} = 0 \Rightarrow x(-4) - y(2) + (10) = 0 \Rightarrow 2x + y = 5$.

Extended and Discovery Exercises for Section 6.7

1. $D = 1[(1)(3) - (1)(2)] - 2[(1)(3) - (1)(1)] + 0[(1)(2) - (1)(1)] = 1 - 4 + 0 = -3$

   $E = 6[(1)(3) - (1)(2)] - 9[(1)(3) - (1)(1)] + 9[(1)(2) - (1)(1)] = 6 - 18 + 9 = -3$

   $F = 1[(9)(3) - (9)(2)] - 2[(6)(3) - (9)(1)] + 0[(6)(2) - (9)(1)] = 9 - 18 + 0 = -9$

   $G = 1[(1)(9) - (1)(9)] - 2[(1)(9) - (1)(6)] + 0[(1)(9) - (1)(6)] = 0 - 6 + 0 = -6$

   $x = \dfrac{E}{D} = \dfrac{-3}{-3} = 1$, $y = \dfrac{F}{D} = \dfrac{-9}{-3} = 3$, $z = \dfrac{G}{D} = \dfrac{-6}{-3} = 2$. The solution is $(1, 3, 2)$.

3. $D = 1[(1)(2) - (1)(0)] - 1[(0)(2) - (1)(1)] + 0[(0)(0) - (1)(1)] = 2 + 1 + 0 = 3$

   $E = 2[(1)(2) - (1)(0)] - 0[(0)(2) - (1)(1)] + 1[(0)(0) - (1)(1)] = 4 + 0 - 1 = 3$

   $F = 1[(0)(2) - (1)(0)] - 1[(2)(2) - (1)(1)] + 0[(0)(2) - (1)(0)] = 0 - 3 + 0 = -3$

   $G = 1[(1)(1) - (1)(0)] - 1[(0)(1) - (1)(2)] + 0[(0)(0) - (1)(2)] = 1 + 2 + 0 = 3$

   $x = \dfrac{E}{D} = \dfrac{3}{3} = 1$, $y = \dfrac{F}{D} = \dfrac{-3}{3} = -1$, $z = \dfrac{G}{D} = \dfrac{3}{3} = 1$. The solution is $(1, -1, 1)$.

5. $D = 1[(1)(2) - (-1)(1)] - (-1)[(0)(2) - (-1)(2)] + 2[(0)(1) - (1)(2)] = 3 + 2 - 4 = 1$

   $E = 7[(1)(2) - (-1)(1)] - 5[(0)(2) - (-1)(2)] + 6[(0)(1) - (1)(2)] = 21 - 10 - 12 = -1$

   $F = 1[(5)(2) - (6)(1)] - (-1)[(7)(2) - (6)(2)] + 2[(7)(1) - (5)(2)] = 4 + 2 - 6 = 0$

   $G = 1[(1)(6) - (-1)(5)] - (-1)[(0)(6) - (-1)(7)] + 2[(0)(5) - (1)(7)] = 11 + 7 - 14 = 4$

   $x = \dfrac{E}{D} = \dfrac{-1}{1} = -1$, $y = \dfrac{F}{D} = \dfrac{0}{1} = 0$, $z = \dfrac{G}{D} = \dfrac{4}{1} = 4$. The solution is $(-1, 0, 4)$.

7. Use cofactors to expand about row 1 of $\begin{bmatrix} x^2 + y^2 & x & y & 1 \\ 4 & 0 & 2 & 1 \\ 4 & 2 & 0 & 1 \\ 4 & -2 & 0 & 1 \end{bmatrix} = 0 \Rightarrow$

   $(x^2 + y^2)[-8] - x(0) + y(0) - (-32) = 0 \Rightarrow x^2 + y^2 - 4 = 0$.

9.   Use cofactors to expand about row 1 of $\begin{bmatrix} x^2+y^2 & x & y & 1 \\ 1 & 0 & 1 & 1 \\ 2 & 1 & -1 & 1 \\ 8 & 2 & 2 & 1 \end{bmatrix} = 0 \Rightarrow$

$(x^2+y^2)(5) - x(15) + y(-5) - 0 = 0 \Rightarrow 5x^2 + 5y^2 - 15x - 5y = 0$.

Checking Basic Concepts for Section 6.7

1. $\det A = a_{11}A_{11} + a_{21}A_{21} + a_{31}A_{31} = a_{11}M_{11} - a_{21}M_{21} + a_{31}M_{31} \Rightarrow$

$\det A = (1)\det\begin{bmatrix} 3 & 1 \\ -2 & 5 \end{bmatrix} - (2)\det\begin{bmatrix} -1 & 2 \\ -2 & 5 \end{bmatrix} + (0)\det\begin{bmatrix} -1 & 2 \\ 3 & 1 \end{bmatrix} = (1)(15-(-2)) - (2)(-5-(-4)) + 0 = 19$. Since

$\det A = 19 \neq 0$, A is invertible.

Chapter 6 Review Exercises

1. $A(b,h) = \dfrac{1}{2}bh \Rightarrow A(3,6) = \dfrac{1}{2}(3)(6) = 9$

3.   The solution is the intersection point $(3, 1)$. The equation of the red line is $y = 2x - 5$ and the equation of the blue line is $x + y = 4$. Substituting the first equation into the second equation we have

$x + (2x-5) = 4 \Rightarrow 3x - 5 = 4 \Rightarrow 3x = 9 \Rightarrow x = 3$ and $3 + y = 4 \Rightarrow y = 1$.

5.   (a) $3x + y = 1 \Rightarrow y = 1 - 3x$ and $2x - 3y = 8 \Rightarrow y = \dfrac{2x-8}{3}$

Graph $Y_1 = 1 - 3X$ and $Y_2 = (2X-8)/3$. The graphs intersect at the point $(1, -2)$. See Figure 5.

       (b)   Substituting $y = 1 - 3x$ into the second equation gives

$2x - 3(1-3x) = 8 \Rightarrow 2x - 3 + 9x = 8 \Rightarrow 11x = 11 \Rightarrow x = 1$. If $x = 1$, then $y = 1 - 3(1) = -2$. The solution is $(1, -2)$.

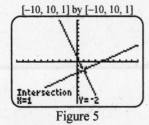
[-10, 10, 1] by [-10, 10, 1]

Figure 5

7.   Multiply the first equation by 2 and add to eliminate the y-variable.

$4x + 2y = 14$

$\underline{x - 2y = -4}$

$5x \quad\quad = 10 \Rightarrow x = 2$

Since $2x + y = 7$, it follows that $y = 3$. The solution is $(2, 3)$. The system is consistent.

9.   Multiply the first equation by 2, the second equation by 3, and add to eliminate the y-variable.

$$12x - 30y = 24$$
$$-12x + 30y = -24$$

$$0 = 0 \;\Rightarrow\; \text{infinitely many solutions of the form} \{(x, y) \mid 2x - 5y = 4\}$$

Since $0 = 0$, the system is consistent and there are infinitely many solutions.

11.   Subtract the equations to eliminate the x-variable.

$$x^2 - 3y = 3$$
$$x^2 + 2y^2 = 5$$

$$-3y - 2y^2 = -2 \Rightarrow 0 = 2y^2 + 3y - 2 \Rightarrow (2y-1)(y+2) = 0 \Rightarrow y = -2, \frac{1}{2}.$$

If $y = -2$ then $x^2 - 3(-2) = 3 \Rightarrow x^2 + 6 = 3 \Rightarrow x^2 = -3$. There is no real solution to this $\Rightarrow y \neq -2$.

If $y = \frac{1}{2}$ then $x^2 - 3\left(\frac{1}{2}\right) = 3 \Rightarrow x^2 = \frac{9}{2} \Rightarrow x = \pm\frac{3}{\sqrt{2}} \Rightarrow x = \pm\frac{3\sqrt{2}}{2}$.

The solutions are $\left(\frac{3\sqrt{2}}{2}, \frac{1}{2}\right)$ and $\left(\frac{-3\sqrt{2}}{2}, \frac{1}{2}\right)$.

13.   See Figure 13.

15.   The solution region is inside the circle or radius 3 and above the line $y = 3 - x$. Their graphs intersect at the points (0, 3) and (3, 0). It does not include the boundary. See Figure 15. One solution to the system is (2, 2).

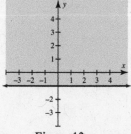

Figure 13   Figure 15

17.   First, subtract the first two equations to eliminate the x-variable.

$$x - y + z = -2$$
$$x + 2y - z = 2$$

$$-3y + 2z = -4$$

Multiply this equation by 2 and multiply the last equation by 3 to eliminate the y-variable.

$$-6y + 4z = -8$$
$$6y + 9z = 21$$

$$13z = 13 \Rightarrow z = 1$$

Substituting $z = 1$ into $2y + 3z = 7$ gives $2y + 3 = 7 \Rightarrow 2y = 4 \Rightarrow y = 2$. Substituting $y = 2$ and $z = 1$ into the original first equation gives $x - 2 + 1 = -2 \Rightarrow x - 1 = -2 \Rightarrow x = -1$.

The solution to the system is $(-1, 2, 1)$.

19.   Add the first two equations to eliminate the x-variable.

$$-x + 2y + 2z = 9$$
$$x + y - 3z = 6$$

$$3y - z = 15$$

Subtract this new equation from the third equation.

$$3y - z = 8$$
$$\underline{3y - z = 15}$$
$$0 = -7$$

Since $0 \neq -7$, there is no solution.

21.   The system is $x + 5y = 6$ and $y = 3$.

Substituting $y = 3$ into the first equation gives $x + 5(3) = 6 \Rightarrow x = -9$. The solution is $(-9, 3)$.

23.   The augmented matrix is in Row-Echelon Form $\Rightarrow x = -2$, $y = 3$, $z = 0 \Rightarrow (-2, 3, 0)$.

25. $\begin{bmatrix} 2 & -1 & 2 & | & 10 \\ 1 & -2 & 1 & | & 8 \\ 3 & -1 & 2 & | & 11 \end{bmatrix} \begin{matrix} (1/2)R_1 \to \\ R_1 - 2R_2 \to \\ 3R_2 - R_3 \to \end{matrix} \begin{bmatrix} 1 & -\frac{1}{2} & 1 & | & 5 \\ 0 & 3 & 0 & | & -6 \\ 0 & -5 & 1 & | & 13 \end{bmatrix} (1/3)R_2 \to \begin{bmatrix} 1 & -\frac{1}{2} & 1 & | & 5 \\ 0 & 1 & 0 & | & -2 \\ 0 & -5 & 1 & | & 13 \end{bmatrix} \begin{matrix} \\ \\ 5R_2 + R_3 \to \end{matrix} \begin{bmatrix} 1 & -\frac{1}{2} & 1 & | & 5 \\ 0 & 1 & 0 & | & -2 \\ 0 & 0 & 1 & | & 3 \end{bmatrix}$

Backward substitution produces $z = 3$; $y = -2$; $x - \frac{1}{2}(-2) + 3 = 5 \Rightarrow x + 1 + 3 = 5 \Rightarrow x = 1$.

The solution is $(1, -2, 3)$.

27.   (a) $a_{12} = 3$ and $a_{22} = 2 \Rightarrow a_{12} + a_{22} = 3 + 2 \Rightarrow a_{12} + a_{22} = 5$

(b) $a_{11} = -2$ and $a_{23} = 4 \Rightarrow a_{11} - 2a_{23} = -2 - 2(4) \Rightarrow a_{11} - 2a_{23} = -10$

29.   (a) $A + 2B = \begin{bmatrix} 1 & -3 \\ 2 & -1 \end{bmatrix} + 2\begin{bmatrix} 3 & 2 \\ -5 & 1 \end{bmatrix} = \begin{bmatrix} 7 & 1 \\ -8 & 1 \end{bmatrix}$

(b) $A - B = \begin{bmatrix} 1 & -3 \\ 2 & -1 \end{bmatrix} - \begin{bmatrix} 3 & 2 \\ -5 & 1 \end{bmatrix} = \begin{bmatrix} -2 & -5 \\ 7 & -2 \end{bmatrix}$

(c) $-4A = -4\begin{bmatrix} 1 & -3 \\ 2 & -1 \end{bmatrix} = \begin{bmatrix} -4 & 12 \\ -8 & 4 \end{bmatrix}$

31. A and B are both 2×2 so AB and BA are also both 2×2.

$AB = \begin{bmatrix} 2 & 0 \\ -5 & 3 \end{bmatrix}\begin{bmatrix} -1 & -2 \\ 4 & 7 \end{bmatrix} = \begin{bmatrix} -2 & -4 \\ 17 & 31 \end{bmatrix}$; $BA = \begin{bmatrix} -1 & -2 \\ 4 & 7 \end{bmatrix}\begin{bmatrix} 2 & 0 \\ -5 & 3 \end{bmatrix} = \begin{bmatrix} 8 & -6 \\ -27 & 21 \end{bmatrix}$

33. A is 2×3 and B is 3×2 so AB is 2×2 and BA is 3×3.

$AB = \begin{bmatrix} 2 & -1 & 3 \\ 2 & 4 & 0 \end{bmatrix}\begin{bmatrix} 1 & 0 \\ -1 & 2 \\ 0 & 3 \end{bmatrix} = \begin{bmatrix} 3 & 7 \\ -2 & 8 \end{bmatrix}$; $BA = \begin{bmatrix} 1 & 0 \\ -1 & 2 \\ 0 & 3 \end{bmatrix}\begin{bmatrix} 2 & -1 & 3 \\ 2 & 4 & 0 \end{bmatrix} = \begin{bmatrix} 2 & -1 & 3 \\ 2 & 9 & -3 \\ 6 & 12 & 0 \end{bmatrix}$

35. B is the inverse of A.

$AB = \begin{bmatrix} 8 & 5 \\ 6 & 4 \end{bmatrix}\begin{bmatrix} 2 & -2.5 \\ -3 & 4 \end{bmatrix} = \begin{bmatrix} 1 & 0 \\ 0 & 1 \end{bmatrix}$ and $BA = \begin{bmatrix} 2 & -2.5 \\ -3 & 4 \end{bmatrix}\begin{bmatrix} 8 & 5 \\ 6 & 4 \end{bmatrix} = \begin{bmatrix} 1 & 0 \\ 0 & 1 \end{bmatrix}$

37. $A | I_2 = \begin{bmatrix} 1 & -2 & | & 1 & 0 \\ -1 & 1 & | & 0 & 1 \end{bmatrix} R_2 + R_1 \to \begin{bmatrix} 1 & -2 & | & 1 & 0 \\ 0 & -1 & | & 1 & 1 \end{bmatrix} (-1)R_2 \to \begin{bmatrix} 1 & -2 & | & 1 & 0 \\ 0 & 1 & | & -1 & -1 \end{bmatrix}$

$R_1 + 2R_2 \to \begin{bmatrix} 1 & 0 & | & -1 & -2 \\ 0 & 1 & | & -1 & -1 \end{bmatrix}$; $A^{-1} = \begin{bmatrix} -1 & -2 \\ -1 & -1 \end{bmatrix}$

39.   (a) $\begin{matrix} x - 3y = 4 \\ 2x - y = 3 \end{matrix} \Rightarrow AX = \begin{bmatrix} 1 & -3 \\ 2 & -1 \end{bmatrix}\begin{bmatrix} x \\ y \end{bmatrix} = \begin{bmatrix} 4 \\ 3 \end{bmatrix} = B$

(b) $X = A^{-1}B \Rightarrow \begin{bmatrix} x \\ y \end{bmatrix} = \begin{bmatrix} -\frac{1}{5} & \frac{3}{5} \\ -\frac{2}{5} & \frac{1}{5} \end{bmatrix}\begin{bmatrix} 4 \\ 3 \end{bmatrix} = \begin{bmatrix} 1 \\ -1 \end{bmatrix}$

41. (a) $AX = B \Rightarrow \begin{bmatrix} 12 & 7 & -3 \\ 8 & -11 & 13 \\ -23 & 0 & 9 \end{bmatrix} \begin{bmatrix} x \\ y \\ z \end{bmatrix} = \begin{bmatrix} 14.6 \\ -60.4 \\ -14.6 \end{bmatrix}$

     (b) See Figure 41. $X = A^{-1}B \Rightarrow X = \begin{bmatrix} -0.5 \\ 1.7 \\ -2.9 \end{bmatrix}$

[−10, 10, 1] by [−10, 10, 1]

```
[A]⁻¹*[B]
         [[-.5 ]
          [1.7 ]
          [-2.9]]
```

Figure 41

43. Expanding about the first column results in

     $\det A = 2\det\begin{bmatrix} 3 & 4 \\ 0 & 5 \end{bmatrix} - 0 + 1\det\begin{bmatrix} 1 & 3 \\ 3 & 4 \end{bmatrix} = (2)(15 - 0) + (1)(4 - 9) = 25$.

45. $\det A = \det\begin{bmatrix} 13 & 22 \\ 55 & -57 \end{bmatrix} = (13)(-57) - (55)(22) = -1951 \neq 0$; A is invertible.

47. The given equations result in the following nonlinear system of equations: $A(l, w) = 77 \Rightarrow lw = 77$ and
     $P(l, w) = 36 \Rightarrow 2l + 2w = 36$. Begin by solving the second equation for l .
     $2l + 2w = 36 \Rightarrow 2l = 36 - 2w \Rightarrow l = 18 - w$. Substitute this into the first equation.
     $lw = 77 \Rightarrow (18 - w)w = 77 \Rightarrow 18w - w^2 = 77 \Rightarrow w^2 - 18w + 77 = 0$. This is a quadratic equation that can be
     solved by factoring. $w^2 - 18w + 77 = 0 \Rightarrow (w - 7)(w - 11) = 0 \Rightarrow w = 7$ or 11.
     Since $l = 18 - w$, if $w = 7$, then $l = 11$ and if $w = 11$ then $l = 7$. For this rectangle $l = 11$ and $w = 7$.

49. (a) Let x represent the amount of the 7% loan and let y represent the amount of the 9% loan. Then the
          system of equations is $x + y = 2000$ and $0.07x + 0.09y = 156$. Multiply the first equation by 0.09 and
          subtract.

$$\begin{array}{r} 0.09x + 0.09y = 180 \\ \underline{0.07x + 0.09y = 156} \\ 0.02x \qquad\quad = 24 \end{array} \Rightarrow x = 1200 \text{ and } y = 2000 - 1200 = 800$$

          The loan amounts are $1200 at 7% and $800 at 9%.

     (b) Graph $Y_1 = 2000 - X$ and $Y_2 = (156 - 0.07X)/0.09$. The graph intersect at (1200, 800). See Figure 49.
          The loan amounts are $1200 at 7% and $800 at 9%.

[0, 2000, 100] by [0, 2000, 100]

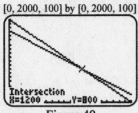

Figure 49

51. Let x be the number of CDs of type A purchased and let y be the number of CDs of type B. Then from the
     table we see that $1x + 2y = 37.47$ and $2x + 3y = 61.95$.

$$AX = B \Rightarrow \begin{bmatrix} 1 & 2 \\ 2 & 3 \end{bmatrix}\begin{bmatrix} x \\ y \end{bmatrix} = \begin{bmatrix} 37.47 \\ 61.95 \end{bmatrix}$$

Using a graphing calculator to solve the system yields that type A CDs cost \$11.49 and type B CD's cost \$12.99.

53.   Enter the vertices as columns in a counterclockwise direction.

$$D = \frac{1}{2}\det\begin{bmatrix} 0 & 5 & 2 \\ 0 & 2 & 5 \\ 1 & 1 & 1 \end{bmatrix} = 10.5 ;$$ The area of the triangle is 10.5 square units.

55.   Since P varies jointly as the square of x and the cube of y, the variation equation $P = kx^2 y^3$ must hold.

If $P = 432$ when $x = 2$ and $y = 3$, then $432 = k(2)^2(3)^3 \Rightarrow 432 = 108k \Rightarrow k = 4$.

Our variation equation becomes $P = 4x^2 y^3$. Thus, when $x = 3$ and $y = 5$, $P = 4(3)^2(5)^3 = 4500$.

Chapters 1-6 Cumulative Review Exercises

1.   Move the decimal point 5 place to the left, $125,000 = 1.25 \times 10^5$.

Move the decimal point 3 places to the left, $4.67 \times 10^{-3} = 0.00467$

3. $D: [-3, 2]$; $R = [-2, 2]$; $f(-0.5) = -2$

5.   (a)   Because $4 - x \not< 0 : 4 - x \geq 0 \Rightarrow -x \geq -4 \Rightarrow x \leq 4$ or $D: \{x | x \leq 4\}$, or $(-\infty, 4]$.

        (b) $f(-1) = \sqrt{4 - (-1)} = \sqrt{5}; f(2a) = \sqrt{4 - (2a)}$

7.   (a)   Using $(0, -1)$ and $(2, 0)$, $m = \frac{0 - (-1)}{2 - 0} = \frac{1}{2}; m = \frac{1}{2}$; y-intercept $(0, -1): -1$.; and x-intercept $(2, 0): 2$.

        (b)   Using slope-intercept form: $f(x) = \frac{1}{2}x - 1$.

        (c)   Graphical: Since the point $(-2, -2)$ is on the graph of $y = f(x)$ it follows that $f(-2) = -2$.

        (d) $f(x) = 0$ at $x = 2$.

9. x-intercept: $y = 0 : -2x + 5(0) = 20 \Rightarrow -2x = 20 \Rightarrow x = -10$. The x-intercept $(-10, 0)$.

y-intercept: $x = 0 : -2(0) + 5y = 20 \Rightarrow 5y = 20 \Rightarrow y = 4$. The y-intercept: $(0, 4)$.

11.   (a) $2(1 - 2x) = 5 - (4 - x) \Rightarrow 2 - 4x = 1 + x \Rightarrow 1 = 5x \Rightarrow x = \frac{1}{5}$

        (b) $2e^x - 1 = 27 \Rightarrow 2e^x = 28 \Rightarrow e^x = 14 \Rightarrow \ln e^x = \ln 14 \Rightarrow x = \ln 14 \Rightarrow x \approx 2.64$

        (c) $\sqrt{2x - 1} = x - 2 \Rightarrow 2x - 1 = x^2 - 4x + 4 \Rightarrow x^2 - 6x + 5 = 0 \Rightarrow (x - 5)(x - 1) = 0 \Rightarrow x = 1, 5$; Checking: $x = 1$ we get $\sqrt{2(1) - 1} = 1 - 2 \Rightarrow \sqrt{1} = -1$ which is false therefore $x = 5$.

        (d) $2x^2 + x = 1 \Rightarrow 2x^2 + x - 1 = 0 \Rightarrow (2x - 1)(x + 1) = 0 \Rightarrow x = -1, \frac{1}{2}$

        (e) $x^3 - 3x^2 + 2x = 0 \Rightarrow x(x^2 - 3x + 2) = 0 \Rightarrow x(x - 2)(x - 1) = 0 \Rightarrow x = 0, 1, 2$

        (f) $x^4 + 8 = 6x^2 \Rightarrow x^4 - 6x^2 + 8 = 0 \Rightarrow (x^2 - 4)(x^2 - 2) = 0 \Rightarrow (x + 2)(x - 2)(x^2 - 2) = 0 \Rightarrow x = -2, 2, \pm\sqrt{2}$

        (g) $\frac{x}{x - 2} = \frac{2x - 1}{x + 1} \Rightarrow x(x + 1) = (2x - 1)(x - 2) \Rightarrow x^2 + x = 2x^2 - 5x + 2 \Rightarrow x^2 - 6x + 2 = 0$

Using the quadratic formula: $\dfrac{6 \pm \sqrt{36 - 4(1)(2)}}{2(1)} = \dfrac{6 \pm \sqrt{28}}{2} = 3 \pm \sqrt{7}$

(h) $|4 - 5x| = 8 \Rightarrow 4 - 5x = -8$ or $4 - 5x = 8$; $4 - 5x = -8 \Rightarrow -5x = -12 \Rightarrow x = \dfrac{12}{5}$ or

$4 - 5x = 8 \Rightarrow -5x = 4 \Rightarrow x = -\dfrac{4}{5}$. Therefore $x = -\dfrac{4}{5}, \dfrac{12}{5}$.

13.   (a) $-3(2 - x) < 4 - (2x + 1) \Rightarrow -6 + 3x < 3 - 2x \Rightarrow 5x < 9 \Rightarrow x < \dfrac{9}{5} \Rightarrow \left(-\infty, \dfrac{9}{5}\right)$ or $\left\{x \,\middle|\, x < \dfrac{9}{5}\right\}$

(b) $-3 \le 4 - 3x < 6 \Rightarrow -7 \le -3x < 2 \Rightarrow \dfrac{7}{3} \ge x > -\dfrac{2}{3} \Rightarrow \left(-\dfrac{2}{3}, \dfrac{7}{3}\right]$ or $\left\{x \,\middle|\, -\dfrac{2}{3} < x \le \dfrac{7}{3}\right\}$

(c)   The solutions to $|4x - 3| \ge 9$ satisfy $x \le s_1$ or $x \ge s_2$ where s_1 and s_2 are the solutions to $|4x - 3| = 9$.

$|4x - 3| = 9$ is equivalent to $4x - 3 = -9 \Rightarrow x = -\dfrac{3}{2}$ and $4x - 3 = 9 \Rightarrow x = 3$.

The interval is $\left(-\infty, -\dfrac{3}{2}\right] \cup [3, \infty)$ or $\left\{x \,\middle|\, x \le -\dfrac{3}{2} \text{ or } x \ge 3\right\}$.

(d)   For $x^2 - 5x + 4 \le 0$ first we solve $x^2 - 5x + 4 = 0 \Rightarrow (x - 4)(x - 1) = 0 \Rightarrow x = 1, 4$. These are the boundary numbers and divide the number line into three sections: $(-\infty, 1]$, $[1, 4]$ and $[4, \infty)$.

Testing a value in each section we get: $x = 0 \Rightarrow 0^2 - 5(0) + 4 \le 0 \Rightarrow 4 \le 0$, which is false;

$x = 2 \Rightarrow 2^2 - 5(2) + 4 \le 0 \Rightarrow 4 - 10 + 4 \le 0 \Rightarrow -2 \le 0$, which is true;

$x = 5 \Rightarrow 5^2 - 5(5) + 4 \le 0 \Rightarrow 25 - 25 + 4 \le 0 \Rightarrow 4 \le 0$, which is false.

Therefore the solution is; $[1, 4]$ or $\{x \,|\, 1 \le x \le 4\}$.

(e)   For $t^3 - t > 0$ first solve $t^3 - t = 0 \Rightarrow t(t^2 - 1) = 0 \Rightarrow t(t + 1)(t - 1) = 0 \Rightarrow t = -1, 0, 1$. These are the boundary numbers and divide the number line into four sections:
$(-\infty, -1)$, $(-1, 0)$, $(0, 1)$ and $(1, \infty)$.

Testing a value in each section we get: $x = -2 \Rightarrow -2^3 - (-2) > 0 \Rightarrow -6 > 0$, which is false;

$x = -\dfrac{1}{2} \Rightarrow \left(-\dfrac{1}{2}\right)^3 - \left(-\dfrac{1}{2}\right) > 0 \Rightarrow -\dfrac{1}{8} + \dfrac{1}{2} > 0 \Rightarrow -\dfrac{3}{8} > 0$, which is true;

$x = \dfrac{1}{2} \Rightarrow \left(\dfrac{1}{2}\right)^3 - \dfrac{1}{2} > 0 \Rightarrow \dfrac{1}{8} - \dfrac{1}{2} > 0 \Rightarrow -\dfrac{3}{8} > 0$, which is false;

$x = 2 \Rightarrow 2^3 - 2 > 0 \Rightarrow 8 - 2 > 0 \Rightarrow 6 > 0$, which is true.

Therefore the solution is: $(-1, 0) \cup (1, \infty)$ or $\{x \,|\, -1 < x < 0 \text{ or } x > 1\}$.

(f)   One boundary number is: $t + 2 = 0 \Rightarrow t = -2$. For the other we solve: $\dfrac{1}{t + 2} - 3 = 0 \Rightarrow$

$1 - 3(t + 2) = 0 \Rightarrow 1 - 3t - 6 = 0 \Rightarrow -3t = 5 \Rightarrow t = -\dfrac{5}{3}$. The two boundary numbers are: -2 and $-\dfrac{5}{3} \Rightarrow$

the possible intervals are: $(-\infty, -2)$, $\left(-2, -\dfrac{5}{3}\right]$, and $\left[-\dfrac{5}{3}, \infty\right)$.

Testing a value in each gives us: $x = -3 \Rightarrow \dfrac{1}{-3 + 2} - 3 \ge 0 \Rightarrow -4 \ge 0$, which is false;

$x = -\dfrac{11}{6} \Rightarrow \dfrac{1}{-\frac{11}{6}+2} - 3 \geq 0 \Rightarrow 3 \geq 0$, which is true;

$x = 0 \Rightarrow \dfrac{1}{0+2} - 3 \geq 0 \Rightarrow -\dfrac{5}{2} \geq 0$, which is false. Therefore the solution is $\left(-2, -\dfrac{5}{3}\right]$ or

$\left\{x \mid -2 < x \leq -\dfrac{5}{3}\right\}$.

15. $-2x^2 + 6x - 1 = 0 \Rightarrow -2x^2 + 6x = 1 \Rightarrow -2(x^2 - 3x) = 1 \Rightarrow$

$-2\left(x^2 - 3x + \dfrac{9}{4}\right) = 1 + \left(-2\left(\dfrac{9}{4}\right)\right) \Rightarrow -2\left(x - \dfrac{3}{2}\right)^2 + \dfrac{7}{2} = 0 \Rightarrow f(x) = -2\left(x - \dfrac{3}{2}\right)^2 + \dfrac{7}{2}$

17. (a) Shift f one unit right and 2 units up. See Figure 17a.

(b) Make graph shorter by taking $\dfrac{1}{2}$ of all y-coordinates of f. See Figure 17b.

(c) Reflect f across both the x-axis and y-axis. Since the graph is symmetric to the y-axis only the reflection across the x-axis will show a result. See Figure 17c.

(d) Make the graph narrower by taking $\dfrac{1}{2}$ of all x-coordinates of f. See Figure 17d.

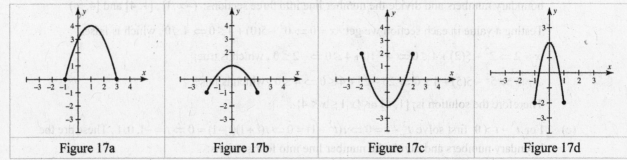

Figure 17a   Figure 17b   Figure 17c   Figure 17d

19. $2x^3 + x^2 - 8x - 4 = (2x^3 - 8x) + (x^2 - 4) = 2x(x^2 - 4) + (x^2 - 4) = (2x + 1)(x^2 - 4) =$
$(2x + 1)(x + 2)(x - 2) \Rightarrow f(x) = (2x + 1)(x + 2)(x - 2)$

21. (a) $(f + g)(2) = f(2) + g(2) = 2 + 3 = 5$

(b) $(g/f)(4) = g(4)/f(4) = 1/0 \Rightarrow$ undefined

(c) $(f \circ g)(3) = f(g(3)) = f(2) = 2$

(d) $(f^{-1} \circ g)(1) = f^{-1}(g(1)) = f^{-1}(4) = 0$

23. (a) $f(2) = 2^2 + 3(2) - 2 = 4 + 6 - 2 = 8$; $g(2) = 2 - 2 = 0$; $(f + g)(2) = 8 + 0 = 8$

(b) $(g \circ f)(1) = g(f(1)) \Rightarrow f(1) = 1^2 + 3(1) - 2 = 1 + 3 - 2 = 2$. Then $g(2) = 2 - 2 = 0 \Rightarrow (g \circ f)(1) = 0$

(c) $f(x) = x^2 + 3x - 2$ and $g(x) = x - 2 \Rightarrow (f - g)(x) = x^2 + 3x - 2 - (x - 2) \Rightarrow (f - g)(x) = x^2 + 2x$

(d) $(f \circ g)(x) = f(g(x)) \Rightarrow (f \circ g)(x) = (x - 2)^2 + 3(x - 2) - 2 \Rightarrow$
$(f \circ g)(x) = x^2 - 4x + 4 + 3x - 6 - 2 \Rightarrow (f \circ g)(x) = x^2 - x - 4$

25. For each unit increase in x, the y-value is multiplied by $\dfrac{2}{3}$. This is an exponential function with $c = 9$ and

$a = \dfrac{2}{3}$, so $f(x) = 9\left(\dfrac{2}{3}\right)^x$.

27.   From the graph when $x = 0$, $y = \dfrac{1}{2}$, $\Rightarrow C = \dfrac{1}{2}$ and when $x = 1$, $y = 1 \Rightarrow$ using $y = Ca^x \Rightarrow 1 = \dfrac{1}{2}a^1 \Rightarrow a = 2$.

So $C = \dfrac{1}{2}$ and $a = 2$.

29.   (a) $\log 100 \Rightarrow 10^x = 100 \Rightarrow x = 2 \Rightarrow \log 100 = 2$

(b) $\log_2 16 \Rightarrow 2^x = 16 \Rightarrow x = 4 \Rightarrow \log_2 16 = 4$

(c) $\ln \dfrac{1}{e^2} \Rightarrow e^x = \dfrac{1}{e^2} \Rightarrow x = -2 \Rightarrow \ln \dfrac{1}{e^2} = -2$

(d) $\log_6 24 - \log_6 4 = \log_6 \dfrac{24}{4} = \log_6 6 \Rightarrow 6^x = 6 \Rightarrow x = 1 \Rightarrow \log_6 24 - \log_6 4 = 1$

31. $\log_2 \dfrac{\sqrt[3]{x^2 - 4}}{\sqrt[3]{x^2 + 4}} = \log_2 \dfrac{(x^2 - 4)^{1/3}}{(x^2 + 4)^{1/2}} = \log_2 (x^2 - 4)^{1/3} - \log_2 (x^2 + 4)^{1/2} =$

$\dfrac{1}{3} \log_2 (x^2 - 4) - \dfrac{1}{2} \log_2 (x^2 + 4) = \dfrac{1}{3} \log_2 [(x+2)(x-2)] - \dfrac{1}{2} \log_2 (x^2 + 4) =$

$\dfrac{1}{3}[\log_2 (x+2) + \log_2 (x-2)] - \dfrac{1}{2} \log_2 (x^2 + 4) = \dfrac{1}{3} \log_2 (x+2) + \dfrac{1}{3} \log_2 (x-2) - \dfrac{1}{2} \log_2 (x^2 + 4)$

33. $f(3,4) = 3^2 + 4^2 = 9 + 16 = 25$

35.   If $z = kx^2 y^{1/2}$, then $7.2 = k(3)^2 (16)^{1/2} \Rightarrow 7.2 = k(9)(4) \Rightarrow 7.2 = 36k \Rightarrow k = 0.2$.

Now $z = 0.2(5)^2 (4)^{1/2} \Rightarrow z = 0.2(25)(2) \Rightarrow z = 10$.

37. $\begin{bmatrix} 1 & -1 & -1 & | & -2 \\ -1 & 1 & -1 & | & 0 \\ 0 & 1 & -2 & | & -6 \end{bmatrix} \begin{matrix} \\ R_1 + R_2 \rightarrow \\ \\ \end{matrix} \begin{bmatrix} 1 & -1 & -1 & | & -2 \\ 0 & 0 & -2 & | & -2 \\ 0 & 1 & -2 & | & -6 \end{bmatrix} \begin{matrix} \\ R_3 \rightarrow \\ R_2 \rightarrow \end{matrix} \begin{bmatrix} 1 & -1 & -1 & | & -2 \\ 0 & 1 & -2 & | & -6 \\ 0 & 0 & -2 & | & -2 \end{bmatrix} (-1/2)R_3 \rightarrow \begin{bmatrix} 1 & -1 & -1 & | & -2 \\ 0 & 1 & -2 & | & -6 \\ 0 & 0 & 1 & | & 1 \end{bmatrix}$

Therefore $z = 1$, substitute this into the second equation $y - 2z = -6 \Rightarrow y - 2(1) = -6 \Rightarrow y - 2 = -6 \Rightarrow y = -4$.
Now substitute $z = 1$ and $y = -4$ into $x - y - z = -2 \Rightarrow x - 1 - (-4) = -2 \Rightarrow x + 3 = -2 \Rightarrow x = -5$. The solution
is: $(-5, -4, 1)$.

39. $\begin{bmatrix} 4 & -5 & | & 1 & 0 \\ 1 & -3 & | & 0 & 1 \end{bmatrix} \begin{matrix} (1/4)R_1 \rightarrow \\ R_1 - 4R_2 \rightarrow \end{matrix} \begin{bmatrix} 1 & -\frac{5}{4} & | & \frac{1}{4} & 0 \\ 0 & 7 & | & 1 & -4 \end{bmatrix} (1/7)R_2 \rightarrow \begin{bmatrix} 1 & -\frac{5}{4} & | & \frac{1}{4} & 0 \\ 0 & 1 & | & \frac{1}{7} & -\frac{4}{7} \end{bmatrix}$

$R_1 + (5/4)R_2 \rightarrow \begin{bmatrix} 1 & 0 & | & \frac{3}{7} & -\frac{5}{7} \\ 0 & 1 & | & \frac{1}{7} & -\frac{4}{7} \end{bmatrix} \Rightarrow A^{-1} = \begin{bmatrix} \frac{3}{7} & -\frac{5}{7} \\ \frac{1}{7} & -\frac{4}{7} \end{bmatrix}$

41.   If $V = \pi r^2 h$, then $12 = \pi(1)^2 h \Rightarrow 12 = \pi h \Rightarrow h = \dfrac{12}{\pi} \Rightarrow h = 3.82$ in.

43.   (a) $t = 0 \Rightarrow D(t) = 16(0)^2 \Rightarrow (0,0)$; $t = 1 \Rightarrow D(t) = 16(1)^2 \Rightarrow (1,16)$

The average rate of change from 0 to 1 is: $\dfrac{16 - 0}{1 - 0} = 16$ ft/sec.

$t = 1 \Rightarrow D(t) = 16(1)^2 \Rightarrow (1, 16)$; $t = 2 \Rightarrow D(t) = 16(2)^2 \Rightarrow (2, 64)$

The average rate of change from 1 to 2 is: $\dfrac{64 - 16}{2 - 1} = 48$ ft/sec.

(b)   During the first second, the average speed is 16 ft/sec. During the next second, the average speed is 48 ft/sec. The object is speeding up.

(c)   The difference quotient is:

$$\frac{16(t+h)^2 - 16t^2}{h} = \frac{16(t^2 + 2th + h^2) - 16t^2}{h} = \frac{16t^2 + 32th + 16h^2 - 16t^2}{h} = 32t + 16h$$

45. $F = \dfrac{k}{d^2} \Rightarrow 150 = \dfrac{k}{(4000)^2} \Rightarrow k = 2,400,000,000$. Now,

$$F = \frac{2,400,000,000}{(10,000)^2} \Rightarrow F = \frac{2,400,000,000}{100,000,000} \Rightarrow F = 24 \text{ lbs.}$$

47.   (a) $C(x) = x(112 - 2x) \Rightarrow C(x) = 112x - 2x^2$

(b) $1470 = 112x - 2x^2 \Rightarrow 2x^2 - 112x + 1470 = 0 \Rightarrow x^2 - 56x + 735 = 0 \Rightarrow (x - 21)(x - 35) = 0 \Rightarrow x = 21, 35$; the cost is \$1470 when either 21 or 35 rooms are rented.

(c)   Graph $C(x) = -2x^2 + 112x$. See Figure 47. The maximum occurs at the vertex $(28, 1568) \Rightarrow \$1568$ when 28 rooms rented.

[0, 70, 10] by [0, 2000, 500]

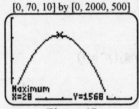

Figure 47

49.   (a)   If $f(x) = \dfrac{5}{9}(x - 32)$, this is subtract 32 and multiply by $\dfrac{5}{9} \Rightarrow$ we divide by $\dfrac{5}{9}$, which is multiply by $\dfrac{9}{5}$,

and add $32 \Rightarrow f^{-1}(x) = \dfrac{9}{5}x + 32$.

(b)   Since $f(x)$ converted Fahrenheit to Celsius, $f^{-1}(x)$ converts Celsius to Fahrenheit.

51. $A = P\left(1 + \dfrac{r}{n}\right)^{nt} \Rightarrow \dfrac{A}{P} = \left(1 + \dfrac{r}{n}\right)^{nt} \Rightarrow \dfrac{A}{P} = \left[\left(1 + \dfrac{r}{n}\right)^{n}\right]^{t} \Rightarrow \ln\left(\dfrac{A}{P}\right) = \ln\left[\left(1 + \dfrac{r}{n}\right)^{n}\right]^{t} \Rightarrow$

$\ln\left(\dfrac{A}{P}\right) = t \ln\left(1 + \dfrac{r}{n}\right)^{n} \Rightarrow \dfrac{\ln\left(\frac{A}{P}\right)}{\ln\left(a + \frac{r}{n}\right)^{n}} = t \Rightarrow t = \dfrac{\ln\left(\frac{A}{P}\right)}{n \ln\left(1 + \frac{r}{n}\right)}$

53.   Let $x =$ amount borrowed at 4% and $y =$ amount borrowed at 3%, then $x + y = 5000$ and $0.04x + 0.03y = 173$. Using elimination, multiply the second equation by 25 and subtract from the first equation:

$$\begin{array}{r} x + y = 5000 \\ x + 0.75y = 4325 \\ \hline 0.25y = 675 \Rightarrow y = 2700 \end{array}$$

Therefore \$2300 was invested at 4% and \$2700 was invested at 3%.

55. $-8 = a(-1)^2 + b(-1) + c \Rightarrow -8 = a - b + c$; $6 = a(1)^2 + b(1) + c \Rightarrow 6 = a + b + c$;

$-4 = a(3)^2 + b(3) + c \Rightarrow -4 = 9a + 3b + c$

Add the first two equations to eliminate the b:

$$\begin{array}{r} -8 = a - b + c \\ 6 = a + b + c \\ \hline -2 = 2a + 2c \end{array}$$

Multiply the first equation by 3 and add the third equation to eliminate the b :

$$-24 = 3a - 3b + 3c$$
$$\underline{-4 = 9a + 3b + \;\; c}$$
$$-28 = 12a \qquad\;\; + 4c$$

Now multiply the first new equation by 2 and subtract the new second equation to eliminate the c :

$$-4 = \qquad 4a + 4c$$
$$\underline{-28 = \quad 12a + 4c}$$
$$24 = -8a \qquad \Rightarrow a = -3$$

Substitute this into $-2 = 2a + 2c \Rightarrow -2 = 2(-3) + 2c \Rightarrow 4 = 2c \Rightarrow c = 2$. Substitute $a = -3$ and $c = 2$ into $6 = a + b + c \Rightarrow 6 = -3 + b + 2 \Rightarrow b = 7$. Therefore $a = -3$, $b = 7$ and $c = 2$.

Chapter 7: Conic Sections

7.1: Parabolas

1. A parabola always opens toward its focus.

3. A parabola with equation $x^2 = 4py$ opens upward if p > 0 and it opens downward if p < 0.

5. A parabola with equation $x^2 = 4py$ has a vertical axis of symmetry.

7.

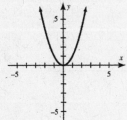

9.

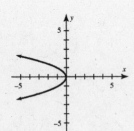

11.

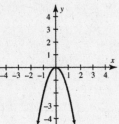

13.

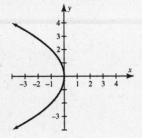

15.

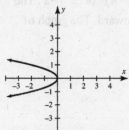

17. Opens upward; e

19. Opens to the left; a

21. Opens to the right, passes through (2, 2); d

23. The equation $16y = x^2$ is in the form $x^2 = 4py$, and the vertex is $V(0, 0)$. Thus, $16 = 4p$ or $p = 4$. The focus is $F(0, 4)$, the equation of the directrix is $y = -4$, and the parabola opens upward. The graph of $y = \frac{1}{16}x^2$ is shown.

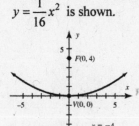

25. The equation $x = \frac{1}{8}y^2$ can be written as $y^2 = 8x$, which is in the form $y^2 = 4px$. The vertex is $V(0, 0)$. Thus, $8 = 4p$ or $p = 2$. The focus is $F(2, 0)$, the equation of the directrix is $x = -2$, and the parabola opens to the right. The graph of $x = \frac{1}{8}y^2$ is shown.

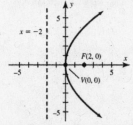

27. The equation $-4x = y^2$ can be written as $y^2 = -4x$, which is in the form $y^2 = 4px$. The vertex is $V(0, 0)$. Thus, $-4 = 4p$ or $p = -1$. The focus is $F(-1, 0)$, the equation of the directrix is $x = 1$, and the parabola opens to the left. The graph of $x = -\frac{1}{4}y^2$ is shown.

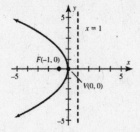

29.   The equation $x^2 = -8y$ is in the form $x^2 = 4py$, and the vertex is $V(0,0)$. Thus, $-8 = 4p$ or $p = -2$. The
      focus is $F(0, -2)$, the equation of the directrix is $y = 2$, and the parabola opens downward. The graph of
      $y = -\frac{1}{8}x^2$ is shown.

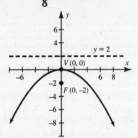

31.   The equation $2y^2 = -8x$ can be written as $y^2 = -4x$, which is in the form $y^2 = 4px$. The vertex is $V(0,0)$.
      Thus, $-4 = 4p$ or $p = -1$. The focus is $F(-1,0)$, the equation of the directrix is $x = 1$, and the parabola
      opens to the left. The graph of $x = -\frac{1}{4}y^2$ is shown.

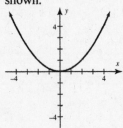

33.   The focus is $F(0,1)$ and the vertex is $V(0,0)$. The distance between these points is 1. Since the focus is
      above the directrix, the parabola opens upward, so $p = 1$. Since the line passing through F and V is vertical,
      the parabola has a vertical axis. Its equation is given by $x^2 = 4py$ or $x^2 = 4y$. The graph of $y = \frac{1}{4}x^2$ is
      shown.

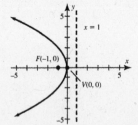

35.   The focus is $F(-3,0)$ and the vertex is $V(0,0)$. The distance between these points is 3. Since the focus is left
      of the directrix, the parabola opens to the left, so $p = -3$. Since the line passing through F and V is
      horizontal, the parabola has a horizontal axis. Its equation is given by $y^2 = 4px$ or $y^2 = -12x$. The graph of
      $x = -\frac{1}{12}y^2$ is shown.

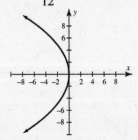

37. If the vertex is $V(0,0)$ and the focus is $F\left(0,\dfrac{3}{4}\right)$, then the parabola opens upward and $p=\dfrac{3}{4}$. Thus,

$x^2=4py \Rightarrow x^2=3y$.

39. If the vertex is $V(0,0)$ and the directrix is $x=2$, then the parabola opens to the left and $p=-2$. Thus,

$y^2=4px \Rightarrow y^2=-8x$.

41. If the vertex is $V(0,0)$ and the focus is $F(1,0)$, then the parabola opens to the right and $p=1$. Thus,

$y^2=4px \Rightarrow y^2=4x$.

43. If the vertex is $V(0,0)$ and the directrix is $x=\dfrac{1}{4}$, then the parabola opens to the left and $p=-\dfrac{1}{4}$. Thus,

$y^2=4px \Rightarrow y^2=-x$.

45. If the vertex is $V(0,0)$ and the parabola has a horizontal axis, the equation is in the form $y^2=4px$. Find the value of p by using the fact that the parabola passes through $(1,-2)$. Thus, $(-2)^2=4p(1) \Rightarrow p=1$. The equation is $y^2=4x$.

47. If the focus is $F(0,-3)$ and the equation of the directrix is $y=3$, the vertex is $V(0,0)$, the parabola opens downward, and $p=-3$. Thus, $x^2=4py \Rightarrow x^2=-12y$.

49. If the focus is $F(-1,0)$ and the equation of the directrix is $x=1$, the vertex is $V(0,0)$, the parabola opens to the left, and $p=-1$. Thus, $y^2=4px \Rightarrow y^2=-4x$.

51.

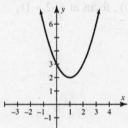

53.

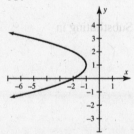

55. Vertex at $(1, 1)$; c

57. Vertex at $(0, 2)$; a

59. The equation $(x-2)^2=8(y+2)$ is in the form $(x-h)^2=4p(y-k)$, with $(h,k)=(2,-2)$ and $p=2$. The parabola opens upward, with vertex at $(2,-2)$, focus at $(2,0)$, and equation of directrix $y=-4$. The graph is

shown.

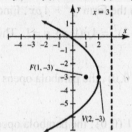

61.   The equation $x = -\dfrac{1}{4}(y+3)^2 + 2$ can be written as $(y+3)^2 = -4(x-2)$, which is in the form

$(y-k)^2 = 4p(x-h)$, with $(h, k) = (2, -3)$ and $p = -1$. The parabola opens to the left, with vertex at

$(2, -3)$, focus at $(1, -3)$, and equation of directrix $x = 3$. The graph is shown.

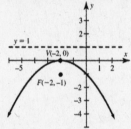

63.   The equation $y = -\dfrac{1}{4}(x+2)^2$ can be written as $(x+2)^2 = -4y$, which is in the form $(x-h)^2 = 4p(y-k)$,

with $(h, k) = (-2, 0)$ and $p = -1$. The parabola opens downward, with vertex at $(-2, 0)$, focus at $(-2, -1)$,

and equation of directrix $y = 1$. The graph is shown.

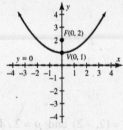

65.   If the focus is at $(0, 2)$ and the vertex at $(0, 1)$, the parabola opens upward and $p = 1$. Substituting in

$(x-h)^2 = 4p(y-k)$, we get $(x-0)^2 = 4(1)(y-1)$ or $x^2 = 4(y-1)$.

67. If the focus is at $(0, 0)$ and the directrix has equation $x = -2$, the vertex is at $(-1,0)$, $p = 1$, and the parabola opens to the right. Substituting in $(y-k)^2 = 4p(x-h)$, we get $(y-0)^2 = 4(1)(x-(-1))$ or $y^2 = 4(x+1)$.

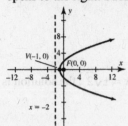

69. If the focus is at $(-1,3)$ and the directrix has equation $y = 7$, the vertex is at $(-1,5)$, $p = -2$, and the parabola opens downward. Substituting in $(x-h)^2 = 4p(y-k)$, we get $(x+1)^2 = 4(-2)(y-5)$ or $(x+1)^2 = -8(y-5)$.

71. Since the parabola has a horizontal axis, the equation is in the form $(y-k)^2 = a(x-h)$. Find the value of a by using the fact that the parabola passes through $(-4,0)$ and the vertex is $V(-2,3)$.

Substituting $x = -4$, $y = 0$, $h = -2$ and $k = 3$ yields $(0-3)^2 = a(-4-(-2)) \Rightarrow a = -\dfrac{9}{2}$.

The equation is $(y-3)^2 = -\dfrac{9}{2}(x+2)$.

73. $-2x = y^2 + 6x + 10 \Rightarrow y^2 = -8x - 10 \Rightarrow (y-0)^2 = -8\left(x + \dfrac{5}{4}\right)$

75. $x = 2y^2 + 4y - 1 \Rightarrow 2y^2 + 4y = x + 1 \Rightarrow y^2 + 2y = \dfrac{1}{2}(x+1) \Rightarrow$

$y^2 + 2y + 1 = \dfrac{1}{2}(x+1) + 1 \Rightarrow (y+1)^2 = \dfrac{1}{2}(x+1+2) \Rightarrow (y+1)^2 = \dfrac{1}{2}(x+3)$

77. $x^2 - 3x + 4 = 2y \Rightarrow x^2 - 3x = 2y - 4 \Rightarrow x^2 - 3x + \dfrac{9}{4} = 2y - 4 + \dfrac{9}{4} \Rightarrow \left(x - \dfrac{3}{2}\right)^2 = 2y - \dfrac{7}{4} \Rightarrow \left(x - \dfrac{3}{2}\right)^2 = 2\left(y - \dfrac{7}{8}\right)$

79. $4y^2 + 4y - 5 = 5x \Rightarrow 4y^2 + 4y = 5x + 5 \Rightarrow y^2 + y = \dfrac{5}{4}(x+1) \Rightarrow$

$y^2 + y + \dfrac{1}{4} = \dfrac{5}{4}(x+1) + \dfrac{1}{4} \Rightarrow \left(y + \dfrac{1}{2}\right)^2 = \dfrac{5}{4}\left(x + 1 + \dfrac{1}{5}\right) \Rightarrow \left(y + \dfrac{1}{2}\right)^2 = \dfrac{5}{4}\left(x + \dfrac{6}{5}\right)$

81. $y = -0.75 \pm \sqrt{-3x}$;

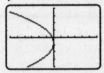

83. $y = 0.5 \pm \sqrt{31(x+13)}$; Note: If a break in the graph appears near the vertex, it should not be there. It is a result of the low resolution of the graphing calculator screen.

85. $y = -1 \pm \sqrt{\dfrac{x}{2.3}}$;

87. $x^2 = 2y$ and $x^2 = y + 1 \Rightarrow 2y = y + 1 \Rightarrow y = 1$; $x^2 = 2y$ when $y = 1 \Rightarrow x^2 = 2 \Rightarrow x = \pm\sqrt{2}$; the solution is $(\pm\sqrt{2}, 1)$.

89. $\dfrac{1}{3}y^2 = -3x$ and $y^2 = x + 1 \Rightarrow -9x = x + 1 \Rightarrow x = -\dfrac{1}{10}$; $y^2 = x + 1$ when

   $x = -\dfrac{1}{10} \Rightarrow y^2 = -\dfrac{1}{10} + 1 \Rightarrow y = \pm\sqrt{09}$; the solution is $\left(-\dfrac{1}{10}, \pm\sqrt{0.9}\right)$.

91. $(y - 1)^2 = x + 1$ and $(y + 2)^2 = -x + 4 \Rightarrow (y - 1)^2 + (y + 2)^2 = 5 \Rightarrow$

   $y^2 - 2y + 1 + y^2 + 4y + 4 = 5 \Rightarrow 2y^2 + 2y = 0 \Rightarrow 2y(y + 1) = 0 \Rightarrow y = 0$ or $y = -1$; $(y - 1)^2 = x + 1$ when

   $y = 0 \Rightarrow 1 = x + 1 \Rightarrow x = 0$, when $y = -1 \Rightarrow 4 = x + 1 \Rightarrow x = 3$, the solution is $(0, 0)$, $(3, -1)$.

93. Substitute the point $(3, 0.75)$ into $x^2 = 4py$ and solve for p; $9 = 4p(0.75) \Rightarrow 9 = 3p \Rightarrow p = 3$. The receiver should be 3 feet from the vertex.

95. (a) Substitute the point $(105, 32)$ into $y = ax^2$ and solve for a; $32 = a(105)^2 \Rightarrow a = \dfrac{32}{11,025}$. The equation

   is $y = \dfrac{32}{11,025}x^2$.

   (b) Rewriting the answer in (a) we have $x^2 = \dfrac{11,025}{32}y$, so $4p = \dfrac{11,025}{32}$ and $p = \dfrac{11,025}{128} \approx 86.1$. The receiver should be located about 86.1 feet from the vertex.

97. (a) Since $y^2 = 100x$, $4p = 100$ and $p = 25$. Thus, the coordinates of the sun are $(25, 0)$.

   (b) The minimum distance occurs when the comet is at the vertex of the parabola, so the minimum distance is 25 million miles.

99. The pipe should be at the focus, so $p = 18$, $k = 4p = 72$ inches or 6 ft.

7.2: Ellipses

1. The endpoints of the major axis of an ellipse are called the vertices of the ellipse.

3. An ellipse with equation $\dfrac{x^2}{a^2} + \dfrac{y^2}{b^2} = 1 (a > b > 0)$ has a horizontal major axis.

5. $\frac{x^2}{4} + \frac{y^2}{9} = 1 \Rightarrow a = 3$ and $b = 2$. $a^2 - b^2 = 3^2 - 2^2 = 5 = c^2 \Rightarrow c = \sqrt{5}$. The foci are $(0, \pm\sqrt{5})$, the endpoints of the major axis (vertices) are $(0, \pm 3)$, while the endpoints of the minor axis are $(\pm 2, 0)$.

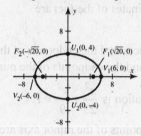

7. $\frac{x^2}{36} + \frac{y^2}{16} = 1 \Rightarrow a = 6$ and $b = 4$. $a^2 - b^2 = 6^2 - 4^2 = 20 = c^2 \Rightarrow c = \sqrt{20}$. The foci are $(\pm\sqrt{20}, 0)$, the endpoints of the major axis (vertices) are $(\pm 6, 0)$, while the endpoints of the minor axis are $(0, \pm 4)$.

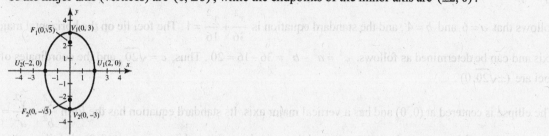

9. $x^2 + 4y^2 = 400 \Rightarrow \frac{x^2}{400} + \frac{y^2}{100} = 1 \Rightarrow a = 20$ and $b = 10$. $a^2 - b^2 = 400 - 100 = 300 = c^2 \Rightarrow c = \sqrt{300}$. The foci are $(\pm\sqrt{300}, 0)$, the endpoints of the major axis (vertices) are $(\pm 20, 0)$, while the endpoints of the minor axis are $(0, \pm 10)$.

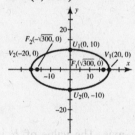

11. $25x^2 + 9y^2 = 225 \Rightarrow \frac{x^2}{9} + \frac{y^2}{25} = 1 \Rightarrow a = 5$ and $b = 3$. $a^2 - b^2 = 25 - 9 = 16 = c^2 \Rightarrow c = 4$. The foci are $(0, \pm 4)$, the endpoints of the major axis (vertices) are $(0, \pm 5)$, while the endpoints of the minor axis are $(\pm 3, 0)$.

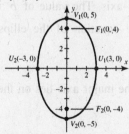

13. Vertices $(0, \pm 6)$; b

15. Vertices $(\pm 4, 0)$; c

17. The ellipse is centered at $(0, 0)$ and has a horizontal major axis. Its standard equation has the form $\dfrac{x^2}{a^2}+\dfrac{y^2}{b^2}=1$. The endpoints of the major axis are $(\pm 6, 0)$ and the endpoints of the minor axis are $(0, \pm 4)$. It follows that $a = 6$ and $b = 4$, and the standard equation is $\dfrac{x^2}{36}+\dfrac{y^2}{16}=1$. The foci lie on the horizontal major axis and can be determined as follows. $c^2 = a^2 - b^2 = 36 - 16 = 20$. Thus, $c = \sqrt{20}$, and the coordinates of the foci are $(\pm\sqrt{20}, 0)$.

19. The ellipse is centered at $(0, 0)$ and has a vertical major axis. Its standard equation has the form $\dfrac{x^2}{b^2}+\dfrac{y^2}{a^2}=1$. The endpoints of the major axis are $(0, \pm 4)$ and the endpoints of the minor axis are $(\pm 2, 0)$. It follows that $a = 4$ and $b = 2$, and the standard equation is $\dfrac{x^2}{4}+\dfrac{y^2}{16}=1$. The foci lie on the vertical major axis and can be determined as follows. $c^2 = a^2 - b^2 = 16 - 4 = 12$. Thus, $c = \sqrt{12}$, and the coordinates of the foci are $(0, \pm\sqrt{12})$.

21. To sketch a graph of an ellipse centered at the origin, it is helpful to plot the vertices and the endpoints of the minor axis. The vertices are $V(\pm 5, 0)$ so $a = 5$, the foci are $F(\pm 4, 0)$ so $c = 4$, and the endpoints of the minor axis are $U(0, \pm 3)$ so $b = 3$. A graph of the ellipse is shown in Figure 21. Its equation is $\dfrac{x^2}{25}+\dfrac{y^2}{9}=1$.

23. The vertices are $V(0, \pm 3)$ so $a = 3$, the foci are $F(0, \pm 2)$ so $c = 2$, and the endpoints of the minor axis are $U(\pm\sqrt{5}, 0)$ so $b = \sqrt{5}$. A graph of the ellipse is shown in Figure 23. Its equation is $\dfrac{x^2}{5}+\dfrac{y^2}{9}=1$.

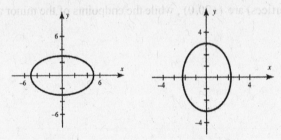

Figure 21   Figure 23

25. Foci of $F(0, \pm 2) \Rightarrow c = 2$ and $V(0, \pm 4) \Rightarrow a = 4$. The major axis lies on the y-axis. The value of b is as follows: $a^2 - b^2 = c^2 \Rightarrow a^2 - c^2 = b^2 \Rightarrow 4^2 - 2^2 = b^2 \Rightarrow b^2 = 12 \Rightarrow b = \sqrt{12}$. The equation of the ellipse is $\dfrac{x^2}{12}+\dfrac{y^2}{16}=1$.

27. Foci of $F(\pm 5, 0) \Rightarrow c = 5$ and $V(\pm 6, 0) \Rightarrow a = 6$. The major axis lies on the x-axis. The value of b is as follows: $a^2 - b^2 = c^2 \Rightarrow a^2 - c^2 = b^2 \Rightarrow 6^2 - 5^2 = b^2 \Rightarrow b^2 = 11 \Rightarrow b = \sqrt{11}$. The equation of the ellipse is $\dfrac{x^2}{36}+\dfrac{y^2}{11}=1$.

29. Horizontal major axis of length $8 \Rightarrow a = 4$. Minor axis of length $6 \Rightarrow b = 3$. The major axis lies on the x-axis. The equation of the ellipse is $\dfrac{x^2}{16}+\dfrac{y^2}{9}=1$.

31. $e = \dfrac{c}{a} = \dfrac{2}{3} \Rightarrow 3c = 2a \Rightarrow c = \dfrac{2}{3}a$. Since the major axis is length $6 \Rightarrow a = 3$. Thus, $c = \dfrac{2}{3}(3) = 2$. Then the value of b is given by the following: $a^2 - b^2 = c^2 \Rightarrow a^2 - c^2 = b^2 \Rightarrow 3^2 - 2^2 = b^2 \Rightarrow b^2 = 5 \Rightarrow b = \sqrt{5}$. The equation of the ellipse is $\dfrac{x^2}{9} + \dfrac{y^2}{5} = 1$.

33. To translate the center from $(0, 0)$ to $(2, -1)$ replace x with $(x - 2)$ and y with $(y + 1)$. This new equation is $\dfrac{(x-2)^2}{4} + \dfrac{(y+1)^2}{3} = 1$.

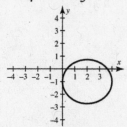

35. To translate the center from $(0, 0)$ to $(-3, -4)$ replace x with $(x + 3)$ and y with $(y + 4)$. This new equation is $\dfrac{(x+3)^2}{2} + \dfrac{(y+4)^2}{9} = 1$.

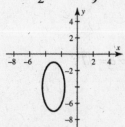

37. The ellipse is centered at $(2, 1)$. The major axis has length $2a = 6$ and the length of the minor axis is $2b = 4$. The major axis is parallel to the y-axis.

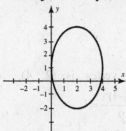

39. The ellipse is centered at $(-1, -2)$. The major axis has length $2a = 10$ and the length of the minor axis is $2b = 8$. The major axis is parallel to the y-axis.

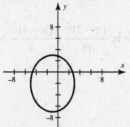

41.   The ellipse is centered at $(-2,0)$. The horizontal major axis has length $2a = 4$ and the vertical minor axis has length $2b = 2$.

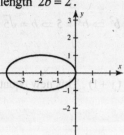

43.   Center at $(2, -4)$; d

45.   Center at $(-1,1)$; c

47.   Center at $(1, 1)$, $a = 5$, $b = 3$, major axis vertical. $c^2 = a^2 - b^2 = 25 - 9 = 16 \Rightarrow c = 4$.
      Foci: $(1, 1 \pm 4)$; veritices: $(1, 1 \pm 5)$; the graph is shown.

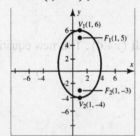

49.   Center at $(-4, 2)$, $a = 4$, $b = 3$, major axis horizontal. $c^2 = a^2 - b^2 = 16 - 9 = 7 \Rightarrow c = \sqrt{7}$.
      Foci: $(-4 \pm \sqrt{7}, 2)$; veritices: $(-4 \pm 4, 2)$; the graph is shown.

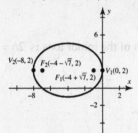

51.   Since the center is $(2, 1)$ and a focus is $(2, 3)$, $c = 2$. Since the vertex is $(2, 4)$, $a = 3$; $b^2 = a^2 - c^2 = 9 - 4 = 5$; the major axis is vertical. The equation is $\dfrac{(x-2)^2}{5} + \dfrac{(y-1)^2}{9} = 1$.

53.   The center is halfway between the vertices at $(0, 2)$; $a = 3$ and $c = 2$; $b^2 = a^2 - c^2 = 9 - 4 = 5$; the major axis is horizontal. The equation is $\dfrac{x^2}{9} + \dfrac{(y-2)^2}{5} = 1$.

55.   Center at $(2, 4)$, $a = 4$ and $b = 2$, major axis parallel to the x-axis; the equation is $\dfrac{(x-2)^2}{16} + \dfrac{(y-4)^2}{4} = 1$.

57. $9x^2 + 18x + 4y^2 - 8y - 23 = 0 \Rightarrow 9(x^2 + 2x) + 4(y^2 - 2y) = 23 \Rightarrow$
      $9(x^2 + 2x + 1) + 4(y^2 - 2y + 1) = 23 + 9 + 4 \Rightarrow 9(x+1)^2 + 4(y-1)^2 = 36 \Rightarrow$
      $\dfrac{(x+1)^2}{4} + \dfrac{(y-1)^2}{9} = 1$; The center is $(-1, 1)$. The vertices are $(-1, 1-3), (-1, 1+3)$ or $(-1, -2), (-1, 4)$.

59. $4x^2 + 8x + y^2 + 2y + 1 = 0 \Rightarrow 4(x^2 + 2x) + (y^2 + 2y) = -1 \Rightarrow$

$4(x^2 + 2x + 1) + (y^2 + 2y + 1) = -1 + 4 + 1 \Rightarrow 4(x+1)^2 + (y+1)^2 = 4 \Rightarrow$

$\dfrac{(x+1)^2}{1} + \dfrac{(y+1)^2}{4} = 1$; The center is $(-1, -1)$. The vertices are $(-1, -1-2), (-1, -1+2)$ or $(-1, -3), (-1, 1)$.

61. $4x^2 + 16x + 5y^2 - 10y + 1 = 0 \Rightarrow 4(x^2 + 4x) + 5(y^2 - 2y) = -1 \Rightarrow$

$4(x^2 + 4x + 4) + 5(y^2 - 2y + 1) = -1 + 16 + 5 \Rightarrow 4(x+2)^2 + 5(y-1)^2 = 20 \Rightarrow \dfrac{(x+2)^2}{5} + \dfrac{(y-1)^2}{4} = 1$; The center

is $(-2, 1)$. The vertices are $(-2-\sqrt{5}, 1), (-2+\sqrt{5}, 1)$.

63. $16x^2 - 16x + 4y^2 + 12y = 51 \Rightarrow 16(x^2 - x) + 4(y^2 + 3y) = 51 \Rightarrow$

$16\left(x^2 - x + \dfrac{1}{4}\right) + 4\left(y^2 + 3y + \dfrac{9}{4}\right) = 51 + 4 + 9 \Rightarrow 16\left(x - \dfrac{1}{2}\right)^2 + 4\left(y + \dfrac{3}{2}\right)^2 = 64 \Rightarrow \dfrac{(x-\frac{1}{2})^2}{4} + \dfrac{(y+\frac{3}{2})^2}{16} = 1$

The center is $\left(\dfrac{1}{2}, -\dfrac{3}{2}\right)$. The vertices are $\left(\dfrac{1}{2}, -\dfrac{3}{2}-4\right), \left(\dfrac{1}{2}, -\dfrac{3}{2}+4\right)$ or $\left(\dfrac{1}{2}, -\dfrac{11}{2}\right), \left(\dfrac{1}{2}, \dfrac{5}{2}\right)$.

65. $x^2 + 2x + y^2 + 6y + 5 = 0 \Rightarrow (x^2 + 2x + \underline{\quad}) + (y^2 + 6y + \underline{\quad}) = -5 \Rightarrow$

$(x^2 + 2x + 1) + (y^2 + 6y + 9) = -5 + 1 + 9 \Rightarrow (x^2 + 2x + 1) + (y^2 + 6y + 9) = 5 \Rightarrow (x+1)^2 + (y+3)^2 = 5$

The center is $(-1, -3)$. The radius is $\sqrt{5}$.

67. $x^2 - 8x + y^2 - 2y + 8 = 0 \Rightarrow (x^2 - 8x + \underline{\quad}) + (y^2 - 2y + \underline{\quad}) = -8 \Rightarrow$

$(x^2 - 8x + 16) + (y^2 - 2y + 1) = -8 + 16 + 1 \Rightarrow (x^2 - 8x + 16) + (y^2 - 2y + 1) = 9 \Rightarrow (x-4)^2 + (y-1)^2 = 9$

The center is $(4, 1)$. The radius is $\sqrt{9} = 3$.

69. $x^2 - 2x + y^2 + 8y - 32 = 0 \Rightarrow (x^2 - 2x + \underline{\quad}) + (y^2 + 8y + \underline{\quad}) = 32 \Rightarrow$

$(x^2 - 2x + 1) + (y^2 + 8y + 16) = 32 + 1 + 16 \Rightarrow (x^2 - 2x + 1) + (y^2 + 8y + 16) = 49 \Rightarrow (x-1)^2 + (y+4)^2 = 49$

The center is $(1, -4)$. The radius is $\sqrt{49} = 7$.

71. $y = \pm\sqrt{10\left(1 - \dfrac{x^2}{15}\right)}$;

73. $y = \pm\sqrt{\dfrac{25 - 4.1x^2}{6.3}}$;

75. $\dfrac{x^2}{4} + \dfrac{y^2}{9} = 1 \Rightarrow 9x^2 + 4y^2 = 36 \Rightarrow 9x^2 + 4(3-x)^2 = 36 \Rightarrow 9x^2 + 4(9 - 6x + x^2) = 36 \Rightarrow$

$13x^2 - 24x = 0$. Then $x(13x - 24) = 0 \Rightarrow x = 0$ or $x = \dfrac{24}{13}$. Since $y = 3 - x$, the corresponding y values are

$3 - 0 = 3$ and $3 - \dfrac{24}{13} = \dfrac{15}{13}$. The solutions are $(0, 3)$ and $\left(\dfrac{24}{13}, \dfrac{15}{13}\right)$.

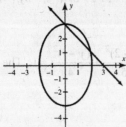

77. $x^2 + y^2 = 9 \Rightarrow x^2 = 9 - y^2$; then $4(9 - y^2) + 16y^2 = 64 \Rightarrow 36 - 4y^2 + 16y^2 = 64 \Rightarrow$

$y^2 = \dfrac{7}{3} \Rightarrow y = \pm\sqrt{\dfrac{7}{3}}$. Substituting in $x^2 + y^2 = 9$ we find $x^2 + \dfrac{7}{3} = 9, x^2 = \dfrac{27}{3} - \dfrac{7}{3} = \dfrac{20}{3}$, so $x = \pm\sqrt{\dfrac{20}{3}}$.

There are four solutions: $\left(\pm\sqrt{\dfrac{20}{3}}, \pm\sqrt{\dfrac{7}{3}}\right)$.

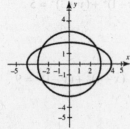

79. $x^2 + y^2 = 9 \Rightarrow y^2 = 9 - x^2$; then $2x^2 + 3(9 - x^2) = 18 \Rightarrow 2x^2 + 27 - 3x^2 = 18 \Rightarrow x^2 = 9 \Rightarrow x = \pm 3$;

Substituting in $x^2 + y^2 = 9$ we find $9 + y^2 = 9$, so $y = 0$. There are two solutions: $(\pm 3, 0)$.

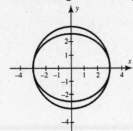

81. $\dfrac{x^2}{2} + \dfrac{y^2}{4} = 1 \Rightarrow 2x^2 + y^2 = 4 \Rightarrow 2(2y - 4) + y^2 = 4 \Rightarrow 4y - 8 + y^2 = 4 \Rightarrow y^2 + 4y - 12 = 0$ Then

$(y + 6)(y - 2) = 0 \Rightarrow y = -6$ or $y = 2$. Since $x = \pm\sqrt{2y - 4}$, the corresponding x values are $\pm\sqrt{2(-6) - 4}$,

which is undefined, and $\pm\sqrt{2(2) - 4} = 0$. The solution is $(0, 2)$.

83. From the first equation $\dfrac{x^2}{2} + \dfrac{y^2}{4} = 1 \Rightarrow 2x^2 + y^2 = 4 \Rightarrow y^2 = 4 - 2x^2$.

From the second equation $\dfrac{x^2}{4} + \dfrac{y^2}{2} = 1 \Rightarrow x^2 + 2y^2 = 4 \Rightarrow y^2 = 2 - \dfrac{1}{2}x^2$. That is $4 - 2x^2 = 2 - \dfrac{1}{2}x^2$.

$4 - 2x^2 = 2 - \dfrac{1}{2}x^2 \Rightarrow \dfrac{3}{2}x^2 = 2 \Rightarrow x^2 = \dfrac{4}{3} \Rightarrow x = \pm\dfrac{2}{\sqrt{3}} = \pm\dfrac{2\sqrt{3}}{3}$

Since $y = \pm\sqrt{4 - 2x^2}$, the y values are $\pm\sqrt{4 - 2\left(\dfrac{2}{\sqrt{3}}\right)^2} = \pm\sqrt{4 - \dfrac{8}{3}} = \pm\sqrt{\dfrac{4}{3}}$.

There are four solutions: $\left(\pm\sqrt{\dfrac{4}{3}}, \pm\sqrt{\dfrac{4}{3}}\right)$.

85. Subtracting the second equation from the first equation yields $(x-2)^2 - x^2 = 0$.

$(x-2)^2 - x^2 = 0 \Rightarrow x^2 - 4x + 4 - x^2 = 0 \Rightarrow -4x + 4 = 0 \Rightarrow -4x = -4 \Rightarrow x = 1$

Since $y = \pm\sqrt{9 - x^2}$, the y values are $\pm\sqrt{9 - (1)^2} = \pm\sqrt{8}$.

The solutions are $(1, -\sqrt{8}), (1, \sqrt{8})$.

87. The system is $(x-1)^2 + (y+1)^2 < 4$ and $(x+1)^2 + y^2 > 1$.

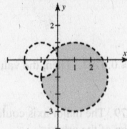

89. The system is $\dfrac{x^2}{4} + \dfrac{y^2}{9} \le 1$ and $x + y \ge 2$.

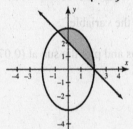

91. The system is $x^2 + y^2 \le 4$ and $x^2 + (y-2)^2 \le 4$.

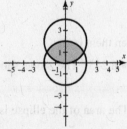

93. The system is $x^2 + y^2 \le 4$ and $(x+1)^2 - y \le 0$.

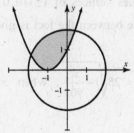

95. The inequality can be written $\dfrac{x^2}{9}+\dfrac{y^2}{4}\le 1$. The shaded region is shown in Figure 95.

Here $a=3$ and $b=2$. The area is $A=\pi ab=\pi(3)(2)=6\pi\approx 18.85\text{ ft}^2$.

97. The shaded region is shown in Figure 97.

Here $a=5$ and $b=4$. The area is $A=\pi ab=\pi(5)(4)=20\pi\approx 62.83\text{ ft}^2$.

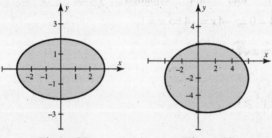

Figure 95   Figure 97

99. $e=\dfrac{c}{a}=0.206\Rightarrow c=0.206a$. Since $a=0.387$, it follows that $c=0.206(0.387)\approx 0.0797$. Then, the value of b is given by the following:

$a^2-b^2=c^2\Rightarrow a^2-c^2=b^2\Rightarrow 0.387^2-0.0797^2=b^2\Rightarrow b^2\approx 0.1434\Rightarrow b\approx 0.379$. The major axis could be located on either the x- or y-axis. We will choose the x-axis. Thus, the equation of the orbit is

$\dfrac{x^2}{0.387^2}+\dfrac{y^2}{0.379^2}=1$. The sun can be located on either of the foci. We will locate the sun at $(0.0797,0)$. To graph the orbit with a graphing calculator, solve the equation for the ellipse for the variable

$y:\dfrac{x^2}{0.387^2}+\dfrac{y^2}{0.379^2}=1\Rightarrow y=\pm 0.379\sqrt{1-\dfrac{x^2}{0.387^2}}$. Graph each of the equations and plot the sun at $(0.0797,$

$0)$. $Y_1=0.379\sqrt{((1-X^2)/0.387^2)}$, $Y_2=-Y_1$. See Figure 99.

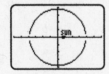

Figure 99

101. The source and the stone are at the two foci of the ellipse, so the distance between them is $2c$.

$c^2=a^2-b^2=4^2-(2.5)^2=16-6.25=9.75\Rightarrow c=\sqrt{9.75}$. Thus, $2c=2\sqrt{9.75}\approx 6.245$. The stone should be 6.245 inches from the source.

103. $c=50$ and $b=40$ for the elliptical floor. $a=\sqrt{50^2+40^2}=10\cdot\sqrt{41}\approx 64.03$. The area of the ellipse is πab, so the area of the floor is $\pi(64.03)(40)\approx 8046.25$ square feet.

105. The ellipse with a mjor axis of 620 feet and minor axis of 513 feet inplies vertices of $(\pm 310,0)$ and $(0,\pm 256.5)$. $c^2=a^2-b^2\Rightarrow c^2=(310)^2-(256.5)^2\Rightarrow c\approx 174.1$. The distance between the foci is given as $2c=2(174.1)=348.2$ feet.

107. The equation of the ellipse is $\dfrac{x^2}{30^2}+\dfrac{y^2}{25^2}=1$. Solving for y we get $y=25\sqrt{1-\dfrac{x^2}{900}}$. When $x=15$,

$y=25\sqrt{1-\dfrac{225}{900}}\approx 21.65$ feet.

109. The minimum height is $4464 - (3960 + 164) = 340$ miles; the maximum height is $4464 - (3960 - 164) = 668$ miles.

111. For Neptune: $e = \dfrac{c}{a} \Rightarrow c = ea \Rightarrow c = 0.009 * 30.10 \Rightarrow c = 0.2709$

     For Pluto: $e = \dfrac{c}{a} \Rightarrow c = ea \Rightarrow c = 0.249 * 39.44 \Rightarrow c = 9.82$

113. Neptune: $\dfrac{(x - 0.271)^2}{30.10^2} + \dfrac{y^2}{30.10^2} = 1$; Pluto: $\dfrac{(x - 9.82)^2}{39.44^2} + \dfrac{y^2}{38.20^2} = 1$

Extended and Discovery Exercises for Section 7.2

1. The slope of the line through $(-2, 6)$ and $(4, -3)$ is $-\dfrac{3}{2}$ and has equation $y - 6 = -\dfrac{3}{2}(x + 2)$.

     $0 - 6 = -\dfrac{3}{2}(x + 2) \Rightarrow x = 2$, x-intercept is $(2, 0)$. $y - 6 = -\dfrac{3}{2}(0 + 2) \Rightarrow y = 3$, y-intercept is $(0, 3)$. The

     equation of the line in intercept form is $\dfrac{x}{2} + \dfrac{y}{3} = 1$.

3. The equation of the line through $(3, -1)$ with slope of -2 is $y + 1 = -2(x - 3)$.

     $0 + 1 = -2(x - 3) \Rightarrow x = 2.5$, x-intercept is $(2.5, 0)$. $y + 1 = -2(0 - 3) \Rightarrow y = 5$, y-intercept is $(0, 5)$.

     The equation of the line in intercept form is $\dfrac{x}{2.5} + \dfrac{y}{5} = 1$.

5. The ellipse $\dfrac{x^2}{25} + \dfrac{y^2}{9} = 1 \Rightarrow a = 5$ and $b = 3$. Therefore, the x-intercept is $(\pm 5, 0)$ and y-intercept is $(0, \pm 3)$.

Checking Basic Concepts for Sections 7.1 and 7.2

1. The equation $x = \frac{1}{2}y^2$ can be written as $y^2 = 2x$, which is in the form $y^2 = 4px$. The vertex is $V(0,0)$.

    Thus, $2 = 4p$ or $p = \frac{1}{2}$. The focus is $F\left(\frac{1}{2}, 0\right)$, the equation of the directrix is $x = -\frac{1}{2}$, and the parabola

    opens to the right. The graph of $x = \frac{1}{2}y^2$ is shown.

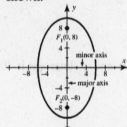

3. $\frac{x^2}{36} + \frac{y^2}{100} = 1 \Rightarrow a = 10$ and $b = 6$. $a^2 - b^2 = 10^2 - 6^2 = 64 = c^2 \Rightarrow c = 8$. The foci are $(0, \pm 8)$, the endpoints

    of the major axis (vertical) are $(0, \pm 10)$, while the endpoints of the minor axis are $(\pm 6, 0)$. The ellipse is

    shown.

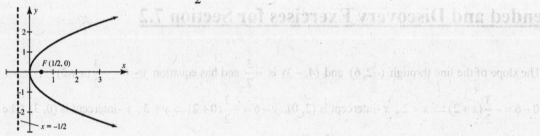

5. For a parabola with vertex at the origin and passing through $(2, 1)$, the equation $x^2 = 4py$ becomes $4 = 4p$,

    so $p = 1$. The filament should be located 1 foot from the vertex of the reflector.

7. $x^2 - 4x + 4y^2 + 8y - 8 = 0 \Rightarrow (x^2 - 4x) + 4(y^2 + 2y) = 8 \Rightarrow$

    $(x^2 - 4x + 4) + 4(y^2 + 2y + 1) = 8 + 4 + 4 \Rightarrow (x - 2)^2 + 4(y + 1)^2 = 16 \Rightarrow$

    $\frac{(x-2)^2}{16} + \frac{(y+1)^2}{4} = 1$; The center is $(2, -1)$. The vertices are $(2-4, -1)$, $(2+4, -1)$ or $(-2, -1)$, $(6, -1)$.

Section 7.3: Hyperbolas

1. The vertices are the endpoints of the transverse axis of a hyperbola.

3. A hyperbola with an equation of the form $\frac{x^2}{a^2} - \frac{y^2}{b^2} = 1$ has a horizontal transverse axis.

5. The transverse axis is horizontal with $a = 3$ and $b = 7$. The vertices are $(\pm 3, 0)$. The asymptotes are

    $y = \pm\frac{7}{3}x$. See Figure 5.

Since $c^2 = a^2 + b^2 \Rightarrow c = \pm\sqrt{9+49} \Rightarrow \sqrt{58}$, the foci are $(\pm\sqrt{58}, 0)$.

7. The transverse axis is vertical with $a = 6$ and $b = 4$. The vertices are $(0, \pm 6)$. The asymptotes are $y = \pm\dfrac{3}{2}x$.

    See Figure 7.

    Since $c^2 = a^2 + b^2 \Rightarrow c = \pm\sqrt{36+16} \Rightarrow \sqrt{52}$, the foci are $(0, \pm\sqrt{52})$.

9. $x^2 - y^2 = 9 \Rightarrow \dfrac{x^2}{9} - \dfrac{y^2}{9} = 1$. The transverse axis is horizontal with $a = 3$ and $b = 3$. The vertices are $(\pm 3, 0)$.

    The asymptotes are $y = \pm x$. See Figure 9.

    Since $c^2 = a^2 + b^2 \Rightarrow c = \pm\sqrt{9+9} \Rightarrow \sqrt{18}$, the foci are $(\pm\sqrt{18}, 0)$.

11. $9y^2 - 16x^2 = 144 \Rightarrow \dfrac{y^2}{16} - \dfrac{x^2}{9} = 1$. The transverse axis is vertical with $a = 4$ and $b = 3$. The vertices are

    $(0, \pm 4)$. The asymptotes are $y = \pm\dfrac{4}{3}x$. See Figure 11.

    Since $c^2 = a^2 + b^2 \Rightarrow c = \pm\sqrt{16+9} \Rightarrow \pm 5$, the foci are $(0, \pm 5)$.

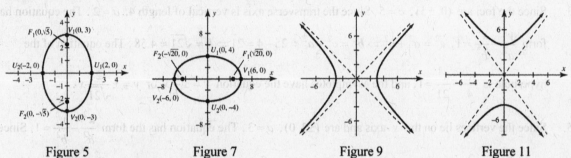

Figure 5   Figure 7   Figure 9   Figure 11

13. Horizontal transverse axis, vertices $(\pm 2, 0)$; d

15. Vertical transverse axis, vertices $(0, \pm 3)$; a

17. Since the foci and the vertices lie on the x-axis, the hyperbola has a horizontal transverse axis with an

    equation of the form $\dfrac{x^2}{a^2} - \dfrac{y^2}{b^2} = 1$. $F(\pm 5, 0) \Rightarrow c = 5$ and $V(\pm 4, 0) \Rightarrow a = 4$.

    $c^2 = a^2 + b^2 \Rightarrow b^2 = c^2 - a^2 = 25 - 16 = 9 \Rightarrow b = 3$. The equation of the hyperbola is $\dfrac{x^2}{16} - \dfrac{y^2}{9} = 1$, and the

    asymptotes have the equation $y = \pm\dfrac{b}{a}x$ or $y = \pm\dfrac{3}{4}x$. The hyperbola is graphed in Figure 17.

19. Since the foci and the vertices lie on the y-axis, the hyperbola has a vertical transverse axis with an equation

    of the form $\dfrac{y^2}{a^2} - \dfrac{x^2}{b^2} = 1$. $F(0, \pm 10) \Rightarrow c = 10$ and $V(0, \pm 6) \Rightarrow a = 6$.

    $c^2 = a^2 + b^2 \Rightarrow b^2 = c^2 - a^2 = 100 - 36 = 64 \Rightarrow b = 8$. The equation of the hyperbola is $\dfrac{y^2}{36} - \dfrac{x^2}{64} = 1$, and the

    asymptotes have the equation $y = \pm\dfrac{a}{b}x$ or $y = \pm\dfrac{3}{4}x$. The hyperbola is graphed in Figure 19.

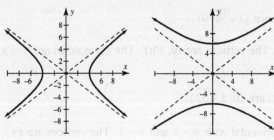

Figure 17   Figure 19

21.   Since the foci and the vertices lie on the y-axis, the hyperbola has a vertical transverse axis with an equation

of the form $\dfrac{y^2}{a^2} - \dfrac{x^2}{b^2} = 1$. $F(0, \pm 13) \Rightarrow c = 13$ and $V(0, \pm 12) \Rightarrow a = 12$.

$c^2 = a^2 + b^2 \Rightarrow b^2 = c^2 - a^2 = 169 - 144 = 25 \Rightarrow b = 5$. The equation of the hyperbola is $\dfrac{y^2}{144} - \dfrac{x^2}{25} = 1$, and the

asymptotes have the equation $y = \pm \dfrac{a}{b} x$ or $y = \pm \dfrac{12}{5} x$.

23.   Since the foci are $(0, \pm 5)$, $c = 5$. Since the transverse axis is vertical of length 4, $a = 2$. The equation has the

form $\dfrac{y^2}{a^2} - \dfrac{x^2}{b^2} = 1$. $c^2 = a^2 + b^2 \Rightarrow b^2 = c^2 - a^2 = 25 - 4 = 21 \Rightarrow b = \sqrt{21} \approx 4.58$. The equation of the

hyperbola is $\dfrac{y^2}{4} - \dfrac{x^2}{21} = 1$, and the asymptotes have the equation $y = \pm \dfrac{a}{b} x$ or $y = \pm \dfrac{2}{\sqrt{21}} x$.

25.   Since the vertices lie on the x-axis and are $(\pm 3, 0)$, $a = 3$. The equation has the form $\dfrac{x^2}{a^2} - \dfrac{y^2}{b^2} = 1$. Since

$y = \pm \dfrac{b}{a} x = \pm \dfrac{2}{3} x$ and $a = 3$, it follows that $b = 2$. The equation of the hyperbola is $\dfrac{x^2}{9} - \dfrac{y^2}{4} = 1$.

27.   Since the endpoints of the conjugate axis are $(0, \pm 3)$, $b = 3$. The vertices $(\pm 4, 0)$ lie on the x-axis so $a = 4$

and the equation of the hyperbola is $\dfrac{x^2}{a^2} - \dfrac{y^2}{b^2} = 1$ or $\dfrac{x^2}{16} - \dfrac{y^2}{9} = 1$. The asymptotes have the equation $y = \pm \dfrac{b}{a} x$

or $y = \pm \dfrac{3}{4} x$.

29.   Since the vertices lie on the x-axis and are $(\pm \sqrt{10}, 0)$, $a^2 = 10$. The equation has the form $\dfrac{x^2}{a^2} - \dfrac{y^2}{b^2} = 1$. The

value of b^2 can be found by substituting $a^2 = 10$, $x = 10$, and $y = 9$ in this equation.

$\dfrac{(10)^2}{10} - \dfrac{(9)^2}{b^2} = 1 \Rightarrow \dfrac{100}{10} - \dfrac{81}{b^2} = 1 \Rightarrow 10 - \dfrac{81}{b^2} = 1 \Rightarrow 9 = \dfrac{81}{b^2} \Rightarrow b^2 = \dfrac{81}{9} \Rightarrow b^2 = 9$

The equation of the hyperbola is $\dfrac{x^2}{10} - \dfrac{y^2}{9} = 1$. The asymptotes have the equation $y = \pm \dfrac{b}{a} x$ or $y = \pm \dfrac{3}{\sqrt{10}} x$.

31.   See Figure 31. The hyperbola has a horizontal transverse axis and its center is $(1, 2)$. Since $a^2 = 16$ and

$b^2 = 4$, it follows that $c^2 = a^2 + b^2 = 16 + 4 = 20$. Thus, $a = 4$, $b = 2$, and $c = \sqrt{20}$. The vertices are located

4 units to the left and right of center and the foci are located $\sqrt{20}$ units to the left and right of center. That is,

the vertices are $(1 \pm 4, 2)$ and the foci are $(1 \pm \sqrt{20}, 2)$. The asymptotes are given by

$y = \pm \dfrac{b}{a}(x - h) + k \Rightarrow y = \pm \dfrac{1}{2}(x - 1) + 2$.

33. See Figure 33. The hyperbola has a vertical transverse axis and its center is $(-2, 2)$. Since $a^2 = 36$ and $b^2 = 4$, it follows that $c^2 = a^2 + b^2 = 36 + 4 = 40$. Thus, $a = 6$, $b = 2$, and $c = \sqrt{40}$. The vertices are located 6 units above and below center and the foci are located $\sqrt{40}$ units above and below center. That is, the vertices are $(-2, 2 \pm 6)$ and the foci are $(-2, 2 \pm \sqrt{40})$. The asymptotes are given by

$$y = \pm \frac{b}{a}(x - h) + k \Rightarrow y = \pm 3(x + 2) + 2.$$

35. See Figure 35. The hyperbola has a horizontal transverse axis and its center is $(0, 1)$. Since $a^2 = 4$ and $b^2 = 1$, it follows that $c^2 = a^2 + b^2 = 4 + 1 = 5$. Thus, $a = 2$, $b = 1$, and $c = \sqrt{5}$. The vertices are located 2 units to the left and right of center and the foci are located $\sqrt{5}$ units to the left and right of center. That is, the vertices are $(\pm 2, 1)$ and the foci are $(\pm \sqrt{5}, 1)$. The asymptotes are given by

$$y = \pm \frac{b}{a}(x - h) + k \Rightarrow y = \pm \frac{1}{2}x + 1.$$

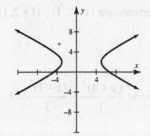

Figure 31

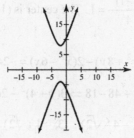

Figure 33

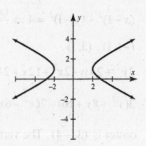

Figure 35

37. Center at $(2, -4)$; b

39. Center at $(2, -1)$; c

41. The hyperbola is centered at $(4, -4)$. The vertical transverse axis has length 8 so $a = 4$. The conjugate axis has length 4, so $b = 2$. Its equation is $\dfrac{(y+4)^2}{16} - \dfrac{(x-4)^2}{4} = 1$ with vertices $(4, -4 \pm 4)$, foci $(4, -4 \pm \sqrt{20})$.

The asymptotes are given by $y = \pm \dfrac{b}{a}(x - h) + k \Rightarrow y = \pm 2(x - 4) - 4$.

43. The hyperbola has a horizontal transverse axis, and its center is $(1, 1)$. Since $a^2 = 4$, $b^2 = 4$, and $c^2 = a^2 + b^2 = 8$, $a = 2$, $b = 2$, and $c = \sqrt{8}$. Thus, the vertices are $(1 \pm 2, 1)$ and the foci are $(1 \pm \sqrt{8}, 1)$. The asymptotes are the lines $y = \pm(x - 1) + 1$. See Figure 43.

45. The hyperbola has a vertical transverse axis, and its center is $(1, -1)$. Since $a^2 = 16$, $b^2 = 9$, and $c^2 = a^2 + b^2 = 25$, $a = 4$, $b = 3$, and $c = 5$. Thus, the vertices are $(1, -1 \pm 4)$ and the foci are $(1, -1 \pm 5)$. The asymptotes are the lines $y = \pm \dfrac{4}{3}(x - 1) - 1$. See Figure 45.

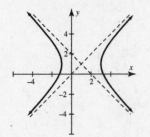

Figure 43

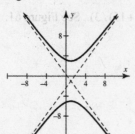

Figure 45

47.   Center $(2,-2) \Rightarrow h=2$ and $k=-2$. Rewrite the coordinates for the given vertex: $(3,-2) \Rightarrow (1+2,-2)$. Rewrite the coordinates for the given focus: $(4,-2) \Rightarrow (2+2,-2)$. Thus, the transverse axis is horizontal with $a=1$ and $c=2$. $b^2 = c^2 - a^2 = 3$. Using the standard equation form, we get the equation:

$$\frac{(x-h)^2}{a^2} - \frac{(y-k)^2}{b^2} = 1 \Rightarrow (x-2)^2 - \frac{(y+2)^2}{3} = 1.$$

49.   Since the vertices $(-1,\pm 1)$ and the foci $(-1,\pm 3)$ have the same x-coordinate, $h=-1$ and the transverse axis is vertical. Rewrite the vertices: $(-1,\pm 1) \Rightarrow (-1, 0\pm 1)$, so $k=0$ and $a=1 \Rightarrow a^2=1$. Rewrite the foci: $(-1,\pm 3) \Rightarrow (-1, 0\pm 3)$, so $k=0$ and $c=3$. So, $b^2 = c^2 - a^2 = 8$. Using the standard equation form, we get the equation: $\dfrac{(y-k)^2}{a^2} - \dfrac{(x-h)^2}{b^2} = 1 \Rightarrow y^2 - \dfrac{(x+1)^2}{8} = 1.$

51. $x^2 - 2x - y^2 + 2y = 4 \Rightarrow (x^2 - 2x + 1) - (y^2 - 2y + 1) = 4 + 1 - 1 \Rightarrow$

$(x-1)^2 - (y-1)^2 = 4 \Rightarrow \dfrac{(x-1)^2}{4} - \dfrac{(y-1)^2}{4} = 1$. The center is $(1, 1)$. The vertices are $(1-2, 1)$, $(1+2, 1)$, $(-1, 1)$, $(3, 1)$.

53. $3y^2 + 24y - 2x^2 + 12x + 24 = 0 \Rightarrow 3(y^2 + 8y) - 2(x^2 - 6x) = -24 \Rightarrow$

$3(y^2 + 8y + 16) - 2(x^2 - 6x + 9) = -24 + 48 - 18 \Rightarrow 3(y+4)^2 - 2(x-3)^2 = 6 \Rightarrow \dfrac{(y+4)^2}{2} - \dfrac{(x-3)^2}{3} = 1$. The center is $(3, -4)$. The vertices are $(3, -4-\sqrt{2})$, $(3, -4+\sqrt{2})$.

55. $x^2 - 6x - 2y^2 + 7 = 0 \Rightarrow (x^2 - 6x + 9) - 2y^2 = -7 + 9 \Rightarrow (x-3)^2 - 2(y-0)^2 = 2 \Rightarrow \dfrac{(x-3)^2}{2} - \dfrac{(y-0)^2}{1} = 1$. The center is $(3, 0)$. The vertices are $(3-\sqrt{2}, 0)$, $(3+\sqrt{2}, 0)$.

57. $4y^2 + 32y - 5x^2 - 10x + 39 = 0 \Rightarrow 4(y^2 + 8y) - 5(x^2 + 2x) = -39 \Rightarrow$

$4(y^2 + 8y + 16) - 5(x^2 + 2x + 1) = -39 + 64 - 5 \Rightarrow 4(y+4)^2 - 5(x+1)^2 = 20 \Rightarrow \dfrac{(y+4)^2}{5} - \dfrac{(x+1)^2}{4} = 1$. The center is $(-1, -4)$. The vertices are $(-1, -4-\sqrt{5})$, $(-1, -4+\sqrt{5})$.

59.   Solve for y: $\dfrac{(y-1)^2}{11} - \dfrac{x^2}{5.9} = 1 \Rightarrow \dfrac{(y-1)^2}{11} = 1 + \dfrac{x^2}{5.9} \Rightarrow (y-1)^2 = 11\left(1 + \dfrac{x^2}{5.9}\right) \Rightarrow$

$y - 1 = \pm\sqrt{11\left(1 + \dfrac{x^2}{5.9}\right)} \Rightarrow y = 1 \pm \sqrt{11\left(1 + \dfrac{x^2}{5.9}\right)}$. Graph $Y_1 = 1 + \sqrt{(11 \cdot (1 + (X^2/5.9)))}$ and

$Y_2 = 1 - \sqrt{(11 \cdot (1 + (X^2/5.9)))}$. See Figure 59.

61.   Solve for y: $3y^2 - 4x^2 = 15 \Rightarrow 3y^2 = 4x^2 + 15 \Rightarrow y^2 = \dfrac{4x^2 + 15}{3} \Rightarrow y = \pm\sqrt{\dfrac{4x^2 + 15}{3}}$.

Graph $Y_1 = \sqrt{((4X^2 + 15)/3)}$ and $Y_2 = \sqrt{((4X^2 + 15)/3)}$. See Figure 61.

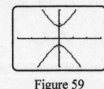
Figure 59

Figure 61

63.   Add both equations together to eliminate the y^2-term:

$$x^2 - y^2 = 4$$

$$\frac{x^2 + y^2 = 9}{2x^2 \qquad = 13} \Rightarrow x^2 = \frac{13}{2} \Rightarrow x = \pm\sqrt{\frac{13}{2}}$$

Substitute $x^2 = \frac{13}{2}$ into the first equation and solve for $y: x^2 - y^2 = 4 \Rightarrow \frac{13}{2} - y^2 = 4 \Rightarrow y^2 = \frac{5}{2} \Rightarrow y = \pm\sqrt{\frac{5}{2}}$.

There are four solutions to the system:

$$\left(\sqrt{\frac{13}{2}}, \sqrt{\frac{5}{2}}\right), \left(-\sqrt{\frac{13}{2}}, \sqrt{\frac{5}{2}}\right), \left(\sqrt{\frac{13}{2}}, -\sqrt{\frac{5}{2}}\right), \text{ and } \left(-\sqrt{\frac{13}{2}}, -\sqrt{\frac{5}{2}}\right).$$

65. Solve the second equation for $y: x + y = 2 \Rightarrow y = 2 - x$. Substitute this result into the first equation and solve

for $x: \dfrac{x^2}{4} - \dfrac{y^2}{9} = 1 \Rightarrow \dfrac{x^2}{4} - \dfrac{(2-x)^2}{9} = 1 \Rightarrow 9x^2 - 4(4 - 4x + x^2) = 36 \Rightarrow$

$9x^2 - 16 + 16x - 4x^2 = 36 \Rightarrow 5x^2 + 16x - 52 = 0 \Rightarrow (5x + 26)(x - 2) = 0 \Rightarrow x = -\dfrac{26}{5} = -5.2 \text{ or } x = 2$.

Substitute for x to find $y: y = 2 - x \Rightarrow y = 2 - (-5.2) = 7.2$ and $y = 2 - x \Rightarrow y = 2 - 2 = 0$. There are two

solutions to the system: $(2, 0)$ and $(-5.2, 7.2)$.

67. Multiply the second equation by 2 and add the equations to eliminate the y^2-term:

$$8x^2 - 6y^2 = 24$$

$$\frac{10x^2 + 6y^2 = 48}{18x^2 \qquad = 72} \Rightarrow x^2 = 4 \Rightarrow x = \pm 2$$

Substitute $x^2 = 4$ into the first equation and solve for y:

$$8x^2 - 6y^2 = 24 \Rightarrow 8(4) - 6y^2 = 24 \Rightarrow 6y^2 = 8 \Rightarrow y^2 = \frac{4}{3} \Rightarrow y = \pm\sqrt{\frac{4}{3}} \Rightarrow y = \pm\frac{2}{\sqrt{3}}.$$

There are four solutions to the system: $\left(-2, -\dfrac{2}{\sqrt{3}}\right), \left(-2, \dfrac{2}{\sqrt{3}}\right), \left(2, -\dfrac{2}{\sqrt{3}}\right), \left(2, \dfrac{2}{\sqrt{3}}\right)$.

69. Solve the second equation for $y: 3x - y = 0 \Rightarrow y = 3x$. Substitute this result into the first equation and solve

for

$x: \dfrac{y^2}{3} - \dfrac{x^2}{4} = 1 \Rightarrow \dfrac{(3x)^2}{3} - \dfrac{x^2}{4} = 1 \Rightarrow \dfrac{9x^2}{3} - \dfrac{x^2}{4} = 1 \Rightarrow 3x^2 - \dfrac{x^2}{4} = 1 \Rightarrow 12x^2 - x^2 = 4 \Rightarrow$

$11x^2 = 4 \Rightarrow x^2 = \dfrac{4}{11} \Rightarrow x = \pm\dfrac{2}{\sqrt{11}}$.

Substitute for x to find y: When $x = \dfrac{2}{\sqrt{11}}, y = 3x \Rightarrow y = 3\left(\dfrac{2}{\sqrt{11}}\right) \Rightarrow y = \dfrac{6}{\sqrt{11}}$.

When $-\dfrac{2}{\sqrt{11}}, y = 3x \Rightarrow y = 3\left(-\dfrac{2}{\sqrt{11}}\right) \Rightarrow y = -\dfrac{6}{\sqrt{11}}$.

The two solutions are: $\left(\dfrac{2}{\sqrt{11}}, \dfrac{6}{\sqrt{11}}\right), \left(-\dfrac{2}{\sqrt{11}}, -\dfrac{6}{\sqrt{11}}\right)$.

71.   (a) $k = 2.82 \times 10^7$ and $D = 42.5 \times 10^6 \Rightarrow \dfrac{k}{\sqrt{D}} = \dfrac{2.82 \times 10^7}{\sqrt{42.5 \times 10^6}} \Rightarrow \dfrac{k}{\sqrt{D}} \approx 4325.68$. Since $V = 2090$ and

            $V < \dfrac{k}{\sqrt{D}}$, the trajectory is elliptic.

         (b)   For $V > \dfrac{k}{\sqrt{D}}$, $V > 4326$. the speed of Explorer IV should be 4326 meters per second or greater so its

            trajectory is hyperbolic.

         (c)   If D is larger, then $\dfrac{k}{\sqrt{D}}$ is smaller, so smaller values for V satisfy $V > \dfrac{k}{\sqrt{D}}$.

Extended and Discovery Exercises for Section 7.3

1.   (a)   Find a and b in the equation $\dfrac{x^2}{a^2} - \dfrac{y^2}{b^2} = 1$. Because the equations of the asymptotes of a hyperbola

            with horizontal transverse axis are $y = \pm \dfrac{b}{a} x$, and the given asymptotes are $y = \pm x$, it follows that

            $\dfrac{b}{a} = 1$ or $a = b$. Since the line $y = x$ intersects the x-axis at a $45°$ angle, the triangle shown in the third

            quadrant is a $45°$-$45°$-$90°$ right triangle and both legs must have length d. Then by the Pythagorean

            theorem, $c^2 = d^2 + d^2 = 2d^2$. That is $c = d\sqrt{2}$. Also, for a hyperbola $c^2 = a^2 + b^2$, and since $a = b$,

            $c^2 = a^2 + a^2 = 2a^2$. That is $c = a\sqrt{2}$. From these two equations, $a\sqrt{2} = d\sqrt{2}$ and so $a = d$. That is,

            $a = b = d = 5 \times 10^{-14}$. Thus the equation of the trajectory of A, where $x > 0$, is given by

            $\dfrac{x^2}{(5 \times 10^{-14})^2} - \dfrac{y^2}{(5 \times 10^{-14})^2} = 1$. Solving for x yields

            $x^2 - y^2 = (5 \times 10^{-14})^2 \Rightarrow x^2 = y^2 + 2.5 \times 10^{-27} \Rightarrow x = \sqrt{y^2 + 25 \times 10^{-27}}$. This equation represents the

            right half of the hyperbola, as shown in the textbook.

         (b)   Since $a = 5 \times 10^{-14}$, the distance from the origin to the vertex is 5×10^{-14}. The distance from N to the

            origin can be found using the Pythagorean theorem. Let h represent this distance, then $h^2 = d^2 + d^2$.

            That is, $h^2 = (5 \times 10^{-14})^2 + (5 \times 10^{-14})^2 \Rightarrow h^2 = 5 \times 10^{-27} \Rightarrow h \approx 7 \times 10^{-14}$. The minimum distance

            between the centers of the alpha partical and the gold nucleus is $5 \times 10^{-14} + 7 \times 10^{-14} = 1.2 \times 10^{-13}$ m.

Checking Basic Concepts for Section 7.3

1.   The center is (0, 0). Since the vertices lie on the x-axis and are $(\pm 4, 0)$, $a = 4$. The equation of one

     asymptote is $y = \dfrac{3}{4} x$ because the line goes through the points (0, 0) and (8, 6). Thus, $a = 4$ and $b = 3$. Using

     the standard equation form, we get the equation: $\dfrac{x^2}{a^2} - \dfrac{y^2}{b^2} = 1 \Rightarrow \dfrac{x^2}{16} - \dfrac{y^2}{9} = 1$.

3. $h = 1$ and $k = 3$. Since the horizontal transverse axis has length 6, $2a = 6 \Rightarrow a = 3$. Since the conjugate axis has length 4, $2b = 4 \Rightarrow b = 2$. $c^2 = a^2 + b^2 = 9 + 4 = 13 \Rightarrow c = \sqrt{13}$. Thus the foci are $(1 \pm \sqrt{13}, 3)$. Using the standard equation form, we get the equation: $\dfrac{(x-h)^2}{a^2} - \dfrac{(y-k)^2}{b^2} = 1 \Rightarrow \dfrac{(x-1)^2}{9} - \dfrac{(y-3)^2}{4} = 1$.

Chapter 7 Review Exercises

1.   The equation is $-x^2 = y$. See Figure 1.

3.   The equation is $\dfrac{x^2}{25} + \dfrac{y^2}{49} = 1$. See Figure 3.

5.   The equation is $\dfrac{y^2}{4} - \dfrac{x^2}{9} = 1$. See Figure 5.

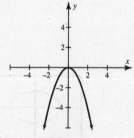

Figure 1

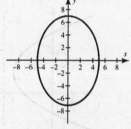

Figure 3

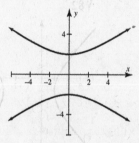

Figure 5

7.   A parabola opening upwards; d

9.   A circle; a

11.   A hyperbola with horizontal transverse axis; e

13.   The parabola opens to the right and $p = 2$. $y^2 = 4px$, so $y^2 = 8x$.

15.   The major axis is horizontal, $a = 5$ and $c = 4$; $b = \sqrt{a^2 - c^2} = \sqrt{25 - 16} = 3$.

      The equation is $\dfrac{x^2}{25} + \dfrac{y^2}{9} = 1$.

17.   The transverse axis is vertical, $c = 10$ and $b = 6$; $a = \sqrt{100 - 36} = 8$.

      The equation is $\dfrac{y^2}{64} - \dfrac{x^2}{36} = 1$.

19. $x^2 = 4py \Rightarrow 4p = -4$, so $p = -1$. The vertex is at the origin and the focus is $(0, -1)$. See Figure 19.

21.   The major axis is horizontal, the center is at the origin, $a = 5$ and $b = 2$. $c^2 = 25 - 4 = 21$, so $c = \sqrt{21}$; the foci are $(\pm\sqrt{21}, 0)$. See Figure 21.

23.   The center is the origin and the transverse axis is horizontal. $c^2 = a^2 + b^2 = 16 + 9 = 25$, so $c = 5$. The foci are $(\pm 5, 0)$. See Figure 23.

25.   The equation represents a circle of radius 3, centered at $(3, -1)$. If we think of the circle as an ellipse, both foci are at $(3, -1)$. See Figure 25.

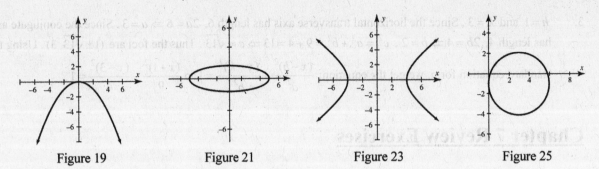

| Figure 19 | Figure 21 | Figure 23 | Figure 25 |

27.   The equation $\dfrac{(x-1)^2}{4} - \dfrac{(y+1)^2}{4} = 1$ represents a hyperbola with horizontal transverse axis and center $(1, -1)$. See Figure 27.

29.   The equation $(y-4)^2 = -8(x-8)$ has the form $(y-k)^2 = 4p(x-h)$ with vertex (h, k) at $(8, 4)$, and $p = -2$. The focus is $(6, 4)$ and the directrix is $x = 10$. See Figure 29.

31. $y = \pm\sqrt{\dfrac{1}{82}(60 - 7.1x^2)}$; See Figure 31.

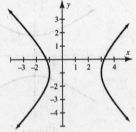   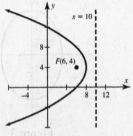   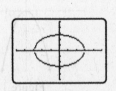

| Figure 27 | Figure 29 | Figure 31 |

33. $-2x = y^2 + 8x + 14 \Rightarrow y^2 = -10x - 14 \Rightarrow (y-0)^2 = -10\left(x + \dfrac{7}{5}\right)$

35. $4x^2 + 8x + 25y^2 - 250y = -529 \Rightarrow 4(x^2 + 2x) + 25(y^2 - 10y) = -529 \Rightarrow$

$4(x^2 + 2x + 1) + 25(y^2 - 10y + 25) = -529 + 4 + 625 \Rightarrow 4(x+1)^2 + 25(y-5)^2 = 100 \Rightarrow$

$\dfrac{(x+1)^2}{25} + \dfrac{(y-5)^2}{4} = 1$; The center is $(-1, 5)$. The vertices are $(-1-5, 5)$, $(-1+5, 5)$ or $(-6, 5)$, $(4, 5)$

37. $x^2 + 4x - 4y^2 + 24y = 36 \Rightarrow (x^2 + 4x + 4) - 4(y^2 - 6y + 9) = 36 + 4 - 36 \Rightarrow$

$(x+2)^2 - 4(y-3)^2 = 4 \Rightarrow \dfrac{(x+2)^2}{4} - \dfrac{(y-3)^2}{1} = 1$. The center is $(-2, 3)$. The vertices are $(-2-2, 3)$,

$(-2+2, 3)$ or $(-4, 3)$, $(0, 3)$.

39.   Clear fractions: $x^2 + y^2 = 4 \Rightarrow y^2 = 4 - x^2$. Substituting in $x^2 + 4y^2 = 8$ gives $x^2 + 4(4 - x^2) = 8$ or $3x^2 = 8$,

so $x = \pm\sqrt{\dfrac{8}{3}}$. Substituting in $x^2 + y^2 = 4$ we find $y^2 = 4 - \dfrac{8}{3} = \dfrac{4}{3}$, so $y = \pm\sqrt{\dfrac{4}{3}}$. There are four solutions:

$\left(\pm\sqrt{\dfrac{8}{3}}, \pm\sqrt{\dfrac{4}{3}}\right)$.

41. The system is $\dfrac{x^2}{9} + \dfrac{y^2}{4} \le 1$ and $x + y \le 3$.

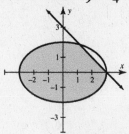

43. (a) $c = \sqrt{500^2 - 70^2} \approx 495.08$, and $a = 500$. The minimum distance is $a - c \approx 4.92$ million miles. The maximum distance is $a + c \approx 995.08$ million miles.

     (b) $2\pi\sqrt{\dfrac{500^2 + 70^2}{2}} \approx 2243$ million miles or 2.243 billion miles.

45. The equation of the ellipse is $\dfrac{x^2}{40^2} + \dfrac{y^2}{30^2} = 1$. Solving for y gives $y = 30\sqrt{1 - \dfrac{x^2}{40^2}}$. When $x = 10$,

     $y = 30\sqrt{1 - \left(\dfrac{10}{40}\right)^2} \approx 29.05$ feet .

Chapter 8: Further Topics in Algebra

8.1: Sequences

1. $a_1 = 2(1)+1 = 3$; $a_2 = 2(2)+1 = 5$; $a_3 = 2(3)+1 = 7$; $a_4 = 2(4)+1 = 9$.
   The first four terms are 3, 5, 7, and 9.

3. $a_1 = 4(-2)^{1-1} = 4$; $a_2 = 4(-2)^{2-1} = -8$; $a_3 = 4(-2)^{3-1} = 16$; $a_4 = 4(-2)^{4-1} = -32$.
   The first four terms are 4, -8, 16, and -32.

5. $a_1 = \dfrac{1}{1^2+1} = \dfrac{1}{2}$; $a_2 = \dfrac{2}{2^2+1} = \dfrac{2}{5}$; $a_3 = \dfrac{3}{3^2+1} = \dfrac{3}{10}$; $a_4 = \dfrac{4}{4^2+1} = \dfrac{4}{17}$.
   The first four terms are $\dfrac{1}{2}, \dfrac{2}{5}, \dfrac{3}{10}$, and $\dfrac{4}{17}$.

7. $a_1 = (-1)^1 \left(\dfrac{1}{2}\right)^1 = -\dfrac{1}{2}$; $a_2 = (-1)^2 \left(\dfrac{1}{2}\right)^2 = \dfrac{1}{4}$; $a_3 = (-1)^3 \left(\dfrac{1}{2}\right)^3 = -\dfrac{1}{8}$; $a_4 = (-1)^4 \left(\dfrac{1}{2}\right)^4 = \dfrac{1}{16}$.
   The first four terms are $-\dfrac{1}{2}, \dfrac{1}{4}, -\dfrac{1}{8}$, and $\dfrac{1}{16}$.

9. $a_1 = (-1)^0 \left(\dfrac{2}{1+2}\right) = \dfrac{2}{3}$; $a_2 = (-1)^1 \left(\dfrac{4}{1+4}\right) = -\dfrac{4}{5}$; $a_3 = (-1)^2 \left(\dfrac{8}{1+8}\right) = \dfrac{8}{9}$; $a_4 = (-1)^3 \left(\dfrac{16}{1+16}\right) = -\dfrac{16}{17}$. The
   first four terms are $\dfrac{2}{3}, -\dfrac{4}{5}, \dfrac{8}{9}$, and $-\dfrac{16}{17}$.

11. $a_1 = 2+1^2 = 3$; $a_2 = 4+2^2 = 8$; $a_3 = 8+3^2 = 17$; $a_4 = 16+4^2 = 32$.
    The first four terms are 3, 8, 17, and 32.

13. The points (1, 2), (2, 4), (3, 3), (4, 5), (5, 3), (6, 6), (7, 4) lie on the graph. Therefore, the terms of the sequence are 2, 4, 3, 5, 3, 6, 4.

15. (a) $a_1 = 1$; $a_2 = 2a_1 = 2(1) = 2$; $a_3 = 2a_2 = 2(2) = 4$; $a_4 = 2a_3 = 2(4) = 8$.
    The first four terms are 1, 2, 4, and 8.
    (b)

17. (a) $a_1 = -3$; $a_2 = a_1 + 3 = -3 + 3 = 0$; $a_3 = a_2 + 3 = 0 + 3 = 3$; $a_4 = a_3 + 3 = 3 + 3 = 6$.
    The first four terms are -3, 0, 3, and 6.
    (b)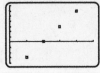

19. (a) $a_1 = 2$; $a_2 = 3a_1 - 1 = 3(2) - 1 = 5$; $a_3 = 3a_2 - 1 = 3(5) - 1 = 14$; $a_4 = 3a_3 - 1 = 3(14) - 1 = 41$. The first
    four terms are 2, 5, 14, and 41.

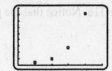

(b)

21. (a) $a_1 = 2$; $a_2 = 5$; $a_3 = a_2 - a_1 = 5 - 2 = 3$; $a_4 = a_3 - a_2 = 3 - 5 = -2$.

    The first four terms are 2, 5, 3, and -2 .

    (b)

23. (a) $a_1 = 2$; $a_2 = a_1^2 = 2^2 = 4$; $a_3 = a_2^2 = 4^2 = 16$; $a_4 = a_3^2 = 16^2 = 256$.

    The first four terms are 2, 4, 16, and 256.

    (b)

25 (a) $a_1 = 1$; $a_2 = a_1 + 2 = 1 + 2 = 3$; $a_3 = a_2 + 3 = 3 + 3 = 6$; $a_4 = a_3 + 4 = 6 + 4 = 10$.

    The first four terms are 1, 3, 6, and 10.

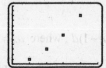

    (b)

27. (a) $a_1 = 2$; $a_2 = 3$; $a_3 = a_2 \cdot a_1 = 2 \cdot 3 = 6$; $a_4 = a_3 \cdot a_2 = 6 \cdot 3 = 18$

    The first four terms are 2, 3, 6, and 18.

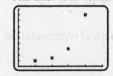

    (b)

29. (a)   Each term can be found by adding 2 to the previous term. A numerical representation for the first eight terms is shown.

n	1	2	3	4	5	6	7	8
a_n	1	3	5	7	9	11	13	15

    (b)   A graphical representation is shown and includes the points from 29a. Notice that the points lie on a line with slope 2 because the sequence is arithmetic.

    (c)   To find the symbolic representation, we will use the formula $a_n = a_1 + (n-1)d$, where $a_n = f(n)$. The common difference of this sequence is $d = 2$ and the first term is $a_1 = 1$. Therefore, a symbolic representation of the sequence is given by $a_n = 1 + (n-1)2$ or $a_n = 2n - 1$.

31. (a)   Each term can be found by subtracting 1.5 to the previous term. A numerical representation for the first eight terms is shown.

n	1	2	3	4	5	6	7	8
a_n	7.5	6	4.5	3	1.5	0	-1.5	-3

(b) A graphical representation is shown and includes the points from Figure 31a. Notice that the points lie on a line with slope -1.5 because the sequence is arithmetic.

(c) To find the symbolic representation, we will use the formula $a_n = a_1 + (n-1)d$, where $a_n = f(n)$. The common difference of this sequence is $d = -1.5$ and the first term is $a_1 = 7.5$. Therefore, a symbolic representation of the sequence is given by $a_n = 7.5 + (n-1)(-1.5)$ or $a_n = -1.5n + 9$.

33. (a) Each term can be found by adding $\dfrac{3}{2}$ to the previous term. A numerical representation for the first eight terms is shown.

n	1	2	3	4	5	6	7	8
a_n	$\frac{1}{2}$	2	$\frac{7}{2}$	5	$\frac{13}{2}$	8	$\frac{19}{2}$	11

(b) A graphical representation is shown and includes the points from 33a. Notice that the points lie on a line with slope $\dfrac{3}{2}$ because the sequence is arithmetic.

(c) To find the symbolic representation, we will use the formula $a_n = a_1 + (n-1)d$, where $a_n = f(n)$. The common difference of this sequence is $d = \dfrac{3}{2}$ and the first term is $a_1 = \dfrac{1}{2}$. Therefore, a symbolic representation of the sequence is given by $a_n = \dfrac{1}{2} + (n-1)\left(\dfrac{3}{2}\right)$ or $a_n = \dfrac{3}{2}n - 1$.

35. (a) Each term can be found by multiplying the previous term by $\dfrac{1}{2}$. A numerical representation for the first eight terms is shown.

n	1	2	3	4	5	6	7	8
a_n	8	4	2	1	$\frac{1}{2}$	$\frac{1}{4}$	$\frac{1}{8}$	$\frac{1}{16}$

(b) A graphical representation is shown and includes the points from 35a. Notice that the points lie on a curve that is decaying exponentially, because the sequence is geometric and the common ratio is less than one in absolute value.

(c) To find the symbolic representation, we will use the formula $a_n = a_1 r^{n-1}$, where $a_n = f(n)$. The common ratio of this sequence is $r = \dfrac{1}{2}$ and the first term is $a_1 = 8$. Therefore, a symbolic representation of the sequence is given by $a_n = 8\left(\dfrac{1}{2}\right)^{n-1}$.

37. (a) Each term can be found by multiplying the previous term by 2. A numerical representation for the first eight terms is shown.

n	1	2	3	4	5	6	7	8
a_n	$\frac{3}{4}$	$\frac{3}{2}$	3	6	12	24	48	96

(b) A graphical representation is shown and includes the points from 37a. Notice that the points lie on a curve that is increasing exponentially, because the sequence is geometric and the common ratio is greater than one in absolute value.

(c)   To find the symbolic representation, we will use the formula $a_n = a_1 r^{n-1}$, where $a_n = f(n)$. The

common ratio of this sequence is $r = 2$ and the first term is $a_1 = \dfrac{3}{4}$. Therefore, a symbolic

representation of the sequence is given by $a_n = \dfrac{3}{4}(2)^{n-1}$.

39.   (a)   Each term can be found by multiplying the previous term by 2. A numerical representation for the first
               eight terms is shown.

n	1	2	3	4	5	6	7	8
a_n	$-\frac{1}{4}$	$-\frac{1}{2}$	-1	-2	-4	-8	-16	-32

       (b)   A graphical representation is shown and includes the points from 39a. Notice that the points lie on a
               curve that is increasing exponentially, because the sequence is geometric and the common ratio is
               greater than one in absolute value.

       (c)   To find the symbolic representation, we will use the formula $a_n = a_1 r^{n-1}$, where $a_n = f(n)$. The

common ratio of this sequence is $r = 2$ and the first term is $a_1 = -\dfrac{1}{4}$. Therefore, a symbolic

representation of the sequence is given by $a_n = -\dfrac{1}{4}(2)^{n-1}$.

41.   Let $a_n = f(n)$, where $f(n) = dn + c$. Then, $f(n) = -2n + c$. Since $f(1) = -2(1) + c = 5 \Rightarrow c = 7$. Thus,
       $a_n = f(n) = -2n + 7$. An alternate solution is to use the formula:
       $a_n = a_1 + (n-1)d = 5 + (n-1)(-2) \Rightarrow a_n = -2n + 7$.

43.   Let $a_n = f(n)$, where $f(n) = dn + c$. Then, $f(n) = 3n + c$. Since $f(3) = 3(3) + c = 1 \Rightarrow c = -8$. Thus,
       $a_n = f(n) = 3n - 8$.

45.   Let $a_n = f(n)$, where $f(n) = dn + c$. Then, $f(n) = 6n + c$. Since $f(1) = 6(1) + c = -5 \Rightarrow c = -11$. Thus,
       $a_n = f(n) = 6n - 11$.

47.   Let $a_n = f(n)$, where $f(n) = dn + c$. Then, $f(n) = -2n + c$. Since $f(4) = -2(4) + c = 8 \Rightarrow c = 16$. Thus,
       $a_n = f(n) = -2n + 16$.

49.   Let $a_n = f(n)$, where $f(n) = dn + c$. $a_2 = -2 \Rightarrow f(2) = -2$ and $a_4 = 8 \Rightarrow f(4) = 8$.

       Thus, $d = \dfrac{f(4) - f(2)}{4 - 2} = \dfrac{8 - (-2)}{2} = 5$. Then, $f(n) = 5n + c$. $f(2) = 5(2) + c = -2 \Rightarrow c = -12$.

       Thus, $a_n = f(n) = 5n - 12$.

51.   Let $a_n = f(n)$, where $f(n) = dn + c$. $a_2 = 5 \Rightarrow f(2) = 5$ and $a_6 = 13 \Rightarrow f(6) = 13$.

       Thus, $d = \dfrac{f(6) - f(2)}{6 - 2} = \dfrac{13 - 5}{4} = 2$. Then, $f(n) = 2n + c$. $f(2) = 2(2) + c = 5 \Rightarrow c = 1$.

       Thus, $a_n = f(n) = 2n + 1$.

53.   Let $a_n = f(n)$, where $f(n) = dn + c$. $a_1 = 8 \Rightarrow f(1) = 8$ and $a_4 = 17 \Rightarrow f(4) = 17$.

Thus, $d = \dfrac{f(4) - f(1)}{4-1} = \dfrac{17-8}{3} = 3$. Then, $f(n) = 3n + c$. $f(1) = 3(1) + c = 8 \Rightarrow c = 5$.

Thus, $a_n = f(n) = 3n + 5$.

55. Let $a_n = f(n)$, where $f(n) = dn + c$. $a_5 = -4 \Rightarrow f(5) = -4$ and $a_8 = -2.5 \Rightarrow f(8) = -2.5$.

Thus, $d = \dfrac{f(8) - f(5)}{8-5} = \dfrac{-2.5 - (-4)}{3} = 0.5$.

Then, $f(n) = 0.5n + c$. $f(5) = 0.5(5) + c = -4 \Rightarrow c = -6.5$. Thus, $a_n = f(n) = 0.5n - 6.5$.

57. Let $a_n = f(n)$, where $f(n) = a_1 r^{n-1}$. Then, $a_n = f(n) = 2\left(\dfrac{1}{2}\right)^{n-1}$.

59. Let $a_n = f(n)$, where $f(n) = a_1 r^{n-1}$. Then, $a_n = f(n) = a_1\left(-\dfrac{1}{4}\right)^{n-1}$.

Since $f(3) = a_1\left(-\dfrac{1}{4}\right)^{3-1} = \dfrac{1}{32} \Rightarrow a_1 = \dfrac{16}{32} = \dfrac{1}{2}$. Thus, $a_n = f(n) = \dfrac{1}{2}\left(-\dfrac{1}{4}\right)^{n-1}$.

61. Let $a_n = f(n)$, where $f(n) = a_1 r^{n-1}$. Then, $a_n = f(n) = -3(2)^{n-1}$.

63. Let $a_n = f(n)$, where $f(n) = a_1 r^{n-1}$. Then, $a_n = f(n) = a_1(5)^{n-1}$.

Since $f(2) = a_1(5)^{2-1} = 10 \Rightarrow a_1 = 2$. Thus, $a_n = f(n) = 2(5)^{n-1}$.

65. Let $a_n = f(n)$, where $f(n) = a_1 r^{n-1}$. $a_2 = 4$ and $a_5 = 32 \Rightarrow 8 = \dfrac{32}{4} = \dfrac{a_5}{a_2} = \dfrac{a_1 r^{5-1}}{a_1 r^{2-1}} = \dfrac{r^4}{r^1} = r^3 \Rightarrow r = 2$. Then,

$f(n) = a_1(2)^{n-1}$. $f(2) = a_1(2)^{2-1} = 4 \Rightarrow a_1 = 2$. Thus, $a_n = f(n) = 2(2)^{n-1}$.

67. Let $a_n = f(n)$, where $f(n) = a_1 r^{n-1}$. $a_3 = 2$ and $a_6 = \dfrac{1}{4} \Rightarrow \dfrac{1}{8} = \dfrac{\frac{1}{4}}{2} = \dfrac{a_6}{a_3} = \dfrac{a_1 r^{6-1}}{a_1 r^{3-1}} = \dfrac{r^5}{r^2} = r^3 \Rightarrow r = \dfrac{1}{2}$. Then,

$f(n) = a_1\left(\dfrac{1}{2}\right)^{n-1}$. $f(3) = a_1\left(\dfrac{1}{2}\right)^{3-1} = 2 \Rightarrow a_1 = 8$. Thus, $a_n = f(n) = 8\left(\dfrac{1}{2}\right)^{n-1}$.

69. Let $a_n = f(n)$, where $f(n) = a_1 r^{n-1}$. $a_1 = -5$ and

$a_3 = -125 \Rightarrow 25 = \dfrac{-125}{-5} = \dfrac{a_3}{a_1} = \dfrac{a_1 r^{3-1}}{a_1 r^{1-1}} = \dfrac{r^2}{r^0} = r^2 \Rightarrow r = -5$. Thus, $a_n = f(n) = -5(-5)^{n-1}$.

71. Let $a_n = f(n)$, where $f(n) = a_1 r^{n-1}$. $a_2 = -1$ and $a_7 = -32 \Rightarrow 32 = \dfrac{-32}{-1} = \dfrac{a_7}{a_2} = \dfrac{a_1 r^{7-1}}{a_1 r^{2-1}} = \dfrac{r^6}{r^1} = r^5 \Rightarrow r = 2$.

Then, $f(n) = a_1(2)^{n-1}$. $f(2) = a_1(2)^{2-1} = -1 \Rightarrow a_1 = -\dfrac{1}{2}$. Thus, $a_n = f(n) = -\dfrac{1}{2}(2)^{n-1}$.

73. Since $f(n) = 4 - 3n^3$ is not a linear function, it does not represent an arithmetic sequence.

75. Since $f(n) = 4n - (3 - n) = 5n - 3$ is a linear function, it represents an arithmetic sequence.

77. Since the plotted points appear to be collinear, the graph represents an arithmetic sequence.

79. Since the common difference is -2, the table represents an arithmetic sequence.

81. Since $f(n) = 4(2)^{n-1}$ is written in the form $f(n) = cr^{n-1}$, it represents a geometric sequence.

83. Since $f(n) = -3(n)^2$ cannot be written in the form $f(n) = cr^{n-1}$, it does not represents a geometric sequence.

85. Since the plotted points appear to be collinear, the graph does not represent a geometric sequence.

87. Since there is no common ratio, the table does not represent a geometric sequence.

89. These terms represent an arithmetic sequence. Each term can be obtained by adding 7 to the previous term.

91. These terms represent a geometric sequence. Each term can be obtained by multiplying the previous term by 4.

93. These terms represent neither an arithmetic nor a geometric sequence. There is no common ratio or difference.

95. This sequence is arithmetic since the points lie on a line (and are evenly spaced). Since the sequence is decreasing the common difference must be negative. The slope of the line passing through these points is -1, so the common difference is $d = -1$.

97. The sequence is either geometric or arithmetic. This sequence must be geometric, since the points do not lie on a line. Since the terms alternate sign, the common ratio r is negative. The absolute value of the terms are dampening to a value of 0. Therefore, $|r| < 1$.

99. The insect population increases rapidly and then levels off at 5000 per acre.

101. (a) The initial density is 500. The population density each successive year is 0.8 of the previous year. Therefore $a_1 = 500$ and $a_n = 0.8a_{n-1}$.

(b) $a_1 = 500$, $a_2 = 0.8(500) = 400$, $a_3 = 0.8(400) = 320$, $a_4 = 0.8(320) = 256$, $a_5 = 0.8(256) = 204.8$ and $a_6 = 0.8(204.8) = 163.84$. The population density is decreasing each year by 20%

(c) The terms of the sequence 500, 400, 320, 256, … are a geometric sequence with $a_1 = 500$ and $r = 0.8$.

Therefore, the n th term is given by $a_n = 500(0.8)^{n-1}$.

103. (a) $a_1 = 8, a_2 = 2.9a_1 - 0.2a_1^2 = 2.9(8) - 0.2(8)^2 = 10.4$,
$a_3 = 2.9a_2 - 0.2a_2^2 = 2.9(10.4) - 0.2(10.4)^2 = 8.528$.

(b)

The population density oscillates above and below 9.5 (approximately).

105. Since 100 million cell phones are thrown away per year we have the following:

$a_1 = 100$, $a_2 = 200$, $a_3 = 300, …, a_n = 100n$, where n is the number of years.

$a_5 = 100(5) = 500$ After 5 years, 500 million cell phones have been thrown out.

107. (a) The first term is $a_1 = 100$ and each successive term can be found by adding $d = 94$ to the previous result. 100, $100 + 94 = 194$, $194 + 94 = 288$, $288 + 94 = 382$, $382 + 94 = 476$

(b)

(c) In an arithmetic sequence with first term a_1 and common difference d, the nth term a_n is given by $a_n = a_1 + (n-1)d$. From part (a) we know that $a_1 = 100$ and $d = 94$ so $a_n = 100 + (n-1)94$.

109. (a) The terms in this sequence can be found by adding the previous two terms. $a_1 = 1$, $a_2 = 1$, $a_3 = 2$, $a_4 = 3$, $a_5 = 5$, $a_6 = 8$, $a_7 = 13$, $a_8 = 21$, $a_9 = 34$, $a_{10} = 55$, $a_{11} = 89$, and $a_{12} = 144$.

(b) $\dfrac{a_2}{a_1} = \dfrac{1}{1} = 1$, $\dfrac{a_3}{a_2} = \dfrac{2}{1} = 2$, $\dfrac{a_4}{a_3} = \dfrac{3}{2} = 1.5$, $\dfrac{a_5}{a_4} = \dfrac{5}{3} \approx 1.6667$, $\dfrac{a_6}{a_5} = \dfrac{8}{5} = 1.6$, $\dfrac{a_7}{a_6} = \dfrac{13}{8} = 1.625$,

$\dfrac{a_8}{a_7} = \dfrac{21}{13} \approx 1.6154$, $\dfrac{a_9}{a_8} = \dfrac{34}{21} = 1.6190$, $\dfrac{a_{10}}{a_9} = \dfrac{55}{34} \approx 1.6176$, $\dfrac{a_{11}}{a_{10}} = \dfrac{89}{55} = 1.6182$, and $\dfrac{a_{12}}{a_{11}} = \dfrac{144}{89} \approx 1.6180$.

These ratios seem to be approaching a number near 1.618. This number is called the golden ratio.

(c) $n = 2$: $a_1 \cdot a_3 - a_2^2 = (1)(2) - (1)^2 = 1 = (-1)^2$;

$$n = 3: a_2 \cdot a_4 - a_3^2 = (1)(3) - (2)^2 = -1 = (-1)^3;$$

$$n = 4: a_3 \cdot a_5 - a_4^2 = (2)(5) - (3)^2 = 1 = (-1)^4$$

111. (a)   The salary of the first employee at the beginning of the nth year is given by:

$$a_n = 30,000 + (n-1)(2000) = 2000n + 28,000.$$

This is a arithmetic sequence with $a_1 = 30,000$ and $d = 2000$.

   (b)   The salary of the second employee at the beginning of the nth year is given by: $b_n = 30,000(1.05)^{n-1}$.

This is a geometric sequence with $b_1 = 30,000$ and $d = 1.05$.

   (c)   At the beginning of the 10th year each salary is:

$$a_{10} = 2000(10) + 28,000 = \$48,000, b_{10} = 30,000(1.05)^{10-1} \approx \$46,540.$$

At the beginning of the 20th year each salary is:

$$a_{20} = 2000(20) + 28,000 = \$68,000, b_{20} = 30,000(1.05)^{20-1} \approx \$75,809.$$

   (d)

With time the geometric sequence overtakes the arithmetic sequence since $r > 1$.

113. Let $a_1 = 2$. Then, $a_2 = \dfrac{1}{2}\left(a_1 + \dfrac{k}{a_1}\right) = \dfrac{1}{2}\left(2 + \dfrac{2}{2}\right) = 1.5$, $a_3 = \dfrac{1}{2}\left(a_2 + \dfrac{k}{a_2}\right) = \dfrac{1}{2}\left(1.5 + \dfrac{2}{1.5}\right) = 1.41\overline{6}$ In a similar

manner, $a_4 \approx 1.414215686$, $a_5 \approx 1.414213562$, and $a_6 \approx 1.414213562$. Since $\sqrt{2} \approx 1.414213562$ this is a very accurate approximation.

115. Let $a_1 = 21$. Then, $a_2 = \dfrac{1}{2}\left(a_1 + \dfrac{k}{a_1}\right) = \dfrac{1}{2}\left(21 + \dfrac{21}{21}\right) = 11$, $a_3 = \dfrac{1}{2}\left(a_2 + \dfrac{k}{a_2}\right) = \dfrac{1}{2}\left(11 + \dfrac{21}{11}\right) = 6.4\overline{5}$ In a similar

manner, $a_4 \approx 4.854033291$, $a_5 \approx 4.59016621$, and $a_6 \approx 4.582581971$. Since $\sqrt{21} \approx 4.582575695$ this is an accurate approximation.

117. By definition $a_n = a_1 + (n-1)d_1$ and $b_n = b_1 + (n-1)d_2$. Then

$$c_n = a_n + b_n = [a_1 + (n-1)d_1] + [b_1 + (n-1)d_2] = (a_1 + b_1) + [(n-1)d_1 + (n-1)d_2] =$$

$$(a_1 + b_1) + (n-1)(d_1 + d_2) = c_1 + (n-1)d \text{ where } c_1 = a_1 + b_1 \text{ and } d = d_1 + d_2.$$

8.2: Series

1. (a)   The counting numbers from 1 to 5 are 1, 2, 3, 4, 5.

   (b)   The series is $1 + 2 + 3 + 4 + 5$.

   (c) $1 + 2 + 3 + 4 + 5 = 15$

3. (a)   The integers counting down from 1 to -3 are $1, 0, -1, -2, -3$.

   (b)   The series is $1 + 0 + (-1) + (-2) + (-3)$.

   (c) $1 + 0 + (-1) + (-2) + (-3) = -5$

5. Since A_n represents the number of AIDS deaths after 2000, the sum from 2005 to 2009 is given by

$$A_5 + A_6 + A_7 + A_8 + A_9$$

7. $S_5 = 3(1) + 3(2) + 3(3) + 3(4) + 3(5) = 3 + 6 + 9 + 12 + 15 = 45$

9. $S_5 = (2(1) - 1) + (2(2) - 1) + (2(3) - 1) + (2(4) - 1) + (2(5) - 1) = 1 + 3 + 5 + 7 + 9 = 25$

11. $S_5 = (1^2 + 1) + (2^2 + 1) + (3^2 + 1) + (4^2 + 1) + (5^2 + 1) = 2 + 5 + 10 + 17 + 26 = 60$

13. $S_5 = \dfrac{1}{1+1} + \dfrac{2}{2+1} + \dfrac{3}{3+1} + \dfrac{4}{4+1} + \dfrac{5}{5+1} = \dfrac{1}{2} + \dfrac{2}{3} + \dfrac{3}{4} + \dfrac{4}{5} + \dfrac{5}{6} = \dfrac{71}{20}$

15. The first term is $a_1 = 3$ and the last term is $a_8 = 17$. To find the sum use:

$$S_n = n\left(\frac{a_1 + a_n}{2}\right) \Rightarrow S_8 = 8\left(\frac{3+17}{2}\right) = 80 .$$ The sum is 80.

17. The first term is $a_1 = 1$ and the last term is $a_{50} = 50$. To find the sum use:

$$S_n = n\left(\frac{a_1 + a_n}{2}\right) \Rightarrow S_{50} = 50\left(\frac{1+50}{2}\right) = 1275 .$$ The sum is 1275.

19. The first term is $a_1 = -7$ and $d = 3$. The last term is 101, but we must determine n. The term of 101 is

$\dfrac{101 - (-7)}{3} = 36$ terms after a_1. Thus, the last term is $a_{37} = 101$. To find the sum use the following:

$$S_n = n\left(\frac{a_1 + a_n}{2}\right) \Rightarrow S_{37} = 37\left(\frac{-7 + 101}{2}\right) = 1739 .$$ The sum is 1739.

21. The number of terms to be added is 40, so $n = 40$. The first term is $a_1 = 5(1) = 5$. The last term is

$a_{40} = 5(40) = 200$. $S_n = n\left(\dfrac{a_1 + a_n}{2}\right) \Rightarrow S_{40} = 40\left(\dfrac{5 + 200}{2}\right) = 4100$. The sum is 4100.

23. $S_{15} = 15\left(\dfrac{a_1 + a_{15}}{2}\right) \Rightarrow 255 = 15\left(\dfrac{3 + a_{15}}{2}\right) \Rightarrow 17 = \dfrac{3 + a_{15}}{2} \Rightarrow 34 = 3 + a_{15} \Rightarrow a_{15} = 31$

25. Since $a_1 = 4$ and $d = 2$, use the formula $S_n = \dfrac{n}{2}(2a_1 + (n-1)d)$.

$$S_{20} = \frac{20}{2}(2(4) + (20-1)2) = 10(8 + 19(2)) = 10(8 + 38) = 10(46) = 460$$

27. Since $a_1 = 10$ and $d = -\dfrac{1}{2}$, use the formula $S_n = \dfrac{n}{2}(2a_1 + (n-1)d)$.

$$S_{20} = \frac{20}{2}\left(2(10) + (20-1)\left(-\frac{1}{2}\right)\right) = 10\left(20 + 19\left(-\frac{1}{2}\right)\right) = 10\left(20 - \frac{19}{2}\right) = 10\left(\frac{21}{2}\right) = 105$$

29. Since $a_1 = 4$ and $a_{20} = 190.2$, use the formula $S_n = n\left(\dfrac{a_1 + a_n}{2}\right)$.

$$S_{20} = 20\left(\frac{4 + 190.2}{2}\right) = 20\left(\frac{194.2}{2}\right) = 20(97.1) = 1942$$

31. Since $a_1 = -2$ and $a_{11} = 50$, the common difference is $d = \dfrac{50 - (-2)}{11 - 1} = \dfrac{52}{10} = 5.2$.

Use the formula $S_n = \dfrac{n}{2}(2a_1 + (n-1)d)$.

$$S_{20} = \frac{20}{2}(2(-2) + (20-1)5.2) = 10(-4 + 19(5.2)) = 10(-4 + 98.8) = 10(94.8) = 948$$

33. Since $a_2 = 6$ and $a_{12} = 31$, the common difference is $d = \dfrac{31 - 6}{12 - 2} = \dfrac{25}{10} = 2.5$ and so $a_1 = 6 - 2.5 = 3.5$.

Use the formula $S_n = \dfrac{n}{2}(2a_1 + (n-1)d)$.

$$S_{20} = \frac{20}{2}(2(3.5) + (20-1)2.5) = 10(7 + 19(2.5)) = 10(7 + 47.5) = 10(54.5) = 545$$

35. The first term is $a_1 = 1$ and the common ratio is $r = 2$. Since there are 8 terms, the sum is

$$S_n = a_1\left(\frac{1-r^n}{1-r}\right) \Rightarrow S_8 = 1\left(\frac{1-2^8}{1-2}\right) = 255.$$

This sum can be verified by adding $1 + 2 + 4 + 8 + 16 + 32 + 64 + 128 = 255$.

37. The first term is $a_1 = 0.5$ and the common ratio is $r = 3$. Since there are 7 terms, the sum is

$$S_n = a_1\left(\frac{1-r^n}{1-r}\right) \Rightarrow S_7 = 0.5\left(\frac{1-(3)^7}{1-3}\right) = 546.5.$$

This sum can be verified by adding $0.5 + 1.5 + 4.5 + 13.5 + 40.5 + 121.5 + 364.5 = 546.5$.

39. The first term is $a_1 = 3(2)^0 = 3$ and the common ratio is $r = 2$. Since there are 20 terms, the sum is

$$S_n = a_1\left(\frac{1-r^n}{1-r}\right) \Rightarrow S_{20} = 3\left(\frac{1-2^{20}}{1-2}\right) = 3,145,725.$$

41. Using the formula $S_n = a_1\left(\frac{1-r^n}{1-r}\right)$ with $a_1 = 1$ and $r = -\frac{1}{2}$, $S_4 = 1\left(\frac{1-(-\frac{1}{2})^4}{1-(-\frac{1}{2})}\right) = 0.625$;

$$S_7 = 1\left(\frac{1-(-\frac{1}{2})^7}{1-(-\frac{1}{2})}\right) = 0.671875 \; ; \; S_{10} = 1\left(\frac{1-(-\frac{1}{2})^{10}}{1-(-\frac{1}{2})}\right) = 0.666015625$$

43. Using the formula $S_n = a_1\left(\frac{1-r^n}{1-r}\right)$ with $a_1 = \frac{1}{3}$ and $r = 2$, $S_4 = \frac{1}{3}\left(\frac{1-2^4}{1-2}\right) = 5 \; ; \; S_7 = \frac{1}{3}\left(\frac{1-2^7}{1-2}\right) = 42.\overline{3}$;

$$S_{10} = \frac{1}{3}\left(\frac{1-2^{10}}{1-2}\right) = 341$$

45. The first term is $a_1 = 1$ and the common ratio is $r = \frac{1}{3}$. The sum is $S = a_1\left(\frac{1}{1-r}\right) \Rightarrow S = 1\left(\frac{1}{1-\frac{1}{3}}\right) = \frac{3}{2}$.

47. The first term is $a_1 = 6$ and the common ratio is $r = -\frac{2}{3}$. The sum is

$$S = a_1\left(\frac{1}{1-r}\right) \Rightarrow S = 6\left(\frac{1}{1-(-\frac{2}{3})}\right) = \frac{18}{5}.$$

49. The first term is $a_1 = 1$ and the common ratio is $r = -\frac{1}{10}$ or -0.1. The sum is

$$S = a_1\left(\frac{1}{1-r}\right) \Rightarrow S = 1\left(\frac{1}{1-(-0.1)}\right) = \frac{1}{1.1} = \frac{10}{11}.$$

51. $\frac{2}{3} = 0.6666666\ldots = 0.6 + 0.06 + 0.006 + 0.0006 + 0.00006 + \cdots$

53. $\frac{9}{11} = 0.81818181\ldots = 0.81 + 0.0081 + 0.000081 + 0.00000081 + \cdots$

55. $\frac{1}{7} = 0.142857142857\ldots = 0.142857 + 0.000000142857 + 0.000000000000142857 + \cdots$

57. The series $0.8 + 0.08 + 0.008 + 0.0008 + \cdots$ is an infinite geometric series with $a_1 = 0.8$ and $r = 0.1$.

$$S = \frac{a_1}{1-r} = \frac{0.8}{1-0.1} = \frac{8}{9}.$$

59. The series $0.45 + 0.0045 + 0.000045 + \cdots$ is an infinite geometric series with $a_1 = 0.45$ and $r = 0.01$.

$$S = \frac{a_1}{1-r} = \frac{0.45}{1-0.01} = \frac{45}{99} = \frac{5}{11}.$$

61. $\sum_{k=1}^{4}(k+1) = (1+1)+(2+1)+(3+1)+(4+1) = 2+3+4+5 = 14$

63. $\sum_{k=1}^{8} 4 = 4+4+4+4+4+4+4+4 = 32$

65. $\sum_{k=1}^{7} k^3 = 1^3+2^3+3^3+4^3+5^3+6^3+7^3 = 1+8+27+64+125+216+343 = 784$

67. $\sum_{k=1}^{5}(k^2-k) = (4^2-4)+(5^2-5) = 12+20 = 32$

69. $1^4+2^4+3^4+4^4+5^4+6^4 = \sum_{k=1}^{6} k^4$

71. $1+\frac{4}{3}+\frac{6}{4}+\frac{8}{5}+\frac{10}{6}+\frac{12}{7}+\frac{14}{8} = \sum_{k=1}^{7}\left(\frac{2k}{k+1}\right)$

73. $2+4+6+8+10+12+14 = \sum_{k=1}^{7} 2k$

75. $24-20-16-12-8-4 = \sum_{k=1}^{6}(28-4k)$

77. $1+\frac{1}{2^2}+\frac{1}{3^2}+\frac{1}{4^2}+\frac{1}{5^2}+\cdots = \sum_{k=1}^{\infty}\left(\frac{1}{k^2}\right)$

79. Because $\sum_{k=6}^{9} k^3 == 6^3+7^3+8^3+9^3$, the summation has 4 terms. We will write $k = 6, 7, 8, 9$ as $n = 1, 2, 3, 4$.

It follows that $n+5 = k$, and $\sum_{n=1}^{4}(n+5)^3$.

81. Because $\sum_{k=9}^{32}(3k-2) = (3(9)-2)+(3(10)-2)+\cdots+(3(32)-2)$, the summation has 24 terms. We will write

$k = 9, 10, 11, \ldots 32$ as $n = 1, 2, 3, 4, \ldots 24$. It follows that $n+8 = k$, and

$\sum_{n=1}^{24}(3(n+8)-2) \Rightarrow \sum_{n=1}^{24}(3n+24-2) \Rightarrow \sum_{n=1}^{24}(3n+22)$.

83. Because $\sum_{k=16}^{52}(k^2-3k) = (16^2-3(16))+(17^2-3(17))+\cdots+(52^2-3(52))$, the summation has 37 terms. We

will write $k = 16, 17, 18, \ldots 52$ as $n = 1, 2, 3, \ldots 37$. It follows that $n+15 = k$, and

$\sum_{n=1}^{37}((n+15)^2-3(n+15)) \Rightarrow \sum_{n=1}^{37}(n^2+30n+225-3n-45) \Rightarrow \sum_{n=1}^{37}(n^2+27n+180)$.

85. $\sum_{k=1}^{60} 9 = 60(9) = 540$

87. $\displaystyle\sum_{k=1}^{15} 5k = 5\sum_{k=1}^{15} k = 5\left[\frac{15(16)}{2}\right] = 600$

89. $\displaystyle\sum_{k=1}^{31} (3k-3) = 3\sum_{k=1}^{31} k - \sum_{k=1}^{31} 3 = 3\left[\frac{31(32)}{2}\right] - 31(3) = 1488 - 93 = 1395$

91. $\displaystyle\sum_{k=1}^{25} k^2 = \frac{(25)(26)(51)}{6} = 5525$

93. $\displaystyle\sum_{k=1}^{16} (k^2 - k) = \sum_{k=1}^{16} k^2 - \sum_{k=1}^{16} k = \frac{(16)(17)(33)}{6} - \frac{(16)(17)}{2} = 1496 - 136 = 1360$

95. $\displaystyle\sum_{K=5}^{24} k = \sum_{K=1}^{24} k - \sum_{K=1}^{4} k = \frac{(24)(25)}{2} - \frac{(4)(5)}{2} = 300 - 10 = 290$

97. $\displaystyle\sum_{k=1}^{n} k = 1 + 2 + 3 + 4 + \cdots + n$. This is an arithmetic series with $a_1 = 1$ and $a_n = n$. The sum is given by

$$S_n = n\left(\frac{a_1 + a_n}{2}\right) = n\left(\frac{1+n}{2}\right) = \frac{n(n+1)}{2}.$$

99. (a)   The arithmetic sequence describing the salary during year n is computed by $a_n = 42,000 + 1800(n-1)$.
The 1st and 15th years salaries are: $a_1 = 42,000 + 1800(1-1) = 42,000$ and
$a_{15} = 42,000 + 1800(15-1) = 67,200$. The total amount earned during the 15 year period is

$$S_{15} = \frac{15(42,000 + 67,200)}{2} = 819,000.$$

   (b)   Verify with a calculator, compute the sum $a_1 + a_2 + a_3 + \cdots + a_{15}$, where $a_n = 42,000 + 1800(n-1)$.

101. Future value is given by $S_n = A_0\left(\dfrac{(1+i)^n - 1}{i}\right)$, where $A_0 = a_1$.

$S_{20} = 2000\left(\dfrac{(1+0.08)^{20} - 1}{0.08}\right) \approx 91,523.93$. If \$2000 is deposited in an account at the end of each year for 20

years and the account pays 8% interest, the future value of this annuity will be \$91,523.93.

103. Future value is given by $S_n = A_0\left(\dfrac{(1+i)^n - 1}{i}\right)$, where $A_0 = a_1$.

$S_5 = 10,000\left(\dfrac{(1+0.11)^5 - 1}{0.11}\right) \approx 62,278.01$. If \$10,000 is deposited in an account at the end of each year for 5

years and the account pays 11% interest, the future value of this annuity will be \$62,278.01.

105. The number of logs in the stack is given by $7 + 8 + 9 + 10 + 11 + 12 + 13 + 14 + 15$. This series is arithmetic with

first term $a_1 = 7$ and the last term $a_9 = 15$ so its sum is given by $S_9 = 9\left(\dfrac{7+15}{2}\right) = 99$. So, there are 99 logs

in the stack.

107. (a)   Since each filter lets half of the impurities through, the series is can be written as

$$0.5(1) + 0.5(0.5) + 0.5(0.25) + \cdots = \sum_{k=1}^{n} 0.5(0.5)^{k-1}.$$

   (b)   The series sums to 1 only if an infinite number of filters are used.

109. The area of the largest square is 1, the area of the next largest square is $\frac{1}{2}$, and each successive square has an area that is $\frac{1}{2}$ the area of the previous square. If there are an infinite number of squares, then the area is represented by the geometric series $1+\frac{1}{2}+\frac{1}{4}+\frac{1}{8}+\frac{1}{16}+\cdots$. This is an infinite geometric series with $a_1 = 1$ and $r = \frac{1}{2}$, whose sum is $S = \frac{1}{1-r} = \frac{1}{1-\frac{1}{2}} = 2$. The area is 2.

111. The value of e^a is approximated by $e^a = 1 + a + \frac{a^2}{2!} + \frac{a^3}{3!} + \cdots + \frac{a^n}{n!}$. Apply the series by letting $a = 1$, since $e^1 = e$. The first 8 terms sum to

$$e^1 \approx 1 + 1 + \frac{1}{2!} + \frac{1}{3!} + \frac{1}{4!} + \frac{1}{5!} + \frac{1}{6!} + \frac{1}{7!} = 1 + 1 + \frac{1}{2} + \frac{1}{6} + \frac{1}{24} + \frac{1}{120} + \frac{1}{720} + \frac{1}{5040} \approx 2.718254.$$ The actual value is $e \approx 2.718282$, so only 8 terms of the series provides an approximation for e that is accurate to 4 decimal places.

113. One can either add the terms directly or apply the formula $S_n = a_1\left(\frac{1-r^n}{1-r}\right)$. We will apply the formula. In this sequence $a_1 = 1$ and $r = \frac{1}{3}$. $S_2 = 1\left(\frac{1-\left(\frac{1}{3}\right)^2}{1-\left(\frac{1}{3}\right)}\right) = \frac{4}{3} \approx 1.3333$; $S_4 = 1\left(\frac{1-\left(\frac{1}{3}\right)^4}{1-\left(\frac{1}{3}\right)}\right) = \frac{40}{27} \approx 1.4815$;

$S_8 = 1\left(\frac{1-\left(\frac{1}{3}\right)^8}{1-\left(\frac{1}{3}\right)}\right) \approx 1.49977$; $S_{16} = 1\left(\frac{1-\left(\frac{1}{3}\right)^{16}}{1-\left(\frac{1}{3}\right)}\right) \approx 1.49999997$. The sum of the infinite series $\sum_{k=1}^{\infty}\left(\frac{1}{3}\right)^{k-1}$ is given by $S = \frac{a_1}{1-r} = \frac{1}{1-\frac{1}{3}} = \frac{3}{2} = 1.5$. As n increases, the partial sums S_2, S_4, S_8, and S_{16} become closer and closer to the value of 1.5.

115. One can either add the terms directly or apply the formula $S_n = a_1\left(\frac{1-r^n}{1-r}\right)$. We will apply the formula. In this sequence $a_1 = 4$ and $r = -\frac{1}{10}$. $S_1 = 4\left(\frac{1-\left(-\frac{1}{10}\right)^1}{1-\left(-\frac{1}{10}\right)}\right) = 4$; $S_2 = 4\left(\frac{1-\left(-\frac{1}{10}\right)^2}{1-\left(-\frac{1}{10}\right)}\right) = 3.6$;

$S_3 = 4\left(\frac{1-\left(-\frac{1}{10}\right)^3}{1-\left(-\frac{1}{10}\right)}\right) = 3.64$; $S_4 = 4\left(\frac{1-\left(-\frac{1}{10}\right)^4}{1-\left(-\frac{1}{10}\right)}\right) = 3.636$; $S_5 = 4\left(\frac{1-\left(-\frac{1}{10}\right)^5}{1-\left(-\frac{1}{10}\right)}\right) = 3.6364$;

$S_6 = 4\left(\frac{1-\left(-\frac{1}{10}\right)^6}{1-\left(-\frac{1}{10}\right)}\right) = 3.63636$; The sum of the infinite series $\sum_{k=1}^{\infty} 4\left(-\frac{1}{10}\right)^{k-1}$ is given by

$S = \frac{a_1}{1-r} = \frac{4}{1-(-0.1)} = \frac{40}{11} = 3.\overline{63}$. As n increases, the partial sums become closer and closer to the value of $3.\overline{63}$.

Extended and Discovery Exercises for Section 8.2

1. A graph of the coordinates representing the series $10 + 15 + 20 + 25 + 30 + 35$ will show that the sum $a_1 + a_n = a_2 + a_{n-1} = a_3 + a_{n-2}$, which can be used to show that

$$a_1 + a_2 + a_3 + a_{n-2} + a_{n-1} + a_n = (a_1 + a_n) + (a_2 + a_{n-1}) + (a_3 + a_{n-2}) = (a_1 + a_n) + (a_1 + a_n) + (a_1 + a_n) \Rightarrow S_n = \frac{n(a_1 + a_n)}{2}$$

Checking Basic Concepts for Sections 8.1 & 8.2

1. A graphical representation of $a_n = -2n + 3$ is shown in Figure 1a. A numerical representation is shown in Figure 1b. The first six terms are 1, –1, –3, –5, –7, –9.

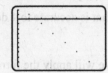

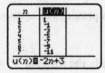

Figure 1a   Figure 1b

3. (a) The series is arithmetic since the common difference between the terms is 4. The sum of the first 10 terms shown is $10\left(\dfrac{1+37}{2}\right) = 190$.

    (b) The series is geometric since the common ratio is $\dfrac{1}{3}$. The sum of the first 6 terms shown is

    $$3\left(\frac{1-(\frac{1}{3})^6}{1-(\frac{1}{3})}\right) = 3\left(\frac{1-(\frac{1}{729})}{1-(\frac{1}{3})}\right) = 3\left(\frac{\frac{728}{729}}{\frac{2}{3}}\right) = 3\left(\frac{364}{243}\right) = \frac{364}{81} \approx 4.494.$$

    (c) The series is an infinite geometric series with a common ratio of $\dfrac{1}{4}$.

    The sum of the series is $\dfrac{2}{1-\frac{1}{4}} = \dfrac{2}{0.75} = \dfrac{8}{3} \approx 2.667$.

    (d) The series is an infinite geometric series with a common ratio of 0.1.

    The sum of the series is $\dfrac{0.9}{1-0.1} = \dfrac{0.9}{0.9} = 1$.

5. (a) $\displaystyle\sum_{k=1}^{15}(k+2) = \sum_{k=1}^{15}k + \sum_{k=1}^{15}2 = \frac{15(16)}{2} + 15(2) = 150$

    (b) $\displaystyle\sum_{k=1}^{21}2k^2 = 2\sum_{k=1}^{21}k^2 = 2\left[\frac{21(22)(43)}{6}\right] = 6622$

8.3: Counting

1. There are $2\cdot2\cdot2\cdot2\cdot2\cdot2\cdot2\cdot2\cdot2\cdot2 = 2^{10} = 1024$ different ways to answer the exam.

3. There are $2\cdot2\cdot2\cdot2\cdot2\cdot4\cdot4\cdot4\cdot4\cdot4\cdot4\cdot4\cdot4\cdot4\cdot4 = 2^5 \cdot 4^{10} = 33,554,432$ different ways.

5. There are 10 digits and 26 letters. There are $10\cdot10\cdot10\cdot26\cdot26\cdot26 = 17,576,000$ different license plates.

7.   There are 36 digits and letters. There are $26 \cdot 26 \cdot 26 \cdot 36 \cdot 36 \cdot 36 = 820,025,856$ different license plates.

9.   There are 3 choices for each of the 5 letters. There are $3 \cdot 3 \cdot 3 \cdot 3 \cdot 3 = 3^5 = 243$ different strings.

11.   There are 5 choices for each of the 5 letters. There are $5 \cdot 5 \cdot 5 \cdot 5 \cdot 5 = 5^5 = 3125$ different strings.

13.   Since there are 2 letters that can only be used once, each string must be 2 letters long. For the first position there are 2 choices and for the second position there is only 1 choice. There are $2 \cdot 1 = 2$ possible strings.

15.   There are $4 \cdot 3 \cdot 2 \cdot 1 = 24$ possible strings.

17.   Each combination can vary between 000 and 999. Thus, there are 1000 possibilities for each lock. That is, there are $1000 \cdot 1000 = 1,000,000$ different combinations in all.

19.   There are 2 settings (on or off) for each of the 12 switches.

      There are $2 \cdot 2 \cdot 2 \cdot 2 \cdot 2 \cdot 2 \cdot 2 \cdot 2 \cdot 2 \cdot 2 \cdot 2 \cdot 2 = 2^{12} = 4096$ different codes for the garage door opener.

21.   For the first letter there are 2 possibilities, whereas there are 26 possibilities for each of the last 3 positions. There are $2 \cdot 26 \cdot 26 \cdot 26 = 35,152$ different call letters possible. There is no shortage of call letters.

23.   There are $2 \cdot 3 \cdot 4 = 24$ different packages that can be purchased.

25.   There are $8 \cdot 10 \cdot 10 \cdot 10 \cdot 10 \cdot 10 \cdot 10 = 8,000,000$ such phone numbers.

27. $6! = 6 \cdot 5 \cdot 4 \cdot 3 \cdot 2 \cdot 1 = 720$

29. $10! = 10 \cdot 9 \cdot 8 \cdot 7 \cdot 6 \cdot 5 \cdot 4 \cdot 3 \cdot 2 \cdot 1 = 3,628,800$

31. $P(n, r) = \dfrac{n!}{(n-r)!} \Rightarrow P(5, 3) = \dfrac{5!}{(5-3)!} = \dfrac{5!}{2!} = \dfrac{120}{2} = 60$

33. $P(n, r) = \dfrac{n!}{(n-r)!} \Rightarrow P(8, 1) = \dfrac{8!}{(8-1)!} = \dfrac{8}{7}! = \dfrac{8 \cdot 7!}{7!} = 8$

35. $P(n, r) = \dfrac{n!}{(n-r)!} \Rightarrow P(7, 3) = \dfrac{7!}{(7-3)!} = \dfrac{7}{4}! = \dfrac{7 \cdot 6 \cdot 5 \cdot 4!}{4!} = 7 \cdot 6 \cdot 5 = 210$

37. $P(n, r) = \dfrac{n!}{(n-r)!} \Rightarrow P(25, 2) = \dfrac{25!}{(25-2)!} = \dfrac{25}{23}! = \dfrac{25 \cdot 24 \cdot 23!}{23!} = 25 \cdot 24 = 600$

39. $P(n, r) = \dfrac{n!}{(n-r)!} \Rightarrow P(10, 4) = \dfrac{10!}{(10-4)!} = \dfrac{10!}{6!} = \dfrac{10 \cdot 9 \cdot 8 \cdot 7 \cdot 6!}{6!} = 10 \cdot 9 \cdot 8 \cdot 7 = 5040$

41. $P(4, 4) = \dfrac{4!}{(4-4)!} = \dfrac{4}{0}! = 4! = 24$

43.   There are $15 \cdot 14 \cdot 13 = 2730$ different arrangements of 3 students from a class of 15.

45.   There are $7 \cdot 6 \cdot 5 = 210$ different routes.

47.   The remaining 4 digits for the 7 digit number are independent events. Since the last 4 digits can be any number from 0 to 9. The total is given by $10 \cdot 10 \cdot 10 \cdot 10 = 10,000$. The first 3 numbers are restricted to 3 different possibilities and there will be $(3) \cdot (10,000)$ or 30,000 phone numbers.

49.   Initially any person can sit at the table. Then the remaining 6 people can sit in $6! = 720$ different ways. Since there is no difference between the people sitting clockwise or counterclockwise around the table, we must divide this result by 2 for the final answer. There are 360 different ways to seat the 7 people.

51. $P(9, 9) = \dfrac{9!}{(9-9)!} = \dfrac{9}{0}! = 9! = 362,880$

53.   Counting February 29th, $p(366, 5) = \dfrac{366!}{(366-5)!} = \dfrac{366!}{361!} = \dfrac{366 \cdot 365 \cdot 364 \cdot 363 \cdot 362 \cdot 361!}{361!} \approx 6.39 \times 10^{12}$.

55.   There are 8 choices for the first and fourth digits and 10 choices for each of the other digits.
      $8 \cdot 10 \cdot 10 \cdot 8 \cdot 10 \cdot 10 \cdot 10 \cdot 10 \cdot 10 \cdot 10 = 6,400,000,000$

57. $C(n, r) = \dfrac{n!}{(n-r)!\, r!} \Rightarrow C(3, 1) = \dfrac{3!}{(3-1)!\,1!} = \dfrac{3!}{2!\,1!} = \dfrac{6}{2} = 3$

59. $C(n, r) = \dfrac{n!}{(n-r)!\, r!} \Rightarrow C(6, 3) = \dfrac{6!}{(6-3)!\,3!} = \dfrac{6!}{3!\,3!} = \dfrac{720}{36} = 20$

61. $C(n, r) = \dfrac{n!}{(n-r)!\, r!} \Rightarrow C(5, 0) = \dfrac{5!}{(5-0)!\,0!} = \dfrac{5!}{5!\,0!} = \dfrac{5!}{5!} = 1$

63. $\dbinom{n}{r} = \dfrac{n!}{(n-r)!\, r!} \Rightarrow \dbinom{8}{2} = \dfrac{8!}{(8-2)!\,2!} = \dfrac{8!}{6!\,2!} = \dfrac{8 \cdot 7 \cdot 6!}{6!\,2!} = \dfrac{56}{2} = 28$

65. $\dbinom{n}{r} = \dfrac{n!}{(n-r)!\, r!} \Rightarrow \dbinom{20}{18} = \dfrac{20!}{(20-18)!\,18!} = \dfrac{20!}{2!\,18!} = \dfrac{20 \cdot 19 \cdot 18!}{2!\,18!} = \dfrac{20 \cdot 19}{2} = 190$

67.   From 39 numbers a player picks 5 numbers. Since order is unimportant, there are $C(39, 5) = 575,757$ different ways of doing this.

69.   Two women can be selected from 5 women in $C(5, 2)$ different ways. Two men can be selected from 3 men in $C(3, 2)$ different ways. The total number of teams is $C(5, 2) \cdot C(3, 2) = 10 \cdot 3 = 30$.

71.   Three questions can be selected from 5 questions in $C(5, 3)$ different ways. Four questions can be selected from 5 questions in $C(5, 4)$ different ways. The total number of possibilities is $C(5, 3) \cdot C(5, 4) = 10 \cdot 5 = 50$.

73.   Three red marbles can be drawn from 10 red marbles in $C(10, 3)$ different ways. Two blue marbles can be drawn from 12 blue marbles in $C(12, 2)$ different ways. There are $C(10, 3) \cdot C(12, 2) = 120 \cdot 66 = 7920$ ways.

75.   Since order is not important, there are $C(24, 3) = 2024$ ways to do this.

77. $P(n, n-1) = \dfrac{n!}{(n-(n-1))!} = \dfrac{n!}{1} = n!$ and $P(n, n) = \dfrac{n!}{(n-n)!} = \dfrac{n}{0}! = \dfrac{n!}{1} = n!$ For example

        $P(7, 6) = 5040 = P(7, 7)$.

8.4: The Binomial Theorem

1. $\dbinom{5}{4} = \dfrac{5!}{1!\,4!} = \dfrac{5 \cdot 4!}{1 \cdot 4!} = 5$

3. $\dbinom{4}{0} = \dfrac{4!}{4!\,0!} = \dfrac{4!}{4! \cdot 1} = 1$

5. $\dbinom{6}{5} = \dfrac{6!}{1!\,5!} = \dfrac{6 \cdot 5!}{1 \cdot 5!} = \dfrac{6}{1} = 6$

7. $\dbinom{3}{3} = \dfrac{3!}{0!\,3!} = \dfrac{3!}{1 \cdot 3!} = 1$

9.   Three a's, two b's $\Rightarrow C(5, 2) = 10$ different strings.

11.   Four a's, four b's $\Rightarrow C(8, 4) = 70$ different strings.

13.   Five a's, zero b's $\Rightarrow C(5, 0) = 1$ string.

15.   Four a's, one $b \Rightarrow C(5, 1) = 5$ different strings.

17. $(x+y)^2 = \dbinom{2}{0} x^2 y^0 + \dbinom{2}{1} x^1 y^1 + \dbinom{2}{2} x^0 y^2 = x^2 + 2xy + y^2$

19. $(m+2)^3 = \binom{3}{0}m^3(2)^0 + \binom{3}{1}m^2(2)^1 + \binom{3}{2}m^1(2)^2 + \binom{3}{3}m^0(2)^3 = m^3 + 6m^2 + 12m + 8$

21. $(2x-3)^3 = \binom{3}{0}(2x)^3(-3)^0 + \binom{3}{1}(2x)^2(-3)^1 + \binom{3}{2}(2x)^1(-3)^2 + \binom{3}{3}(2x)^0(-3)^3 = 8x^3 - 36x^2 + 54x - 27$

23. $(p-q)^6 = \binom{6}{0}p^6(-q)^0 + \binom{6}{1}p^5(-q)^1 + \binom{6}{2}p^4(-q)^2 + \binom{6}{3}p^3(-q)^3 + \binom{6}{4}p^2(-q)^4 + $

 $\binom{6}{5}p^1(-q)^5 + \binom{6}{6}p^0(-q)^6 = p^6 - 6p^5q + 15p^4q^2 - 20p^3q^3 + 15p^2q^4 - 6pq^5 + q^6$

25. $(2m+3n)^3 = \binom{3}{0}(2m)^3(3n)^0 + \binom{3}{1}(2m)^2(3n)^1 + \binom{3}{2}(2m)^1(3n)^2 + \binom{3}{3}(2m)^0(3n)^3 = $

 $8m^3 + 36m^2n + 54mn^2 + 27n^3$

27. $(1-x^2)^4 = \binom{4}{0}(1)^4(-x^2)^0 + \binom{4}{1}(1)^3(-x^2)^1 + \binom{4}{2}(1)^2(-x^2)^2 + \binom{4}{3}(1)^1(-x^2)^3 + \binom{4}{4}(1)^0(-x^2)^4$

 $= 1 - 4x^2 + 6x^4 - 4x^6 + x^8$

29. $(2p^3-3)^3 = \binom{3}{0}(2p^3)^3(-3)^0 + \binom{3}{1}(2p^3)^2(-3)^1 + \binom{3}{2}(2p^3)^1(-3)^2 + \binom{3}{3}(2p^3)^0(-3)^3 = $

 $8p^9 - 36p^6 + 54p^3 - 27$

31. Using Pascal's triangle, the coefficients are 1, 2, and 1.
 $(x+y)^2 = 1x^2y^0 + 2x^1y^1 + 1x^0y^2 = x^2 + 2xy + y^2$

33. Using Pascal's triangle, the coefficients are 1, 4, 6, 4, and 1.
 $(3x+1)^4 = 1(3x)^4(1)^0 + 4(3x)^3(1)^1 + 6(3x)^2(1)^2 + 4(3x)^1(1)^3 + 1(3x)^0(1)^4 = $

 $81x^4 + 108x^3 + 54x^2 + 12x + 1$

35. Using Pascal's triangle, the coefficients are 1, 5, 10, 10, 5, and 1.
 $(2-x)^5 = 1(2)^5(-x)^0 + 5(2)^4(-x)^1 + 10(2)^3(-x)^2 + 10(2)^2(-x)^3 + 5(2)^1(-x)^4 + 1(2)^0(-x)^5 = $

 $\quad 32 - 80x + 80x^2 - 40x^3 + 10x^4 - x^5$

37. Using Pascal's triangle, the coefficients are 1, 4, 6, 4, and 1.
 $(x^2+2)^4 = 1(x^2)^4(2)^0 + 4(x^2)^3(2)^1 + 6(x^2)^2(2)^2 + 4(x^2)^1(2)^3 + 1(x^2)^0(2)^4 = x^8 + 8x^6 + 24x^4 + 32x^2 + 16$

39. Using Pascal's triangle, the coefficients are 1, 4, 6, 4, and 1.
 $(4x-3y)^4 = 1(4x)^4(-3y)^0 + 4(4x)^3(-3y)^1 + 6(4x)^2(-3y)^2 + 4(4x)^1(-3y)^3 + 1(4x)^0(-3y)^4 = $

 $\quad 256x^4 - 768x^3y + 864x^2y^2 - 432xy^3 + 81y^4$

41. Using Pascal's triangle, the coefficients are 1, 6, 15, 20, 15, 6, and 1.
 $(m+n)^6 = m^6 + 6m^5n + 15m^4n^2 + 20m^3n^3 + 15m^2n^4 + 6mn^5 + n^6$

43. Using Pascal's triangle, the coefficients are 1, 3, 3, and 1.
 $(2x^3-y^2)^3 = 1(2x^3)^3(-y^2)^0 + 3(2x^3)^2(-y^2)^1 + 3(2x^3)^1(-y^2)^2 + 1(2x^3)^0(-y^2)^3 = 8x^9 - 12x^6y^2 + 6x^3y^4 - y^6$

45. The fourth term of $(a+b)^9$ is $\binom{9}{6}a^{9-3}b^3 = \binom{9}{6}a^6b^3 = 84a^6b^3$.

47. The fifth term of $(x+y)^8$ is $\binom{8}{4}x^{8-4}y^4 = \binom{8}{4}x^4y^4 = 70x^4y^4$.

49. The fourth term of $(2x+y)^5$ is $\binom{5}{2}(2x)^{5-3}y^3 = \binom{5}{2}(2x)^2 y^3 = 40x^2 y^3$.

51. The sixth term of $(3x-2y)^6$ is $\binom{6}{1}(3x)^{6-5}(-2y)^5 = \binom{6}{1}(3x)(-2y)^5 = -576xy^5$.

Checking Basic Concepts for Sections 8.3 and 8.4

1. There are $2\cdot2\cdot2\cdot2\cdot2\cdot2\cdot2\cdot2 = 2^8 = 256$ different ways to answer the quiz.

3. There are $26\cdot36\cdot36\cdot36\cdot36\cdot36 = 26\cdot36^5 = 1,572,120,576$ different license plates of this kind.

8.5: Mathematical Induction

1. $3+6+9+\cdots+3n=\dfrac{3n(n+1)}{2}$

    (i)   Show that the statement is true for $n=1$: $3(1)=\dfrac{3(1)(2)}{2}\Rightarrow 3=3$

    (ii) Assume that S_k is true: $3+6+9+\cdots+3k=\dfrac{3k(k+1)}{2}$

          Show that S_{k+1} is true: $3+6+\cdots+3(k+1)=\dfrac{3(k+1)(k+2)}{2}$

          Add $3(k+1)$ to each side of S_k: $3+6+9+\cdots+3k+3(k+1)=\dfrac{3k(k+1)}{2}+3(k+1)=$

          $\dfrac{3k(k+1)+6(k+1)}{2}=\dfrac{(k+1)(3k+6)}{2}=\dfrac{3(k+1)(k+2)}{2}$

    Since S_k implies S_{k+1}, the statement is true for every positive integer n.

3. $5+10+15+\cdots+5n=\dfrac{5n(n+1)}{2}$

    (i)   Show that the statement is true for $n=1$: $5(1)=\dfrac{5(1)(2)}{2}\Rightarrow 5=5$

    (ii) Assume that S_k is true: $5+10+15+\cdots+5k=\dfrac{5k(k+1)}{2}$

          Show that S_{k+1} is true: $5+10+\cdots+5(k+1)=\dfrac{5(k+1)(k+2)}{2}$

          Add $5(k+1)$ to each side of S_k: $5+10+15+\cdots+5k+5(k+1)=\dfrac{5k(k+1)}{2}+5(k+1)=$

          $\dfrac{5k(k+1)+10(k+1)}{2}=\dfrac{(k+1)(5k+10)}{2}=\dfrac{5(k+1)(k+2)}{2}$

    Since S_k implies S_{k+1}, the statement is true for every positive integer n.

5. $3+3^2+3^3+\cdots+3^n=\dfrac{3(3^n-1)}{2}$

    (i)   Show that the statement is true for $n=1$: $3^1=\dfrac{3(3^1-1)}{2}\Rightarrow 3=3$

    (ii) Assume that S_k is true: $3+3^2+3^3+\cdots+3^k=\dfrac{3(3^k-1)}{2}$

          Show that S_{k+1} is true: $3+3^2+3^3+\cdots+3^{k+1}=\dfrac{3(3^{k+1}-1)}{2}$

          Add 3^{k+1} to each side of S_k: $3+3^2+3^3+\cdots+3^k+3^{k+1}=\dfrac{3(3^k-1)}{2}+3^{k+1}=$

          $\dfrac{3(3^k-1)+2(3^{k+1})}{2}=\dfrac{3^{k+1}-3+2(3^{k+1})}{2}=\dfrac{3(3^{k+1})-3}{2}=\dfrac{3(3^{k+1}-1)}{2}$

    Since S_k implies S_{k+1}, the statement is true for every positive integer n.

7. $1^3 + 2^3 + 3^3 + \cdots + n^3 = \dfrac{n^2(n+1)^2}{4}$

   (i) Show that the statement is true for $n = 1$: $1^3 = \dfrac{1^2(1+1)^2}{4} \Rightarrow 1 = 1$

   (ii) Assume that S_k is true: $1^3 + 2^3 + 3^3 + \cdots + k^3 = \dfrac{k^2(k+1)^2}{4}$

   Show that S_{k+1} is true: $1^3 + 2^3 + \cdots + (k+1)^3 = \dfrac{(k+1)^2(k+2)^2}{4}$

   Add $(k+1)^3$ to each side of S_k: $1^3 + 2^3 + 3^3 + \cdots + k^3 + (k+1)^3 = \dfrac{k^2(k+1)^2}{4} + (k+1)^3 =$

   $\dfrac{k^2(k+1)^2 + 4(k+1)^3}{4} = \dfrac{(k+1)^2(k^2 + 4k + 4)}{4} = \dfrac{(k+1)^2(k+2)^2}{4}$

   Since S_k implies S_{k+1}, the statement is true for every positive integer n.

9. $\dfrac{1}{1 \cdot 2} + \dfrac{1}{2 \cdot 3} + \cdots + \dfrac{1}{n(n+1)} = \dfrac{n}{n+1}$

   (i) Show that the statement is true for $n = 1$: $\dfrac{1}{1+(1+1)} = \dfrac{1}{1+1} \Rightarrow \dfrac{1}{2} = \dfrac{1}{2}$

   (ii) Assume that S_k is true: $\dfrac{1}{1 \cdot 2} + \dfrac{1}{2 \cdot 3} + \cdots + \dfrac{1}{k(k+1)} = \dfrac{k}{k+1}$

   Show that S_{k+1} is true: $\dfrac{1}{1 \cdot 2} + \dfrac{1}{2 \cdot 3} + \cdots + \dfrac{1}{(k+1)(k+2)} = \dfrac{k+1}{k+2}$

   Add $\dfrac{1}{(k+1)(k+2)}$ to each side of S_k: $\dfrac{1}{1 \cdot 2} + \dfrac{1}{2 \cdot 3} + \cdots + \dfrac{1}{(k+1)(k+2)} =$

   $\dfrac{k}{k+1} + \dfrac{1}{(k+1)(k+2)} = \dfrac{k(k+2)+1}{(k+1)(k+2)} = \dfrac{k^2 + 2k + 1}{(k+1)(k+2)} = \dfrac{(k+1)(k+1)}{(k+1)(k+2)} = \dfrac{k+1}{k+2}$

   Since S_k implies S_{k+1}, the statement is true for every positive integer n.

11. $\dfrac{4}{5} + \dfrac{4}{5^2} + \dfrac{4}{5^3} + \cdots + \dfrac{4}{5^n} = 1 - \dfrac{1}{5^n}$

   (i) Show that the statement is true for $n = 1$: $\dfrac{4}{5^1} = 1 - \dfrac{1}{5^1} \Rightarrow \dfrac{4}{5} = \dfrac{4}{5}$

   (ii) Assume that S_k is true: $\dfrac{4}{5} + \dfrac{4}{5^2} + \dfrac{4}{5^3} + \cdots + \dfrac{4}{5^k} = 1 - \dfrac{1}{5^k}$

   Show that S_{k+1} is true: $\dfrac{4}{5} + \dfrac{4}{5^2} + \cdots + \dfrac{4}{5^{k+1}} = 1 - \dfrac{1}{5^{k+1}}$

   Add $\dfrac{4}{5^{k+1}}$ to each side of S_k: $\dfrac{4}{5} + \dfrac{4}{5^2} + \dfrac{4}{5^3} + \cdots + \dfrac{4}{5^k} + \dfrac{4}{5^{k+1}} = 1 - \dfrac{1}{5^k} + \dfrac{4}{5^{k+1}} =$

   $1 - \dfrac{1}{5^k} \cdot \dfrac{5}{5} + \dfrac{4}{5^{k+1}} = 1 - \dfrac{5}{5^{k+1}} + \dfrac{4}{5^{k+1}} = 1 - \dfrac{1}{5^{k+1}}$

   Since S_k implies S_{k+1}, the statement is true for every positive integer n.

13. $\dfrac{1}{1 \cdot 4} + \dfrac{1}{4 \cdot 7} + \cdots + \dfrac{1}{(3n-2)(3n+1)} = \dfrac{n}{3n+1}$

(i)   Show that the statement is true for $n=1$: $\dfrac{1}{1\cdot 4}=\dfrac{1}{3(1)+1}\Rightarrow \dfrac{1}{4}=\dfrac{1}{4}$

(ii) Assume that S_k is true: $\dfrac{1}{1\cdot 4}+\cdots+\dfrac{1}{(3k-2)(3k+1)}=\dfrac{k}{3k+1}$

Show that S_{k+1} is true: $\dfrac{1}{1\cdot 4}+\cdots+\dfrac{1}{[3(k+1)-2][3(k+1)+1]}=\dfrac{k+1}{3(k+1)+1}$

Add $\dfrac{1}{[3(k+1)-2][3(k+1)+1]}$ to each side of S_k: $\dfrac{1}{1\cdot 4}+\cdots+\dfrac{1}{[3(k+1)-2][3(k+1)+1]}$

$=\dfrac{k}{3k+1}+\dfrac{1}{[3(k+1)-2][3(k+1)+1]}=\dfrac{k}{3k+1}+\dfrac{1}{(3k+1)(3k+4)}=\dfrac{k(3k+4)+1}{(3k+1)(3k+4)}=$

$\dfrac{3k^2+4k+1}{(3k+1)(3k+4)}=\dfrac{(3k+1)(k+1)}{(3k+1)(3k+4)}=\dfrac{k+1}{3k+4}=\dfrac{k+1}{3(k+1)+1}$

Since S_k implies S_{k+1}, the statement is true for every positive integer n.

15.   When $n=1$, $3^1<6(1)\Rightarrow 3<6$, When $n=2$, $3^2<6(2)\Rightarrow 9<12$.

When $n=3$, $3^3>6(3)\Rightarrow 27>18$. For all $n\geq 3$, $3^n>6n$. The only values are 1 and 2.

17.   When $n=1$, $2^1>1^2\Rightarrow 2>1$. When $n=2$, $2^2=2^2\Rightarrow 4=4$. When $n=3$, $2^3<3^2\Rightarrow 8<9$.

When $n=4$, $2^4=4^2\Rightarrow 16=16$. For all $n\geq 5$, $2^n>n^2$. The only values are 2, 3, and 4.

19. $(a^m)^n=a^{mn}$

(i)   Show that the statement is true for $n=1$: $(a^m)^1=a^{m\cdot 1}\Rightarrow a^m=a^m$

(ii) Assume that S_k is true: $(a^m)^k=a^{mk}$

Show that S_{k+1} is true: $(a^m)^{k+1}=a^{m(k+1)}$

Multiply each side of S_k by a^m: $(a^m)^k\cdot(a^m)^1=a^{mk}\cdot a^m\Rightarrow (a^m)^{k+1}=a^{mk+m}\Rightarrow (a^m)^{k+1}=a^{m(k+1)}$

Since S_k implies S_{k+1}, the statement is true for every positive integer n.

21. $2^n>2n$, if $n\geq 3$
(i)   Show that the statement is true for $n=3$: $2^3>2(3)\Rightarrow 8>6$

(ii) Assume that S_k is true: $2^k>2k$

Show that S_{k+1} is true: $2^{k+1}>2(k+1)$

Multiply each side of S_k by 2: $2^k\cdot 2>2k\cdot 2\Rightarrow 2^{k+1}>2(k+1)$

Since S_k implies S_{k+1}, the statement is true for every positive integer $n\geq 3$.

23. $a^n>1$, if $a>1$

(i)   Show that the statement is true for $n=1$: $a^1>1\Rightarrow a>1$, which is true by the given restriction.

(ii) Assume that S_k is true: $a^k>1$

Show that S_{k+1} is true: $a^{k+1}>1$

Multiply each side of S_k by a: $a^k\cdot a>1\cdot a\Rightarrow a^{k+1}>a$

Because $a>1$, we may substitute 1 for a in the expression. That is $a^{k+1}>1$

Since S_k implies S_{k+1}, the statement is true for every positive integer n.

25. $a^n<a^{n-1}$, if $0<a<1$

(i)   Show that the statement is true for $n = 1$: $a^1 < a^0 \Rightarrow a < 1$, which is true by the given restriction.

(ii)   Assume that S_k is true: $a^k < a^{k-1}$

Show that S_{k+1} is true: $a^{k+1} < a^k$

Multiply each side of S_k by a: $a^k \cdot a < a^{k-1} \cdot a \Rightarrow a^{k+1} < a^k$

Since S_k implies S_{k+1}, the statement is true for every positive integer n.

27. $n! > 2^n$, if $n \geq 4$

(i)   Show that the statement is true for $n = 4$: $4! > 2^4 \Rightarrow 24 > 16$

(ii)   Assume that S_k is true: $k! > 2^k$

Show that S_{k+1} is true: $(k+1)! > 2^{k+1}$

Multiply each side of S_k by $k+1$: $(k+1)k! > 2^k(k+1) \Rightarrow (k+1)! > 2^k(k+1)$

Because $(k+1) > 2$ for all $k \geq 4$, we may substitute 2 for $(k+1)$ in the expression.

That is $(k+1)! > 2^k(2)$ or $(k+1)! > 2^{k+1}$.

Since S_k implies S_{k+1}, the statement is true for every positive integer $n \geq 4$.

29.   The number of handshakes is $\dfrac{n^2 - n}{2}$ if $n \geq 2$.

(i)   Show that the statement is true for $n = 2$: The number of handshakes for 2 people is $\dfrac{2^2 - 2}{2} = \dfrac{2}{2} = 1$,

which is true.

(ii)   Assume that S_k is true: The number of handshakes for k people is $\dfrac{k^2 - k}{2}$.

Show that S_{k+1} is true: The number of handshakes for $k+1$ people is

$$\frac{(k+1)^2 - (k+1)}{2} = \frac{k^2 + 2k + 1 - k - 1}{2} = \frac{k^2 + k}{2}.$$

When a person joins a group of k people, each person must shake hands with the new person.

Since there are a total of k people that will shake hands with the new person, the total number of

handshakes for $k+1$ people is $\dfrac{k^2 - k}{2} + k = \dfrac{k^2 - k + 2k}{2} = \dfrac{k^2 + k}{2}$.

Since S_k implies S_{k+1}, the statement is true for every positive integer $n \geq 2$.

31.   The first figure has perimeter $P = 3$. When a new figure is generated, each side if the previous figure

increases in length by a factor of $\dfrac{4}{3}$. Thus, the second figure has perimeter $P = 3\left(\dfrac{4}{3}\right)$, the third figure has

perimeter $P = 3\left(\dfrac{4}{3}\right)^2$, and so on. In general, the nth figure has perimeter $P = 3\left(\dfrac{4}{3}\right)^{n-1}$.

33.   With 1 ring, 1 move is required. With 2 rings, 3 moves are required. Note that $3 = 2 + 1$. With 3 rings, 7

moves are required. Note that $7 = 2^2 + 2 + 1$.

With n rings $2^{n-1} + 2^{n-2} + \cdots + 2^1 + 1 = 2^n - 1$ moves are required.

(i)   Show that the statement is true for $n = 1$: The number of moves for 1 ring is $2^1 - 1 = 1$, which is true.

(ii)   Assume that S_k is true: The number of moves for k rings is $2^k - 1$.

Show that S_{k+1} is true: The number of moves for $k+1$ rings is $2^{k+1} - 1$.

Assume $k+1$ rings are on the first peg. Since S_k is true, the top f rings can be moved to the second peg in 2^k-1 moves. Now move the bottom ring to the third peg. Since S_k is true, move the k rings from the second peg on top of the ring on the third peg in 2^k-1 moves. The total number of moves is
$$(2^k-1)+1+(2^k-1)=2\cdot2^k-1=2^{k+1}-1$$

Since S_k implies S_{k+1}, the statement is true for every positive integer n.

8.6: Probability

1. Yes. The number is between 0 and 1.

3. No. The number is greater 1.

5. Yes. The number is between 0 and 1, inclusive.

7. No. This number is less than 0.

9. A head when tossing a fair coin $\Rightarrow \dfrac{1}{2}$.

11. Rolling a 2 with a fair die $\Rightarrow \dfrac{1}{6}$.

13. Guessing the correct answer for a true-false question $\Rightarrow \dfrac{1}{2}$.

15. Drawing a king from a standard deck of 52 cards $\Rightarrow \dfrac{4}{52}=\dfrac{1}{13}$.

17. The access code can vary between 0000 and 9999. This is 10,000 possibilities, so the probability of guessing the ATM code at random is $\dfrac{1}{10,000}$.

19. (a) The probability that their favorite is pepperoni is 0.43, so the probability it is not pepperoni is $1-0.43=0.57$ or 57%.

    (b) $19\%+14\%=33\%$ or 0.33

21. The set $A\cup B$ is the set of elements that belong to either A or B. $A\cup B=\{10,25,26,35\}$

    The set $A\cap B$ is the set of elements that belong to both A and B. $A\cap B=\{25,26\}$

23. The set $A\cup B$ is the set of elements that belong to either A or B. $A\cup B=\{1,3,5,7,9,11\}$

    The set $A\cap B$ is the set of elements that belong to both A and B. $A\cap B=\varnothing$

25. The set $A\cup B$ is the set of elements that belong to either A or B.

    $A\cup B=\{\text{Tossing a heads, Tossing a tails}\}$

    The set $A\cap B$ is the set of elements that belong to both A and B. $A\cap B=\varnothing$

27. The probability of one tail is $\dfrac{1}{2}$. The events of tossing a coin are independent so the probability of tossing two tails is $\dfrac{1}{2}\cdot\dfrac{1}{2}=\dfrac{1}{4}$.

29. The probability of rolling either a 5 or a 6 is $\dfrac{2}{6}=\dfrac{1}{3}$. The probability of obtaining a 5 or 6 on three consecutive roles is $\dfrac{1}{3}\cdot\dfrac{1}{3}\cdot\dfrac{1}{3}=\dfrac{1}{27}$.

31. To obtain a sum of 2, both die must show a 1. The probability of rolling a 1 with one die is $\frac{1}{6}$. Since the

     events are independent, the probability of obtaining a 1 on two dice is $\frac{1}{6} \cdot \frac{1}{6} = \frac{1}{36}$.

33. The probability of rolling a die and not obtaining a 6 is $\frac{5}{6}$. The probability of rolling a die four times and not

     obtaining a 6 is $\frac{5}{6} \cdot \frac{5}{6} \cdot \frac{5}{6} \cdot \frac{5}{6} = \frac{625}{1296} \approx 0.482$.

35. The probability of drawing the first ace is $\frac{4}{52}$, the second ace $\frac{3}{51}$, the third ace $\frac{2}{50}$, and the fourth ace $\frac{1}{49}$.

     The probability of drawing four aces is $\frac{4}{52} \cdot \frac{3}{51} \cdot \frac{2}{50} \cdot \frac{1}{49} = \frac{24}{6,497,400} = \frac{1}{270,725}$.

37. There are $\binom{13}{3}$ ways to draw 3 hearts, and there are $\binom{13}{2}$ ways to draw 2 diamonds. there are $\binom{52}{5}$ different

     poker hands. Thus, the probability of drawing 3 hearts and 2 diamonds is

     $$P(E) = \frac{n(E)}{n(S)} = \frac{\binom{13}{3} \cdot \binom{13}{2}}{\binom{52}{5}} = \frac{286 \cdot 78}{2,598,960} = 0.0086 \text{, or a 0.86\% chance.}$$

39. There are 4 strings out of 20 that are defective. Therefore, there is a $\frac{4}{20} = 0.2$ probability or 20% chance of

     drawing a defective string and rejecting the box.

41. (a)
     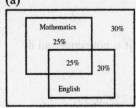

     (b) The percentage of students needing help with mathematics, English, or both is $25 + 25 + 20 = 70\%$. The
          probability of needing help in mathematics, English, or both is 0.7.

     (c) If M represents the event of a student needing help with mathematics, and E represents the event that
          a student needs help with English, then

          $P(M \text{ or } E) = P(M \cup E) = P(M) + P(E) - P(M \cap E) = 0.5 + 0.45 - 0.25 = 0.7$.

43. (a) $18\% = \frac{18}{100} = \frac{9}{50}$

     (b) $1 - \frac{18 + 11 + 11 + 11 + 9 + 4}{100} = \frac{36}{100} = \frac{9}{25}$

     (c) $1 - \frac{18 + 11}{100} = \frac{71}{100}$

45. The probability is $\frac{634}{100,000} \approx 0.00634$.

47. (a) The probability is $\frac{223,508}{679,590} \approx 0.329$.

(b)   This is the complement of the event in part (a), which is $1 - \dfrac{223,508}{679,590} = \dfrac{476,082}{679,590} \approx 0.671$.

(c)   The probability is $\dfrac{65,947 + 64,573}{679,590} \approx 0.192$.

49. $P(\text{Pair or Sum 6}) = P(\text{Pair}) + P(\text{Sum 6}) - P(\text{Pair and Sum 6}) = \dfrac{6}{36} + \dfrac{5}{36} - \dfrac{1}{36} = \dfrac{10}{36} = \dfrac{5}{18}$

51.   There are four 2's, four 3's and four 4's in a standard deck. The probability is $\dfrac{12}{52} = \dfrac{3}{13}$.

53.   (a)   Since the events of rolling a 4, 5, or 6 are mutually exclusive, the probability of either a 4, 5, or 6 is $0.2 + 0.2 + 0.3 = 0.7$.

       (b)   The events of rolling a 6 followed by a second 6 are independent events. The probability of two consecutive 6's is $(0.3)(0.3) = 0.09$.

55.   (a)   The only way to obtain a sum of 12 is for both dice to show a 6. Since these events are independent, the probability of two 6's is $(0.3)(0.3) = 0.09$ or 9%.

       (b)   There are two ways to obtain a sum of 11. The red die shows a 5 and the blue die a 6, or vice versa. The roll of the two dice are independent. The probability of the red die showing a 5 is 0.2 and the probability of the blue die showing a 6 is 0.3. The probability of a sum of 11 is $(0.2)(0.3) = 0.06$. Similarly, the probability of the red die showing a 6 and the blue die showing a 5 is 0.06. Since these events are mutually exclusive, the probability of a sum of 11 is $0.06 + 0.06 = 0.12$ or 12%.

57.   There are ten possibilities for each of the three digits. By the fundamental counting principle, there are $10 \cdot 10 \cdot 10 = 10^3 = 1000$ different ways to pick these numbers. There is only one winning number so the probability is $\dfrac{1}{1000}$.

59.   (a)   There are a total of $22 + 18 + 10 = 50$ marbles in the jar. Since 22 of them are red, there is a probability of $\dfrac{22}{50} = 0.44$ of drawing a red ball.

       (b)   Since there is a 0.44 probability of drawing a red ball, there is a $1 - 0.44 = 0.56$ probability of not drawing a red ball.

       (c)   Drawing a blue or green ball is equivalent to not drawing a red ball. The probability is the same as in part (b), which was 0.56.

61.   Since one card, a queen, has been drawn, there are 3 queens left in the set of 51 cards. So the probability of drawing a queen is $\dfrac{3}{51}$.

63.   Since there are 4 kings in a total of 12 face cards, the probability of drawing a king is $\dfrac{4}{12} = \dfrac{1}{3}$.

65.   Since 2 red out of 10 red marbles and 4 blue out of 23 blue marbles have already been drawn, 6 marbles out of 33 marbles have been removed from the jar. There are $23 - 4 = 19$ blue marbles left out of $33 - 6 = 27$ marbles in the jar. The probability of drawing a blue marble next is $\dfrac{19}{27}$.

67.   Let E_1 denote the event that it is cloudy and E_2 denote the event that it is windy. $P(E_1) = 0.30$ and $P(E_1 \text{ and } E_2) = P(E_1 \cap E_2) = 0.12$. $P(E_1 \cap E_2) = P(E_1) \cdot P(E_2, \text{ given that } E_1 \text{ has occurred}) \Rightarrow$ $0.12 = 0.30 \cdot P(E_2, \text{ given that } E_1 \text{ has occured}) \Rightarrow 0.4 = P(E_2, \text{ given that } E_1 \text{ has occurred})$. The probability that it will be windy given that the day is cloudy is 40%.

69.   The possibilities for the result that the first die is a 2 and the sum of the two dice is 7 or more, can be represented by $\{(2, 5), (2, 6)\}$. The sample space can be represented by $\{(2, 1), (2, 2), (2, 3), (2, 4), (2, 5),$ and $(2, 6)\}$. Out of a total of 6, 2 outcomes satisfy the conditions, so the probability is $\frac{2}{6} = \frac{1}{3}$.

71.   (a)   Let D represent the event that part is defective. Since 18 of the 235 parts are defective, $P(D) = \frac{18}{235}$.

       (b)   Let A represent the event that part is type A. Since 7 of the 18 parts are type A, $P(A, \text{given } D) = \frac{7}{18}$.

       (c) $P(D \text{ and } A) = P(D \cap A) = P(D) \cdot P(A, \text{given } D) = \frac{18}{235} \cdot \frac{7}{18} = \frac{7}{235}$.

73.   (a)   Let O represent the event that the number is odd. Since 8 of the 15 numbers are odd, $P(O) = \frac{8}{15}$.

       (b)   Let E represent the event that the number is even. Since 7 of the 15 numbers are even, $P(E) = \frac{7}{15}$.

       (c)   Let x represent the event that the number is prime. Since 6 of the 15 numbers are prime,
              $P(M) = \frac{6}{15} = \frac{2}{5}$.

       (d)   Since 5 of the 6 prime numbers are odd, $P(M \cap O) = \frac{5}{15} = \frac{1}{3}$.

       (e)   Since 1 of the 6 prime numbers is even, $P(M \cap E) = \frac{1}{15}$.

Extended and Discovery Exercises for Section 8.6

1.   (a) $P_{k,j} = \dfrac{\binom{2k}{j}\binom{4-2k}{2-j}}{\binom{4}{2}} \Rightarrow P_{00} = \dfrac{\binom{0}{0}\binom{4}{2}}{\binom{4}{2}} = 1;\ P_{01} = \dfrac{\binom{0}{1}\binom{4}{1}}{\binom{4}{2}} = 0;\ P_{02} = \dfrac{\binom{0}{2}\binom{4}{0}}{\binom{4}{2}} = 0;\ P_{10} = \dfrac{\binom{2}{0}\binom{2}{2}}{\binom{4}{2}} = \dfrac{1}{6};$

       $P_{11}\dfrac{\binom{2}{1}\binom{2}{1}}{\binom{4}{2}} = \dfrac{2}{3};\ P_{12} = \dfrac{\binom{2}{2}\binom{2}{0}}{\binom{4}{2}} = \dfrac{1}{6};\ P_{20} = \dfrac{\binom{4}{0}\binom{0}{2}}{\binom{4}{2}} = 0;\ P_{21} = \dfrac{\binom{4}{1}\binom{0}{1}}{\binom{4}{2}} = 0;\ P_{22} = \dfrac{\binom{4}{2}\binom{0}{0}}{\binom{4}{2}} = 1.$ Thus,

       $$P = \begin{bmatrix} P_{00} & P_{01} & P_{02} \\ P_{10} & P_{11} & P_{12} \\ P_{20} & P_{21} & P_{22} \end{bmatrix} = \begin{bmatrix} 1 & 0 & 0 \\ \frac{1}{6} & \frac{2}{3} & \frac{1}{6} \\ 0 & 0 & 1 \end{bmatrix}.$$

       (b)   The sum of the probabilities in each row is 1. The greatest probabilities lie along the diagonal. A mother cell is most likely to produce a daughter cell like itself. *Answers may vary.*

3.   The quantity $\dfrac{a_1 + a_n}{2}$ represents not only the average of the two terms a_1 and a_n, but it also represents the average of all of the terms $a_1, a_2, a_3, \ldots, a_n$ in the series. This is true whether n is odd or even. The total sum is equal to n times the average of the terms.

Checking Basic Concepts for Section 8.5 and 8.6

1. $4 + 8 + 12 + \cdots + 4n = 2n(n+1)$

   (i)   Show that the statement is true for $n = 1$: $4(1) = 2(1)(1+1) \Rightarrow 4 = 4$.

   (ii) Assume that S_k is true: $4 + 8 + 12 + \cdots + 4k = 2k(k+1)$.

   Show that S_{k+1} is true: $4 + 8 + \cdots + 4(k+1) = 2(k+1)(k+2)$

   Add $4(k+1)$ to each side of S_k: $4 + 8 + 12 + \cdots + 4k + 4(k+1) = 2k(k+1) + 4(k+1) =$

   $2k^2 + 6k + 4 = 2(k+1)(k+2)$

   Since S_k implies S_{k+1}, the statement is true for every positive integer n.

3. The probability of one head is $\dfrac{1}{2}$. The events of tossing a coin are independent so the probability of tossing

   four heads is $\dfrac{1}{2} \cdot \dfrac{1}{2} \cdot \dfrac{1}{2} \cdot \dfrac{1}{2} = \dfrac{1}{16}$.

5. There are $\binom{4}{4}$ different ways to draw four aces and $\binom{4}{1}$ different ways to draw a queen. There are a total of

   $\binom{52}{5}$ different 5-card hands in a set of 52 cards. Thus, the probability of drawing four aces and a queen is

   $\dfrac{\binom{4}{4} \cdot \binom{4}{1}}{\binom{52}{5}} = \dfrac{1 \cdot 4}{2{,}598{,}960} = 0.0000015$.

Chapter 8 Review Exercises

1. $a_1 = -3(1) + 2 = -1$; $a_2 = -3(2) + 2 = -4$; $a_3 = -3(3) + 2 = -7$; $a_4 = -3(4) + 2 = -10$ The first four terms are -1, -4, -7, and -10.

3. $a_1 = 0$; $a_2 = 2a_1 + 1 = 2(0) + 1 = 1$; $a_3 = 2a_2 + 1 = 2(1) + 1 = 3$; $a_4 = 2a_3 + 1 = 2(3) + 1 = 7$. The first four terms are 0, 1, 3, and 7.

5. The points (1, 5), (2, 3), (3, 1), (4, 2), (5, 4), and (6, 6) lie on the graph. Therefore, the terms of this sequence are 5, 3, 1, 2, 4, 6.

7. (a) Each term can be found by adding -2. to the previous term, beginning with 3. It is an arithmetic sequence. A numerical representation for the first eight terms is shown.

n	1	2	3	4	5	6	7	8
a_n	3	1	−1	−3	−5	−7	−9	−11

   (b) A graphical representation for the first eight terms is shown. Notice that the points lie on a line with slope -2 because the sequence is arithmetic with $d = -2$.

(c)   To find the symbolic representation, we will use the formula $a_n = a_1 + (n-1)d$, where $a_n = f(n)$. The common difference of the sequence is $d = -2$. and the first term is $a_1 = 3$. Therefore a symbolic representation is given by $a_n = 3 + (n-1)(-2)$ or $a_n = -2n + 5$.

9.   Let $a_n = f(n)$, where $f(n) = dn + c$. Then, $f(n) = 4n + c$. Since $f(3) = 4(3) + c = -3 \Rightarrow c = -15$. Thus $a_n = f(n) = 4n - 15$.

11. Since f is a linear function, the sequence is arithmetic.

13. Since f. can be written in the form $f(n) = cr^{n-1}$, the sequence is geometric.

15. $S_5 = (4(1)+1) + (4(2)+1) + (4(3)+1) + (4(4)+1) + (4(5)+1) = 5 + 9 + 13 + 17 + 21 = 65$

17. The first term is $a_1 = -2$ and the last term is $a_9 = 22$.

     To find the sum use $S_n = n\left(\dfrac{a_1 + a_n}{2}\right) \Rightarrow S_9 = 9\left(\dfrac{-2+22}{2}\right) = 90$. The sum is 90.

19. The first term is $a_1 = 1$ and the common ratio is $r = 3$. Since there are 8 terms, the sum is

     $$S_n = a_1\left(\frac{1-r^n}{1-r}\right) \Rightarrow S_8 = 1\left(\frac{1-3^8}{1-3}\right) = 3280.$$

21. The first term is $a_1 = 4$ and the common ratio is $r = -\dfrac{1}{3}$. The sum is $S_n = a_1\left(\dfrac{1}{1-r}\right) \Rightarrow S = 4\left(\dfrac{1}{1+\frac{1}{3}}\right) = 3$.

23. $\displaystyle\sum_{k=1}^{5}(5k+1) = (5+1) + (10+1) + (15+1) + (20+1) + (25+1) = 6 + 11 + 16 + 21 + 26$

25. $1^3 + 2^3 + 3^3 + 4^3 + 5^3 + 6^3 = \displaystyle\sum_{k=1}^{6}k^3$

27. Since $a_1 = 5$ and $d = -3$, use the formula $S_n = \dfrac{n}{2}(2a_1 + (n-1)d)$.

     $S_{30} = \dfrac{30}{2}(2(5) + (30-1)(-3)) = 15(10 + 29(-3)) = 15(-77) = -1155$

29. $\dfrac{2}{11} = 0.18181818\ldots = 0.18 + 0.0018 + 0.000018 + 0.00000018 + \cdots$

31. $P(n,r) = \dfrac{n!}{(n-r)!} \Rightarrow P(6,3) = \dfrac{6!}{(6-3)!} = \dfrac{6!}{3!} = \dfrac{6 \cdot 5 \cdot 4 \cdot 3!}{3!} = 6 \cdot 5 \cdot 4 = 120$

33. $1 + 3 + 5 + \cdots + (2n-1) = n^2$. The

     (i)   Show that the statement is true for $n = 1$: $2(1) - 1 = 1^2 \Rightarrow 1 = 1$

     (ii) Assume that S_k is true: $1 + 3 + 5 + \cdots + (2k-1) = k^2$.

           Show that S_{k+1} is true: $1 + 3 + \cdots + (2(k+1)-1) = (k+1)^2$

           Add $2k+1$ to each side of S_k: $1 + 3 + 5 + \cdots + (2k-1) + (2k+1) = k^2 + 2k + 1 = (k+1)^2$

     Since S_k implies S_{k+1}, the statement is true for every positive integer n.

35. Since 1, 2, or 3 are three outcomes out of six possible outcomes, the probability is $\dfrac{3}{6} = \dfrac{1}{2}$.

37. There are $P(5,5) = 5! = 120$ different arrangements.

39. There are $4^{20} \approx 1.1 \times 10^{12}$ different ways to answer the exam.

41.   There are 50 choices for each number in the combination. So there are $50^4 = 6,250,000$ possible combinations.

43.   (a)   The initial height is 4 feet. On the first rebound it reaches 90% of 4 or 3.6 feet. On the second rebound it attains a height of 90% of 3.6 or 3.24 feet. Each term in this sequence is found by multiplying the previous term by 0.9. Thus, the first five terms are 4, 3.6, 3.24, 2.916, 2.6244. This is a geometric sequence.

      (b)   Plot the points (1, 4), (2, 3.6), (3, 3.24), (4, 2.916), and (5, 2.6244) as shown.

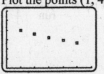

      (c)   The first term is 4 and the common ratio is 0.9. Thus, $a_n = 4(0.9)^{n-1}$.

45.   From a group of 6 people, we must select a set of 3 people. This can be done $C(6,3) = 20$ different ways.

47.   They can be selected in $C(10,6) = 210$ different ways.

49.   There are $C(16,2)$ different ways to select 2 batteries from a pack of 16. In order to avoid testing a defective battery, we must pick 2 batteries from the 14 good ones. This can be done in $C(14,2)$ different ways. The probability that the box is not rejected is $\dfrac{C(14,2)}{C(16,2)} = \dfrac{91}{120} \approx 0.758$.

51.   (a)   The total number of marbles in the jar is $13 + 27 + 20 = 60$. Since 27 of the marbles are blue, there is a probability of $\dfrac{27}{60} = 0.45$ of drawing a blue marble.

      (b)   Since there is a 0.45 probability of drawing a blue marble, there is a $1 - 0.45 = 0.55$ probability of drawing a marble that is not blue.

      (c)   There is a $\dfrac{13}{60} \approx 0.217$ probability of drawing a red marble.

53.   Initially, the population density grows slowly, then it increases rapidly. After some time, it levels off near 4000 thousand (4,000,000) per acre.

Chapters 1-8 Cumulative Review Exercises

1. $34,500 = 3.45 \times 10^4$; $1.52 \times 10^{-4} = 0.000152$

3. $d\sqrt{(1-(-4))^2 + (-2-2)^2} = \sqrt{5^2 + (-4)^2} = \sqrt{25+16} = \sqrt{41}$

5.   (a) $f(-3) = \sqrt{1-(-3)} = \sqrt{4} = 2$; $f(a+1) = \sqrt{1-(a+1)} = \sqrt{-a}$

      (b)   For f to be defined, $1-x \geq 0$. Thus, the domain is $D = \{x \mid x \leq 1\}$, or $(-\infty, 1]$.

7. $\dfrac{f(-1) - f(-2)}{-1 - (-2)} = \dfrac{((-1)^3 - 4) - ((-2)^3 - 4)}{1} = -5 - (-12) = 7$

9.   A line passing through the points $(2, -4)$ and $(-3, 2)$ has slope $m = \dfrac{2-(-4)}{-3-2} = \dfrac{6}{-5} = -\dfrac{6}{5}$.

Using the point $(2,-4)$, the equation is $y=-\dfrac{6}{5}(x-2)-4 \Rightarrow y=-\dfrac{6}{5}x-\dfrac{8}{5}$.

11. (a) The graph passes through the points $(0,-1)$ and $(4,2)$. The slope is $m=\dfrac{2-(-1)}{4-0}=\dfrac{3}{4}$.

The y-intercept is $(0,-1)$. The x-intercept can be found by noting that a 1-unit rise from the point $(0,-1)$ would require a $\dfrac{4}{3}$-unit run to return to a line with slope $\dfrac{3}{4}$. That is, $m=\dfrac{\text{rise}}{\text{run}}=\dfrac{1}{\frac{4}{3}}=\dfrac{3}{4}$. The x-intercept is $\left(\dfrac{4}{3},0\right)$.

(b) Since the slope is $\dfrac{3}{4}$ and the $b=-1$, the formula for f is $f(x)=\dfrac{3}{4}x-1$.

(c) $\dfrac{3}{4}x-1=0 \Rightarrow \dfrac{3}{4}x=1 \Rightarrow x=\dfrac{4}{3}$

13. (a) $4(x-2)+1=3-\dfrac{1}{2}(2x+3) \Rightarrow 4x-8+1=3-x-\dfrac{3}{2} \Rightarrow 5x=\dfrac{17}{2} \Rightarrow x=\dfrac{17}{10}$

(b) $6x^2=13x+5 \Rightarrow 6x^2-13x-5=0 \Rightarrow (3x+1)(2x-5)=0 \Rightarrow x=-\dfrac{1}{3}$ or $x=\dfrac{5}{2}$

(c) By the quadratic formula, $x=\dfrac{-b\pm\sqrt{b^2-4ac}}{2a} \Rightarrow x=\dfrac{-(-1)\pm\sqrt{(-1)^2-4(1)(-3)}}{2(1)}=\dfrac{1\pm\sqrt{13}}{2}$

(d) $x^3+x^2=4x+4 \Rightarrow x^3+x^2-4x-4=0 \Rightarrow x^2(x+1)-4(x+1)=0 \Rightarrow$
$(x^2-4)(x+1)=0 \Rightarrow (x+2)(x-2)(x+1)=0 \Rightarrow x=-2,-1,$ or 2

(e) $x^4-4x^2+3=0 \Rightarrow (x^2-3)(x^2-1)=0 \Rightarrow x^2=3$ or $x^2=1 \Rightarrow x=\pm\sqrt{3}$ or ± 1

(f) $\dfrac{1}{x-3}=\dfrac{4}{x+5} \Rightarrow (x-3)(x+5)\cdot\dfrac{1}{x-3}=\dfrac{4}{x+5}\cdot(x-3)(x+5) \Rightarrow x+5=4(x-3) \Rightarrow$
$x+5=4x-12 \Rightarrow -3x=-17 \Rightarrow x=\dfrac{17}{3}$

(g) $3e^{2x}-5=23 \Rightarrow 3e^{2x}=28 \Rightarrow e^{2x}=\dfrac{28}{3} \Rightarrow \ln e^{2x}=\ln\dfrac{28}{3} \Rightarrow 2x=\ln\dfrac{28}{3} \Rightarrow x=\dfrac{\ln(\frac{28}{3})}{2}\approx 1.117$

(h) $2\log(x+1)-1=2 \Rightarrow 2\log(x+1)=3 \Rightarrow \log(x+1)=\dfrac{3}{2} \Rightarrow 10^{\log(x+1)}=10^{3/2} \Rightarrow$
$x+1=10^{3/2} \Rightarrow x=10^{3/2}-1\approx 30.623$

(i) $\sqrt{x+3}+4=x+1 \Rightarrow \sqrt{x+3}=x-3 \Rightarrow (\sqrt{x+3})^2=(x-3)^2 \Rightarrow$
$x+3=x^2-6x+9 \Rightarrow x^2-7x+6=0 \Rightarrow (x-1)(x-6)=0 \Rightarrow x=1$ or $x=6$. Note, 1 is extraneous. The only solution is 6.

(j) $|3x-1|=5 \Rightarrow 3x-1=-5$ or $3x-1=5 \Rightarrow 3x=-4$ or $3x=6 \Rightarrow x=-\dfrac{4}{3}$ or 2

15. f is continuous.

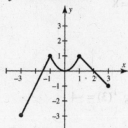

17.   (a)   The solution to $f(x) = 0$ is the x-intercepts. The solutions are -2, -1, 1 and 2.

       (b)   The graph of f is above the x-axis on the interval $(-\infty, -2) \cup (-1, 1) \cup (2, \infty)$ or
              $\{x \mid x < -2 \text{ or } -1 < x < 1 \text{ or } x > 2\}$.

       (c)   The graph of f is on or below the x-axis on the interval $[-2, -1] \cup [1, 2]$ or
              $\{x \mid -2 \le x \le -1 \text{ or } 1 \le x \le 2\}$.

19.   The x-coordinate of the vertex is $x = -\dfrac{b}{2a} = -\dfrac{9}{2(-3)} = \dfrac{3}{2}$. The y-coordinate of the vertex is $f\left(\dfrac{3}{2}\right)$.

       $y = f\left(\dfrac{3}{2}\right) = -3\left(\dfrac{3}{2}\right)^2 + 9\left(\dfrac{3}{2}\right) + 1 = -\dfrac{27}{4} + \dfrac{27}{2} + 1 = -\dfrac{27}{4} + \dfrac{54}{4} + \dfrac{4}{4} = \dfrac{31}{4}$. The vertex is $\left(\dfrac{3}{2}, \dfrac{31}{4}\right)$.

21.   (a)   The graph of f is increasing on $(-\infty, -2] \cup [1, \infty)$ and decreasing on $[-2, 1]$ increasing on
              $\{x \mid x \le -2 \text{ or } x \ge 1\}$ or $(-\infty, -2] \cup [1, \infty)$ decreasing on $\{x \mid -2 \le x \le 1\}$ or $[-2, 1]$.

       (b)   The zeros of f are the x-intercepts which are approximately -3.3, 0, and 1.8.

       (c)   The turning points are approximately $(-2, 2)$ and $(1, -0.7)$.

       (d)   There is a local minimum of -0.7 and a local maximum of 2.

23.   (a) $\dfrac{6x^4 - 2x^2 + 1}{2x^2} = \dfrac{6x^4}{2x^2} - \dfrac{2x^2}{2x^2} + \dfrac{1}{2x^2} = 3x^2 - 1 + \dfrac{1}{2x^2}$

       $\begin{array}{r} 2x^3 - 5x^2 + 5x - 6 \\ x+1 \overline{)\ 2x^4 - 3x^3 + 0x^2 - x + 2} \end{array}$ The solution is : $2x^3 - 5x^2 + 5x - 6 + \dfrac{8}{x+1}$

              $\underline{2x^4 + 2x^3}$
              $\quad -5x^3 + 0x^2 - x + 2$
              $\quad \underline{-5x^3 - 5x^2}$
              $\qquad\quad 5x^2 - x + 2$
              $\qquad\quad \underline{5x^2 + 5x}$
              $\qquad\qquad\quad -6x + 2$
              $\qquad\qquad\quad \underline{-6x - 6}$
       (b) $\qquad\qquad\qquad\qquad 8$

25.   Since $3i$ is a zero its conjugate, $-3i$, is also a zero.
       The complete factored form is
       $f(x) = 3(x - (-1))(x - (-3i))(x - 3i)$ or $f(x) = 3(x + 1)(x + 3i)(x - 3i)$.

       To find the expanded form, first multiply $(x + 3i)(x - 3i) = x^2 + 9$. Continuing to multiply out gives
       $f(x) = 3(x + 1)(x^2 + 9) = 3(x^3 + x^2 + 9x + 9)$ or $f(x) = 3x^3 + 3x^2 + 27x + 27$.

27. $x = \dfrac{-b \pm \sqrt{b^2 - 4ac}}{2a} \Rightarrow x = \dfrac{-2 \pm \sqrt{2^2 - 4(1)(5)}}{2(1)} = \dfrac{-2 \pm \sqrt{-16}}{2} = \dfrac{-2 \pm 4i}{2} = -1 \pm 2i$

29. $\sqrt[5]{(x+1)^3} = (x+1)^{3/5}$. When $x = 31$, $(x+1)^{3/5} = (31+1)^{3/5} = 32^{3/5} = (\sqrt[5]{32})^3 = 2^3 = 8$

31. (a) $(f+g)(2) = f(2) + g(2) = 3 + 0 = 3$

   (b) $(fg)(0) = f(0) \cdot g(0) = -1(1) = -1$

   (c) $(g \circ f)(1) = g(f(1)) = g(0) = 1$

   (d) Note, for $g^{-1}(3)$, read the graph backward. $(g^{-1} \circ f)(-2) = g^{-1}(f(-2)) = g^{-1}(3) = -4$

33. Let $y = f(x)$: $y = \dfrac{x}{x+1}$. Interchange x and y: $x = \dfrac{y}{y+1}$. Then solve for y.

$x = \dfrac{y}{y+1} \Rightarrow x(y+1) = y \Rightarrow xy + x = y \Rightarrow x = y - xy \Rightarrow x = y(1-x) \Rightarrow \dfrac{x}{1-x} = y$ The inverse function is

$f^{-1}(x) = \dfrac{x}{1-x}$ or $f^{-1}(x) = -\dfrac{x}{x-1}$.

35. Since each unit increase in x results in $f(x)$ increasing by a factor of 3, the function is exponential with base 3. Since $f(0) = 2$, the initial value is 2 and the function is $f(x) = 2(3^x)$.

37. $A = 500\left(1 + \dfrac{0.06}{4}\right)^{15.4} \approx \$1221.16$

39. (a) A quadratic function is defined for all real inputs. The domain is $D = (-\infty, \infty)$ or $\{x \mid -\infty < x < \infty\}$. The graph of f is a parabola opening upward with vertex $(1, 0)$. The range is $R = [0, \infty)$ or $\{x \mid x \geq 0\}$.

   (b) An exponential function is defined for all real inputs. The domain is $D = (-\infty, \infty)$ or $\{x \mid -\infty < x < \infty\}$. The function $f(x) = 10^x$ has only positive outputs. The range is $R = (0, \infty)$ or $\{x \mid x > 0\}$.

   (c) The natural logarithm function is defined for only positive real inputs. The domain is x or $\{x \mid x > 0\}$. The natural logarithm function can output any real value. The range is $R = (-\infty, \infty)$ or $\{x \mid -\infty < x < \infty\}$.

   (d) The reciprocal function is defined for all real inputs except 0. The domain is $D = (-\infty, 0) \cup (0, \infty)$ or $\{x \mid x \neq 0\}$.
   The reciprocal function can output any real value except 0. The range is $R = (-\infty, 0) \cup (0, \infty)$ or $\{x \mid x \neq 0\}$.

41. $2\log x + 3\log y - \dfrac{1}{3}\log z = \log x^2 + \log y^3 - \log z^{1/3} = \log \dfrac{x^2 y^3}{z^{1/3}} = \log \dfrac{x^2 y^3}{\sqrt[3]{z}}$

43. (a) Multiplying the first equation by -1 and adding the two equations will eliminate the variable x.

$$-2x - 3y = -4$$
$$\underline{2x - 5y = -12}$$
$$-8y = -16$$ Thus, $y = 2$. And so $2x + 3(2) = 4 \Rightarrow x = -1$. The solution is $(-1, 2)$.

   (b) Multiplying the first equation by 2 and adding the two equations will eliminate both variables.

$$-4x + y = \;\;2$$
$$\underline{4x + y = -2}$$
$$0 = 0$$ This is an identity. The system is dependent with solutions. $\{(x, y) \mid 4x - y = -2\}$

   (c) Adding the two equations will eliminate the variable y.

$$x^2 + y^2 = 16$$
$$\underline{2x^2 - y^2 = 11}$$
$$3x^2 = 27 \quad \text{Thus, } x = \pm 3. \text{ And so } (\pm 3)^2 + y^2 = 16 \Rightarrow y^2 = 7 \Rightarrow y = \pm\sqrt{7}.$$

There are four solutions, $(\pm 3, \pm\sqrt{7})$.

(d)   Multiply the first equation by 2 and subtract the second equations to eliminate the variable x.

$$2x + 2y - 4z = -12$$
$$\underline{2x - y - 3z = -18}$$
$$3y - z = \quad 6$$

Subtract the third equation from this *new* equation to eliminate both y and z.

$$3y - z = 6$$
$$\underline{3y - z = 6}$$
$$0 = 6$$

This is an identity which means there are an infinite number of solutions to the system.

Solving the third equation for y yields $3y - z = 6 \Rightarrow 3y = z + 6 \Rightarrow y = \dfrac{z+6}{3}$. Substituting this result for

y in the first equation yields $x + \dfrac{z+6}{3} - 2z = -6 \Rightarrow x = 2z - \dfrac{z+6}{3} - 6 \Rightarrow$

$$x + \frac{z+6}{3} - 2z = -6 \Rightarrow x = 2z - \frac{z+6}{3} - 6 \Rightarrow x = \frac{5z - 24}{3}.$$

The solutions set is $\left\{ (x, y, z) \mid x = \dfrac{5z-24}{3},\ y = \dfrac{z+6}{3},\ \text{and } z = z \right\}$.

45.   For the graphs of parts (a) and (b), see Figures 45a and 45b.

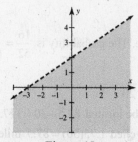

Figure 45a

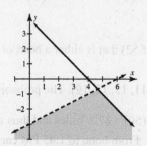

Figure 45b

47. $\begin{bmatrix} 1 & -2 & | & 1 & 0 \\ -3 & 4 & | & 0 & 1 \end{bmatrix} R_2 + 3R_1 \rightarrow \begin{bmatrix} 1 & -2 & | & 1 & 0 \\ 0 & -2 & | & 3 & 1 \end{bmatrix} R_1 - R_2 \rightarrow \begin{bmatrix} 1 & 0 & | & -2 & 1 \\ 0 & -2 & | & 3 & 1 \end{bmatrix} -\frac{1}{2}R_2 \rightarrow \begin{bmatrix} 1 & 0 & | & -2 & -1 \\ 0 & 1 & | & -\frac{3}{2} & -\frac{1}{2} \end{bmatrix}$

That is $A^{-1} = \begin{bmatrix} -2 & -1 \\ -1.5 & -0.5 \end{bmatrix}$.

49. $\det\left(\begin{bmatrix} -1 & 4 \\ 2 & 3 \end{bmatrix} \right) = -1(3) - 2(4) = -11$

$\det\left(\begin{bmatrix} 2 & 3 & -1 \\ 3 & -1 & 5 \\ 0 & 0 & -2 \end{bmatrix} \right) = 0(3(5) - (-1)(-1)) - 0(2(5) - 3(-1)) + -2(2(-1) - 3(3)) =$

$-2(-11) = 22$

51.   If the focus is at $\left(\frac{3}{4}, 0\right)$ and the vertex at $(0, 0)$, the parabola opens to the right and $p = \frac{3}{4}$. Substituting in

$(y-h)^2 = 4p(x-k)$, we get $(y-0)^2 = 4\left(\frac{3}{4}\right)(x-0)$ or $y^2 = 3x$.

53.   Since the foci and the vertices lie on the y-axis, the hyperbola has a vertical transverse axis with an equation

of the form $\frac{y^2}{a^2} - \frac{x^2}{b^2} = 1$. $F(0, \pm13) \Rightarrow c = 13$ and $V(0, \pm5) \Rightarrow a = 5$.

$c^2 = a^2 + b^2 \Rightarrow b^2 = c^2 - a^2 = 169 - 25 = 144 \Rightarrow b = 12$. The equation of the hyperbola is $\frac{y^2}{25} - \frac{x^2}{144} = 1$.

55.   Let $a_n = f(n)$, where $f(n) = dn + c$. $a_1 = 4 \Rightarrow f(1) = 4$ and $a_3 = 12 \Rightarrow f(3) = 12$.

Thus, $d = \frac{f(3) - f(1)}{3 - 1} = \frac{12 - 4}{2} = 4$. Then, $f(n) = 4n + c$. $f(1) = 4(1) + c = 4 \Rightarrow c = 0$.

Thus, $a_n = f(n) = 4n$.

57.   (a)   The first term is $a_1 = 2$ and the last term is $a_{25} = 74$. There are 25 terms. To find the sum use:

$S_n = n\left(\frac{a_1 + a_n}{2}\right) \Rightarrow S_{25} = 25\left(\frac{2 + 74}{2}\right) = 950$. The sum is 950.

      (b)   The first term is $a_1 = 0.2$ and the common ratio is $r = \frac{1}{10}$ or 0.1. The sum is

$S = a_1\left(\frac{1}{1-r}\right) \Rightarrow S = 0.2\left(\frac{1}{1-(0.1)}\right) = \frac{0.2}{0.9} = \frac{2}{9}$.

59.   There are $26 \cdot 26 \cdot 26 \cdot 10 \cdot 10 \cdot 10 \cdot 10 = 175,760,000$ such license plates.

61. $(2x-1)^4 = \binom{4}{0}(2x)^4(-1)^0 + \binom{4}{1}(2x)^3(-1)^1 + \binom{4}{2}(2x)^2(-1)^2 + \binom{4}{3}(2x)^1(-1)^3 + \binom{4}{4}(2x)^0(-1)^4$

$= 16x^4 - 32x^3 + 24x^2 - 8x + 1$

63.   The card must be one of the 16 cards (out of 52) that is either a heart or an ace. The probability is $\frac{16}{52} = \frac{4}{13}$.

65.   There are 8 primes from 1 to 20 (2, 3, 5, 7, 11, 13, 17, 19). The probability is $\frac{8}{20} = \frac{2}{5}$.

67.   From noon to 1:45 PM car A travelled $1.75(50) = 87.5$ miles and thus would be located $87.5 + 30 = 117.5$

miles north of the original location of car B. From noon to 1:45 PM car B travelled $1.75(50) = 87.5$ miles and

thus would be located 87.5 miles east of its original location. The paths of the two cars form a right triangle

with legs of length 117.5 and 87.5. The length of the hypotenuse is equal to the distance between the cars.

$c^2 = 117.5^2 + 87.5^2 \Rightarrow c^2 = 21,462.5 \Rightarrow c \approx 146.5$ miles

69.   Let x represent the time spent jogging at 7 miles per hour and let y represent the time spent jogging at 9

miles per hour. Then the system of equations to be solved is $x + y = 1.3$ and $7x + 9y = 10.5$. Solving the first

equation for y gives $y = 1.3 - x$. Substituting this expression in the second equation and solving for x yields

$7x + 9(1.3 - x) = 10.5 \Rightarrow -2x + 11.7 = 10.5 \Rightarrow -2x = -1.2 \Rightarrow x = 0.6$. From the first equation

$0.6 + y = 1.3 \Rightarrow y = 0.7$. The jogger ran for 0.6 hours at 7 mph and 0.7 hours at 9 mph.

71. Let x represent the number of hours they work together. The first person mows $\frac{x}{5}$ of the lawn in x hours and the second person mows $\frac{x}{4}$ of the lawn in x hours. Together they mow $\frac{x}{5} + \frac{x}{4}$ of the lawn in x hours.

   The job is complete when the fraction of the lawn mowed is 1. That is, we must solve $\frac{x}{5} + \frac{x}{4} = 1$.

   The LCD is 20. The first step is to multiply each side of the equation by the LCD.

   $$20 \cdot \left(\frac{x}{5} + \frac{x}{4} \right) = 1 \cdot 20 \Rightarrow 4x + 5x = 20 \Rightarrow 9x = 20 \Rightarrow x = \frac{20}{9} \approx 2.22 \text{ hours.}$$

73. (a) The cost of 1 ticket is $1(405 - 1(5)) = 400$. The cost for 2 tickets is $2(405 - 2(5)) = \$790$. The cost for 3 tickets is $3(405 - 3(5)) = \$1170$. Following this pattern, the cost for x tickets is $C(x) = x(405 - 5x)$.

   (b) $x(405 - 5x) = 7000 \Rightarrow 405x - 5x^2 = 7000 \Rightarrow 5x^2 - 405x + 7000 = 0 \Rightarrow$
   $x^2 - 81x + 1400 = 0 \Rightarrow (x - 25)(x - 56) = 0 \Rightarrow x = 25$ or $x = 56$. It costs \$7000 to buy either 25 or 56 tickets.

   (c) The maximum for the function occurs at the vertex of its graph - a parabola. For $C(x) = -5x^2 + 405x$,
   $x = -\frac{b}{2a} = -\frac{405}{2(-5)} = 40.5$. Since the number of tickets must be an integer, the maximum occurs at either 40 or 41 tickets being purchased. In either case, the maximum expenditure is
   $40(405 - 5(40)) = \$8200$.

75. Substitute the point (1.5, 0.5) into $x^2 = 4py$ and solve for p; $1.5^2 = 4p(0.5) \Rightarrow 2.25 = 2p \Rightarrow p = 1.125$.
   The receiver should be 1.125 feet from the vertex.

77. There are 49 marbles that are not blue and there are a total of 77 marbles. The probability of drawing a marble that is not blue is $\frac{49}{77} = \frac{7}{11}$.

Chapter R: Basic Concepts from Algebra and Geometry

R.1: Formulas from Geometry

1. $A = LW = (15)(7) = 105 \text{ ft}^2$; $P = 2L + 2W = 2(15) + 2(7) = 30 + 14 = 44 \text{ ft}$

3. $A = LW = (100)(35) = 3500 \text{ m}^2$; $P = 2L + 2W = 2(100) + 2(35) = 200 + 70 = 270 \text{ m}$

5. $A = LW = (3x)(y) = 3xy$ square units ; $P = 2L + 2W = 2(3x) + 2(y) = 6x + 2y$ units

7. Since the width is half of the length, $L = 2W$.
    $A = LW = (2W)(W) = 2W^2$ square units ; $P = 2L + 2W = 2(2W) + 2(W) = 6W$ units

9. Since the length equals the width plus 5, $L = W + 5$. $A = LW = (W + 5)(W) = W(W + 5)$ square units ;
    $P = 2L + 2W = 2(W + 5) + 2(W) = 4W + 10$ units

11. $A = \frac{1}{2}bh = \left(\frac{1}{2}\right)(8)(5) = 20 \text{ cm}^2$

13. $A = \frac{1}{2}bh = \left(\frac{1}{2}\right)(5)(8) = 20 \text{ in.}^2$

15. $A = \frac{1}{2}bh = \left(\frac{1}{2}\right)(10.1)(730) = 3686.5 \text{ m}^2$

17. $A = \frac{1}{2}bh = \left(\frac{1}{2}\right)(2x)(6x) = 6x^2$ square units

19. $A = \frac{1}{2}bh = \left(\frac{1}{2}\right)(z)(5z) = \frac{5}{2}z^2$ square units

21. $C = 2\pi r = 2\pi(4) = 8\pi \approx 25.1 \text{ m}$; $A = \pi r^2 = \pi(4)^2 = 16\pi \approx 50.3 \text{ m}^2$

23. $C = 2\pi r = 2\pi(19) = 38\pi \approx 119.4 \text{ in.}$; $A = \pi r^2 = \pi(19)^2 = 361\pi \approx 1134.1 \text{ in.}^2$

25. $C = 2\pi r = 2\pi(2x) = 4\pi x$ units ; $A = \pi r^2 = \pi(2x)^2 = 4\pi x^2$ square units

27. $c^2 = a^2 + b^2 \Rightarrow c^2 = (60)^2 + (11)^2 \Rightarrow c^2 = 3721 \Rightarrow c = \sqrt{3721} \Rightarrow c = 61 \text{ ft}$
    $P = 60 + 11 + 61 = 132 \text{ ft}$

29. $c^2 = a^2 + b^2 \Rightarrow b^2 = c^2 - a^2 \Rightarrow b^2 = (13)^2 - (5)^2 \Rightarrow b^2 = 144 \Rightarrow b = \sqrt{144} \Rightarrow b = 12 \text{ cm}$
    $P = 5 + 12 + 13 = 30 \text{ cm}$

31. $c^2 = a^2 + b^2 \Rightarrow a^2 = c^2 - b^2 \Rightarrow a^2 = (10)^2 - (7)^2 \Rightarrow a^2 = 51 \Rightarrow a = \sqrt{51} \Rightarrow a \approx 7.1 \text{ mm}$
    $P \approx 7.1 + 7 + 10 = 24.1 \text{ mm}$

33. $A = \frac{1}{2}bh = \left(\frac{1}{2}\right)(6)(3) = 9 \text{ ft}^2$

35. $c^2 = a^2 + b^2 \Rightarrow b^2 = c^2 - a^2 \Rightarrow b^2 = (15)^2 - (11)^2 \Rightarrow b^2 = 104 \Rightarrow b = \sqrt{104} \Rightarrow b \approx 10.2 \text{ in.}$
    $A = \frac{1}{2}bh = \left(\frac{1}{2}\right)(11)(\sqrt{104}) \approx 56.1 \text{ in.}^2$

37. $V = LWH = (4)(3)(2) = 24 \text{ ft}^3$; $S = 2LW + 2WH + 2LH = 2(4)(3) + 2(3)(2) + 2(4)(2) = 52 \text{ ft}^2$

39. Since 1 foot is 12 inches, $V = LWH = (4.5)(4)(12) = 216 \text{ in.}^3$

$S = 2LW + 2WH + 2LH = 2(4.5)(4) + 2(4)(12) + 2(4.5)(12) = 240 \text{ in.}^2$

41. $V = LWH = (3x)(2x)(x) = 6x^3$ cubic units

$S = 2LW + 2WH + 2LH = 2(3x)(2x) + 2(2x)(x) + 2(3x)(x) = 22x^2$ square units

43. $V = LWH = (x)(2y)(3z) = 6xyz$ cubic units

$S = 2LW + 2WH + 2LH = 2(x)(2y) + 2(2y)(3z) + 2(x)(3z) = 4xy + 12yz + 6xz$ square units

45. Since the length is twice W and the height is half W, $L = 2W$ and $H = 0.5W$

$V = LWH = (2W)(W)(0.5W) = W^3$

47. $V = \frac{4}{3}\pi r^3 \Rightarrow V = \frac{4}{3}\pi(3)^3 = 36\pi \approx 113.1 \text{ ft}^3$; $S = 4\pi r^2 \Rightarrow S = 4\pi(3)^2 = 36\pi \approx 113.1 \text{ ft}^2$

49. $V = \frac{4}{3}\pi r^3 \Rightarrow V = \frac{4}{3}\pi(3.2)^3 \approx 43.7\pi \approx 137.3 \text{ m}^3$; $S = 4\pi r^2 \Rightarrow S = 4\pi(3.2)^2 \approx 41.0\pi \approx 128.8 \text{ m}^2$

51. $V = \pi r^2 h \Rightarrow V = \pi(0.5)^2(2) = 0.5\pi \approx 1.6 \text{ ft}^3$; $S_{side} = 2\pi rh \Rightarrow S_{side} = 2\pi(0.5)(2) = 2\pi \approx 6.3 \text{ ft}^2$

$S_{total} = 2\pi rh + 2\pi r^2 \Rightarrow S_{total} = 2\pi(0.5)(2) + 2\pi(0.5)^2 = 2.5\pi \approx 7.9 \text{ ft}^2$

53. Since h is twice r, $h = 2(12) = 24$, thus $V = \pi r^2 h \Rightarrow V = \pi(12)^2(24) = 3456\pi \approx 10,857.3 \text{ mm}^3$

$S_{side} = 2\pi rh \Rightarrow S_{side} = 2\pi(12)(24) = 576\pi \approx 1809.6 \text{ mm}^2$

$S_{total} = 2\pi rh + 2\pi r^2 \Rightarrow S_{total} = 2\pi(12)(24) + 2\pi(12)^2 = 864\pi \approx 2714.3 \text{ mm}^2$

55. $V = \frac{1}{3}\pi r^2 h \Rightarrow V = \frac{1}{3}\pi(5)^2(6) \approx 157.1 \text{ cm}^3$;

$S = \pi r\sqrt{r^2 + h^2} \Rightarrow S = \pi(5)\sqrt{(5)^2 + (6)^2} \approx 122.7 \text{ cm}^2$

57. Since 24 inches is 2 feet,

$V = \frac{1}{3}\pi r^2 h \Rightarrow V = \frac{1}{3}\pi(2)^2(3) \approx 12.6 \text{ ft}^3$; $S = \pi r\sqrt{r^2 + h^2} \Rightarrow S = \pi(2)\sqrt{(2)^2 + (3)^2} \approx 22.7 \text{ ft}^2$

59. Since h is three times r, $h = 3(2.4) = 7.2$.

$V = \frac{1}{3}\pi r^2 h \Rightarrow V = \frac{1}{3}\pi(2.4)^2(7.2) \approx 43.4 \text{ ft}^3$;

$S = \pi r\sqrt{r^2 + h^2} \Rightarrow S = \pi(2.4)\sqrt{(24)^2 + (72)^2} \approx 57.2 \text{ ft}^2$

61. $\frac{x}{4} = \frac{5}{3} \Rightarrow 3x = 20 \Rightarrow x = \frac{20}{3} \approx 6.7$

63. $\frac{x}{7} = \frac{9}{6} \Rightarrow 6x = 63 \Rightarrow x = \frac{63}{6} = \frac{21}{2} = 10.5$

R.2: Integer Exponents

1. No. $2^3 = 8$ and $3^2 = 9$

3. $\frac{1}{7^n}$

5. 5^{m-n}

7. 2^{mk}

9. 5000

11. 2^3

13. 4^4

15. 3^0

17. $5^3 = 125$

19. $-2^4 = -16$

21. $5^0 = 1$

23. $\left(\dfrac{2}{3}\right)^3 = \dfrac{2^3}{3^3} = \dfrac{2 \cdot 2 \cdot 2}{3 \cdot 3 \cdot 3} = \dfrac{8}{27}$

25. $\left(-\dfrac{1}{2}\right)^4 = \left(-\dfrac{1}{2}\right) \cdot \left(-\dfrac{1}{2}\right) \cdot \left(-\dfrac{1}{2}\right) \cdot \left(-\dfrac{1}{2}\right) = \dfrac{1}{16}$

27. $4^{-3} = \dfrac{1}{4^3} = \dfrac{1}{64}$

29. $\dfrac{1}{2^{-4}} = 2^4 = 16$

31. $\left(\dfrac{3}{4}\right)^{-3} = \left(\dfrac{4}{3}\right)^3 = \left(\dfrac{4}{3}\right) \cdot \left(\dfrac{4}{3}\right) \cdot \left(\dfrac{4}{3}\right) = \dfrac{64}{27}$

33. $\dfrac{3^{-2}}{2^{-3}} = \dfrac{2^3}{3^2} = \dfrac{8}{9}$

35. $6^3 \cdot 6^{-4} = 6^{3+(-4)} = 6^{-1} = \dfrac{1}{6}$

37. $2x^2 \cdot 3x^{-3} \cdot x^4 = 6x^{2+(-3)+4} = 6x^3$

39. $10^0 \cdot 10^6 \cdot 10^2 = 10^{0+6+2} = 10^8 = 100{,}000{,}000$

41. $5^{-2} \cdot 5^3 \cdot 2^{-4} \cdot 2^3 = 5^{-2+3} \cdot 2^{-4+3} = 5^1 \cdot 2^{-1} = 5 \cdot \dfrac{1}{2} = \dfrac{5}{2}$

43. $2a^3 \cdot b^2 \cdot a^{-4} \cdot 4b^{-5} = 2a^{3+(-4)} \cdot 4b^{2+(-5)} = 8a^{-1}b^{-3} = \dfrac{8}{ab^3}$

45. $\dfrac{5^4}{5^2} = 5^{4-2} = 5^2 = 25$

47. $\dfrac{a^{-3}}{a^2 \cdot a} = a^{-3-(2+1)} = a^{-6} = \dfrac{1}{a^6}$

49. $\dfrac{24x^3}{6x} = \dfrac{24}{6}x^{3-1} = 4x^2$

51. $\dfrac{12a^2b^3}{18a^4b^2} = \dfrac{12}{18}a^{2-4} \cdot b^{3-2} = \dfrac{2}{3}a^{-2}b^1 = \dfrac{2b}{3a^2}$

53. $\dfrac{21x^{-3}y^4}{7x^4y^{-2}} = \dfrac{21}{7}x^{-3-4} \cdot y^{4-(-2)} = 3x^{-7}y^6 = \dfrac{3y^6}{x^7}$

55. $(5^{-1})^3 = 5^{-1(3)} = 5^{-3} = \dfrac{1}{5^3} = \dfrac{1}{125}$

57. $(y^4)^{-2} = y^{4(-2)} = y^{-8} = \dfrac{1}{y^8}$

59. $(4y^2)^3 = 4^3 \cdot y^{2 \cdot 3} = 64y^6$

61. $\left(\dfrac{4}{x}\right)^3 = \dfrac{4^3}{x^3} = \dfrac{64}{x^3}$

63. $\left(\dfrac{2x}{z^4}\right)^{-5} = \left(\dfrac{z^4}{2x}\right)^5 = \dfrac{(z^4)^5}{(2x)^5} = \dfrac{z^{4 \cdot 5}}{2^5 \cdot x^5} = \dfrac{z^{20}}{32x^5}$

65. $\dfrac{2}{(ab)^{-1}} = 2(ab)^1 = 2ab$

67. $\dfrac{2^{-3}}{2t^{-2}} = \dfrac{t^2}{2(2^3)} = \dfrac{t^2}{16}$

69. $\dfrac{6a^2b^{-3}}{4ab^{-2}} = \dfrac{6}{4} \cdot a^{2-1}b^{-3-(-2)} = \dfrac{3}{2} \cdot ab^{-1} = \dfrac{3a}{2b}$

71. $\dfrac{5r^2st^{-3}}{25rs^{-2}t^2} = \dfrac{5}{25} \cdot r^{2-1}s^{1-(-2)}t^{-3-2} = \dfrac{1}{5} \cdot rs^3t^{-5} = \dfrac{rs^3}{5t^5}$

73. $(3x^2y^{-3})^{-2} = \dfrac{1}{(3x^2y^{-3})^2} = \dfrac{1}{3^2(x^2)^2(y^{-3})^2} = \dfrac{1}{9x^{2 \cdot 2}y^{-3 \cdot 2}} = \dfrac{1}{9x^4y^{-6}} = \dfrac{y^6}{9x^4}$

75. $\dfrac{(d^3)^{-2}}{(d^{-2})^3} = \dfrac{d^{3 \cdot (-2)}}{d^{-2 \cdot 3}} = \dfrac{d^{-6}}{d^{-6}} = d^{-6-(-6)} = d^0 = 1$

77. $\left(\dfrac{3t^2}{2t^{-1}}\right)^3 = \dfrac{(3t^2)^3}{(2t^{-1})^3} = \dfrac{3^3(t^2)^3}{2^3(t^{-1})^3} = \dfrac{27t^{2 \cdot 3}}{8t^{-1 \cdot 3}} = \dfrac{27t^6}{8t^{-3}} = \dfrac{27}{8} \cdot t^{6-(-3)} = \dfrac{27}{8} \cdot t^9 = \dfrac{27t^9}{8}$

79. $\dfrac{(-m^2n^{-1})^{-2}}{(mn)^{-1}} = \dfrac{(mn)^1}{(-m^2n^{-1})^2} = \dfrac{mn}{(-m^2)^2(n^{-1})^2} = \dfrac{mn}{m^{2 \cdot 2}n^{-1 \cdot 2}} = \dfrac{mn}{m^4n^{-2}} = m^{1-4}n^{1-(-2)} = m^{-3}n^3 = \dfrac{n^3}{m^3}$

81. $\left(\dfrac{2a^3}{6b}\right)^4 = \left(\dfrac{a^3}{3b}\right)^4 = \dfrac{(a^3)^4}{3^4b^4} = \dfrac{a^{3 \cdot 4}}{81b^4} = \dfrac{a^{12}}{81b^4}$

83. $\dfrac{8x^{-3}y^{-2}}{4x^{-2}y^{-4}} = \dfrac{8}{4}x^{-3-(-2)}y^{-2-(-4)} = 2x^{-1}y^2 = \dfrac{2y^2}{x}$

85. $\dfrac{(r^2t^2)^{-2}}{(r^3t)^{-1}} = \dfrac{r^3t}{(r^2t^2)^2} = \dfrac{r^3t}{(r^2)^2(t^2)^2} = \dfrac{r^3t}{r^{2 \cdot 2}t^{2 \cdot 2}} = \dfrac{r^3t}{r^4t^4} = r^{3-4}t^{1-4} = r^{-1}t^{-3} = \dfrac{1}{rt^3}$

87. $\dfrac{4x^{-2}y^3}{(2x^{-1}y)^2} = \dfrac{4x^{-2}y^3}{2^2(x^{-1})^2y^2} = \dfrac{4x^{-2}y^3}{4x^{-1 \cdot 2}y^2} = \dfrac{4x^{-2}y^3}{4x^{-2}y^2} = \left(\dfrac{4x^{-2}}{4x^{-2}}\right)y^{3-2} = 1y = y$

89. $\left(\dfrac{15r^2t}{3r^{-3}t^4}\right)^3 = \left(\dfrac{15}{3}r^{2-(-3)}t^{1-4}\right)^3 = (5r^5t^{-3})^3 = 5^3(r^5)^3(t^{-3})^3 = 125r^{5 \cdot 3}t^{-3 \cdot 3} = 125r^{15}t^{-9} = \dfrac{125r^{15}}{t^9}$

R.3: Polynomial Expressions

1. $3x^3 + 5x^3 = 8x^3$

3. $5y^7 - 8y^7 = -3y^7$

5. $5x^2 + 8x + x^2 = 6x^2 + 8x$

7. $9x^2 - x + 4x - 6x^2 = 3x^2 + 3x$

9. $x^2 + 9x - 2 + 4x^2 + 4x = 5x^2 + 13x - 2$

11. $7y + 9x^2y - 5y + x^2y = 2y + 10x^2y = 10x^2y + 2y$

13. The degree is 2. The leading coefficient is 5.

15. The degree is 3. The leading coefficient is $-\dfrac{2}{5}$.

17. The degree is 5. The leading coefficient is 1.

19. $(5x + 6) + (-2x + 6) = 3x + 12$

21. $(2x^2 - x + 7) + (-2x^2 + 4x - 9) = 3x - 2$

23. $(4x) + (1 - 4.5x) = -0.5x + 1$

25. $(x^4 - 3x^2 - 4) + \left(-8x^4 + x^2 - \dfrac{1}{2}\right) = -7x^4 - 2x^2 - \dfrac{9}{2}$

27. $(2z^3 + 5z - 6) + (z^2 - 3z + 2) = 2z^3 + z^2 + 2z - 4$

29. $-(7x^3) = -7x^3$

31. $-(19z^5 - 5z^2 + 3z) = -19z^5 + 5z^2 - 3z$

33. $-(z^4 - z^2 - 9) = -z^4 + z^2 + 9$

35. $(5x - 3) - (2x + 4) = 5x - 3 - 2x - 4 = 3x - 7$

37. $(x^2 - 3x + 1) - (-5x^2 + 2x - 4) = x^2 - 3x + 1 + 5x^2 - 2x + 4 = 6x^2 - 5x + 5$

39. $(4x^4 + 2x^2 - 9) - (x^4 - 2x^2 - 5) = 4x^4 + 2x^2 - 9 - x^4 + 2x^2 + 5 = 3x^4 + 4x^2 - 4$

41. $(x^4 - 1) - (4x^4 + 3x + 7) = x^4 - 1 - 4x^4 - 3x - 7 = -3x^4 - 3x - 8$

43. $5x(x - 5) = 5x^2 - 25x$

45. $-5(3x + 1) = -15x - 5$

47. $5(y + 2) = 5y + 10$

49. $-2(5x + 9) = -10x - 18$

51. $(y - 3)6y = 6y^2 - 18y$

53. $-4(5x - y) = -20x + 4y$

55. $(y + 5)(y - 7) = y^2 - 7y + 5y - 35 = y^2 - 2y - 35$

57. $(3 - 2x)(3 + x) = 9 + 3x - 6x - 2x^2 = -2x^2 - 3x + 9$

59. $(-2x + 3)(x - 2) = -2x^2 + 4x + 3x - 6 = -2x^2 + 7x - 6$

61. $\left(x - \dfrac{1}{2}\right)\left(x + \dfrac{1}{4}\right) = x^2 + \dfrac{1}{4}x - \dfrac{1}{2}x - \dfrac{1}{8} = x^2 - \dfrac{1}{4}x - \dfrac{1}{8}$

63. $(x^2 + 1)(2x^2 - 1) = 2x^4 - x^2 + 2x^2 - 1 = 2x^4 + x^2 - 1$

65. $(x+y)(x-2y) = x^2 - 2xy + xy - 2y^2 = x^2 - xy - 2y^2$

67. $3x(2x^2 - x - 1) = 6x^3 - 3x^2 - 3x$

69. $-x(2x^4 - x^2 + 10) = -2x^5 + x^3 - 10x$

71. $(2x^2 - 4x + 1)(3x^2) = 6x^4 - 12x^3 + 3x^2$

73. $(x+1)(x^2 + 2x - 3) = x^3 + 2x^2 - 3x + x^2 + 2x - 3 = x^3 + 3x^2 - x - 3$

75. $(2-3x)(5-2x)(x^2-1) \Rightarrow [(2-3x)(5-2x)](x^2-1) = (10 - 19x + 6x^2)(x^2 - 1) =$
    $10x^2 - 19x^3 + 6x^4 - 10 + 19x - 6x^2 = 6x^4 - 19x^3 + 4x^2 + 19x - 10$

77. $(x^2 + 2)(3x - 2) = 3x^3 - 2x^2 + 6x - 4$

79. $(x-7)(x+7) = x^2 - 7^2 = x^2 - 49$

81. $(3x+4)(3x-4) = (3x)^2 - 4^2 = 9x^2 - 16$

83. $(2x-3y)(2x+3y) = (2x)^2 - (3y)^2 = 4x^2 - 9y^2$

85. $(x+4)^2 = x^2 + 2(4)x + 4^2 = x^2 + 8x + 16$

87. $(2x+1)^2 = 4x^2 + 2(2x) + 1 = 4x^2 + 4x + 1$

89. $(x-1)^2 = x^2 - 2(x) + 1 = x^2 - 2x + 1$

91. $(2-3x)^2 = 4 - 2(6x) + 9x^2 = 4 - 12x + 9x^2$

93. $3x(x+1)(x-1) = 3x(x^2 - 1) = 3x^3 - 3x$

95. $(2-5x^2)(2+5x^2) = 2^2 - (5x^2)^2 = 4 - 25x^4$

R.4: Factoring Polynomials

1. $10x - 15 = 5(2x - 3)$

3. $2x^3 - 5x = x(2x^2 - 5)$

5. $8x^3 - 4x^2 + 16x = 4x(2x^2 - x + 4)$

7. $5x^4 - 15x^3 + 15x^2 = 5x^2(x^2 - 3x + 3)$

9. $15x^3 + 10x^2 - 30x = 5x(3x^2 + 2x - 6)$

11. $6r^5 - 8r^4 + 12r^3 = 2r^3(3r^2 - 4r + 6)$

13. $8x^2 y^2 - 24x^2 y^3 = 8x^2 y^2(1 - 3y)$

15. $18mn^2 - 12m^2 n^3 = 6mn^2(3 - 2mn)$

17. $-4a^2 - 2ab + 6ab^2 = -2a(2a + b - 3b^2)$

19. $x^3 + 3x^2 + 2x + 6 = x^2(x+3) + 2(x+3) = (x+3)(x^2+2)$

21. $6x^3 - 4x^2 + 9x - 6 = 2x^2(3x-2) + 3(3x-2) = (3x-2)(2x^2+3)$

23. $z^3 - 5z^2 + z - 5 = z^2(z-5) + 1(z-5) = (z-5)(z^2+1)$

25. $y^4 + 2y^3 - 5y^2 - 10y = y(y^3 + 2y^2 - 5y - 10) = y[y^2(y+2) - 5(y+2)] = y(y+2)(y^2-5)$

27. $2x^3 - 3x^2 + 2x - 3 = x^2(2x-3) + 1(2x-3) = (x^2+1)(2x-3)$

29. $2x^4 - x^3 + 4x - 2 = x^3(2x-1) + 2(2x-1) = (x^3+2)(2x-1)$

31. $ab - 3a + 2b - 6 = a(b-3) + 2(b-3) = (a+2)(b-3)$

33. $x^2 + 7x + 10 = (x+2)(x+5)$

35. $x^2 + 8x + 12 = (x+2)(x+6)$

37. $z^2 + z - 42 = (z-6)(z+7)$

39. $z^2 + 11z + 24 = (z+3)(z+8)$

41. $24x^2 + 14x - 3 = (4x+3)(6x-1)$

43. $6x^2 - x - 2 = (2x+1)(3x-2)$

45. $1 + x - 2x^2 = (1-x)(1+2x)$

47. $20 + 7x - 6x^2 = (5-2x)(4+3x)$

49. $5x^3 + x^2 - 6x = x(5x^2 + x - 6) = x(x-1)(5x+6)$

51. $3x^3 + 12x^2 + 9x = 3x(x^2 + 4x + 3) = 3x(x+3)(x+1)$

53. $2x^2 - 14x + 20 = 2(x^2 - 7x + 10) = 2(x-5)(x-2)$

55. $60t^4 + 230t^3 - 40t^2 = 10t^2(6t^2 + 23t - 4) = 10t^2(t+4)(6t-1)$

57. $4m^3 + 10m^2 - 6m = 2m(2m^2 + 5m - 3) = 2m(m+3)(2m-1)$

59. $x^2 - 25 = (x-5)(x+5)$

61. $4x^2 - 25 = (2x-5)(2x+5)$

63. $36x^2 - 100 = 4(9x^2 - 25) = 4(3x-5)(3x+5)$

65. $64z^2 - 25z^4 = z^2(64 - 25z^2) = z^2(8-5z)(8+5z)$

67. $16x^4 - y^4 = (4x^2 - y^2)(4x^2 + y^2) = (2x-y)(2x+y)(4x^2 + y^2)$

69. The sum of two squares does not factor using real numbers.

71. $4 - r^2t^2 = (2-rt)(2+rt)$

73. $(x-1)^2 - 16 = ((x-1)-4)((x-1)+4) = (x-5)(x+3)$

75. $4 - (z+3)^2 = (2-(z+3))(2+(z+3)) = (-1-z)(5+z) = -(z+1)(z+5)$

77. $x^2 + 2x + 1 = (x+1)^2$

79. $4x^2 + 20x + 25 = (2x+5)^2$

81. $x^2 - 12x + 36 = (x-6)^2$

83. $9z^3 - 6z^2 + z = z(9z^2 - 6z + 1) = z(3z-1)^2$

85. $9y^3 + 30y^2 + 25y = y(9y^2 + 30y + 25) = y(3y+5)^2$

87. $4x^2 - 12xy + 9y^2 = (2x-3y)^2$

89. $9a^3b - 12a^2b + 4ab = ab(9a^2 - 12a + 4) = ab(3a-2)^2$

91. $x^3 - 1 = (x-1)(x^2 + 1x + 1^2) = (x-1)(x^2 + x + 1)$

93. $y^3 + z^3 = (y+z)(y^2 - yz + z^2)$

95. $8x^3 - 27 = (2x)^3 - 3^3 = (2x-3)((2x)^2 + 2x(3) + 3^2) = (2x-3)(4x^2 + 6x + 9)$

97. $x^4 + 125x = x(x^3 + 5^3) = x(x+5)(x^2 - 5x + 5^2) = x(x+5)(x^2 - 5x + 25)$

99. $8r^6 - t^3 = (2r^2 - t)((2r^2)^2 + 2r^2t + t^2) = (2r^2 - t)(4r^4 + 2r^2t + t^2)$

101. $10m^9 - 270n^6 = 10(m^9 - 27n^6) = 10(m^3 - 3n^2)((m^3)^2 + 3m^3n^2 + (3n^2)^2) = 10(m^3 - 3n^2)(m^6 + 3m^3n^2 + 9n^4)$

103. $16x^2 - 25 = (4x)^2 - 5^2 = (4x - 5)(4x + 5)$

105. $x^3 - 64 = x^3 - 4^3 = (x - 4)(x^2 + 4x + 4^2) = (x - 4)(x^2 + 4x + 16)$

107. $x^2 + 16x + 64 = (x + 8)(x + 8) = (x + 8)^2$

109. $5x^2 - 38x - 16 = (x - 8)(5x + 2)$

111. $x^4 + 8x = x(x^3 + 8) = x(x + 2)(x^2 - 2x + 4)$

113. $64x^3 + 8y^3 = 8(8x^3 + y^3) = 8(2x + y)(4x^2 - 2xy + y^2)$

115. $3x^2 - 5x - 8 = (x + 1)(3x - 8)$

117. $7a^3 + 20a^2 - 3a = a(7a^2 + 20a - 3) = a(a + 3)(7a - 1)$

119. $2x^3 - x^2 + 6x - 3 = x^2(2x - 1) + 3(2x - 1) = (x^2 + 3)(2x - 1)$

121. $2x^4 - 5x^3 - 25x^2 = x^2(2x^2 - 5x - 25) = x^2(2x + 5)(x - 5)$

123. $2x^4 + 5x^2 + 3 = (2x^2 + 3)(x^2 + 1)$

125. $x^3 + 3x^2 + x + 3 = x^2(x + 3) + (x + 3) = (x + 3)(x^2 + 1)$

127. $5x^3 - 5x^2 + 10x - 10 = 5x^2(x - 1) + 10(x - 1) = (x - 1)(5x^2 + 10) = 5(x^2 + 2)(x - 1)$

129. $ax + bx - ay - by = x(a + b) - y(a + b) = (a + b)(x - y)$

131. $18x^2 + 12x + 2 = 2(9x^2 + 6x + 1) = 2(3x + 1)^2$

133. $-4x^3 + 24x^2 - 36x = -4x(x^2 - 6x + 9) = -4x(x - 3)^2$

135. $27x^3 - 8 = (3x - 2)(9x^2 + 6x + 4)$

137. $-x^4 - 8x = -x(x^3 + 8) = -x(x + 2)(x^2 - 2x + 4)$

139. $x^4 - 2x^3 - x + 2 = x^3(x - 2) - 1(x - 2) = (x - 2)(x^3 - 1) = (x - 2)(x - 1)(x^2 + x + 1)$

141. $r^4 - 16 = (r^2 - 4)(r^2 + 4) = (r + 2)(r - 2)(r^2 + 4)$

143. $25x^2 - 4a^2 = (5x - 2a)(5x + 2a)$

145. $(2x^4 - 2y^4) = 2(x^4 - y^4) = 2(x^2 - y^2)(x^2 + y^2) = 2(x - y)(x + y)(x^2 + y^2)$

147. $9x^3 + 6x^2 - 3x = 3x(3x^2 + 2x - 1) = 3x(3x - 1)(x + 1)$

149. $(z - 2)^2 - 9 = (z - 2 - 3)(z - 2 + 3) = (z - 5)(z + 1)$

151. $3x^5 - 27x^3 + 3x^2 - 27 = 3(x^5 - 9x^3 + x^2 - 9) = 3[x^3(x^2 - 9) + (x^2 - 9)] =$
      $3(x^2 - 9)(x^3 + 1) = 3(x + 3)(x - 3)(x + 1)(x^2 - x + 1)$

153. $(x + 2)^2(x + 4)^4 + (x + 2)^3(x + 4)^3 = (x + 2)^2(x + 4)^3[x + 4 + x + 2] =$
      $(x + 2)^2(x + 4)^3(2x + 6) = 2(x + 2)^2(x + 4)^3(x + 3)$

155. $(6x + 1)(8x - 3)^4 - (6x + 1)^2(8x - 3)^3 = (6x + 1)(8x - 3)^3[8x - 3 - (6x + 1)] =$
      $(6x + 1)(8x - 3)^2(2x - 4) = 2(6x + 1)(8x - 3)^3(x - 2)$

157. $4x^2(5x - 1)^5 + 2x(5x - 1)^6 = 2x(5x - 1)^5[2x + 5x - 1] = 2x(5x - 1)^5(7x - 1)$

R.5: Rational Expressions

1. $\dfrac{10x^3}{5x^2} = 2x$

3. $\dfrac{(x-5)(x+5)}{x-5} = x+5$

5. $-\dfrac{4-t}{t-4} = \dfrac{-(4-t)}{t-4} = \dfrac{t-4}{t-4} = 1$

7. $\dfrac{4m-n}{-4m+n} = \dfrac{-(-4m+n)}{-4m+n} = -1$

9. $\dfrac{5-y}{y-5} = \dfrac{-(y-5)}{y-5} = -1$

11. $\dfrac{x^2-16}{x-4} = \dfrac{(x-4)(x+4)}{x-4} = x+4$

13. $\dfrac{x+3}{2x^2+5x-3} = \dfrac{x+3}{(2x-1)(x+3)} = \dfrac{1}{2x-1}$

15. $-\dfrac{z+2}{4z+8} = -\dfrac{z+2}{4(z+2)} = -\dfrac{1}{4}$

17. $\dfrac{x^2+2x}{x^2+3x+2} = \dfrac{x(x+2)}{(x+1)(x+2)} = \dfrac{x}{x+1}$

19. $\dfrac{a^3+b^3}{a+b} = \dfrac{(a+b)(a^2-ab+b^2)}{a+b} = a^2-ab+b^2$

21. $\dfrac{5}{8} \cdot \dfrac{4}{15} = \dfrac{1}{6}$

23. $\dfrac{5}{6} \cdot \dfrac{3}{10} \cdot \dfrac{8}{3} = \dfrac{2}{3}$

25. $\dfrac{4}{7} \div \dfrac{8}{7} = \dfrac{4}{7} \cdot \dfrac{7}{8} = \dfrac{1}{2}$

27. $\dfrac{1}{2} \div \dfrac{3}{4} \div \dfrac{5}{6} = \left(\dfrac{1}{2} \cdot \dfrac{4}{3}\right) \div \dfrac{5}{6} = \dfrac{2}{3} \cdot \dfrac{6}{5} = \dfrac{4}{5}$

29. $\dfrac{3}{8} + \dfrac{5}{8} = \dfrac{8}{8} = 1$

31. $\dfrac{3}{7} - \dfrac{4}{7} = -\dfrac{1}{7}$

33. $\dfrac{2}{3} + \dfrac{5}{11} = \dfrac{22}{33} + \dfrac{15}{33} = \dfrac{37}{33}$

35. $\dfrac{4}{5} - \dfrac{1}{10} = \dfrac{8}{10} - \dfrac{1}{10} = \dfrac{7}{10}$

37. $\dfrac{1}{3} + \dfrac{3}{4} - \dfrac{3}{7} = \dfrac{28}{84} + \dfrac{63}{84} - \dfrac{36}{84} = \dfrac{55}{84}$

39. $\dfrac{1}{x^2} \cdot \dfrac{3x}{2} = \dfrac{3}{2x}$

41. $\dfrac{5x}{3} \div \dfrac{10x}{6} = \dfrac{5x}{3} \cdot \dfrac{6}{10x} = \dfrac{2}{2} = 1$

43. $\dfrac{x+1}{2x-5} \cdot \dfrac{x}{x+1} = \dfrac{x}{2x-5}$

45. $\dfrac{(x-5)(x+3)}{3x-1} \cdot \dfrac{x(3x-1)}{(x-5)} = \dfrac{x+3}{1} \cdot \dfrac{x}{1} = x(x+3)$

47. $\dfrac{x^2-2x-35}{2x^3-3x^2} \cdot \dfrac{x^3-x^2}{2x-14} = \dfrac{(x-7)(x+5)}{x^2(2x-3)} \cdot \dfrac{x^2(x-1)}{2(x-7)} = \dfrac{x+5}{2x-3} \cdot \dfrac{x-1}{2} = \dfrac{(x-1)(x+5)}{2(2x-3)}$

49. $\dfrac{6b}{b+2} \div \dfrac{3b^4}{2b+4} = \dfrac{6b}{b+2} \cdot \dfrac{2(b+2)}{3b^4} = \dfrac{2}{1} \cdot \dfrac{2}{b^3} = \dfrac{4}{b^3}$

51. $\dfrac{3a+1}{a^7} \div \dfrac{a+1}{3a^8} = \dfrac{3a+1}{a^7} \cdot \dfrac{3a^8}{a+1} = \dfrac{3a+1}{1} \cdot \dfrac{3a}{a+1} = \dfrac{3a(3a+1)}{a+1}$

53. $\dfrac{x+5}{x^3-x} \div \dfrac{x^2-25}{x^3} = \dfrac{x+5}{x(x+1)(x-1)} \cdot \dfrac{x^3}{(x-5)(x+5)} = \dfrac{1}{(x+1)(x-1)} \cdot \dfrac{x^2}{x-5} = \dfrac{x^2}{(x-5)(x+1)(x-1)}$

55. $\dfrac{x-2}{x^3-x} \div \dfrac{x^2-2x}{x^2-1} = \dfrac{x-2}{x(x+1)(x-1)} \cdot \dfrac{(x+1)(x-1)}{x(x-2)} = \dfrac{1}{x^2}$

57. $\dfrac{x^2-3x+2}{x^2+5x+6} \div \dfrac{x^2+x-2}{x^2+2x-3} = \dfrac{(x-2)(x-1)}{(x+3)(x+2)} \cdot \dfrac{(x+3)(x-1)}{(x+2)(x-1)} = \dfrac{(x-2)(x-1)}{(x+2)^2}$

59. $\dfrac{x^2-4}{x^2+x-2} \div \dfrac{x-1}{x-2} = \dfrac{(x+2)(x-2)}{(x+2)(x-1)} \cdot \dfrac{x-1}{x-2} = 1$

61. $\dfrac{3y}{x^2} \div \dfrac{y^2}{x} \div \dfrac{y}{5x} = \left(\dfrac{3y}{x^2} \cdot \dfrac{x}{y^2}\right) \div \dfrac{y}{5x} = \dfrac{3}{xy} \cdot \dfrac{5x}{y} = \dfrac{15}{y^2}$

63. $\dfrac{x-3}{x-1} \div \dfrac{x^2}{x-1} \div \dfrac{x-3}{x} = \left(\dfrac{x-3}{x-1} \cdot \dfrac{x-1}{x^2}\right) \div \dfrac{x-3}{x} = \dfrac{x-3}{x^2} \cdot \dfrac{x}{x-3} = \dfrac{1}{x}$

65. Since, $12 = 2 \cdot 2 \cdot 3$ and $18 = 2 \cdot 3 \cdot 3$, the LCM is $2 \cdot 2 \cdot 3 \cdot 3 = 36$.

67. Since $5a^3 = 5 \cdot a \cdot a \cdot a$ and $10a = 2 \cdot 5 \cdot a$, the LCM is $2 \cdot 5 \cdot a \cdot a \cdot a = 10a^3$.

69. Since $z^2 - 4z = z(z-4)$ and $(z-4)^2 = (z-4)(z-4)$, the LCM is $z(z-4)(z-4) = z(z-4)^2$.

71. Since $x^2 - 6x + 9 = (x-3)(x-3)$ and $x^2 - 5x + 6 = (x-2)(x-3)$, the LCM is

    $(x-2)(x-3)(x-3) = (x-2)(x-3)^2$.

73. The factored denominators are $x+1$ and 7. The LCD is $7(x+1)$.

75. The factored denominators are $x+4$ and $(x+4)(x-4)$. The LCD is $(x+4)(x-4)$.

77. The factored denominators are 2 and $2x+1$ and $2(x-2)$. The LCD is $2(2x+1)(x-2)$.

79. $\dfrac{4}{x+1} + \dfrac{3}{x+1} = \dfrac{7}{x+1}$

81. $\dfrac{2}{x^2-1} - \dfrac{x+1}{x^2-1} = \dfrac{2-(x+1)}{x^2-1} = \dfrac{2-x-1}{x^2-1} = \dfrac{-x+1}{(x-1)(x+1)} = \dfrac{-(x-1)}{(x+1)(x-1)} = -\dfrac{1}{x+1}$

83. $\dfrac{x}{x+4} - \dfrac{x+1}{x(x+4)} = \dfrac{x^2}{x(x+4)} - \dfrac{x+1}{x(x+4)} = \dfrac{x^2-(x+1)}{x(x+4)} = \dfrac{x^2-x-1}{x(x+4)}$

85. $\dfrac{2}{x^2} - \dfrac{4x-1}{x} = \dfrac{2}{x^2} - \dfrac{x(4x-1)}{x^2} = \dfrac{2-(4x^2-x)}{x^2} = \dfrac{-4x^2+x+2}{x^2}$

87. $\dfrac{5b}{3a} - \dfrac{7b}{5a} = \dfrac{5(5b)}{5(3a)} - \dfrac{3(7b)}{3(5a)} = \dfrac{25b}{15a} - \dfrac{21b}{15a} = \dfrac{4b}{15a}$

89. $\dfrac{3}{2n-1} - \dfrac{3}{1-2n} = \dfrac{3}{2n-1} + \dfrac{3}{2n-1} = \dfrac{6}{2n-1}$

91. $\dfrac{x+3}{x-5} + \dfrac{5}{x-3} = \dfrac{(x+3)(x-3)}{(x-5)(x-3)} + \dfrac{5(x-5)}{(x-5)(x-3)} = \dfrac{x^2-9+5x-25}{(x-5)(x-3)} = \dfrac{x^2+5x-34}{(x-5)(x-3)}$

93. $\dfrac{3}{x-5} - \dfrac{1}{x-3} - \dfrac{2x}{x-5} = \dfrac{3(x-3)}{(x-5)(x-3)} - \dfrac{x-5}{(x-5)(x-3)} - \dfrac{2x(x-3)}{(x-5)(x-3)} = \dfrac{-2(x^2-4x+2)}{(x-5)(x-3)}$

95. $\dfrac{x}{x^2-9} + \dfrac{5x}{x-3} = \dfrac{x}{(x-3)(x+3)} + \dfrac{5x(x+3)}{(x-3)(x+3)} = \dfrac{x+5x^2+15x}{(x-3)(x+3)} = \dfrac{x(5x+16)}{(x-3)(x+3)}$

97. $\dfrac{b}{2b-4} - \dfrac{b-1}{b-2} = \dfrac{b}{2(b-2)} - \dfrac{2(b-1)}{2(b-2)} = \dfrac{b-(2b-2)}{2(b-2)} = \dfrac{-(b-2)}{2(b-2)} = -\dfrac{1}{2}$

99. $\dfrac{2x}{x-5} + \dfrac{2x-1}{3x^2-16x+5} = \dfrac{2x(3x-1)}{(x-5)(3x-1)} + \dfrac{2x-1}{(x-5)(3x-1)} = \dfrac{6x^2-2x+2x-1}{(x-5)(3x-1)} = \dfrac{6x^2-1}{(x-5)(3x-1)}$

101. $\dfrac{x}{(x-1)^2} - \dfrac{1}{(x-1)(x+3)} = \dfrac{x(x+3)}{(x-1)^2(x+3)} - \dfrac{x-1}{(x-1)^2(x+3)} = \dfrac{x^2+3x-x+1}{(x-1)^2(x+3)} = \dfrac{x^2+2x+1}{(x-1)^2(x+3)} = \dfrac{(x+1)^2}{(x-1)^2(x+3)}$

103. $\dfrac{x}{x^2-5x+4} + \dfrac{2}{x^2-2x-8} = \dfrac{x}{(x-4)(x-1)} + \dfrac{2}{(x-4)(x+2)} =$

$\dfrac{x(x+2)}{(x-4)(x+2)(x-1)} + \dfrac{2(x-1)}{(x-4)(x+2)(x-1)} = \dfrac{x^2+2x+2x-2}{(x-4)(x+2)(x-1)} = \dfrac{x^2+4x-2}{(x-4)(x+2)(x-1)}$

105. $\dfrac{x}{x^2-4} - \dfrac{1}{x^2+4x+4} = \dfrac{x}{(x-2)(x+2)} - \dfrac{1}{(x+2)(x+2)} =$

$\dfrac{x(x+2)}{(x+2)^2(x-2)} - \dfrac{x-2}{(x+2)^2(x-2)} = \dfrac{x^2+2x-x+2}{(x+2)^2(x-2)} = \dfrac{x^2+x+2}{(x+2)^2(x-2)}$

107. $\dfrac{3x}{x-y} - \dfrac{3y}{x^2-2xy+y^2} = \dfrac{3x}{x-y} - \dfrac{3y}{(x-y)(x-y)} = \dfrac{3x(x-y)}{(x-y)^2} - \dfrac{3y}{(x-y)^2} = \dfrac{3x^2-3xy-3y}{(x-y)^2} = \dfrac{3(x^2-xy-y)}{(x-y)^2}$

109. $x + \dfrac{1}{x-1} - \dfrac{1}{x+1} = \dfrac{x(x-1)(x+1)}{(x-1)(x+1)} + \dfrac{x+1}{(x-1)(x+1)} - \dfrac{x-1}{(x-1)(x+1)} = \dfrac{x^3-x+x+1-x+1}{(x-1)(x+1)} = \dfrac{x^3-x+2}{(x-1)(x+1)}$

111. $\dfrac{6}{t-1} + \dfrac{2}{t-2} + \dfrac{1}{t} = \dfrac{6t(t-2)}{t(t-1)(t-2)} + \dfrac{2t(t-1)}{t(t-1)(t-2)} + \dfrac{(t-1)(t-2)}{t(t-1)(t-2)} = \dfrac{6t^2-12t+2t^2-2t+t^2-3t+2}{t(t-1)(t-2)} = \dfrac{9t^2-17t+2}{t(t-1)(t-2)}$

113. The factored denominators are x and x^2. The LCD is x^2.

$\dfrac{1}{x} + \dfrac{3}{x^2} = 0 \Rightarrow \dfrac{1(x^2)}{x} + \dfrac{3(x^2)}{x^2} = 0(x^2) \Rightarrow x+3 = 0 \Rightarrow x = -3$

115. The factored denominators are x and $2x-1$. The LCD is $x(2x-1)$.

$$\frac{1}{x} + \frac{3x}{2x-1} = 0 \Rightarrow \frac{1(x)(2x-1)}{x} + \frac{3x(x)(2x-1)}{2x-1} = 0(x)(2x-1) \Rightarrow 2x-1+3x^2 = 0 \Rightarrow$$

$$3x^2 + 2x - 1 = 0 \Rightarrow (3x-1)(x+1) = 0 \Rightarrow x = \frac{1}{3} \text{ or } x = -1$$

117. The factored denominators are $(3-x)(3+x)$ and $3-x$. The LCD is $(3-x)(3+x)$.

$$\frac{2x}{9-x^2} + \frac{1}{3-x} = 0 \Rightarrow \frac{2x(3-x)(3+x)}{9-x^2} + \frac{1(3-x)(3+x)}{3-x} = 0(3-x)(3+x) \Rightarrow$$

$$2x + 3 + x = 0 \Rightarrow 3x + 3 = 0 \Rightarrow x = -1$$

119. The factored denominators are $2x$, $2x^2$ and x^3. The LCD is $2x^3$.

$$\frac{1}{2x} + \frac{1}{2x^2} - \frac{1}{x^3} = 0 \Rightarrow \frac{1(2x^3)}{2x} + \frac{1(2x^3)}{2x^2} - \frac{1(2x^3)}{x^3} = 0(2x^3) \Rightarrow x^2 + x - 2 = 0 \Rightarrow$$

$$(x+2)(x-1) = 0 \Rightarrow x = -2 \text{ or } x = 1$$

121. The factored denominators are x, $x+5$ and $x-5$. The LCD is $x(x+5)(x-5)$.

$$\frac{1}{x} - \frac{2}{x+5} + \frac{1}{x-5} = 0 \Rightarrow$$

$$\frac{1(x)(x+5)(x-5)}{x} - \frac{2(x)(x+5)(x-5)}{x+5} + \frac{1(x)(x+5)(x-5)}{x-5} = 0(x)(x+5)(x-5) \Rightarrow$$

$$(x+5)(x-5) - 2x(x-5) + x(x+5) = 0 \Rightarrow x^2 - 25 - 2x^2 + 10x + x^2 + 5x = 0 \Rightarrow$$

$$15x - 25 = 0 \Rightarrow x = \frac{25}{15} = \frac{5}{3}$$

123. $\dfrac{1+\dfrac{1}{x}}{1-\dfrac{1}{x}} = \dfrac{1+\dfrac{1}{x}}{1-\dfrac{1}{x}} \cdot \dfrac{x}{x} = \dfrac{x+1}{x-1}$

125. $\dfrac{\dfrac{1}{x-5}}{\dfrac{4}{x} - \dfrac{1}{x-5}} = \dfrac{\dfrac{1}{x-5}}{\dfrac{4}{x} - \dfrac{1}{x-5}} \cdot \dfrac{x(x-5)}{x(x-5)} = \dfrac{x}{4(x-5)-x} = \dfrac{x}{4x-20-x} = \dfrac{x}{3x-20}$

127. $\dfrac{\dfrac{1}{x} + \dfrac{2-x}{x^2}}{\dfrac{3}{x^2} - \dfrac{1}{x}} = \dfrac{\dfrac{1}{x} + \dfrac{2-x}{x^2}}{\dfrac{3}{x^2} - \dfrac{1}{x}} \cdot \dfrac{x^2}{x^2} = \dfrac{x+2-x}{3-x} = \dfrac{2}{3-x}$

129. $\dfrac{\dfrac{1}{x-3} + \dfrac{2}{x-3}}{2 - \dfrac{1}{x-3}} = \dfrac{\dfrac{1}{x-3} + \dfrac{2}{x-3}}{2 - \dfrac{1}{x-3}} \cdot \dfrac{(x-3)(x+3)}{(x-3)(x+3)} = \dfrac{x-3+2(x+3)}{2(x-3)(x+3)-(x+3)} = \dfrac{3(x+1)}{(x+3)(2x-7)}$

131. $\dfrac{\dfrac{4}{x-5}}{\dfrac{1}{x+5} + \dfrac{1}{x}} = \dfrac{\dfrac{4}{x-5}}{\dfrac{1}{x+5} + \dfrac{1}{x}} \cdot \dfrac{x(x-5)(x+5)}{x(x-5)(x+5)} = \dfrac{4x(x+5)}{x(x-5)+(x-5)(x+5)} = \dfrac{4x(x+5)}{(x-5)(2x+5)}$

133. $\dfrac{\dfrac{1}{2a} - \dfrac{1}{2b}}{\dfrac{1}{2a^2} - \dfrac{1}{2b^2}} = \dfrac{\dfrac{1}{2a} - \dfrac{1}{2b}}{\dfrac{1}{2a^2} - \dfrac{1}{2b^2}} \cdot \dfrac{2a^2b^2}{2a^2b^2} = \dfrac{ab^2 - a^2b}{2b^2 - 2a^2} = \dfrac{ab(b-a)}{2(b+a)(b-a)} = \dfrac{ab}{2(a+b)}$

R.6: Radical Notation and Rational Exponents

1. $-\sqrt{25} = -\sqrt{5^2} = -5$ and $\sqrt{25} = \sqrt{5^2} = 5$

3. $-\sqrt{\dfrac{16}{25}} = -\sqrt{\left(\dfrac{4}{5}\right)^2} = -\dfrac{4}{5}$ and $\sqrt{\dfrac{16}{25}} = \sqrt{\left(\dfrac{4}{5}\right)^2} = \dfrac{4}{5}$

5. $-\sqrt{11} \approx -3.32$ and $\sqrt{11} \approx 3.32$

7. $\sqrt{144} = \sqrt{12^2} = 12$

9. $\sqrt{23} \approx 4.80$

11. $\sqrt{\dfrac{4}{49}} = \sqrt{\left(\dfrac{2}{7}\right)^2} = \dfrac{2}{7}$

13. Since $b < 0$, $\sqrt{b^2} = -b$.

15. $\sqrt[3]{27} = \sqrt[3]{3^3} = 3$

17. $\sqrt[3]{-8} = \sqrt[3]{(-2)^3} = -2$

19. $\sqrt[3]{\dfrac{1}{27}} = \sqrt[3]{\left(\dfrac{1}{3}\right)^3} = \dfrac{1}{3}$

21. $\sqrt[3]{b^9} = \sqrt[3]{(b^3)^3} = b^3$

23. $\sqrt{9} = 3$

25. $-\sqrt{5} \approx -2.24$

27. $\sqrt{0.36} = 0.6$

29. $\sqrt{\dfrac{16}{25}} = \dfrac{\sqrt{16}}{\sqrt{25}} = \dfrac{4}{5}$

31. $\sqrt[4]{-16}$ Not possible

33. $\sqrt[3]{27} = 3$

35. $\sqrt[3]{-64} = -4$

37. $\sqrt[3]{5} \approx 1.71$

39. $-\sqrt[3]{x^9} = -\sqrt[3]{(x^3)^3} = -x^3$

41. $\sqrt[3]{(2x)^6} = \sqrt[3]{((2x)^2)^3} = (2x)^2 = 4x^2$

43. $\sqrt[4]{81} = 3$

45. $\sqrt[5]{-7} \approx -1.48$

47. $6^{1/2} = \sqrt{6}$

49. $(xy)^{1/2} = \sqrt{xy}$

51. $y^{-1/5} = \dfrac{1}{\sqrt[5]{y}}$

53. The expression can be written $27^{2/3} = \sqrt[3]{27^2}$ or $(\sqrt[3]{27})^2$ and evaluated as $(\sqrt[3]{27})^2 = 3^2 = 9$.

55. The expression can be written $(-1)^{4/3} = \sqrt[3]{(-1)^4}$ or $(\sqrt[3]{-1})^4$ and evaluated as $(\sqrt[3]{-1})^4 = (-1)^4 = 1$.

57. The expression can be written $8^{-1/3} = \dfrac{1}{8^{1/3}} = \dfrac{1}{\sqrt[3]{8}}$ and evaluated as $\dfrac{1}{\sqrt[3]{8}} = \dfrac{1}{2}$.

59. The expression can be written $13^{-3/5} = \dfrac{1}{13^{3/5}} = \dfrac{1}{\sqrt[5]{13^3}}$ or $\dfrac{1}{(\sqrt[5]{13})^3}$. The result is not an integer.

61. $\sqrt{t} = t^{1/2}$

63. $\sqrt[3]{(x+1)} = (x+1)^{1/3}$

65. $\dfrac{1}{\sqrt{x+1}} = (x+1)^{-1/2}$

67. $\sqrt{a^2 - b^2} = \left(a^2 - b^2\right)^{1/2}$

69. $\dfrac{1}{\sqrt[3]{x^7}} = x^{-7/3}$

71. $16^{1/2} = \sqrt{16} = 4$

73. $256^{1/4} = \sqrt[4]{256} = 4$

75. $32^{1/5} = \sqrt[5]{32} = 2$

77. $(-8)^{4/3} = (\sqrt[3]{-8})^4 = (-2)^4 = 16$

79. $2^{1/2} \cdot 2^{2/3} = 2^{1/2+2/3} = 2^{7/6} \approx 2.24$

81. $\left(\dfrac{4}{9}\right)^{1/2} = \dfrac{4^{1/2}}{9^{1/2}} = \dfrac{\sqrt{4}}{\sqrt{9}} = \dfrac{2}{3}$

83. $\dfrac{4^{2/3}}{4^{1/2}} = 4^{2/3-1/2} = 4^{1/6} \approx 1.26$

85. $4^{-1/2} = \dfrac{1}{4^{1/2}} = \dfrac{1}{\sqrt{4}} = \dfrac{1}{2}$

87. $(-8)^{-1/3} = \dfrac{1}{(-8)^{1/3}} = \dfrac{1}{\sqrt[3]{-8}} = \dfrac{1}{-2} = -\dfrac{1}{2}$

89. $\left(\dfrac{1}{16}\right)^{-1/4} = 16^{1/4} = \sqrt[4]{16} = 2$

91. $(2^{1/2})^3 = 2^{1/2 \cdot 3} = 2^{3/2} \approx 2.83$

93. $(x^2)^{3/2} = x^{2 \cdot 3/2} = x^3$

95. $(x^2 y^8)^{1/2} = x^{2 \cdot 1/2} \cdot y^{8 \cdot 1/2} = xy^4$

97. $\sqrt[3]{x^3 y^6} = (x^3 y^6)^{1/3} = x^{3 \cdot 1/3} \cdot y^{6 \cdot 1/3} = xy^2$

99. $\sqrt{\dfrac{y^4}{x^2}} = \left(\dfrac{y^4}{x^2}\right)^{1/2} = \dfrac{y^{4 \cdot 1/2}}{x^{2 \cdot 1/2}} = \dfrac{y^2}{x}$

101. $\sqrt{y^3} \cdot \sqrt[3]{y^2} = (y^3)^{1/2} \cdot (y^2)^{1/3} = y^{3 \cdot 1/2} \cdot y^{2 \cdot 1/3} = y^{3/2} \cdot y^{2/3} = y^{3/2+2/3} = y^{13/6}$

103. $\left(\dfrac{x^6}{27}\right)^{2/3} = \dfrac{x^{6 \cdot 2/3}}{27^{2/3}} = \dfrac{x^4}{(\sqrt[3]{27})^2} = \dfrac{x^4}{3^2} = \dfrac{x^4}{9}$

105. $\left(\dfrac{x^2}{y^6}\right)^{-1/2} = \left(\dfrac{y^6}{x^2}\right)^{1/2} = \dfrac{y^{6\cdot 1/2}}{x^{2\cdot 1/2}} = \dfrac{y^3}{x}$

107. $\sqrt{\sqrt{y}} = (y^{1/2})^{1/2} = y^{1/2 \cdot 1/2} = y^{1/4}$

109. $(a^{-1/2})^{4/3} = a^{-1/2 \cdot 4/3} = a^{-2/3} = \dfrac{1}{a^{2/3}}$

111. $(a^3 b^6)^{1/3} = a^{3 \cdot 1/3} \cdot b^{6 \cdot 1/3} = ab^2$

113. $\dfrac{(k^{1/2})^{-3}}{(k^2)^{1/4}} = \dfrac{k^{-3/2}}{k^{1/2}} = k^{-3/2 - 1/2} = k^{-4/2} = k^{-2} = \dfrac{1}{k^2}$

115. $\sqrt{b} \cdot \sqrt[4]{b} = b^{1/2} \cdot b^{1/4} = b^{1/2 + 1/4} = b^{3/4}$

117. $\sqrt{z} \cdot \sqrt[3]{z^2} \cdot \sqrt[4]{z^3} = z^{1/2} \cdot z^{2/3} \cdot z^{3/4} = z^{1/2 + 2/3 + 3/4} = z^{23/12}$

119. $p^{1/2}(p^{3/2} + p^{1/2}) = p^{1/2 + 3/2} + p^{1/2 + 1/2} = p^2 + p$

121. $\sqrt[3]{x}(\sqrt{x} - \sqrt[3]{x^2}) = x^{1/3}(x^{1/2} - x^{2/3}) = x^{1/3 + 1/2} - x^{1/3 + 2/3} = x^{5/6} - x$

R.7: Radical Expressions

1. $\sqrt{3} \cdot \sqrt{3} = \sqrt{3 \cdot 3} = \sqrt{9} = 3$

3. $\sqrt{2} \cdot \sqrt{50} = \sqrt{2 \cdot 50} = \sqrt{100} = 10$

5. $\sqrt[3]{4} \cdot \sqrt[3]{16} = \sqrt[3]{4 \cdot 16} = \sqrt[3]{64} = 4$

7. $\sqrt{\dfrac{9}{25}} = \dfrac{\sqrt{9}}{\sqrt{25}} = \dfrac{3}{5}$

9. $\sqrt{\dfrac{1}{2}} \cdot \sqrt{\dfrac{1}{8}} = \sqrt{\dfrac{1 \cdot 1}{2 \cdot 8}} = \sqrt{\dfrac{1}{16}} = \dfrac{\sqrt{1}}{\sqrt{16}} = \dfrac{1}{4}$

11. $\sqrt{\dfrac{x}{2}} \cdot \sqrt{\dfrac{x}{8}} = \sqrt{\dfrac{x \cdot x}{2 \cdot 8}} = \sqrt{\dfrac{x^2}{16}} = \dfrac{\sqrt{x^2}}{\sqrt{16}} = \dfrac{x}{4}$

13. $\dfrac{\sqrt{45}}{\sqrt{5}} = \sqrt{\dfrac{45}{5}} = \sqrt{9} = 3$

15. $\sqrt[4]{9} \cdot \sqrt[4]{9} = \sqrt[4]{9 \cdot 9} = \sqrt[4]{81} = 3$

17. $\dfrac{\sqrt[5]{64}}{\sqrt[5]{-2}} = \sqrt[5]{\dfrac{64}{-2}} = \sqrt[5]{-32} = -2$

19. $\dfrac{\sqrt{a^2 b}}{\sqrt{b}} = \sqrt{\dfrac{a^2 b}{b}} = \sqrt{a^2} = a$

21. $\sqrt[3]{\dfrac{x^3}{8}} = \dfrac{\sqrt[3]{x^3}}{\sqrt[3]{8}} = \dfrac{x}{2}$

23. $\sqrt{4x^4} = \sqrt{4} \cdot \sqrt{(x^2)^2} = 2x^2$

25. $\sqrt[4]{16x^4 y} = \sqrt[4]{16} \cdot \sqrt[4]{x^4} \cdot \sqrt[4]{y} = 2x\sqrt[4]{y}$

27. $\sqrt{3x} \cdot \sqrt{12x} = \sqrt{3 \cdot 12 \cdot x \cdot x} = \sqrt{36x^2} = \sqrt{36} \cdot \sqrt{x^2} = 6x$

29. $\sqrt[3]{8x^6y^3z^9} = \sqrt[3]{8} \cdot \sqrt[3]{(x^2)^3} \cdot \sqrt[3]{y^3} \cdot \sqrt[3]{(z^3)^3} = 2x^2yz^3$

31. $\sqrt[4]{\dfrac{3}{4}} \cdot \sqrt[4]{\dfrac{27}{4}} = \sqrt[4]{\dfrac{3}{4} \cdot \dfrac{27}{4}} = \sqrt[4]{\dfrac{81}{16}} = \dfrac{\sqrt[4]{81}}{\sqrt[4]{16}} = \dfrac{3}{2}$

33. $\sqrt[4]{25z} \cdot \sqrt[4]{25z} = \sqrt[4]{625z^2} = \sqrt[4]{625} \cdot \sqrt[4]{z^2} = 5\sqrt{z}$

35. $\sqrt[5]{\dfrac{7a}{b^2}} \cdot \sqrt[5]{\dfrac{b^2}{7a^6}} = \sqrt[5]{\dfrac{7ab^2}{7a^6b^2}} = \sqrt[5]{\dfrac{1}{a^5}} = \dfrac{1}{a}$

37. $\sqrt{200} = \sqrt{100 \cdot 2} = \sqrt{100} \cdot \sqrt{2} = 10\sqrt{2}$

39. $\sqrt[3]{81} = \sqrt[3]{27 \cdot 3} = \sqrt[3]{27} \cdot \sqrt[3]{3} = 3\sqrt[3]{3}$

41. $\sqrt[4]{64} = \sqrt[4]{16 \cdot 4} = \sqrt[4]{16} \cdot \sqrt[4]{4} = 2\sqrt[4]{4} = 2\sqrt[4]{2^2} = 2\sqrt{2}$

43. $\sqrt[5]{-64} = \sqrt[5]{-2^6} = \sqrt[5]{-2^5 \cdot 2} = \sqrt[5]{-2^5} \cdot \sqrt[5]{2} = -2\sqrt[5]{2}$

45. $\sqrt{8n^3} = \sqrt{(2n^2 \cdot 2n)} = \sqrt{(2n)^2} \cdot \sqrt{2n} = 2n\sqrt{2n}$

47. $\sqrt{12a^2b^5} = \sqrt{(2ab^2)^2 \cdot 3b} = \sqrt{(2ab^2)^2} \cdot \sqrt{3b} = 2ab^2\sqrt{3b}$

49. $\sqrt[3]{-125x^4y^5} = \sqrt[3]{(-5xy)^3 \cdot xy^2} = \sqrt[3]{(-5xy)^3} \cdot \sqrt[3]{xy^2} = -5xy\sqrt[3]{xy^2}$

51. $\sqrt[3]{5t} \cdot \sqrt[3]{125t} = \sqrt[3]{625t^2} = \sqrt[3]{5^4t^2} = \sqrt[3]{5^3 \cdot 5t^2} = \sqrt[3]{5^3} \cdot \sqrt[3]{5t^2} = 5\sqrt[3]{5t^2}$

53. $\sqrt[4]{\dfrac{9t^5}{r^8}} \cdot \sqrt[4]{\dfrac{9r}{5t}} = \sqrt[4]{\dfrac{81rt^5}{5r^8t}} = \sqrt[4]{\dfrac{81t^4}{5r^7}} = \dfrac{\sqrt[4]{(3t)^4}}{\sqrt[4]{r^4 \cdot 5r^3}} = \dfrac{3t}{r\sqrt[4]{5r^3}}$

55. $\sqrt{3} \cdot \sqrt[3]{3} = 3^{1/2} \cdot 3^{1/3} = 3^{1/2+1/3} = 3^{5/6} = \sqrt[6]{3^5}$

57. $\sqrt[4]{8} \cdot \sqrt[3]{4} = \sqrt[4]{2^3} \cdot \sqrt[3]{2^2} = 2^{3/4} \cdot 2^{2/3} = 2^{3/4+2/3} = 2^{17/12} = 2^{12/12+5/12} = 2 \cdot 2^{5/12} = 2\sqrt[12]{2^5}$

59. $\sqrt[4]{x^3} \cdot \sqrt[3]{x} = x^{3/4} \cdot x^{1/3} = x^{3/4+1/3} = x^{13/12} = x^{12/12} \cdot x^{1/12} = x\sqrt[12]{x}$

61. $\sqrt[4]{rt} \cdot \sqrt[3]{r^2t} = (rt)^{1/4} \cdot (r^2t)^{1/3} = r^{1/4}t^{1/4} \cdot r^{2/3}t^{1/3} = r^{1/4+2/3}t^{1/4+1/3} = r^{11/12}t^{7/12} = \sqrt[12]{r^{11}t^7}$

63. $2\sqrt{3} + 7\sqrt{3} = 9\sqrt{3}$

65. $\sqrt{x} + \sqrt{x} - \sqrt{y} = 2\sqrt{x} - \sqrt{y}$

67. $2\sqrt[3]{6} - 7\sqrt[3]{6} = -5\sqrt[3]{6}$

69. $3\sqrt{28} + 3\sqrt{7} = 3\sqrt{4 \cdot 7} + 3\sqrt{7} = 3 \cdot 2\sqrt{7} + 3\sqrt{7} = 9\sqrt{7}$

71. $\sqrt{44} - 4\sqrt{11} = \sqrt{4 \cdot 11} - 4\sqrt{11} = 2\sqrt{11} - 4\sqrt{11} = -2\sqrt{11}$

73. $2\sqrt[3]{16} + \sqrt[3]{2} - \sqrt{2} = 2\sqrt[3]{8 \cdot 2} + \sqrt[3]{2} - \sqrt{2} = 2 \cdot 2\sqrt[3]{2} + \sqrt[3]{2} - \sqrt{2} = 5\sqrt[3]{2} - \sqrt{2}$

75. $\sqrt[3]{xy} - 2\sqrt[3]{xy} = -\sqrt[3]{xy}$

77. $\sqrt{4x+8} + \sqrt{x+2} = \sqrt{4(x+2)} + \sqrt{x+2} = 2\sqrt{x+2} + \sqrt{x+2} = 3\sqrt{x+2}$

79. $\dfrac{15\sqrt{8}}{4} - \dfrac{2\sqrt{2}}{5} = \dfrac{15 \cdot 2\sqrt{2}}{4} \cdot \dfrac{5}{5} - \dfrac{2\sqrt{2}}{5} \cdot \dfrac{4}{4} = \dfrac{150\sqrt{2}}{20} - \dfrac{8\sqrt{2}}{20} = \dfrac{150\sqrt{2}-8\sqrt{2}}{20} = \dfrac{142\sqrt{2}}{20} = \dfrac{71\sqrt{2}}{10}$

81. $2\sqrt[4]{64} - \sqrt[4]{324} + \sqrt[4]{4} = 2\sqrt[4]{16 \cdot 4} - \sqrt[4]{81 \cdot 4} + \sqrt[4]{4} = 4\sqrt[4]{4} - 3\sqrt[4]{4} + \sqrt[4]{4} = 2\sqrt[4]{4} = 2\sqrt{2}$

83. $2\sqrt{3z} + 3\sqrt{12z} + 3\sqrt{48z} = 2\sqrt{3z} + 3\sqrt{4 \cdot 3z} + 3\sqrt{16 \cdot 3z} = 2\sqrt{3z} + 6\sqrt{3z} + 12\sqrt{3z} = 20\sqrt{3z}$

85. $\sqrt[4]{81a^5b^5} - \sqrt[4]{ab} = \sqrt[4]{(3ab)^4 \cdot ab} - \sqrt[4]{ab} = 3ab\sqrt[4]{ab} - \sqrt[4]{ab} = (3ab-1)\sqrt[4]{ab}$

87. $5\sqrt[3]{\dfrac{n^4}{125}} - 2\sqrt[3]{n} = 5\sqrt[3]{\dfrac{n^3}{125} \cdot n} - 2\sqrt[3]{n} = 5 \cdot \dfrac{n}{5}\sqrt[3]{n} - 2\sqrt[3]{n} = n\sqrt[3]{n} - 2\sqrt[3]{n} = (n-2)\sqrt[3]{n}$

89. $(3+\sqrt{7})(3-\sqrt{7}) = 3^2 - (\sqrt{7})^2 = 9 - 7 = 2$

91. $(\sqrt{x}+8)(\sqrt{x}-8) = (\sqrt{x})^2 - 8^2 = x - 64$

93. $(\sqrt{ab}-\sqrt{c})(\sqrt{ab}+\sqrt{c}) = (\sqrt{ab})^2 - (\sqrt{c})^2 = ab - c$

95. $(\sqrt{x}-7)(\sqrt{x}+8) = (\sqrt{x})^2 + 8\sqrt{x} - 7\sqrt{x} - 56 = x + \sqrt{x} - 56$

97. $\dfrac{4}{\sqrt{3}} = \dfrac{4}{\sqrt{3}} \cdot \dfrac{\sqrt{3}}{\sqrt{3}} = \dfrac{4\sqrt{3}}{3}$

99. $\dfrac{5}{3\sqrt{5}} = \dfrac{5}{3\sqrt{5}} \cdot \dfrac{\sqrt{5}}{\sqrt{5}} = \dfrac{5\sqrt{5}}{3 \cdot 5} = \dfrac{5\sqrt{5}}{15} = \dfrac{\sqrt{5}}{3}$

101. $\sqrt{\dfrac{b}{12}} = \dfrac{\sqrt{b}}{\sqrt{12}} = \dfrac{\sqrt{b}}{\sqrt{12}} \cdot \dfrac{\sqrt{12}}{\sqrt{12}} = \dfrac{\sqrt{12b}}{12} = \dfrac{\sqrt{4 \cdot 3b}}{12} = \dfrac{2\sqrt{3b}}{12} = \dfrac{\sqrt{3b}}{6}$

103. $\dfrac{1}{3-\sqrt{2}} = \dfrac{1}{3-\sqrt{2}} \cdot \dfrac{3+\sqrt{2}}{3+\sqrt{2}} = \dfrac{3+\sqrt{2}}{9-2} = \dfrac{3+\sqrt{2}}{7}$

105. $\dfrac{\sqrt{2}}{\sqrt{5}+2} = \dfrac{\sqrt{2}}{\sqrt{5}+2} \cdot \dfrac{\sqrt{5}-2}{\sqrt{5}-2} = \dfrac{\sqrt{10}-2\sqrt{2}}{5-4} = \dfrac{\sqrt{10}-2\sqrt{2}}{1} = \sqrt{10} - 2\sqrt{2}$

107. $\dfrac{1}{\sqrt{7}-\sqrt{6}} = \dfrac{1}{\sqrt{7}-\sqrt{6}} \cdot \dfrac{\sqrt{7}+\sqrt{6}}{\sqrt{7}+\sqrt{6}} = \dfrac{\sqrt{7}+\sqrt{6}}{7-6} = \dfrac{\sqrt{7}+\sqrt{6}}{1} = \sqrt{7} + \sqrt{6}$

109. $\dfrac{\sqrt{z}}{\sqrt{z}-3} = \dfrac{\sqrt{z}}{\sqrt{z}-3} \cdot \dfrac{\sqrt{z}+3}{\sqrt{z}+3} = \dfrac{z+3\sqrt{z}}{z-9}$

111. $\dfrac{\sqrt{a}+\sqrt{b}}{\sqrt{a}-\sqrt{b}} = \dfrac{\sqrt{a}+\sqrt{b}}{\sqrt{a}-\sqrt{b}} \cdot \dfrac{\sqrt{a}+\sqrt{b}}{\sqrt{a}+\sqrt{b}} = \dfrac{a+2\sqrt{ab}+b}{a-b}$

Appendix C: Partial Fractions

1. Multiply $\dfrac{5}{3x(2x+1)} = \dfrac{A}{3x} + \dfrac{B}{2x+1}$ by $3x(2x+1) \Rightarrow 5 = A(2x+1) + B(3x)$.

    Let $x = 0 \Rightarrow 5 = A(1) \Rightarrow A = 5$. Let $x = -\dfrac{1}{2} \Rightarrow 5 = B\left(-\dfrac{3}{2}\right) \Rightarrow B = -\dfrac{10}{3}$.

    The expression can be written $\dfrac{5}{3x} + \dfrac{-10}{3(2x+1)}$.

3. Multiply $\dfrac{4x+2}{(x+2)(2x-1)} = \dfrac{A}{x+2} + \dfrac{B}{2x-1}$ by $(x+2)(2x-1) \Rightarrow 4x+2 = A(2x-1) + B(x+2)$.

    Let $x = -2 \Rightarrow -6 = A(-5) \Rightarrow A = \dfrac{6}{5}$. Let $x = \dfrac{1}{2} \Rightarrow 4 = B\left(\dfrac{5}{2}\right) \Rightarrow B = \dfrac{8}{5}$.

    The expression can be written $\dfrac{6}{5(x+2)} + \dfrac{8}{5(2x-1)}$.

5. Factoring $\dfrac{x}{x^2+4x-5}$ results in $\dfrac{x}{(x+5)(x-1)}$.

    Multiply $\dfrac{x}{(x+5)(x-1)} = \dfrac{A}{x+5} + \dfrac{B}{x-1}$ by $(x+5)(x-1) \Rightarrow x = A(x-1) + B(x+5)$.

    Let $x = -5 \Rightarrow -5 = A(-6) \Rightarrow A = \dfrac{5}{6}$. Let $x = 1 \Rightarrow 1 = B(6) \Rightarrow B = \dfrac{1}{6}$.

    The expression can be written $\dfrac{5}{6(x+5)} + \dfrac{1}{6(x-1)}$.

7. Multiply $\dfrac{2x}{(x+1)(x+2)^2} = \dfrac{A}{x+1} + \dfrac{B}{x+2} + \dfrac{C}{(x+2)^2}$ by $(x+1)(x+2)^2 \Rightarrow$

    $2x = A(x+2)^2 + B(x+1)(x+2) + C(x+1)$. Let $x = -1 \Rightarrow -2 = A(1) \Rightarrow A = -2$.

    Let $x = -2 \Rightarrow -4 = C(-1) \Rightarrow C = 4$. Let $x = 0$ with $A = -2$ and $C = 4 \Rightarrow$

    $\quad 0 = -2(4) + B(2) + 4(1) \Rightarrow 4 = 2B \Rightarrow B = 2$. The expression can be written $\dfrac{-2}{x+1} + \dfrac{2}{x+2} + \dfrac{4}{(x+2)^2}$.

9. Multiply $\dfrac{4}{x(1-x)} = \dfrac{A}{x} + \dfrac{B}{1-x}$ by $x(1-x) \Rightarrow 4 = A(1-x) + B(x)$.

    Let $x = 0 \Rightarrow 4 = A(1) \Rightarrow A = 4$. Let $x = 1 \Rightarrow 4 = B(1) \Rightarrow B = 4$.

    The expression can be written $\dfrac{4}{x} + \dfrac{4}{1-x}$.

11. Multiply $\dfrac{4x^2-x-15}{x(x+1)(x-1)} = \dfrac{A}{x} + \dfrac{B}{x+1} + \dfrac{C}{x-1}$ by $x(x+1)(x-1) \Rightarrow$

    $4x^2 - x - 15 = A(x-1)(x+1) + B(x)(x-1) + C(x)(x+1)$. Let $x = 0 \Rightarrow -15 = A(-1) \Rightarrow A = 15$. Let
    $x = 1 \Rightarrow -12 = C(1)(2) \Rightarrow C = -6$. Let $x = -1 \Rightarrow -10 = B(-1)(-2) \Rightarrow B = -5$.

The expression can be written $\dfrac{15}{x} + \dfrac{-5}{x+1} + \dfrac{-6}{x-1}$.

13. By long division $\dfrac{x^2}{x^2+2x+1} = 1 + \dfrac{-2x-1}{(x+1)^2}$.

Multiply $\dfrac{-2x-1}{(x+1)^2} = \dfrac{A}{x+1} + \dfrac{B}{(x+1)^2}$ by $(x+1)^2 \Rightarrow -2x-1 = A(x+1) + B$.

Let $x = -1 \Rightarrow 1 = B$. Let $x = 0$ with $B = 1 \Rightarrow -1 = A+1 \Rightarrow A = -2$.

The expression can be written $1 + \dfrac{-2}{x+1} + \dfrac{1}{(x+1)^2}$.

15. By long division $\dfrac{2x^5 + 3x^4 - 3x^3 - 2x^2 + x}{2x^2+5x+2} = x^3 - x^2 + \dfrac{x}{2x^2+5x+2} = x^3 - x^2 + \dfrac{x}{(2x+1)(x+2)}$.

Multiply $\dfrac{x}{(2x+1)(x+2)} = \dfrac{A}{2x+1} + \dfrac{B}{x+2}$ by $(2x+1)(x+2) \Rightarrow x = A(x+2) + B(2x+1)$.

Let $x = -\dfrac{1}{2} \Rightarrow -\dfrac{1}{2} = A\left(\dfrac{3}{2}\right) \Rightarrow A = -\dfrac{1}{3}$. Let $x = -2 \Rightarrow -2 = B(-3) \Rightarrow B = \dfrac{2}{3}$.

The expression can be written $x^3 - x^2 + \dfrac{-1}{3(2x+1)} + \dfrac{2}{3(x+2)}$.

17. By long division $\dfrac{x^3+4}{9x^3-4x} = \dfrac{1}{9} + \dfrac{\frac{4}{9}x+4}{9x^3-4x} = \dfrac{1}{9} + \dfrac{\frac{4}{9}x+4}{x(3x+2)(3x-2)}$.

Multiply $\dfrac{\frac{4}{9}x+4}{x(3x+2)(3x-2)} = \dfrac{A}{x} + \dfrac{B}{3x+2} + \dfrac{C}{3x-2}$ by $x(3x+2)(3x-2) \Rightarrow$

$\dfrac{4}{9}x+4 = A(3x+2)(3x-2) + B(x)(3x-2) + C(x)(3x+2)$. Let $x = 0 \Rightarrow 4 = A(-4) \Rightarrow A = -1$

Let $x = -\dfrac{2}{3} \Rightarrow -\dfrac{8}{27} + 4 = B\left(-\dfrac{2}{3}\right)(-4) \Rightarrow \dfrac{100}{27} = \dfrac{8}{3}B \Rightarrow B = \dfrac{25}{18}$.

Let $x = \dfrac{2}{3} \Rightarrow \dfrac{8}{27} + 4 = C\left(\dfrac{2}{3}\right)(4) \Rightarrow \dfrac{116}{27} = \dfrac{8}{3}C \Rightarrow C = \dfrac{29}{18}$.

The expression can be written $\dfrac{1}{9} + \dfrac{-1}{x} + \dfrac{25}{18(3x+2)} + \dfrac{29}{18(3x-2)}$.

19. Multiply $\dfrac{-3}{x^2(x^2+5)} = \dfrac{A}{x} + \dfrac{B}{x^2} + \dfrac{Cx+D}{x^2+5}$ by $x^2(x^2+5) \Rightarrow$

$-3 = A(x)(x^2+5) + B(x^2+5) + (Cx+D)(x^2) \Rightarrow -3 = Ax^3 + 5Ax + Bx^2 + 5B + Cx^3 + Dx^2$.

Equate coefficients. For x^3: $0 = A+C$.

For x^2: $0 = B+D$. For x: $0 = 5A \Rightarrow A = 0$. For the constants: $-3 = 5B \Rightarrow B = -\dfrac{3}{5}$.

Substitute $A = 0$ in the first equation. $C = 0$. Substitute $B = -\dfrac{3}{5}$ in the second equation. $D = \dfrac{3}{5}$.

The expression can be written $\dfrac{-3}{5x^2} + \dfrac{3}{5(x^2+5)}$.

21. Multiply $\dfrac{3x-2}{(x+4)(3x^2+1)} = \dfrac{A}{x+4} + \dfrac{Bx+C}{3x^2+1}$ by $(x+4)(3x^2+1) \Rightarrow$

$3x-2 = A(3x^2+1) + (Bx+C)(x+4) \Rightarrow 3x-2 = 3Ax^2 + A + Bx^2 + 4Bx + Cx + 4C$.

Let $x = -4 \Rightarrow -14 = 49A \Rightarrow A = -\dfrac{2}{7}$. Equate coefficients.

For x^2: $0 = 3A + B \Rightarrow 0 = -\dfrac{6}{7} + B \Rightarrow B = \dfrac{6}{7}$. For x: $3 = 4B + C \Rightarrow 3 = \dfrac{24}{7} + C \Rightarrow C = -\dfrac{3}{7}$.

The expression can be written $\dfrac{-2}{7(x+4)} + \dfrac{6x-3}{7(3x^2+1)}$.

23. Multiply $\dfrac{1}{x(2x+1)(3x^2+4)} = \dfrac{A}{x} + \dfrac{B}{2x+1} + \dfrac{Cx+D}{3x^2+4}$ by $x(2x+1)(3x^2+4) \Rightarrow$

$1 = A(2x+1)(3x^2+4) + B(x)(3x^2+4) + (Cx+D)(x)(2x+1)$.

Let $x = 0 \Rightarrow 1 = A(1)(4) \Rightarrow A = \dfrac{1}{4}$. Let $x = -\dfrac{1}{2} \Rightarrow 1 = B\left(-\dfrac{1}{2}\right)\left(\dfrac{19}{4}\right) \Rightarrow B = -\dfrac{8}{19}$.

Multiply the right side out. $1 = A(6x^3 + 3x^2 + 8x + 4) + 3Bx^3 + 4Bx + 2Cx^3 + Cx^2 + 2Dx^2 + Dx \Rightarrow$

$1 = 6Ax^3 + 3Ax^2 + 8Ax + 4A + 3Bx^3 + 4Bx + 2Cx^3 + Cx^2 + 2Dx^2 + Dx$. Equate coefficients.

For x^3: $0 = 6A + 3B + 2C \Rightarrow 0 = 6\left(\dfrac{1}{4}\right) + 3\left(-\dfrac{8}{19}\right) + 2C \Rightarrow 0 = \dfrac{9}{38} + 2C \Rightarrow C = -\dfrac{9}{76}$.

For x^2: $0 = 3A + C + 2D \Rightarrow 0 = \dfrac{3}{4} - \dfrac{9}{76} + 2D \Rightarrow 0 = \dfrac{48}{76} + 2D \Rightarrow D = -\dfrac{24}{76}$.

The expression can be written $\dfrac{1}{4x} + \dfrac{-8}{19(2x+1)} + \dfrac{-9x-24}{76(3x^2+4)}$.

25. Multiply $\dfrac{3x-1}{x(2x^2+1)^2} = \dfrac{A}{x} + \dfrac{Bx+C}{2x^2+1} + \dfrac{Dx+E}{(2x^2+1)^2}$ by $x(2x^2+1)^2 \Rightarrow$

$3x-1 = A(2x^2+1)^2 + (Bx+C)(x)(2x^2+1) + (Dx+E)(x)$.

Let $x = 0 \Rightarrow -1 = A(1) \Rightarrow A = -1$. Multiply the right side out.

$3x-1 = A(4x^4 + 4x^2 + 1) + 2Bx^4 + Bx^2 + Cx + 2Cx^3 + Dx^2 + Ex \Rightarrow$

$3x-1 = 4Ax^4 + 4Ax^2 + A + 2Bx^4 + Bx^2 + Cx + 2Cx^3 + Dx^2 + Ex$.

Equate coefficients. For x^4: $0 = 4A + 2B \Rightarrow 0 = -4 + 2B \Rightarrow B = 2$. For x^3: $0 = 2C \Rightarrow C = 0$.

For x^2: $0 = 4A + B + D \Rightarrow 0 = -4 + 2 + D \Rightarrow D = 2$. For x: $3 = C + E \Rightarrow 3 = 0 + E \Rightarrow E = 3$. The expression

can be written $\dfrac{-1}{x} + \dfrac{2x}{2x^2+1} + \dfrac{2x+3}{(2x^2+1)^2}$.

27. Multiply $\dfrac{-x^4 - 8x^2 + 3x - 10}{(x+2)(x^2+4)^2} = \dfrac{A}{x+2} + \dfrac{Bx+C}{x^2+4} + \dfrac{Dx+E}{(x^2+4)^2}$ by $(x+2)(x^2+4)^2 \Rightarrow$

$-x^4 - 8x^2 + 3x - 10 = A(x^2+4)^2 + (Bx+C)(x+2)(x^2+4) + (Dx+E)(x+2)$.

Let $x = -2 \Rightarrow -64 = A(64) \Rightarrow A = -1$. Multiply the right side out.

$-x^4 - 8x^2 + 3x - 10 = Ax^4 + 8Ax^2 + 16A + Bx^4 + 2Bx^3 + 4Bx^2 + 8Bx + Cx^3 + 2Cx^2 +$.

$4Cx + 8C + Dx^2 + 2Dx + Ex + 2E$.

Equate coefficients.

For x^4: $-1 = A + B \Rightarrow -1 = -1 + B \Rightarrow B = 0$. For x^3: $0 = 2B + C \Rightarrow 0 = 0 + C \Rightarrow C = 0$.

For x^2: $-8 = 8A + 4B + 2C + D \Rightarrow -8 = -8 + 0 + 0 + D \Rightarrow D = 0$.

For x: $3 = 8B + 4C + 2D + E \Rightarrow 3 = 0 + 0 + 0 + E \Rightarrow E = 3$.

The expression can be written $\dfrac{-1}{x+2} + \dfrac{3}{(x^2+4)^2}$.

29. By long division . $\dfrac{5x^5 + 10x^4 - 15x^3 + 4x^2 + 13x - 9}{x^3 + 2x^2 - 3x} = 5x^2 + \dfrac{4x^2 + 13x - 9}{x^3 + 2x^2 - 3x} = 5x^2 + \dfrac{4x^2 + 13x - 9}{x(x+3)(x-1)}$.

Multiply $\dfrac{4x^2 + 13x - 9}{x(x+3)(x-1)} = \dfrac{A}{x} + \dfrac{B}{x+3} + \dfrac{C}{x-1}$ by $x(x+3)(x-1) \Rightarrow$

$4x^2 + 13x - 9 = A(x+3)(x-1) + B(x)(x-1) + C(x)(x+3)$ Let $x = 0 \Rightarrow -9 = A(-3) \Rightarrow A = 3$. Let

$x = -3 \Rightarrow -12 = B(-3)(-4) \Rightarrow B = -1$. Let $x = 1 \Rightarrow 8 = C(4) \Rightarrow C = 2$.

The expression can be written $5x^2 + \dfrac{3}{x} + \dfrac{-1}{x+3} + \dfrac{2}{x-1}$.

Appendix D: Percent Change and Exponential Functions

1. (a) $\dfrac{B-A}{A} \times 100 \Rightarrow \dfrac{1000-500}{500} \times 100 \Rightarrow \dfrac{500}{500} \times 100 \Rightarrow 1 \times 100 \Rightarrow 100\%$

   (b) $\dfrac{A-B}{B} \times 100 \Rightarrow \dfrac{500-1000}{1000} \times 100 \Rightarrow -\dfrac{500}{1000} \times 100 \Rightarrow -\dfrac{1}{2} \times 100 \Rightarrow -50\%$

3. (a) $\dfrac{B-A}{A} \times 100 \Rightarrow \dfrac{1.3-1.27}{1.27} \times 100 \Rightarrow \dfrac{0.03}{1.27} \times 100 \Rightarrow 0.0236 \times 100 \Rightarrow 2.36\%$

   (b) $\dfrac{A-B}{B} \times 100 \Rightarrow \dfrac{1.27-1.30}{1.30} \times 100 \Rightarrow \dfrac{-0.03}{1.30} \times 100 \Rightarrow -0.0231 \times 100 \Rightarrow -2.31\%$

5. (a) $\dfrac{B-A}{A} \times 100 \Rightarrow \dfrac{65-45}{45} \times 100 \Rightarrow \dfrac{20}{45} \times 100 \Rightarrow 0.4444 \times 100 \Rightarrow 44.44\%$

   (b) $\dfrac{A-B}{B} \times 100 \Rightarrow \dfrac{45-65}{65} \times 100 \Rightarrow \dfrac{-20}{65} \times 100 \Rightarrow -0.3077 \times 100 \Rightarrow -30.77\%$

7. (a) Let $A = 1500$ and $r = 1.2$. The increase is $rA = 1.2 \cdot 1500 = \$1800$.

   (b) The final value of the account is $A + rA = 1500 + 1800 = \$3300$

   (c) The account increased in value by a factor of $1 + r = 1 + 1.2 = 2.2$.

9. (a) Let $A = 4000$ and $r = -0.55$. The decrease is $rA = -0.5 \cdot 4000 = -\$2200$.

   (b) The final value of the account is $A + rA = 4000 + (-2200) = -\$1800$

   (c) The account decreased in value by a factor of $1 + r = 1 + (-0.55) = 0.45$.

11. (a) Let $A = 7500$ and $r = -0.60$. The decrease is $rA = -0.60 \cdot 7500 = -\$4500$.

    (b) The final value of the account is $A + rA = 7500 + (-4500) = \$3000$

    (c) The account decreased in value by a factor of $1 + r = 1 + (-0.60) = 0.4$.

13. $A = 4000\left(1 + \dfrac{0.055}{12}\right)^{12} = 4225.63$

    effective annual interest for monthly compounding $= \dfrac{4225.63 - 4000}{4000} \times 100 = 5.641\%$

15. $A = 4000e^{0.055(1)} = 4226.16$

    effective annual interest for continuous compounding $= \dfrac{4226.16 - 4000}{4000} \times 100 = 5.654\%$

17. Effective interest rate compounded continuously $= \left(e^{0.0425(1)} - 1\right) \times 100 = 4.34\%$

19. $E = \left(1 + \dfrac{r}{n}\right)^n - 1 = \left(1 + \dfrac{0.04}{12}\right)^{12} - 1 = 0.0407 = 4.07\%$

21. $E = \left(e^{0.09(1)} - 1\right) = .0942 = 9.42\%$

23. The initial value is $C = 9500$, the rate of decrease is $r = -0.35$, and the decay factor is
    $a = 1 + (-0.35) = 0.65$. The sample of insects contains $f(x) = 9500(0.65)^x$ insects after x weeks.

25. The initial value is $C = 2500$, the rate of increase is $r = 0.05$, and the growth factor is $a = 1 + (0.05) = 1.05$. The sample of fish contains $f(x) = 2500(1.05)^x$ fish after x months.

27. The initial value is $C = 1000$, the rate of decrease is $r = -0.065$, and the decay factor is $a = 1 + (-0.065) = 0.935$. The mutual fund contains $f(x) = 1000(0.935)^x$ dollars after x years.

29. For $f(x) = 8(1.12)^x$ the initial value is $C = 8$ and the growth factor is $a = 1.12$. Because $a = 1 + r$, it follows that $r = a - 1 \Rightarrow r = 1.12 - 1 = 0.12$ or 12%.

31. For $f(x) = 1.5(0.35)^x$ the initial value is $C = 1.5$ and the decay factor is $a = 0.35$. Because $a = 1 + r$, it follows that $r = a - 1 \Rightarrow r = 0.35 - 1 = -0.65$ or -65%.

33. For $f(x) = 0.55^x$ the initial value is $C = 1$ and the decay factor is $a = 0.55$. Because $a = 1 + r$, it follows that $r = a - 1 \Rightarrow r = 0.55 - 1 = -0.45$ or -45%.

35. For $f(x) = 7e^x$ the initial value is $C = 7$ and the growth factor is $a = e$. Because $a = 1 + r$, it follows that $r = e - 1 \Rightarrow r \approx 1.718$ or 171.8%.

37. For $f(x) = 6(3^{-x})$ the initial value is $C = 6$ and the decay factor is $a = 3^{-1} = \dfrac{1}{3}$. Because $a = 1 + r$, it follows that $r = \dfrac{1}{3} - 1 \Rightarrow r = -\dfrac{2}{3}$ or -66.7%.

39. The population of the city is 35,000 and is increasing by a factor of 9.8% every 2 years. Thus $A_0 = 35,000$, $b = 1.098$, $k = 2$, and $f(t) = 35,000(1.098)^{t/2}$. After 8 years the population is $f(5) = 35,000(1.098)^{5/2} \approx 44,215$.

41. The sample of bacteria is 1000 and is increasing by a factor of 3 every 7 hours. Thus $A_0 = 1000$, $b = 3$, $k = 7$, and $f(t) = 1000(3)^{t/7}$. After 11 hours the bacteria population is $f(11) = 1000(3)^{11/7} \approx 5620$.

43. The intensity of light is I_0 and is decreasing by a factor of $\dfrac{1}{3}$ every 2 millimeters. Thus $A_0 = I_0$, $b = 1 - \dfrac{1}{3} = \dfrac{2}{3}$, $k = 2$, and $f(t) = I_0\left(\dfrac{2}{3}\right)^{t/2}$. After 4.3 millimeters the intensity is $f(4.3) = I_0\left(\dfrac{2}{3}\right)^{4.3/2} \approx 0.418 I_0$.

45. The initial amount is 5000 and is increasing by a factor of 4 every 35 years. Thus $A_0 = 5000$, $b = 4$, $k = 35$, and $f(t) = 5000(4)^{t/35}$. After 8 years the amount is $f(8) = 5000(4)^{8/35} = \$6864.10$.

47. $T = \dfrac{70}{7} = 10$ years, $A = 2000e^{0.07(10)} = \$4027.51$

49. $T = \dfrac{70}{20} = 3.5$ years, $A = 500e^{0.2(3.5)} = \$1006.88$

51. $T = \dfrac{70}{25} = 2.8$ years, $A = 1500e^{0.25(2.8)} = \$3020.63$

53. $R = \dfrac{70}{40} = 1.75\%$

55. $R = \dfrac{70}{35} = 2\%$

57. $R = \dfrac{70}{70} = 1\%$

59. The sample of bacteria is increasing by a factor of 6% every 8 hours. Thus $A_0 = 1$, $b = 1.06$, $k = 8$, and $f(t) = 1 \cdot 1.06^{t/8}$. After 3 hours the percent increase is $f(3) = 1.06^{3/8} \approx 1.02209$. Thus, the increase is about 2.21%.

61. The number of cell phone subscribers increases by 25% in one year. Thus, $A_0 = 1$, $b = 1.25$, $k = 1$, and $f(t) = 1 \cdot 1.25^{t}$. $f(1) = 1 \cdot 1.25^{1} = 1.25$ and we know that after one year the percent of cell phone users has increased to 125%.

Now after the second year the number of cell phone subscribers has decreased 20%. Thus, $A_0 = 1.25$, $b = 0.8$, $k = 1$, and $f(t) = 1.25 \cdot 0.8^{t}$. $f(1) = 1.25 \cdot 0.8^{1} = 1$. Therefore, the number of subscribers at the beginning of the first year and the end of the second year is the same. Therefore, they are equal.

63. The initial amount is 9.81 and is decreasing by a factor of 9% in one year. Thus $A_0 = 9.81$, $b = 1 - 0.09 = 0.91$, $k = 1$, and $f(t) = 9.81 \cdot 0.91^{t}$. After 3 years the new wage is $f(1) = 9.81 \cdot 0.91^{3} = \$7.39$.

65. The initial amount of the element is 100% and has decreased to 40% after 2 years. Thus $A_0 = 1$, $b = 0.4$, $k = 2$, and $f(t) = 1 \cdot 0.4^{t/2}$. After 8 years the remaining percentage is $f(8) = 1 \cdot 0.4^{8/2} = 0.0256 = 2.56\%$.